Universitext

Emmanuele DiBenedetto

Degenerate Parabolic Equations

With 12 Figures

Springer-Verlag
New York Berlin Heidelberg London Paris
Tokyo Hong Kong Barcelona Budapest

Emmanuele DiBenedetto
Northwestern University
USA
and
University of Rome II
Italy

AMS Subject Classifications (1991): 35K65

Library of Congress Cataloging-in-Publication Data
DiBenedetto, Emmanuele.
Degenerate parabolic equations/Emmanuele DiBenedetto.
p. cm. – (Universitext)
Includes bibliographical references.
ISBN 0-387-94020-0 (New York: acid-free). – ISBN 3-540-94020-0
(Berlin: acid-free)
1. Differential equations, Parabolic. I. Title.
QA377.D62 1993
515′.353 – dc20 93-285

Printed on acid-free paper.

Production managed by Natalie Johnson; manufacturing supervised by Jacqui Ashri.
Photocomposed using the author's TEX files.
Printed and bound by R.R. Donnelley & Sons, Harrisonburg, VA.
Printed in the United States of America.

9 8 7 6 5 4 3 2 1

ISBN 0-387-94020-0 Springer-Verlag New York Berlin Heidelberg
ISBN 3-540-94020-0 Springer-Verlag Berlin Heidelberg New York

Preface

1. Elliptic equations: Harnack estimates and Hölder continuity

Considerable progress was made in the early 1950s and mid-1960s in the theory of elliptic equations, due to the discoveries of DeGiorgi [33] and Moser [81,82]. Consider local weak solutions of

$$(1.1)\qquad \begin{cases} u \in W^{1,2}_{loc}(\Omega), & \Omega \text{ a domain in } \mathbf{R}^N \\ (a_{ij}u_{x_i})_{x_j} = 0 & \text{in } \Omega, \end{cases}$$

where the coefficients $x \to a_{ij}(x)$, $i,j = 1,2,\dots,N$ are assumed to be only bounded and measurable and satisfying the ellipticity condition

$$(1.2)\qquad a_{ji}\xi_i\xi_j \ge \lambda|\xi|^2, \quad \text{a.e. } \Omega, \quad \forall\xi \in \mathbf{R}^N, \quad \text{for some } \lambda > 0.$$

DeGiorgi established that local solutions are Hölder continuous and Moser proved that non-negative solutions satisfy the Harnack inequality. Such inequality can be used, in turn, to prove the Hölder continuity of solutions. Both authors worked with *linear* p.d.e.'s. However the linearity has no bearing in the proofs. This permits an extension of these results to quasilinear equations of the type

$$(1.3)\qquad \begin{cases} u \in W^{1,p}_{loc}(\Omega), & p > 1 \\ \operatorname{div} \mathbf{a}(x,u,Du) + b(x,u,Du) = 0, & \text{in } \Omega, \end{cases}$$

with structure conditions

$$(1.4)\qquad \begin{cases} \mathbf{a}(x,u,Du)\cdot Du \ge \lambda|Du|^p - \varphi(x), & \text{a.e. } \Omega_T,\ p>1 \\ |\mathbf{a}(x,u,Du)| \le \Lambda|Du|^{p-1} + \varphi(x), \\ |b(x,u,Du)| \le \Lambda|Du|^{p-1} + \varphi(x). \end{cases}$$

Here $0<\lambda\le\Lambda$ are two given constants and $\varphi\in L^{\infty}_{loc}(\Omega)$ is non-negative. As a prototype we may take

$$\text{(1.5)}\qquad \operatorname{div}\left|Du\right|^{p-2}Du=0,\qquad \text{in}\quad \Omega,\qquad p>1\,.$$

The modulus of ellipticity of (1.5) is $|Du|^{p-2}$. Therefore at points where $|Du|=0$, the p.d.e. is degenerate if $p>2$ and it is singular if $1<p<2$.

By using the methods of DeGiorgi, Ladyzhenskaja and Ural'tzeva [66] established that weak solutions of (1.4) are Hölder continuous, whereas Serrin [92] and Trudinger [96], following the methods of Moser, proved that non-negative solutions satisfy a Harnack principle. The generalisation is twofold, i.e., the principal part $\mathbf{a}(x,u,Du)$ is permitted to have a *non-linear dependence* with respect to u_{x_i}, $i=1,2,\ldots,N$, and a *non-linear growth* with respect to $|Du|$. The latter is of particular interest since the equation in (1.5) might be either degenerate or singular.

2. Parabolic equations: Harnack estimates and Hölder continuity

The first parabolic version of the Harnack inequality is due to Hadamard [50] and Pini [86] and applies to non-negative solutions of the heat equation. It takes the following form. Let u be a non-negative solution of the heat equation in the cylindrical domain $\Omega_T\equiv\Omega\times(0,T)$, $0<T<\infty$, and for $(x_o,t_o)\in\Omega_T$ consider the cylinder

$$\text{(2.1)}\qquad Q_\rho\equiv B_\rho(x_o)\times(t_o-\rho^2,\,t_o],\quad B_\rho(x_o)\equiv\{|x-x_o|<\rho\}\,.$$

There exists a constant γ depending only upon N, such that if $Q_{2\rho}\subset\Omega_T$, then

$$\text{(2.2)}\qquad u(x_o,t_o)\ge\gamma\sup_{B_\rho(x_o)}u(x,t_o-\rho^2)\,.$$

The proof is based on local representations by means of heat potentials. A striking result of Moser [83] is that (2.2) continues to hold for non-negative weak solutions of

$$\text{(2.3)}\qquad \begin{cases} u\in V^{1,2}(\Omega_T)\equiv L^\infty\left(0,T;L^2(\Omega)\right)\cap L^2\left(0,T;W^{1,2}(\Omega)\right),\\ u_t-\left(a_{ij}(x,t)u_{x_i}\right)_{x_j}=0,\qquad \text{in }\Omega_T\end{cases}$$

where $a_{ij}\in L^\infty(\Omega_T)$ satisfy the analog of the ellipticity condition (1.2). As before, it can be used to prove that weak solutions are locally Hölder continuous in Ω_T. Since the linearity of (2.3) is immaterial to the proof, one might expect, as in the elliptic case, an extension of these results to quasilinear equations of the type

$$\text{(2.4)}\qquad \begin{cases} u\in V^{1,p}(\Omega_T)\equiv L^\infty\left(0,T;L^2(\Omega)\right)\cap L^p\left(0,T;W^{1,p}(\Omega)\right),\\ u_t-\operatorname{div}\mathbf{a}(x,t,u,Du)=b(x,t,u,Du),\qquad \text{in }\Omega_T,\end{cases}$$

where the structure condition is as in (1.4). Surprisingly however, Moser's proof could be extended only for the case $p=2$, i.e., for equations whose principal part has a *linear growth* with respect to $|Du|$. This appears in the work of Aronson and Serrin [7] and Trudinger [97]. The methods of DeGiorgi also could not be extended. Ladyzenskaja et al. [67] proved that solutions of (2.4) are Hölder continuous, provided the principal part has exactly a *linear growth* with respect to $|Du|$. Analogous results were established by Kruzkov [60,61,62] and by Nash [84] by entirely different methods.

Thus it appears that unlike the elliptic case, the degeneracy or singularity of the principal part plays a peculiar role, and for example, for the non-linear equation

$$u_t - \operatorname{div} \left|Du\right|^{p-2} Du = 0, \qquad \text{in } \Omega_T, \quad p > 1, \tag{2.5}$$

one could not establish whether non-negative weak solutions satisfy the Harnack estimate or whether a solution is locally Hölder continuous.

3. Parabolic equations and systems

These issues have remained open since the mid-1960s. They were revived however with the contributions of N.N. Ural'tzeva [100] in 1968 and K. Uhlenbeck [99] in 1977. Consider the system

$$\begin{cases} \mathbf{u} \equiv (u_1, u_2, \dots, u_n), & u_i \in W^{1,p}_{loc}(\Omega),\ p > 1,\ i = 1, 2, \dots n, \\ \operatorname{div} |D\mathbf{u}|^{p-2} Du_i = 0, & \text{in } \Omega\,. \end{cases} \tag{3.1}$$

When $p > 2$, Ural'tzeva and Uhlenbeck prove that local solutions of (3.1) are of class $C^{1,\alpha}_{loc}(\Omega)$, for some $\alpha \in (0,1)$. The parabolic version of (3.1) is

$$\begin{cases} \mathbf{u} \equiv (u_1, u_2, \dots, u_n), & u_i \in V^{1,p}(\Omega_T),\ i = 1, 2, \dots n, \\ u_t - \operatorname{div} |D\mathbf{u}|^{p-2} Du_i = 0, & \text{in } \Omega_T\,. \end{cases} \tag{3.2}$$

Besides their intrinsic mathematical interest, this kind of system arises from geometry [99], quasiregular mappings [2,17,55,89] and fluid dynamics [5,8,56,57,74,75]. In particular Ladyzenskaja [65] suggests systems of the type of (3.2) as a model of motion of non-newtonian fluids. In such a case $\mathbf{u}$ is the velocity vector. Non-newtonian here means that the stress tensor at each point of the fluid is *not* linearly proportional to the matrix of the space-gradient of the velocity.

The function $w = |D\mathbf{u}|^2$ is formally a subsolution of

$$\frac{\partial}{\partial t} w - \left(a_{\ell,k} w^{\frac{p-2}{2}} w_{x_k} \right)_{x_\ell} \le 0 \quad \text{in } \Omega_T, \tag{3.3}$$

where

$$a_{\ell,k} \equiv \left\{ \delta_{\ell,k} + (p-2) \frac{u_{i,x_\ell} u_{i,x_k}}{|D\mathbf{u}|^2} \right\}.$$

This is a *parabolic* version of a similar finding observed in [99,100] for elliptic systems. Therefore a *parabolic* version of the Ural'tzeva and Uhlenbeck result requires some understanding of the *local* behaviour of solutions of the porous media equation

$$u_t - \Delta u^m = 0, \qquad u \ge 0, \quad m > 0, \tag{3.4}$$

and its quasilinear versions. Such an equation is degenerate at those points of Ω_T where $u=0$ if $m>1$ and singular if $0<m<1$.

The porous medium equation has a life of its own. We only mention that questions of regularity were first studied by Caffarelli and Friedman. It was shown in [21] that *non-negative* solutions of the *Cauchy problem* associated with (3.4) are Hölder continuous. The result is not *local*.

A more *local* point of view was adopted in [20,35,90]. However these contributions could only establish that the solution is continuous with a *logarithmic* modulus of continuity.

In the mid-1980s, some progress was made in the theory of degenerate p.d.e.'s of the type of (2.5), for $p>2$. It was shown that the solutions are locally Hölder continuous (see [39]). Surprisingly, the same techniques can be suitably modified to establish the *local* Hölder continuity of *any local* solution of quasilinear porous medium-type equations. These modified methods, in turn, are crucial in proving that weak solutions of the systems (3.2) are of class $C^{1,\alpha}_{loc}(\Omega_T)$.

Therefore understanding the local structure of the solutions of (2.5) has implications to the theory of systems and the theory of equations with degeneracies quite different than (2.5).

4. Main results

In these notes we will discuss these issues and present results obtained during the past five years or so. These results follow, one way or another, from a single unifying idea which we call *intrinsic rescaling*. The diffusion process in (2.5) evolves in a time scale determined instant by instant by the solution itself, so that, loosely speaking, it can be regarded as the heat equation in its own intrinsic time-configuration. A precise description of this fact as well as its effectiveness is linked to its technical implementations.

We collect in Chap. I notation and standard material to be used as we proceed. Degenerate or singular p.d.e. of the type of (2.4) are introduced in Chap. II. We make precise their functional setting and the meaning of solutions and we derive *truncated* energy estimates for them. In Chaps. III and VI, we state and prove theorems regarding the local and global Hölder continuity of weak solutions of (2.4) both for $p>2$ and $1<p<2$ and discuss some open problems. In the singular case $1<p<2$, we introduce in Chap. IV a novel iteration technique quite different than that of DeGiorgi [33] or Moser [83].

These theorems assume the solutions to be locally or globally bounded. A theory of boundedness of solutions is developed in Chap. V and it includes equations with lower order terms exhibiting the Hadamard *natural* growth condition. The sup-estimates we prove appear to be dramatically different than those in the linear theory. Solutions are locally bounded only if they belong to $L^r_{loc}(\Omega_T)$ for some $r\geq 1$ satisfying

$$\lambda_r \equiv N(p-2)+rp>0 \tag{4.1}$$

and such a condition is sharp. In Chap. XII we give a counterexample that shows that if (4.1) is violated, then (2.5) has unbounded solutions.

The Hölder estimates and the L^∞-bounds are the basis for an organic theory of local and global behaviour of solutions of such degenerate and/or singular equations.

In Chaps. VI and VII we present an *intrinsic* version of the Harnack estimate and attempt to trace their connection with Hölder continuity. The natural parabolic cylinders associated with (2.5) are

$$(4.2) \qquad Q_\rho \equiv B_\rho(x_o) \times (t_o - \rho^p, , t_o], \qquad (x_o, t_o) \in \Omega_T .$$

We show by counterexamples that the Harnack estimate (2.2) cannot hold for non-negative solutions of (2.5), in the geometry of (4.2). It does hold however in a time-scale intrinsic to the solution itself. These Harnack inequalities reduce to (2.2) when $p = 2$. In the degenerate case $p > 2$ we establish a *global* Harnack type estimate for non-negative solutions of (1.5) in the whole strip $\Sigma_T \equiv \mathbf{R}^N \times (0, T)$. We show that such an estimate is equivalent to a growth condition on the solution as $|x| \to \infty$. If $\max\{1; \frac{2N}{N+1}\} < p < 2$, a surprising result is that the Harnack estimate holds in an *elliptic* form, i.e., holds over a ball B_ρ at a given time level. This is in contrast to the behaviour of non-negative solutions of the heat equation as pointed out by Moser [83] by a counterexample. These Harnack estimates in either the degenerate or singular case have been established *only* for non-negative solutions of the homogeneous equation (2.5). The proofs rely on some sort of non-linear versions of 'fundamental solutions'. It is natural to ask whether they hold for quasilinear equations. This is a challenging open problem and parallels the Hadamard [50] and Pini [86] approach via *fundamental* solutions, versus the *'non-linear'* approach of Moser [83].

The number p is required to be larger than $2N/(N+1)$ and such a condition is sharp for a Harnack estimate to hold. The case $1 < p \leq 2N/(N+1)$ is not fully understood and it seems to suggest questions similar to those of the limiting Sobolev exponent for elliptic equations (see Brézis [19]) and questions in differential geometry. Here we only mention that as $p \searrow 1$, (2.5) tends *formally* to a p.d.e. of the type of motion by mean curvature.

Hölder and Harnack estimates as well as precise sup-bounds coalesce in the theory of the Cauchy problem associated with (2.4). This is presented in Chap. XI for the degenerate case $p > 2$ and in Chap. XII for the singular case $1 < p < 2$. When $p > 2$, we identify the optimal growth of the initial datum as $|x| \to \infty$ for a solution, local or global in time, to exist. This is the analog of the theory of Tychonov [98], Tacklind [94] and Widder [105] for the heat equation. When $1 < p < 2$ it turns out that any non-negative initial datum $u_o \in L^1_{loc}(\mathbf{R}^N)$ yields a *unique* solution global in time. In general

$$|Du| \notin L^p_{loc}\left(\mathbf{R}^N \times \mathbf{R}^+\right), \qquad 1 < p \leq \frac{2N}{N+1}.$$

Therefore the main difficulty of the theory is to make precise the meaning of solution. We introduce in Chap. XII a new notion of non-negative weak solutions and establish the existence and *uniqueness* of such solutions. We show by a counterexample that these might be discontinuous. Thus, in view of the possible singularities, the notion of solution is dramatically different than the notion of *'viscosity'* solution. Issues of solutions of variable sign as well as their local and global behaviour are open.

In Chaps. VIII-X, we turn to systems of the type (3.2) and prove that

$$(4.3) \qquad u^{(i)}_{x_j} \in C^\alpha_{loc}(\Omega_T), \quad i = 1, 2, \ldots, n, \quad j = 1, 2, \ldots, N,$$

provided $p > 2N/(N+1)$. Analogous estimates are derived for all $p > 1$ for solutions in $L^r_{loc}(\Omega_T)$, where $r \geq 1$ satisfies (4.1). Again such a condition is sharp for (4.3) to hold. Near the lateral boundary of Ω_T we establish C^α estimates *for all* $\alpha \in (0,1)$, provided $p > \max\left\{1; \frac{2N}{N+2}\right\}$. Estimates in the class $C^{1,\alpha}$ near the boundary are still lacking even in the elliptic case.

A similar spectrum of results could be developed for equations of the type (3.4). We have avoided doing this to keep the theory as organic and unified as possible.

We have chosen not to present existence theorems for boundary value problems associated with (2.4) or (3.2). Theorems of this kind are mostly based on Galerkin approximations and appear in the literature in a variety of forms. We refer, for example, to [67] or [73]. Given the a priori estimates presented here these can be obtained alternatively by a limiting process in a family of approximating problems and an application of Minty's Lemma. These notes can be ideally divided in four parts:

1. Hölder continuity and boundedness of solutions (Chapters I-V)
2. Harnack type estimates (Chapters VI-VII)
3. Systems (Chapters VIII-X)
4. Non-negative solutions in a strip Σ_T (Chapters XI-XII).

These parts are technically linked but they are conceptually independent, in the sense that they deal with issues that have developed in independent directions. We have attempted to present them in such a way that they can be approached independently.

The motivation in writing these notes, beyond the specific degenerate and singular p.d.e., is to present a body of ideas and techniques that are surprisingly flexible and adaptable to a variety of parabolic equations bearing, in one way or another, a degeneracy or singularity.

Acknowledgments

The book is an outgrowth of my notes for the Lipschitz Vorlesungen that I delivered in the summer of 1990 at the Institut für Angewandte Math. of the University of Bonn, Germany. I would like to thank the Reinische Friedrich Wilhelm Universität and the grantees of the Sonderforschungsbereich 256 for their kind hospitality and support.

I have used preliminary drafts and portions of the manuscript as a basis for lecture series delivered in the Spring of 1989 at Ist. Naz. Alta Matematica, Rome Italy, in July 1992 at the Summer course of the Universidad Complutense de Madrid Spain and in the Winter 1992 at the Korean National Univ. Seoul Korea. My thanks to all the participants for their critical input and to these institutions for their support.

I like to thank Y.C. Kwong for a critical reading of a good portion of the manuscript and for valuable suggestions. I have also benefited from the input of M. Porzio who read carefully the first draft of the first four Chapters, V. Vespri and Chen Ya-Zhe who have read various portions of the script and my students J. Park and M. O'Leary for their input.

Contents

Preface

I. Notation and function spaces

II. Weak solutions and local energy estimates

III. Hölder continuity of solutions of degenerate parabolic equations

IV. Hölder continuity of solutions of singular parabolic equations

V. Boundedness of weak solutions

VI. Harnack estimates: the case $p>2$

VII. Harnack estimates and extinction profile for singular equations

VIII. Degenerate and singular parabolic systems

IX. Parabolic p-systems: Hölder continuity of $D\mathbf{u}$

X. Parabolic p-systems: boundary regularity

XI. Non-negative solutions in Σ_T. The case $p>2$

XII. Non-negative solutions in Σ_T. The case $1<p<2$

I

Notation and function spaces

1. Some notation

Let Ω be a bounded domain in $\mathbf{R}^N$ of boundary $\partial\Omega$ and for $0 < T < \infty$ let Ω_T denote the cylindrical domain $\Omega \times (0, T]$. Also let,

$$S_T \equiv \partial\Omega \times [0, T], \qquad \Gamma \equiv S_T \cup (\Omega \times \{0\})$$

denote the lateral boundary and the parabolic boundary of Ω_T respectively.

If Ω is a sphere of radius $\rho > 0$ centered at some $x_o \in \mathbf{R}^N$, we denote it by $B_\rho(x_o) \equiv \{|x - x_o| < \rho\}$, and if x_o coincides with the origin, we let $B_\rho(0) \equiv B_\rho$.

The boundary $\partial\Omega$ will be assumed to satisfy the property of *positive geometric density*, i.e.,

$$(1.1) \qquad \begin{cases} \text{there exists } \alpha^* \in (0,1) \text{ and } \rho_o > 0 \text{ such that } \forall x_o \in \partial\Omega, \\ \text{for every ball } B_\rho(x_o) \text{ centered at } x_o \text{ and radius } \rho \le \rho_o \\ |\Omega \cap B_\rho(x_o)| \le (1 - \alpha^*)|B_\rho(x_o)|, \end{cases}$$

where $|\Sigma|$ denotes the Lebesgue measure of a measurable set Σ.

At times it will be necessary to assume that $\partial\Omega$ is of class $C^{1,\lambda}$ for some $\lambda \in (0,1)$. That is, there exist a positive number ρ_o such that for all $x_o \in \partial\Omega$, the portion of $\partial\Omega$ within the ball $B_{\rho_o}(x_o)$ can be represented, in a local system of coordinates, as the graph of a $C^{1,\lambda}$ function $\phi^{(x_o)}$ such that $\phi^{(x_o)}(x_o) = 0$. We set

$$(1.2) \qquad |||\partial\Omega|||_{1+\lambda} \equiv \sup_{x_o \in \partial\Omega} \left[D\phi^{(x_o)} \right]_{\lambda, \bar{B}_{\rho_o}(x_o)}.$$

Here for a smooth function ϕ defined on a compact subset $\mathcal{K}$ of $\mathbf{R}^n$, for some positive integer n

$$[\phi]_{\lambda,\mathcal{K}} \equiv \sup_{x\in\mathcal{K}} |\phi(x)| + \sup_{(x,y)\in\mathcal{K}} \frac{|\phi(x)-\phi(y)|}{|x-y|^\lambda}. \tag{1.3}$$

The boundary $\partial\Omega$ is *piecewise smooth* if it satisfies (1.1) and is the union of finitely many portions of $(N\!-\!1)$–dimensional hypersurfaces of class $C^{1,\lambda}$.

If $\partial\Omega$ is piecewise smooth, we say that a certain quantity, say C or γ, *depends upon the structure of* $\partial\Omega$ if it can be calculated apriori only in terms of the numbers α^*, ρ_o in the definition (1.1), the number of the components making up $\partial\Omega$ and the $|||\cdot|||_{1+\lambda}$ norm of each of the components making up $\partial\Omega$.

If $f \in L^q(\Omega), 1 \le q \le \infty$, denote by $\|f\|_{q,\Omega}$ the $L^q(\Omega)$-norm of f. The function f is in $L^q_{loc}(\Omega)$ if $\|f\|_{q,\mathcal{K}}$ is finite for all compact subsets $\mathcal{K}$ of Ω.

Let $q, r \ge 1$. A function f defined and measurable in Ω_T belongs to

$$L^{q,r}(\Omega_T) \equiv L^r(0,T; L^q(\Omega))$$

if

$$\|f\|_{q,r;\Omega_T} \equiv \left(\int_0^T \left(\int_\Omega |f|^q dx \right)^{\frac{r}{q}} d\tau \right)^{\frac{1}{r}} < \infty.$$

Also $f \in L^{q,r}_{loc}(\Omega_T)$, if for every compact subset $\mathcal{K}$ of Ω and every subinterval $[t_1, t_2] \subset (0,T]$

$$\int_{t_1}^{t_2} \left(\int_{\mathcal{K}} |f|^q dx \right)^{\frac{r}{q}} d\tau < \infty.$$

Whenever $q = r$ we set $L^{q,r}(\Omega_T) \equiv L^q(\Omega_T)$, $L^{q,r}_{loc}(\Omega_T) \equiv L^q_{loc}(\Omega_T)$ and $\|f\|_{q,q;\Omega_T} \equiv \|f\|_{q,\Omega_T}$. These definitions are extended in the obvious way when either q or r are infinity.

If $f \in C^1(\Omega_T)$, we denote by $Df \equiv (f_{x_1}, f_{x_2}, \dots, f_{x_N})$ the gradient of f with respect to the space variables only.

The spaces $W^{1,p}(\Omega)$ and $W^{1,p}_o(\Omega)$, $p \ge 1$, are defined by

$W^{1,p}(\Omega)$ is the completion of $C^\infty(\Omega)$ under the norm

$$\|v\|_{W^{1,p}(\Omega)} \equiv \|v\|_{p,\Omega} + \|Dv\|_{p,\Omega}, \; v \in C^\infty(\Omega) \cap L^p(\Omega).$$

$W^{1,p}_o(\Omega)$ is the completion of $C^\infty_o(\Omega)$ under the norm

$$\|v\|_{W^{1,p}_o(\Omega)} \equiv \|Dv\|_{p,\Omega}, \; v \in C^\infty_o(\Omega).$$

Equivalently $W^{1,p}(\Omega)$ is the Banach space of functions $v \in L^p(\Omega)$ whose generalised derivatives v_{x_i} belong to $L^p(\Omega)$ for all $i = 1, 2, \dots, N$.

A function $v \in L^p_{loc}(\Omega)$ is in $W^{1,p}_{loc}(\Omega)$ if for every compact subset $\mathcal{K} \subset \Omega$, $v \in W^{1,p}(\mathcal{K})$.

We let $W^{1,\infty}(\Omega)$ denote the space of functions $v \in L^\infty(\Omega)$ whose distributional derivatives v_{x_i} are in $L^\infty(\Omega)$, for $i=1,2,\ldots,N$. The space $W^{1,\infty}_{loc}(\Omega)$ is defined analogously.

2. Basic facts about $W^{1,p}(\Omega)$ and $W_o^{1,p}(\Omega)$

We collect here a few facts that will be of frequent use in what follows. The first is about the Gagliardo–Nirenberg multiplicative embedding inequality.

THEOREM 2.1. *Let $v \in W_o^{1,p}(\Omega)$, $p \ge 1$. For every fixed number $s \ge 1$ there exists a constant C depending only upon N, p and s such that*

$$\|v\|_{q,\Omega} \le C\|Dv\|^{\alpha}_{p,\Omega}\|v\|^{1-\alpha}_{s,\Omega}, \tag{2.1}$$

where $\alpha \in [0,1]$, $p,q \ge 1$, are linked by

$$\alpha = \left(\frac{1}{s}-\frac{1}{q}\right)\left(\frac{1}{N}-\frac{1}{p}+\frac{1}{s}\right)^{-1}, \tag{2.2}$$

and their admissible range is:

(2.2-i) *If* $N=1$,

$$q \in [s,\infty], \qquad \alpha \in \left[0, \frac{p}{p+s(p-1)}\right];$$

(2.2-ii) *if* $1 \ge p < N$, $\alpha \in [0,1]$ *and*

$$q \in \left[s, \frac{Np}{N-p}\right] \quad \text{if} \quad s \le \frac{Np}{N-p},$$

$$q \in \left[\frac{Np}{N-p}, s\right] \quad \text{if} \quad s \ge \frac{Np}{N-p};$$

(2.2-iii) *if* $p \ge N > 1$, $q \in [s,\infty)$ *and*

$$\alpha \in \left[0, \frac{Np}{Np+s(p-N)}\right).$$

COROLLARY 2.1. *Let $v \in W_o^{1,p}(\Omega)$, and assume $p \in [1,N)$. There exists a constant γ depending only upon N, and p, such that*

$$\|v\|_{q,\Omega} \le \gamma\|Dv\|_{p,\Omega}, \qquad \text{where} \qquad q = \frac{Np}{N-p}. \tag{2.1$'$}$$

PROOF: We may take $\alpha=1$ and $s=1$ in (2.2–ii).

If $\partial\Omega$ is piecewise smooth, functions v in $W^{1,p}(\Omega)$ are defined up to $\partial\Omega$ via their traces. We will denote by $v|_{\partial\Omega}$ the trace on $\partial\Omega$ of a function $v \in W^{1,p}(\Omega)$.

THEOREM 2.2. *Assume that $\partial\Omega$ is piecewise smooth. There exists a constant C depending only upon N, p and the structure of $\partial\Omega$ such that*

$$\|v\|_{q,\partial\Omega} \le C\|v\|_{W^{1,p}(\Omega)}, \tag{2.3}$$

where

$$q \in \left[1, \frac{(N-1)p}{N-p}\right], \qquad \text{if} \quad 1 < p < N, \tag{2.3-i}$$

$$q \in [1, \infty), \qquad \text{if} \quad p = N. \tag{2.3-ii}$$

If $\partial\Omega$ is piecewise smooth, the space $W_o^{1,p}(\Omega)$ can be defined equivalently as the set of functions $v \in W^{1,p}(\Omega)$ whose trace on $\partial\Omega$ is zero.

Remark 2.1. The embedding inequalities of Theorem 2.1 and Corollary 2.1 continue to hold for functions v in $W^{1,p}(\Omega)$, not necessarily vanishing on $\partial\Omega$ in the sense of the traces, provided we assume further that $\partial\Omega$ is piecewise smooth and that

$$\int_\Omega v(x)dx = 0. \tag{2.4}$$

In such a case the constant C depends upon s, p, q, α, N and the structure of $\partial\Omega$. However it does not depend on the *size* of Ω, i.e., it does not change under dilations of Ω.

Let k be any real number and for a function $v \in W^{1,p}(\Omega)$ consider the truncations of v given by

$$(v-k)_+ \equiv \max\{(v-k)\ ;\ 0\}, \quad (v-k)_- \equiv \max\{-(v-k)\ ;\ 0\}. \tag{2.5}$$

LEMMA 2.1. *Let $v \in W^{1,p}(\Omega)$. Then for all $k \in \mathbf{R}$, $(v \mp k)_\pm \in W^{1,p}(\Omega)$. Assume in addition that the trace of v on $\partial\Omega$ is essentially bounded and*

$$\|v\|_{\infty,\partial\Omega} \le k_o, \quad \textit{for some} \quad k_o > 0.$$

Then for all $k \ge k_o$, $(v-k)_\pm \in W_o^{1,p}(\Omega)$.

COROLLARY 2.2. *Let $v_i \in W^{1,p}(\Omega)$, $i = 1, 2, \dots, n \in \mathbf{N}$. Then*

$$w \equiv \min\{v_1, v_2, \dots, v_n\} \in W^{1,p}(\Omega).$$

PROOF: Assume first $n = 2$. Then

$$\min\{v_1, v_2\} = \frac{v_1 - (v_2 - v_1)_+}{2} + \frac{v_2 - (v_1 - v_2)_+}{2}.$$

The general case is proved by induction.

If v is a continuous function defined in Ω and $k < l$ is a pair of real numbers, we set

$$(2.6)\qquad \begin{cases} [v>l] & \equiv \{x\in\Omega \,|\, v(x)>l\}\,, \\ [v<k] & \equiv \{x\in\Omega \,|\, v(x)<k\}\,, \\ [k<v<l] & \equiv \{x\in\Omega \,|\, k<v(x)<l\}\,. \end{cases}$$

LEMMA 2.2. *Let $v\in W^{1,1}(B_\rho(x_o))\cap C(B_\rho(x_o))$ for some $\rho>0$ and some $x_o\in\mathbf{R}^N$ and let k and l be any pair of real numbers such that $k<l$.*

There exists a constant γ depending only upon N,p and independent of k,l,v,x_o,ρ, such that

$$(2.7)\qquad (l-k)\big|\,[v>l]\,\big| \le \gamma\frac{\rho^{N+1}}{\big|\,[v<k]\,\big|}\int_{[k<v<l]}|Dv|dx.$$

Remark 2.2. The conclusion of the lemma continues to hold for functions $v\in W^{1,1}(\Omega)\cap C(\Omega)$ provided Ω is *convex.* We will use it in the case Ω is a hemisphere or a cube.

Remark 2.3. The continuity is not necessary to the conclusion of Lemma 2.2. The function v has been assumed to be continuous to give an unambiguous meaning to the definitions (2.6). If v is only in $W^{1,1}(\Omega)$, one could fix an arbitrary representative out of the equivalence class v say $\tilde{v}$ and define (2.6) accordingly. The conclusion of the Lemma is independent of the choice of $\tilde{v}$.

2-(i). Poincaré–type inequalities

Inequality (2.7) is due to De Giorgi [33] and it is a particular case of a more general Poincaré–type inequality. The embedding (2.1)′ of Corollary 2.1 gives a majorisation of the $L^q(\Omega)$–norm of u solely in terms of the $L^p(\Omega)$–norm of its gradient. This is possible because one knows that u vanishes on $\partial\Omega$ in the sense of the traces. A Poincaré–type inequality bounds some integral norm of a function $u\in W^{1,p}(\Omega)$ in terms *only* of some integral norm of its gradient, provided some information is available on the set where u vanishes.

PROPOSITION 2.1. *Let Ω be a bounded convex set in $\mathbf{R}^N$ and let $\varphi\in C\left(\overline{\Omega}\right)$ satisfy*

$$(2.8)\qquad \begin{cases} 0\le\varphi\le 1, & \forall x\in\Omega, \\ \textit{the sets } [\varphi>k] \textit{ are convex}, & \forall k\in(0,1). \end{cases}$$

Let $v\in W^{1,p}(\Omega)$, $p\ge 1$, and assume that the set

$$\mathcal{E}\equiv[v=0]\cap[\varphi=1]$$

has positive measure. There exists a constant C depending only upon N and p and independent of v and φ, such that

$$(2.9)\qquad \left(\int_\Omega \varphi|v|^p\,dx\right)^{\frac{1}{p}} \le C\frac{(\operatorname{diam}\Omega)^N}{|\mathcal{E}|^{\frac{N-1}{N}}}\left(\int_\Omega \varphi|Dv|^p\,dx\right)^{\frac{1}{p}}.$$

PROOF: We first prove (2.9) for $p=1$. For every $z\in\mathcal{E}$ and $x\in\Omega$,

$$|v(x)| = |v(x)-v(z)| = \Big|\int_0^{|x-z|}\frac{\partial}{\partial\rho}v\big(z+\tfrac{x-z}{|x-z|}\rho\big)\,d\rho\Big| \le \int_0^{|x-z|}\Big|D\,v\big(z+\tfrac{x-z}{|x-z|}\rho\big)\Big|\,d\rho.$$

Multiply this inequality by $\varphi(x)$ and integrate in dx over Ω and in dz over $\mathcal{E}$. This gives

$$|\mathcal{E}|\int_\Omega \varphi|v|dx \le \int_{\mathcal{E}} dz\int_\Omega dx\int_0^{|x-z|}\varphi(x)\Big|D\,v\big(z+\tfrac{x-z}{|x-z|}\rho\big)\Big|\,d\rho. \tag{2.10}$$

By virtue of the assumption (2.8)

$$\varphi(x)\Big|D\,v\big(z+\tfrac{x-z}{|x-z|}\rho\big)\Big| \le \varphi|Dv|\big(z+\tfrac{x-z}{|x-z|}\rho\big).$$

We put this inequality in (2.10) and compute the integral in dx on the right hand side, in polar coordinates with pole at z and radial variable $r=|x-z|$. We let ω be the angular variables and denote by $\mathcal{R}(\omega)$ the polar representation of $\partial\Omega$ with pole at z.

$$\begin{aligned}|\mathcal{E}|\int_\Omega\varphi|v|dx &\le \int_{\mathcal{E}}dz\left(\int_0^{\operatorname{diam}\Omega}r^{N-1}dr\right)\int_{|\omega|=1}d\omega\int_0^{\mathcal{R}(\omega)}\frac{\varphi|Du|(x)}{|x-z|^{N-1}}\,r^{N-1}dr\\ &\le \gamma\,(\operatorname{diam}\Omega)^N\int_{\mathcal{E}}dz\int_\Omega\frac{\varphi|Dv|(x)}{|x-z|^{N-1}}\,dx\\ &\le \gamma\,(\operatorname{diam}\Omega)^N\sup_{x\in\Omega}\int_{\mathcal{E}}\frac{dz}{|x-z|^{N-1}}\int_\Omega\varphi|Dv|\,dx.\end{aligned}$$

Next, for all $x\in\Omega$, and for all $\delta>0$,

$$\begin{aligned}\int_{\mathcal{E}}\frac{dz}{|x-z|^{N-1}} &= \int_{\mathcal{E}\cap\{|x-z|<\delta\}}\frac{dz}{|x-z|^{N-1}}+\int_{\mathcal{E}\cap\{|x-z|\ge\delta\}}\frac{dz}{|x-z|^{N-1}}\\ &\le \gamma(N,p)\left\{\delta+\frac{|\mathcal{E}|}{\delta^{N-1}}\right\}.\end{aligned}$$

Minimising with respect to the parameter δ gives

$$\int_\Omega\varphi|v|\,dx \le \gamma\frac{(\operatorname{diam}\Omega)^N}{|\mathcal{E}|^{\frac{N-1}{N}}}\int_\Omega\varphi|Dv|\,dx, \tag{2.11}$$

for a constant $\gamma = \gamma(N,p)$. By replacing v with $|v|^p$ in (2.11), we obtain

$$\begin{aligned}\int_\Omega \varphi|v|^p\,dx &\le p\gamma\frac{(\operatorname{diam}\Omega)^N}{|\mathcal{E}|^{1-\frac{1}{N}}}\int_\Omega \varphi|v|^{p-1}|Dv|\,dx\\ &\le \frac{1}{2}\int_\Omega \varphi|v|^p\,dx + \gamma(p)\left[\frac{(\operatorname{diam}\Omega)^N}{|\mathcal{E}|^{1-\frac{1}{N}}}\right]^p\int_\Omega \varphi\,|Dv|^p\,dx.\end{aligned}$$

Remark 2.4. Inequality (2.7) follows by applying (2.9) with $\varphi \equiv 1$ and $p = 1$ to the function

$$w = \begin{cases}\min\{v,l\} - k & \text{if } v > k\\ 0 & \text{if } v \le k.\end{cases}$$

By Corollary 2.2 such a function is in $W^{1,1}(\Omega)$.

3. Parabolic spaces and embeddings

We introduce spaces of functions, depending on $(x,t) \in \Omega_T$, that exhibit different regularity in the space and time variables. These are spaces where typically solutions of parabolic equations in divergence form are found.

Let $m, p \ge 1$ and consider the Banach spaces

$$V^{m,p}(\Omega_T) \equiv L^\infty(0,T;L^m(\Omega)) \cap L^p\left(0,T;W^{1,p}(\Omega)\right)$$

and

$$V_o^{m,p}(\Omega_T) \equiv L^\infty(0,T;L^m(\Omega)) \cap L^p\left(0,T;W_o^{1,p}(\Omega)\right),$$

both equipped with the norm, $v \in V^{m,p}(\Omega_T)$,

$$\|v\|_{V^{m,p}(\Omega_T)} \equiv \operatorname*{ess\,sup}_{0<t<T}\|v(\cdot,t)\|_{m,\Omega} + \|Dv\|_{p,\Omega_T}.$$

When $m = p$, we set $V_o^{p,p}(\Omega_T) \equiv V_o^p(\Omega_T)$ and $V^{p,p}(\Omega_T) \equiv V^p(\Omega_T)$. Both spaces are embedded in $L^q(\Omega_T)$ for some $q > p$. In a precise way we have

PROPOSITION 3.1. *There exists a constant γ depending only upon N, p, m such that for every $v \in V_o^{m,p}(\Omega_T)$*

$$\iint_{\Omega_T}|v(x,t)|^q\,dxdt \le \gamma^q\left(\iint_{\Omega_T}|Dv(x,t)|^p\,dxdt\right)\left(\operatorname*{ess\,sup}_{0<t<T}\int_\Omega |v(x,t)|^m\,dx\right)^{\frac{p}{N}}, \tag{3.1}$$

where

$$q = p\frac{N+m}{N}.$$

Moreover

$$\|v\|_{q,\Omega_T} \le \gamma \|v\|_{V^{m,p}(\Omega_T)}. \tag{3.2}$$

The multiplicative inequality (3.1) and the embedding (3.2) continue to hold for functions $v \in V^{m,p}(\Omega_T)$ *such that*

$$\int_\Omega v(x,t)dx = 0, \qquad \textit{for a.e.} \quad t \in (0,T),$$

provided $\partial\Omega$ *is piecewise smooth. In such a case the constant* γ *depends also on the structure of* $\partial\Omega$.

PROPOSITION 3.2. *Assume that* $\partial\Omega$ *is piecewise smooth. There exists a constant* γ *depending only upon* N, p, m *and the structure of* $\partial\Omega$*, such that for every* $v \in V^{m,p}(\Omega_T)$,

$$\|v\|_{q,\Omega_T} \le \gamma \left(1 + \frac{T}{|\Omega|^{\frac{N(p-m)+mp}{Nm}}}\right)^{\frac{1}{q}} \|v\|_{V^{m,p}(\Omega_T)}, \quad q = p\frac{N+m}{N}. \tag{3.3}$$

PROOF OF PROPOSITION 3.1: Assume first that $N(p-m)+mp>0$. Write the embedding inequality (2.1) for the function $x \to v(x,t)$ for a.e. $t \in (0,T)$ and for the choice of the parameters $s=m$ and

$$\alpha = \frac{p}{q}\,; \qquad q = p\frac{N+m}{N}\,; \qquad N(p-m)+mp>0.$$

Taking the q^{th}power in the resulting inequality and then integrating over $(0,T)$ proves (3.1). If $N(p-m)+mp \le 0$, we must have $p < N$. Therefore applying Corollary 2.1,

$$\begin{aligned}
\iint_{\Omega_T} |v|^q \, dx\, dt &= \int_0^T \int_\Omega |v|^p |v|^{m\frac{p}{N}} dx\, dt \\
&\le \int_0^T \left(\int_\Omega |v|^{\frac{pN}{N-p}} dx\right)^{\frac{N-p}{N}} \left(\int_\Omega |v|^m dx\right)^{\frac{p}{N}} dt \\
&\le \left(\iint_{\Omega_T} |Dv|^p dx\, dt\right)\left(\operatorname*{ess\,sup}_{0<t<T} \int_\Omega |v(x,t)|^m dx\right)^{\frac{p}{N}}.
\end{aligned}$$

To prove (3.2), we rewrite (3.1) as

$$\|v\|_{q,\Omega_T} \le C\|Dv\|_{p,\Omega_T}^{\frac{p}{q}} \left(\operatorname*{ess\,sup}_{0<t<T} \|v(\cdot,t)\|_{m,\Omega} \right)^{1-\frac{p}{q}},$$

and apply Young's inequality.

PROOF OF PROPOSITION 3.2: If $v \in V^{m,p}(\Omega_T)$, consider the function

$$w(\cdot,t) = v(\cdot,t) - \frac{1}{|\Omega|} \int_{\Omega} v(x,t)dx, \qquad \text{a.e.} \quad t \in (0,T),$$

which has zero integral average over Ω for a.e. $t \in (0,T)$. By Remark 2.1, $x \to w(x,t)$ satisfies the embedding inequality (2.1) for a.e. $t \in (0,T]$ and with constant C depending also upon the structure of $\partial\Omega$.

Proceeding as before, we arrive at (3.2) for w. For a.e. $t \in (0,T)$,

$$\|Dw\|_{p,\Omega_T} = \|Dv\|_{p,\Omega_T}, \qquad \|w(\cdot,t)\|_{m,\Omega} \le 2\|v(\cdot,t)\|_{m,\Omega}.$$

Moreover

$$\|w\|_{q,\Omega_T} \ge \|v\|_{q,\Omega_T} - |\Omega|^{\frac{1}{q}-\frac{1}{m}} \left(\int_0^T \left(\int_{\Omega} |v|^m dx \right)^{\frac{q}{m}} dt \right)^{\frac{1}{q}}.$$

Therefore

$$\|v\|_{q,\Omega_T} \le C\|Dv\|_{p,\Omega_T} + 2C \operatorname*{ess\,sup}_{0<t<T} \|v(\cdot,t)\|_{m,\Omega}$$
$$+ |\Omega|^{\frac{1}{q}-\frac{1}{m}} \left(\int_0^T \left(\int_{\Omega} |v|^m dx \right)^{\frac{q}{m}} dt \right)^{\frac{1}{q}}.$$

The last term is majorised by

$$\left(\frac{T}{|\Omega|^{\frac{N(p-m)+mp}{Nm}}} \right)^{\frac{1}{q}} \operatorname*{ess\,sup}_{0<t<T} \|v(\cdot,t)\|_{m,\Omega},$$

and the proposition follows.

We will use the following Corollaries obtained from the previous Propositions by taking $m=p$ and by applying the Hölder inequality.

COROLLARY 3.1. *Let $p>1$. There exists a constant γ depending only upon N and p, such that for every $v \in V_o^p(\Omega_T)$,*

$$\|v\|_{p,\Omega_T}^p \le \gamma \big| |v|>0 \big|^{\frac{p}{N+p}} \|v\|_{V^p(\Omega_T)}^p. \tag{3.4}$$

COROLLARY 3.2. *Let $p>1$. There exists a constant γ depending only upon N, p and the structure of $\partial\Omega$, such that for every $v \in V^p(\Omega_T)$,*

$$\|v\|_{p,\Omega_T}^p \le \gamma \left(1 + \frac{T}{|\Omega|^{\frac{p}{N}}} \right)^{\frac{N}{N+p}} \big| |v|>0 \big|^{\frac{p}{N+p}} \|v\|_{V^p(\Omega_T)}^p. \tag{3.5}$$

The next two Propositions hold in the case $m=p$.

PROPOSITION 3.3. *There exists a constant γ depending only upon N and p such that for every $v\in V_o^p(\Omega_T)$,*

$$\|v\|_{q,r;\Omega_T} \le \gamma \|v\|_{V^p(\Omega_T)}, \tag{3.6}$$

where the numbers $q, r \ge 1$ are linked by

$$\frac{1}{r}+\frac{N}{pq}=\frac{N}{p^2}, \tag{3.7}$$

and their admissible range is

$$\begin{cases} q\in(p,\infty], \quad r\in[p^2,\infty); & \text{if } N=1, \\ q\in\left[p,\frac{Np}{N-p}\right], \quad r\in[p,\infty]; & \text{if } 1\le p<N, \\ q\in[p,\infty), \quad r\in\left(\frac{p^2}{N},\infty\right]; & \text{if } 1<N\le p. \end{cases} \tag{3.8}$$

PROOF: Let $v\in V_0^p(\Omega_T)$ and let $r\ge 1$ to be chosen. From (2.1) with $s=p$ it follows that

$$\left(\int_0^T \|v(\cdot,\tau)\|_{q,\Omega}^r d\tau\right)^{\frac{1}{r}} \tag{3.9}$$

$$\le \gamma \left(\int_0^T \|Dv(\cdot,\tau)\|_p^{\alpha r} d\tau\right)^{\frac{1}{r}} \operatorname*{ess\,sup}_{0\le\tau\le T} \|v(\cdot,\tau)\|_{p,\Omega_T}^{1-\alpha}.$$

Choose $\alpha r=p$. Then conditions (2.2)-(2.2-iii) imply (3.7)-(3.8), and the Proposition follows.

The next Proposition holds for functions $v\in V^p(\Omega_T)$ not necessarily vanishing on the lateral boundary of Ω_T.

PROPOSITION 3.4. *There exists a constant γ depending only upon N, p, m and the structure of $\partial\Omega$, such that for every $v\in V^p(\Omega_T)$,*

$$\|v\|_{q,r;\Omega_T} \le \gamma \left(1+\frac{T}{|\Omega|^{\frac{p}{N}}}\right)^{\frac{1}{r}} \|v\|_{V^p(\Omega_T)} \tag{3.9}$$

where q and r satisfy (3.7) and (3.8).

PROOF: Apply (2.1) to the function

$$w(\cdot,t)=v(\cdot,t)-\frac{1}{|\Omega|}\int_\Omega v(x,t)dx, \qquad \text{a.e.} \quad t\in(0,T],$$

which has zero average in Ω for a.e. $t\in(0,T)$. Proceeding as in the proof of Proposition 3.3 we arrive at (3.6) for w where γ now also depends upon the structure of $\partial\Omega$. From this

$$\|v\|_{q,r;\Omega_T} \le \gamma \|v\|_{V^p(\Omega_T)} + \gamma |\Omega|^{\frac{1}{q}-1} \left(\int_0^T \left(\int_\Omega |v(x,\tau)| dx \right)^r d\tau \right)^{\frac{1}{r}}$$

$$\le \gamma \|v\|_{V^p(\Omega_T)} + \gamma \left(\frac{T}{|\Omega|^{p/N}} \right)^{\frac{1}{r}} \operatorname*{ess\,sup}_{0 \le \tau \le T} \|v(\cdot,\tau)\|_{p,\Omega}.$$

We conclude this section by stating a *parabolic* version of Lemma 2.1 and Corollary 2.2 concerning the truncated functions $(v-k)_\pm$.

LEMMA 3.1. *Let $v \in V^{m,p}(\Omega_T)$. Then for all $k \in \mathbf{R}$, $(v-k)_\pm \in V^{m,p}(\Omega_T)$. Assume in addition that the trace of $x \to v(x,t)$ on $\partial\Omega$ is essentially bounded and*

$$\operatorname*{ess\,sup}_{0<t<T} \|v(\cdot,t)\|_{\infty,\partial\Omega} \le k_o, \quad \textit{for some } k_o > 0.$$

Then for all $k \ge k_o$, $(v \mp k)_\pm \in V_o^{m,p}(\Omega_T)$.

COROLLARY 3.3. *Let $v_i \in L^p\left(0,T; W^{1,p}(\Omega)\right)$, $i = 1,2,\dots,n \in \mathbf{N}$. Then*

$$w \equiv \min\{v_1, v_2, \dots, v_n\} \in L^p\left(0,T; W^{1,p}(\Omega)\right).$$

3-(i). Steklov averages

Let v be a function in $L^1(\Omega_T)$ and for $0 < h < T$ introduce the Steklov averages $v_h(\cdot,t)$ defined for all $0<t<T$ by

$$v_h \equiv \begin{cases} \frac{1}{h} \int_t^{t+h} v(\cdot,\tau) d\tau, & t \in (0, T-h], \\ 0, & t > T-h\,; \end{cases}$$

$$v_{\bar{h}} \equiv \begin{cases} \frac{1}{h} \int_{t-h}^{t} v(\cdot,\tau) d\tau, & t \in (h, T], \\ 0, & t < h. \end{cases}$$

LEMMA 3.2. *Let $v \in L^{q,r}(\Omega_T)$. Then, as $h \to 0$, v_h converges to v in $L^{q,r}(\Omega_{T-\varepsilon})$ for every $\varepsilon \in (0,T)$. If $v \in C\left(0,T; L^q(\Omega)\right)$, then as $h \to 0$, $v_h(\cdot,t)$ converges to $v(\cdot,t)$ in $L^q(\Omega)$ for every $t \in (0, T-\varepsilon)$, $\forall \varepsilon \in (0,T)$.*

A similar statement holds for $v_{\bar{h}}$. The proof of the lemma is straightforward from the theory of L^p spaces.

4. Auxiliary lemmas

4-(i). *Fast geometric convergence*

We state and prove two lemmas concerning the geometric convergence of sequences of numbers.

LEMMA 4.1. *Let* $\{Y_n\}$, $n=0,1,2,\ldots$, *be a sequence of positive numbers, satisfying the recursive inequalities*

$$Y_{n+1} \le C b^n Y_n^{1+\alpha} \tag{4.1}$$

where $C, b, > 1$ *and* $\alpha > 0$ *are given numbers. If*

$$Y_o \le C^{-1/\alpha} b^{-1/\alpha^2}, \tag{4.2}$$

then $\{Y_n\}$ *converges to zero as* $n \to \infty$.

The proof is by induction.

LEMMA 4.2. *Let* $\{Y_n\}$ *and* $\{Z_n\}$, $n=0,1,2,\ldots$, *be sequences of positive numbers, satisfying the recursive inequalities*

$$\begin{cases} Y_{n+1} \le C b^n \left(Y_n^{1+\alpha} + Z_n^{1+\kappa} Y_n^{\alpha} \right), \\ Z_{n+1} \le C b^n \left(Y_n + Z_n^{1+\kappa} \right) \end{cases} \tag{4.3}$$

where $C, b > 1$ *and* $\kappa, \alpha > 0$ *are given numbers. If*

$$Y_o + Z_o^{1+\kappa} \le (2C)^{-\frac{1+\kappa}{\sigma}} b^{-\frac{1+\kappa}{\sigma^2}}, \quad \textit{where} \quad \sigma = \min\{\kappa; \alpha\}, \tag{4.4}$$

then $\{Y_n\}$ *and* $\{Z_n\}$ *tend to zero as* $n \to \infty$.

PROOF: Set $M_n = Y_n + Z_n^{1+\kappa}$ and rewrite the second of (4.3) as

$$Z_{n+1}^{1+\kappa} \le C^{1+\kappa} b^{(1+\kappa)n} M_n^{1+\kappa}. \tag{4.5}$$

Consider the term in braces in the first of (4.3). If $Z_n^{1+\kappa} \le Y_n$, such a term is majorised by $2M_n^{1+\alpha}$. If $Z_n^{1+\kappa} \ge Y_n$, then the same term can be majorised by

$$Y_n^{1+\alpha} + \left(Z_n^{1+\kappa} \right)^{1+\alpha} \le M_n^{1+\alpha}.$$

Combining this with (4.5) we deduce that in either case

$$M_{n+1} \le 2 C^{1+\kappa} b^{(1+\kappa)n} M_n^{1+\min\{\kappa,\alpha\}}.$$

The proof is concluded by induction as in Lemma 4.1.

4-(ii). An interpolation lemma

LEMMA 4.3. *Let* $\{Y_n\}$, $n=0,1,2\ldots$, *be a sequence of equibounded positive numbers satisfying the recursive inequalities*

$$Y_n \le C b^n Y_{n+1}^{1-\alpha}, \tag{4.6}$$

where $C, b>1$ *and* $\alpha\in(0,1)$ *are given constants. Then*

$$Y_o \le \left(\frac{2C}{b^{1-\frac{1}{\alpha}}}\right)^{\frac{1}{\alpha}}. \tag{4.7}$$

Remark 4.1. The Lemma turns the *qualitative* information of equiboundedness of the sequence $\{Y_n\}$ into a *quantitative* apriori estimate for Y_o.

PROOF OF LEMMA 4.3: From (4.6), by Young's inequality

$$Y_n \le \varepsilon Y_{n+1} + \left(\frac{C}{\varepsilon^{1-\alpha}}\right)^{\frac{1}{\alpha}} b^{\frac{n}{\alpha}}, \qquad \forall\varepsilon\in(0,1), \quad n=0,1,2,\ldots .$$

By iteration

$$Y_o \le \varepsilon^n Y_n + \left(\frac{C}{\varepsilon^{1-\alpha}}\right)^{\frac{1}{\alpha}} \sum_{i=0}^{n-1} \left(b^{\frac{1}{\alpha}}\varepsilon\right)^i .$$

Choose $b^{\frac{1}{\alpha}}\varepsilon=\frac{1}{2}$ so that the sum on the right hand side can be majorised with a series convergent to 2. Letting $n\to\infty$ proves the Lemma.

4-(iii). An algebraic lemma

We conclude this section by recording two algebraic inequalities needed in what follows.

LEMMA 4.4. *Let* $p\ge 2$. *Then* $\forall a,b\in\mathbf{R}^m$, $m\in\mathbf{N}$

$$\langle |a|^{p-2}a-|b|^{p-2}b, a-b\rangle \ge \gamma_o|a-b|^p \tag{4.8}$$

where γ_0 *depends only upon* p, m.

Let $1<p<2$. Then $\forall a,b\in\mathbf{R}^m$

$$\langle |a|^{p-2}a-|b|^{p-2}b, a-b\rangle \le \gamma_1|a-b|^p, \tag{4.9}$$

where γ_1 depends only upon p, m.

PROOF:

$$I(p) = \langle |a|^{p-2}a - |b|^{p-2}b, a-b\rangle$$
$$= \left\langle \int_0^1 \frac{d}{ds}|sa+(1-s)b|^{p-2}(sa+(1-s)b)ds, a-b \right\rangle$$
$$= \int_0^1 |sa+(1-s)b|^{p-2}|a-b|^2 ds$$
$$+ (p-2)\int_0^1 |sa+(1-s)b|^{p-4}|\langle sa+(1-s)b, a-b\rangle|^2 ds.$$

If $p \geq 2$, $I(p) \geq |a-b|^2 \int_0^1 |sa+(1-s)b|^{p-2}ds$. If $|a| \geq |b-a|$, we have

$$|sa+(1-s)b| \geq \big||a|-(1-s)|a-b|\big| \geq s|a-b|$$

and (4.8) follows. If $|a| < |b-a|$,

$$|a-b|^2 \int_0^1 |sa+(1-s)b|^{p-2}dx$$
$$\geq |a-b|^2 \int_0^1 \frac{(|sa+(1-s)b|^2)^{p/2}}{(2-s)|a-b|^2}ds$$
$$\geq \frac{1}{2}\left(\int_0^1 |sa+(1-s)b|^2 ds\right)^{p/2}$$
$$= \frac{1}{2}\frac{1}{3^{p/2}}\left(|a|^2+|b|^2+\langle a,b\rangle\right)^{p/2}$$
$$\geq \gamma_0|a-b|^p.$$

Remark 4.2. The reverse inequality to (4.8) is, in general, false. Indeed, if $a \in \mathbf{R}^+$, $b=a+1$

$$I(p) = (p-1)\xi^{p-2}; \quad \text{for some } \xi \in (a, a+1).$$

Letting $a \to \infty$ shows that the inequality $I(p) \leq \gamma|a-b|^p$ cannot hold with γ independent of a and b.

Next, if $1 < p < 2$,

$$I(p) \leq (p-1)|a-b|^2 \int_0^1 |sa+(1-s)b|^{p-2}ds.$$

If $|a| \geq |a-b|$, since $p < 2$,

$$|sa + (1-s)b|^{p-2} \le \big||a| - (1-s)|a-b|\big|^{p-2} \le s^{p-2}|a-b|^{p-2}$$

and (4.9) follows since s^{p-2} is integrable. If $|a| < |b-a|$, let s_* be defined by

$$(1-s_*)|a-b| = |a|.$$

Then estimate

$$\begin{aligned} I(p) &\le (p-1)|a-b|^2 \int_0^1 \big||a| - (1-s)|a-b|\big|^{p-2} ds \\ &\le |a-b|^p \left| \int_0^{s_*} \frac{d}{ds} (|a| - (1-s)|a-b|)^{p-1} ds \right| \\ &+ |a-b| \left| \int_{s_*}^1 \frac{d}{ds} (|a| - (1-s)|a-b|)^{p-1} ds \right| \\ &\le \gamma_1 |a-b|^p. \end{aligned}$$

Remark 4.3. The reverse inequality is false, in general, with γ independent of a, b.

5. Bibliographical notes

For the theory of Sobolev spaces we refer to the monographs of Adams [1] and Mazja [76]. The embedding theorems 2.1 and 2.2 are special cases of more general embedding theorems. No attempt has been made to state them in the most general setting and under the best assumptions on the regularity of $\partial\Omega$. For a proof of Lemma 2.1 we refer to Mazja [76] and Stampacchia [93]. Lemma 2.2 is due to DeGiorgi [33]. Also the statement in Remark 2.2 follows from the proof in [33]. The parabolic spaces $V^{m,p}(\Omega_T)$ and $V_o^{m,p}(\Omega_T)$ are standard in the theory of parabolic partial differential equations and we refer for example to [67,73]. The embedding theorems 3.2 and 3.3 are a modification of similar statements and proofs in [67]. The lemmas on rapid geometric convergence are stated in [67]; we have given a different proof. The interpolation inequality of Lemma 4.3 is taken from Campanato [22,23]. Lemma 4.4 is taken from [27].

II

Weak solutions and local energy estimates

1. Quasilinear degenerate or singular equations

We introduce a class of quasilinear parabolic equations with the principal part in divergence form, that are either degenerate or singular due to the vanishing of the gradient $|Du|$ of their solutions.

$$u_t - \operatorname{div} \mathbf{a}(x,t,u,Du) = b(x,t,u,Du) \quad \text{in } \mathcal{D}'(\Omega_T). \tag{1.1}$$

The functions $\mathbf{a} : \Omega_T \times \mathbf{R}^{N+1} \to \mathbf{R}^N$ and $b : \Omega_T \times \mathbf{R}^{N+1} \to \mathbf{R}$ are only assumed to be measurable and satisfying the structure conditions

$$\begin{aligned}
&(\mathbf{A}_1) \qquad \mathbf{a}(x,t,u,Du)\cdot Du \ge C_o|Du|^p - \varphi_o(x,t),\\
&(\mathbf{A}_2) \qquad |\mathbf{a}(x,t,u,Du)| \le C_1|Du|^{p-1} + \varphi_1(x,t),\\
&(\mathbf{A}_3) \qquad |b(x,t,u,Du)| \le C_2|Du|^p + \varphi_2(x,t)
\end{aligned}$$

for $p > 1$ and a.e. $(x,t) \in \Omega_T$. Here C_i, $i = 0,1,2$, are given positive constants and φ_i, $i = 0,1,2$, are given non-negative functions, defined in Ω_T and subject to the condition

$$(\mathbf{A}_4) \qquad \varphi_o, \quad \varphi_1^{\frac{p}{p-1}}, \quad \varphi_2 \in L^{\hat{q},\hat{r}}(\Omega_T)$$

where $\hat{q}, \hat{r} \ge 1$ satisfy

$$(\mathbf{A}_5) \qquad \frac{1}{\hat{r}} + \frac{N}{p\hat{q}} = 1 - \kappa_1,$$

and

(A$_5$–i) $$\hat{q} \in (1, \infty], \quad \hat{r} \in \left[\frac{1}{1-\kappa_1}, \frac{p}{p(1-\kappa_1)-1}\right),$$
$$\kappa_1 \in \left(0, \frac{p-1}{p}\right) \qquad \text{if} \quad N = 1;$$

(A$_5$–ii) $$\hat{q} \in \left[\frac{N}{p(1-\kappa_1)}, \infty\right], \quad \hat{r} \in \left[\frac{1}{1-\kappa_1}, \infty\right],$$
$$\kappa_1 \in (0,1) \quad \text{if } N > 1, \quad 1 < p \le N;$$

(A$_5$–iii) $$\hat{q} \in \left[\frac{N}{p(1-\kappa_1)}, \infty\right], \quad \hat{r} \in \left[\frac{1}{1-\kappa_1}, \infty\right],$$
$$\kappa_1 \in \left(\frac{p-N}{p}, 1\right), \quad \text{if } 1 < N < p.$$

A measurable function u is a *local* weak sub(super)-solution of (1.1) in Ω_T if

(1.2) $$u \in C_{loc}\left(0,T; L^2_{loc}(\Omega)\right) \cap L^p_{loc}\left(0,T; W^{1,p}_{loc}(\Omega)\right),$$

and for every compact subset $\mathcal{K}$ of Ω and for every subinterval $[t_1, t_2]$ of $(0, T]$

(1.3) $$\int_{\mathcal{K}} u\varphi dx \bigg|_{t_1}^{t_2} + \int_{t_1}^{t_2}\int_{\mathcal{K}} \{-u\varphi_t + \mathbf{a}(x,\tau,u,Du)\cdot D\varphi\}\, dx d\tau$$
$$\le (\ge) \int_{t_1}^{t_2}\int_{\mathcal{K}} b(x,\tau,u,Du)\varphi\, dx d\tau,$$

for all locally bounded testing functions

(1.4) $$\varphi \in W^{1,2}_{loc}\left(0,T; L^2(\mathcal{K})\right) \cap L^p_{loc}\left(0,T; W^{1,p}_o(\mathcal{K})\right), \qquad \varphi \ge 0.$$

The local boundedness of the testing functions φ is required to guarantee the convergence of the integral on the right hand side of (1.3).

A function u that is both a local subsolution and a local supersolution of (1.1) is a local solution.

Remark 1.1. If $p = 2$, then (1.1) is non-degenerate. In such a case it is known that locally bounded weak solutions are locally Hölder continuous; moreover the assumptions $(\mathbf{A}_1)$–$(\mathbf{A}_5)$ are optimal for a Hölder modulus to hold.

It would be technically convenient to have a formulation of weak solution that involves u_t. Unfortunately solutions of (1.1), whenever they exist, possess a modest degree of regularity in the time variable and, in general, u_t has a meaning only in the sense of distributions.

The following notion of local weak sub(super)-solution involves the discrete time derivative of u and is equivalent to (1.3).

Fix $t \in (0,T)$ and let h be a small positive number such that $0 < t < t+h < T$. In (1.3) take $t_1 = t$, $t_2 = t+h$ and choose a testing function φ independent of the variable $\tau \in (t, t+h)$. Dividing by h and recalling the definition of Steklov averages we obtain

$$(1.5)\quad \int_{\mathcal{K}\times\{t\}} \{u_{h,t}\varphi + [\mathbf{a}(x,\tau,u,Du)]_h \cdot D\varphi - [b(x,\tau,u,Du)]_h\,\varphi\}\,dx \le (\ge) 0,$$

$$\text{for all } 0<t<T-h \text{ and for all } \varphi \in W_o^{1,p}(\mathcal{K}) \cap L^\infty_{loc}(\Omega), \quad \varphi \ge 0.$$

To recover (1.3), fix a subinterval $0<t_1<t_2<T$, choose h so small that $t_2+h \le T$ and in (1.5) take a testing function as in (1.4). Such a choice is admissible, since the testing functions in (1.5) are independent of the variable $\tau \in (t, t+h)$ but may be dependent upon t. Integrating over $[t_1, t_2]$ and letting $h \to 0$ with the aid of Lemma 3.2 of Chap. I gives (1.3).

1-(i). Subsolutions and parabolic equations

The structure conditions $(\mathbf{A}_1)$–$(\mathbf{A}_5)$ are not sufficient to characterize parabolic p.d.e.'s. For example the *'principal part'*

$$\mathbf{a}(x,t,u,Du) \equiv Du - Du/|Du|$$

satisfies $(\mathbf{A}_1)$–$(\mathbf{A}_5)$ with $p=2$. However its *'modulus of ellipticity'* changes type at $|Du|=1$. In what follows we assume that (1.1) is weakly parabolic in the sense that it satisfies $(\mathbf{A}_1)$–$(\mathbf{A}_5)$ and in addition, whenever u is a weak solution of (1.1),

$(\mathbf{A}_6)$ for all $k \in \mathbf{R}$ the truncated functions $(u-k)_\pm$ are weak subsolutions of (1.1) in the sense of (1.3).

with $\mathbf{a}(x,t,u,Du)$ replaced by

$$\pm\mathbf{a}(x,t,k \pm (u-k)_\pm, \pm D(u-k)_\pm)$$

and $b(x,t,u,Du)$ replaced by

$$\pm b(x,t,k \pm (u-k)_\pm, \pm D(u-k)_\pm).$$

To clarify the connection between subsolutions and parabolic structures, we derive some sufficient conditions on $\mathbf{a}(x,t,u,Du)$ for $(\mathbf{A}_6)$ to be verified. Let u be a local weak solution of (1.1), and in (1.5) take the testing function

$$\frac{(u_h)_+}{(u_h)_+ + \varepsilon}\,\varphi, \qquad \varepsilon > 0 \qquad \text{and } \varphi \text{ satisfying (1.4).}$$

We integrate in dt over $[t_1, t_2] \subset (0,T)$ and let first $h \to 0$ and then $\varepsilon \to 0$ to obtain

$$\int_\Omega u_+\varphi(x,t)\Big|_{t_1}^{t_2} + \int_{t_1}^{t_2}\int_\Omega \{-u_+\varphi_t + \mathbf{a}(x,t,u_+,Du_+)\cdot D\varphi\}\,dxd\tau$$

$$(1.6) \qquad = \int_{t_1}^{t_2}\int_\Omega b(x,t,u_+,Du_+)\,\varphi\,dxd\tau$$

$$- \liminf_{\varepsilon\searrow 0}\,\varepsilon\int_{t_1}^{t_2}\int_\Omega \frac{[\mathbf{a}(x,t,u_+,Du_+)\cdot Du_+]}{(u_+ + \varepsilon)^2}\,\varphi\,dxd\tau.$$

LEMMA 1.1. *Assume that*

$$\mathbf{a}(x,t,u,\eta)\cdot\eta \geq 0, \qquad \forall\eta\in\mathbf{R}^N.$$

Then (1.1) is weakly parabolic.

PROOF: It suffices to verify $(\mathbf{A}_6)$ for $(u-k)_+$ and $k=0$. This is the content of (1.6).

A more general condition for $(\mathbf{A}_6)$ to hold can be given by the notion of monotonicity. We say that $\eta\to\mathbf{a}(x,t,u,\eta)$ is monotone if[(1)]

$$(1.7) \qquad \big\langle \mathbf{a}(x,t,u,\eta_1) - \mathbf{a}(x,t,u,\eta_2), \eta_1 - \eta_2\big\rangle \geq 0, \quad \forall\eta_i\in\mathbf{R}^N,\ i=1,2.$$

LEMMA 1.2. *Assume that* $\eta\to\mathbf{a}(x,t,u,\eta)$ *is monotone and*

$$(1.8) \qquad \operatorname{div}\mathbf{a}(x,t,u,0)\in L^1_{loc}(\Omega_T).$$

Then $(\mathbf{A}_6)$ *holds.*

PROOF: Write the last integral on the right hand side of (1.6) as

$$\varepsilon\int_{t_1}^{t_2}\int_\Omega \frac{[(\mathbf{a}(x,t,u,Du_+) - \mathbf{a}(x,t,u,0))\cdot Du_+]}{(u_+ + \varepsilon)^2}\,\varphi\,dxd\tau$$

$$- \int_{t_1}^{t_2}\int_\Omega \operatorname{div}\mathbf{a}(x,t,u,0)\frac{u_+}{u_+ + \varepsilon}\,\varphi\,dxd\tau$$

$$- \int_{t_1}^{t_2}\int_\Omega \mathbf{a}(x,t,u,0)\cdot D\varphi\,\frac{u_+}{u_+ + \varepsilon}\,dxd\tau.$$

We let $\varepsilon\searrow 0$ and discard the non–negative contribution of the first integral. The sum of the last two terms tends to zero since

(1) The monotonicity assumption (1.7) is natural in the existence theory. It permits one to apply Minty's Lemma [78] to identify the weak limit of the principal part of the p.d.e. when (1.1) is approximated by a sequence of regularised problems.

$$\lim_{\varepsilon\searrow 0}\int_{t_1}^{t_2}\!\!\int_{\Omega}\operatorname{div}\mathbf{a}(x,t,u,0)\frac{u_+}{u_++\varepsilon}\varphi\,dxd\tau \tag{1.9}$$

$$=\int_{t_1}^{t_2}\!\!\int_{\Omega}\operatorname{div}\mathbf{a}(x,t,u_+,0)\varphi\,dxd\tau$$

$$=-\int_{t_1}^{t_2}\!\!\int_{\Omega}\mathbf{a}(x,t,u_+,0)\cdot D\varphi\,dxd\tau.$$

One checks that the assumptions of the lemma are verified for example by equations with principal part

$$\left(|Du|^{p-2}a_{ij}(x,t)u_{x_i}+\psi_o(x,t)\,u_{x_j}/|Du|+\psi_j(x,t)\right)_{x_j},$$

where ψ_o is bounded, non-negative and $\psi_{j,x_j}\in L^1(\Omega_T)$ and the matrix (a_{ij}) is only measurable and positive definite.

Remark 1.1. The *'regularity'* assumption (1.8) is only needed to justify the limit in (1.9). It can be dispensed with when working with a sequence of approximating solutions.

2. Boundary value problems

We will give regularity results for weak solutions of (1.1) up to the lateral boundary S_T, provided u satisfies appropriate Dirichlet or Neumann boundary conditions. We also prove that weak solutions are Hölder continuous up to $t=0$ if the initial datum is Hölder continuous.

Since the arguments are local in nature, for these results to hold, the prescribed boundary data have to be taken only locally. However, for simplicity of presentation we will state them globally, in terms of boundary value problems.

2-(i). *The Dirichlet problem*

Consider formally the Dirichlet problem

$$\begin{cases} u_t-\operatorname{div}\mathbf{a}(x,t,u,Du)=b(x,t,u,Du), & \text{in } \Omega_T,\\ u(\cdot,t)\big|_{\partial\Omega}=g(\cdot,t), & \text{a.e. } t\in(0,T),\\ u(\cdot,0)=u_o(\cdot), \end{cases} \tag{2.1}$$

where the structure conditions $(\mathbf{A}_1)$–$(\mathbf{A}_5)$ are retained. On the Dirichlet data g and u_o we assume

(D) g is continuous on $\overline{S}_T$ with modulus of continuity, say $\omega_g(\cdot)$,

(U$_o$) u_o is continuous in $\overline{\Omega}$ with modulus of continuity, say $\omega_o(\cdot)$.

A weak sub(super)-solution of the Dirichlet problem (2.1) is a measurable function u, satisfying

$$u \in C\left(0,T;L^2(\Omega)\right)\cap L^p\left(0,T;W^{1,p}(\Omega)\right), \tag{2.2}$$

and for all $t\in(0,T]$

$$\begin{aligned}\int_\Omega u\varphi(x,t)\,dx \iint_{\Omega_t} \{-u\varphi_t + \mathbf{a}(x,\tau,u,Du)\cdot D\varphi\}\,dxd\tau \\ \le (\ge) \int_\Omega u_o\varphi(x,0)\,dx + \iint_{\Omega_t} b(x,\tau,u,Du)\varphi\,dxd\tau,\end{aligned} \tag{2.3}$$

for all bounded testing functions

$$\varphi \in W^{1,2}\left(0,T;L^2(\Omega)\right)\cap L^p\left(0,T;W_o^{1,p}(\Omega)\right), \quad \varphi\ge 0. \tag{2.4}$$

In addition the second of (2.1) holds in the sense that $u \le (\ge)\, g$ on $\partial\Omega$ in the sense of the traces of functions in $W^{1,p}(\Omega)$ for a.e. $t\in(0,T)$.

A function u that is both a sub-solution and a super-solution of (2.1) is a solution of the Dirichlet problem.

The formulation can be rephrased in terms of Steklov averages as in the previous section, namely

$$\int_{\Omega\times\{t\}} \{u_{h,t}\varphi + [\mathbf{a}(x,\tau,u,Du)]_h\cdot D\varphi - [b(x,\tau,u,Du)]_h\,\varphi\}\,dx \le (\ge)\, 0, \tag{2.5}$$

for all $0<t<T-h$ and for all $\varphi\in W_o^{1,p}(\Omega)\cap L^\infty(\Omega)$, $\quad \varphi\ge 0$.

Moreover the initial datum is taken in the sense of $L^2(\Omega)$, i.e.,

$$(u_h(\cdot,0)-u_o)_{+(-)} \longrightarrow 0 \quad \text{in } L^2(\Omega). \tag{2.6}$$

2-(ii). *Variational boundary data*

Assume that $\partial\Omega$ is piecewise smooth, so that the outward unit normal, which we denote with $\mathbf{n}$, is defined a.e. on $\partial\Omega$ and consider formally the Neumann problem

$$\begin{cases} u_t - \operatorname{div}\mathbf{a}(x,t,u,Du) = b(x,t,u,Du), & \text{in } \Omega_T,\\ \mathbf{a}(x,t,u,Du)\cdot\mathbf{n} = \psi(x,t,u), & \text{on } S_T,\\ u(\cdot,0) = u_o(\cdot). \end{cases} \tag{2.7}$$

We retain the structure conditions $(\mathbf{A}_1)$–$(\mathbf{A}_5)$ and the assumption $(\mathbf{U_o})$ on the initial datum. We assume that $\psi(\cdot,t,u(\cdot,t))$ admits, for a.e. $t\in(0,T)$, an extension into Ω which we denote by $\hat{\psi}(\cdot,t,u(\cdot,t))$, such that

$$\text{(N)}\qquad \begin{cases} |\hat{\psi}| \le \psi_o u + \psi_1, \text{ a.e. } \Omega_T, \\ |\hat{\psi}_u| \le \psi_o, \\ |\hat{\psi}_{x_i}| \le \psi_1,\ i=1,2,\dots,N, \end{cases}$$

where ψ_1, ψ_o are given non-negative functions satisfying

$$\text{(N–i)}\qquad \psi_1^{\frac{p}{p-1}}, \psi_o^{\frac{p}{p-1}} \in L^{\hat{q},\hat{r}}(\Omega_T), \text{ where } \hat{q} \text{ and } \hat{r} \text{ satisfy } (\mathbf{A}_5).$$

To give a notion of weak sub(super)-solution we let $\mathcal{K}$ be an arbitrary compact subset of $\mathbf{R}^N$ and consider testing functions

$$\text{(2.8)}\qquad \varphi \in W^{1,2}\left(0,T;L^2(\mathcal{K})\right) \cap L^p_{loc}\left(0,T;W^{1,p}_o(\mathcal{K})\right), \qquad \varphi \ge 0.$$

A function u,

$$\text{(2.9)}\qquad u \in C\left(0,T;L^2(\Omega)\right) \cap L^p\left(0,T;W^{1,p}(\Omega)\right),$$

is a weak sub(super)-solution of the Neumann problem (2.7) if for every compact subset $\mathcal{K}$ of $\mathbf{R}^N$ and for every subinterval $[t_1,t_2]$ of $[0,T]$

$$\text{(2.10)}\qquad \int_{\mathcal{K}\cap\Omega} u\varphi dx \Big|_{t_1}^{t_2} + \int_{t_1}^{t_2}\int_{\mathcal{K}\cap\Omega} \{-u\varphi_t + \mathbf{a}(x,\tau,u,Du)\cdot D\varphi\}\, dx d\tau$$
$$\le (\ge) \int_{t_1}^{t_2}\int_{\mathcal{K}\cap\Omega} b(x,\tau,u,Du)\varphi dx d\tau + \int_{t_1}^{t_2}\int_{\mathcal{K}\cap\partial\Omega} \psi(x,t,u)\varphi\, d\sigma d\tau,$$

where $d\sigma$ denotes the surface measure on $\partial\Omega$. We remark that the testing functions φ vanish, in the sense on the traces, on the boundary of $\mathcal{K}$ and *not* on the boundary of Ω. The variational datum is reflected in the boundary integral on the right hand side of (2.10). The formulation in terms of Steklov averages is

$$\text{(2.11)}\qquad \int_{[\mathcal{K}\cap\Omega]\times\{t\}} \{u_{h,t}\varphi + [\mathbf{a}(x,\tau,u,Du)]_h \cdot D\varphi - [b(x,\tau,u,Du)]_h\, \varphi\}\, dx$$
$$\le (\ge) \int_{[\mathcal{K}\cap\partial\Omega]\times\{t\}} [\psi(x,t,u)]_h\, \varphi\, d\sigma,$$
$$\text{for all } 0 < t < T-h \text{ and for all } \varphi \in W^{1,p}_o(\mathcal{K}), \quad \varphi \ge 0,$$

and the initial datum is taken in the sense of (2.6).

3. Local integral inequalities

We will derive some integral inequalities in the interior of Ω_T, which will be the main tools in establishing local Hölder estimates for the solutions. Analogous estimates near the lateral boundary S_T as well as near $t=0$ will be derived in the next section.

Let K_ρ denote the N-dimensional cube centered at the origin and wedge 2ρ, i.e.,

$$K_\rho \equiv \left\{ x \in \mathbf{R}^N \middle| \max_{1 \le i \le N} |x_i| < \rho \right\}.$$

If $x_o \in \mathbf{R}^N$, we let $[x_o + K_\rho]$ denote the cube of centre x_o and wedge 2ρ which is congruent to K_ρ, i.e.,

$$[x_o + K_\rho] \equiv \left\{ x \in \mathbf{R}^N \middle| \max_{1 \le i \le N} |x_i - x_{o,i}| < \rho \right\}.$$

Let θ be a given positive number and consider the cylinder

$$Q(\theta, \rho) \equiv K_\rho \times \{-\theta, 0\},$$

and if $(x_o, t_o) \in \mathbf{R}^{N+1}$, we let $[(x_o, t_o) + Q(\theta, \rho)]$ denote the cylinder with *'vertex'* at (x_o, t_o) congruent to $Q(\theta, \rho)$, i.e.,

$$[(x_o, t_o) + Q(\theta, \rho)] \equiv \left\{ x \in \mathbf{R}^N \middle| \max_{1 \le i \le N} |x_i - x_{o,i}| < \rho \right\} \times \{t_o - \theta, t_o\}.$$

We will refer to these as cubes and cylinders of *'radius'* ρ and height θ.

Fix $(x_o, t_o) \in \Omega_T$ and let ρ and θ be so small that $[(x_o, t_o) + Q(\theta, \rho)] \in \Omega_T$. Let ζ denote a piecewise smooth cutoff function in $[(x_o, t_o) + Q(\theta, \rho)]$ such that

(3.1) $$\zeta \in [0,1], |D\zeta| < \infty, \text{ and } \zeta(x,t) = 0, \text{ for } x \text{ outside } [x_o + K_\rho].$$

Assume that

(3.2) $$u \in L^\infty_{loc}(\Omega_T),$$

and construct the truncated functions $(u-k)_\pm$. We will choose levels k satisfying

(3.3) $$\operatorname*{ess\,sup}_{[(x_o,t_o)+Q(\theta,\rho)]} |(u-k)_\pm| \equiv H_k^\pm \le \delta,$$

where δ is a positive parameter to be chosen later.

Remark 3.1. Suppose (3.3) is written for $(u-k)_+$ and assume the number δ is small. Then the levels k are forced to be near the essential sup of u in $[(x_o, t_o) + Q(\theta, \rho)]$. Likewise if (3.3) is written for $(u-k)_-$, then k has to be close to the essential inf of u within $[(x_o, t_o) + Q(\theta, \rho)]$.

Roughly speaking, the function u is Hölder continuous if, within $[(x_o, t_o) + Q(\theta, \rho)]$, it is *close* in some integral norm to its integral average. Accordingly the sets where the function is near its supremum or near its infimum, within $[(x_o, t_o) + Q(\theta, \rho)]$, have relatively small measure. Our energy inequalities reflect this through the sets

$$A^{\pm}_{k,\rho}(\tau) \equiv \{x \in [x_o + K_\rho] \mid (u(x,\tau)-k)_\pm > 0\}. \tag{3.4}$$

In estimating the contribution of the *lower order* terms $\varphi_i,\ i=0,1,2$, it is convenient to introduce the numbers q, r, κ constructed starting from $\hat q, \hat r, \kappa_1$ as follows:

$$q = \frac{\hat q p(1+\kappa)}{\hat q - 1}; \qquad r = \frac{\hat r p(1+\kappa)}{\hat r - 1}; \qquad \kappa = \frac{p}{N}\kappa_1. \tag{3.5}$$

It is seen from $(\mathbf{A}_5\text{–}\mathbf{i})$–$(\mathbf{A}_5\text{–}\mathbf{iii})$ that they satisfy

$$\frac{1}{r} + \frac{N}{pq} = \frac{N}{p^2}, \tag{3.6}$$

and their admissible range is

$$\begin{cases} q \in (p,\infty), \quad r \in (p^2,\infty); & \text{if } N=1,\\ q \in \left(p, \frac{Np}{N-p}\right), \quad r \in (p,\infty); & \text{if } 1<p<N,\\ q \in (p,\infty), \quad r \in \left(\frac{p^2}{N},\infty\right); & \text{if } 1<N\le p. \end{cases} \tag{3.7}$$

The statement that a constant γ depends only upon the data means that it can be determined a priori only in terms of the numbers N, p, q, r, κ, the constants $C_i,\ i=0,1,2$, and the norms

$$\|\varphi_o, \varphi_1^{\frac{p}{p-1}}, \varphi_2\|_{\hat q, \hat r; \Omega_T}.$$

3-(i). Local energy estimates

PROPOSITION 3.1. *Let u be a locally bounded weak solution of (1.1) in Ω_T. There exist constants γ and δ_o that can be determined a priori only in terms of the data such that for every cylinder $[(x_o,t_o)+Q(\theta,\rho)] \subset \Omega_T$ and for every level k satisfying (3.3) for $\delta \le \delta_o$*

(3.8)

$$\begin{aligned} &\sup_{t_o-\theta<t<t_o} \int_{[x_o+K_\rho]} (u-k)_\pm^2 \zeta^p(x,t)\,dx + \gamma^{-1} \iint_{[(x_o,t_o)+Q(\theta,\rho)]} |D(u-k)_\pm \zeta|^p\,dx d\tau \\ &\quad \le \int_{[x_o+K_\rho]} (u-k)_\pm^2 \zeta^p(x,t_o-\theta)\,dx + \gamma \iint_{[(x_o,t_o)+Q(\theta,\rho)]} (u-k)_\pm^p |D\zeta|^p\,dx d\tau \\ &\quad + \gamma \iint_{[(x_o,t_o)+Q(\theta,\rho)]} (u-k)_\pm^2 \zeta^{p-1}\zeta_t\,dx d\tau + \gamma \left\{ \int_{t_o-\theta}^{t_o} |A^{\pm}_{k,\rho}(\tau)|^{\frac{r}{q}}\,d\tau \right\}^{\frac{p(1+\kappa)}{r}}. \end{aligned}$$

PROOF: After a translation we may assume that (x_o,t_o) coincides with the origin and it will suffice to prove (3.8) for the cube $Q(\theta,\rho)$. In the weak formulation (1.5) take the testing functions

$$\varphi = \pm(u_h - k)_\pm \zeta^p$$

and integrate over $(-\theta, t)$, $t \in (-\theta, 0)$. Estimating the various terms separately we have first

$$\int_{-\theta}^{t}\int_{K_\rho} \frac{\partial}{\partial\tau} u_h \varphi dx d\tau = \frac{1}{2}\int_{-\theta}^{t}\int_{K_\rho} \frac{\partial}{\partial\tau}(u_h - k)_\pm^2 \zeta^p dx d\tau.$$

Therefore integrating by parts and letting $h \to 0$ with the aid of Lemma 3.2 of Chap. I,

$$\int_{-\theta}^{t}\int_{K_\rho} \frac{\partial}{\partial\tau} u_h \varphi \, dx d\tau \longrightarrow$$

$$\frac{1}{2}\int_{K_\rho}(u-k)_\pm^2 \zeta^p(x,t)\, dx - \frac{1}{2}\int_{K_\rho}(u-k)_\pm^2 \zeta^p(x,-\theta)\, dx$$

$$-\frac{p}{2}\int_{-\theta}^{t}\int_{K_\rho}(u-k)_\pm^2 \zeta^{p-1}\zeta_t \, dx d\tau.$$

In estimating the remaining parts we first let $h \to 0$ and then use the structure conditions $(\mathbf{A}_1)$–$(\mathbf{A}_5)$. To simplify the notation, set

$$Q^t \equiv K_\rho \times (-\theta, t), \qquad t \in (-\theta, 0).$$

Then

$$(3.9) \qquad \pm \iint_{Q^t} [\mathbf{a}(x,t,u,Du)]_h \cdot D\left((u_h - k)_\pm \zeta^p\right) dx d\tau \longrightarrow$$

$$\iint_{Q^t} \mathbf{a}(x,t,u,Du) \cdot \left[\pm D(u-k)_\pm \zeta^p \pm p(u-k)_\pm \zeta^{p-1} D\zeta\right] dx d\tau$$

$$\geq C_o \iint_{Q^t} |D(u-k)_\pm|^p \zeta^p dx d\tau - \iint_{Q^t} \varphi_o \zeta^p \chi\left[(u-k)_\pm > 0\right] dx d\tau$$

$$- pC_1 \iint_{Q^t} |D(u-k)_\pm|^{p-1}(u-k)_\pm \zeta^{p-1}|D\zeta| dx d\tau$$

$$- p \iint_{Q^t} \varphi_1 (u-k)_\pm \zeta^{p-1}|D\zeta| dx d\tau,$$

where $\chi(\Sigma)$ denotes the characteristic function of the set Σ. By Young's inequality

(i)
$$pC_1 \iint_{Q^t} |D(u-k)_\pm|^{p-1}(u-k)_\pm \zeta^{p-1}|D\zeta| dx d\tau$$
$$\le \frac{C_o}{2} \iint_{Q^t} |D(u-k)_\pm|^p \zeta^p dx d\tau + \gamma(C_o) \iint_{Q^t} (u-k)_\pm^p |D\zeta|^p dx d\tau,$$

(ii)
$$p \iint_{Q^t} \varphi_1 (u-k)_\pm \zeta^{p-1}|D\zeta| dx d\tau \le \iint_{Q^t} (u-k)_\pm^p |D\zeta|^p dx d\tau$$
$$+ \gamma \iint_{Q^t} \varphi_1^{\frac{p}{p-1}} \chi\left[(u-k)_\pm > 0\right] dx d\tau.$$

Combining this in (3.9) we arrive at

$$\iint_{Q^t} \mathbf{a}(x,\tau,u,Du) \cdot D\left((u-k)_\pm \zeta^p\right) dx d\tau$$
$$\ge \frac{C_o}{2} \iint_{Q^t} |D(u-k)_\pm \zeta|^p dx d\tau - \gamma \iint_{Q^t} (u-k)_\pm^p |D\zeta|^p dx d\tau$$
$$- \gamma \iint_{Q^t} \left(\varphi_o + \varphi_1^{\frac{p}{p-1}}\right) \chi\left[(u-k)_\pm > 0\right] dx d\tau.$$

Finally

(3.10)
$$\iint_{Q^t} |b(x,\tau,u,Du)(u-k)_\pm \zeta^p| dx d\tau$$
$$\le C_2 \iint_{Q^t} |D(u-k)_\pm|^p (u-k)_\pm \zeta^p dx d\tau + \iint_{Q^t} \varphi_2 (u-k)_\pm \zeta^p dx d\tau.$$

Now if we impose on the levels k the restriction

(3.11)
$$\operatorname*{ess\,sup}_{Q(\theta,\rho)} |(u-k)_\pm| \le \delta_o \equiv \frac{C_o}{4C_2},$$

we deduce from (3.10)

$$
\begin{aligned}
(3.10') \qquad & \iint_{Q^t} \left| b(x,\tau,u,Du)(u-k)_\pm \zeta^p \right| dx d\tau \\
& \leq \frac{C_o}{4} \iint_{Q^t} |D(u-k)_\pm|^p \zeta^p dx d\tau + \iint_{Q^t} \varphi_2 (u-k)_\pm \zeta^p dx d\tau \\
& \leq \frac{C_o}{4} \iint_{Q^t} |D(u-k)_\pm \zeta|^p dx d\tau + \delta_o \iint_{Q^t} \varphi_2 \chi\left[(u-k)_\pm > 0\right] dx d\tau \\
& + \gamma \iint_{Q^t} (u-k)_\pm^p |D\zeta|^p dx d\tau.
\end{aligned}
$$

Combining these estimates and recalling that $t \in (-\theta, 0)$ is arbitrary, we obtain

$$
\begin{aligned}
& \sup_{-\theta<t<0} \int_{K_\rho} (u-k)_\pm^2 \zeta^p(x,t)\, dx + \frac{C_o}{2} \iint_{Q(\theta,\rho)} |D(u-k)_\pm \zeta|^p dx d\tau \\
& \quad \leq \int_{K_\rho} (u-k)_\pm^2 \zeta^p(x,-\theta) dx \\
& \quad + \gamma \iint_{Q(\theta,\rho)} (u-k)_\pm^p |D\zeta|^p dx d\tau + \gamma \iint_{Q(\theta,\rho)} (u-k)_\pm^2 \zeta^{p-1} \zeta_t \, dx d\tau \\
& \quad + \gamma \iint_{Q(\theta,\rho)} \left(\varphi_o + \varphi_1^{\frac{p}{p-1}} + \varphi_2 \right) \chi\left[(u-k)_\pm > 0\right] dx d\tau.
\end{aligned}
$$

By Hölder's inequality

$$
\begin{aligned}
& \iint_{Q(\theta,\rho)} \left(\varphi_o + \varphi_1^{\frac{p}{p-1}} + \varphi_2 \right) \chi\left[(u-k)_\pm > 0\right] dx d\tau \\
& \quad \leq \|\varphi_o + \varphi_1^{\frac{p}{p-1}} + \varphi_2\|_{\hat{q},\hat{r},\Omega_T} \left\{ \int_{-\theta}^{0} |A_{k,\rho}^\pm(\tau)|^{\frac{\hat{q}-1}{\hat{q}} \frac{\hat{r}}{\hat{r}-1}} d\tau \right\}^{\frac{\hat{r}-1}{\hat{r}}}.
\end{aligned}
$$

Therefore recalling the definition (3.5) of the numbers q, r, κ, inequality (3.8) follows.

Remark 3.2. The proof shows that the number δ_o in (3.11) has to be chosen small according to the constant C_2. If in (1.1) $b(x,t,u,Du) \equiv 0$, then δ_o can be taken to be infinite and no restriction is imposed on the levels k.

Remark 3.3. If the lower order terms φ_i are all zero, then in (3.8) the last term can be discarded.

3-(ii). Local logarithmic estimates

Introduce the logarithmic function

$$\text{(3.12)}\quad \Psi\left(H_k^\pm,(u-k)_\pm,c\right)\equiv \ln^+\left\{\frac{H_k^\pm}{H_k^\pm-(u-k)_\pm+c}\right\},\quad 0<c<H_k^\pm,$$

where $H_k^\pm$ is defined in (3.3) via the levels k, and for $s>0$

$$\ln^+ s\equiv\max\{\ln s\,;\,0\}.$$

In the cylinder $[(x_o,t_o)+Q(\theta,\rho)]$ we take a cutoff function satisfying (3.1) and

$$\text{(3.13)}\qquad \zeta \text{ is independent of } t\in(t_o-\theta,t_o].$$

PROPOSITION 3.2. *Let u be a locally bounded weak solution of (1.1) in Ω_T. There exist constants γ and δ_o that can be determined a priori only in terms of the data, such that for every cylinder $(x_o,t_o)+Q(\theta,\rho)\in\Omega_T$ and for every level k satisfying (3.3) for $\delta\le\delta_o$*

$$\text{(3.14)}\qquad \sup_{t_o-\theta<t<t_o}\int_{[x_o+K_\rho]}\Psi^2\left(H_k^\pm,(u-k)_\pm,c\right)(x,t)\zeta^p(x)\,dx$$
$$\le\int_{[x_o+K_\rho]}\Psi^2\left(H_k^\pm,(u-k)_\pm,c\right)(x,t_o-\theta)\zeta^p(x)\,dx$$
$$+\gamma\iint_{[(x_o,t_o)+Q(\theta,\rho)]}\Psi\left|\Psi_u\left(H_k^\pm,(u-k)_\pm,c\right)\right|^{2-p}|D\zeta|^p\,dxd\tau$$
$$+\frac{\gamma}{c^2}\left(1+\ln\frac{H_k^\pm}{c}\right)\left\{\int_{t_o-\theta}^{t_o}|A_{k,\rho}^\pm(\tau)|^{\frac{r}{q}}d\tau\right\}^{\frac{p(1+\kappa)}{r}}.$$

PROOF: As before, we may take $(x_o,t_o)\equiv(0,0)$ and will work within the cylinder Q^t introduced earlier. Also, to simplify the symbolism let us set

$$\text{(3.15)}\qquad \Psi\left(H_k^\pm,(u-k)_\pm,c\right)=\psi(u).$$

In (1.5) take the testing function

$$\varphi=\frac{\partial}{\partial u_h}\left[\psi^2(u_h)\right]\zeta^p=\left[\psi^2(u_h)\right]'\zeta^p.$$

By direct calculation

$$\left[\psi^2(u_h)\right]''=2(1+\psi)\psi'^2\in L^\infty_{loc}(\Omega_T)$$

which implies that such a φ is an admissible testing function in (1.5). Since $\psi(u_h)$ vanishes on the set where $(u_h-k)_\pm=0$,

$$\iint_{Q^t} \frac{\partial}{\partial\tau} u_h \left[\psi^2\right]' \zeta^p dx d\tau = \iint_{Q^t} \frac{\partial}{\partial\tau} \psi^2 \zeta^p dx d\tau$$

$$= \int_{K_\rho \times \{t\}} \psi^2(u_h)\zeta^p dx - \int_{K_\rho \times \{-\theta\}} \psi^2(u_h)\zeta^p dx.$$

Therefore letting $h \to 0$

$$\iint_{Q^t} \frac{\partial}{\partial\tau} u_h \left[\psi^2\right]' \zeta^p dx d\tau \longrightarrow \int_{K_\rho \times \{t\}} \Psi^2\left(H_k^\pm, (u-k)_\pm, c\right) \zeta^p dx$$

$$- \int_{K_\rho \times \{-\theta\}} \Psi^2\left(H_k^\pm, (u-k)_\pm, c\right) \zeta^p dx.$$

To estimate the remaining terms we let $h \to 0$ first and then make use of the structure conditions $(\mathbf{A}_1)$–$(\mathbf{A}_5)$.

$$\iint_{Q^t} \mathbf{a}(x,\tau,u,Du) \cdot D\varphi dx d\tau \geq 2C_o \iint_{Q^t} (1+\psi)\,\psi'^2 |Du|^p \zeta^p dx d\tau$$

$$- 2\iint_{Q^t} (1+\psi)\,\psi'^2 \varphi_o(x,\tau)\zeta^p dx d\tau$$

$$- 2pC_1 \iint_{Q^t} |Du|^{p-1} \psi\psi' \zeta^{p-1} |D\zeta| dx d\tau$$

$$- 2pC_1 \iint_{Q^t} \psi\psi' \varphi_1(x,\tau)\zeta^{p-1}|D\zeta| dx d\tau.$$

From this, by repeated application of Young's inequality

$$\iint_{Q^t} \mathbf{a}(x,\tau,u,Du) \cdot D\varphi dx d\tau \geq C_o \iint_{Q^t} (1+\psi)\,\psi'^2 |Du|^p \zeta^p dx d\tau$$

$$- 2\iint_{Q^t} (1+\psi)\,\psi'^2 \varphi_o(x,\tau)\zeta^p dx d\tau$$

$$- \gamma(p) \iint_{Q^t} \psi\,(\psi')^{2-p}\,|D\zeta|^p dx d\tau$$

$$- \gamma(p) \iint_{Q^t} \psi\,(\psi')^2\,\varphi_1^{\frac{p}{p-1}} \zeta^p dx d\tau.$$

For the lower order terms we have

$$\iint_{Q^t} |b(x,\tau,u,Du)\psi\psi'\zeta^p| dx d\tau \le C_2 \iint_{Q^t} |Du|^p (1+\psi)\psi'^2\psi'^{-1}\zeta^p dx d\tau$$
$$+ \iint_{Q^t} \varphi_2 \psi\psi'\zeta^p dx d\tau.$$

Next we observe that by virtue of (3.3) and the definition (3.12) and (3.15) of ψ

$$\psi'^{-1} = H_k^{\pm} - (u-k)_{\pm} + c < 2\delta$$

and

$$\psi \le \ln\left(\frac{H_k^{\pm}}{c}\right), \qquad \psi' \le \frac{1}{c}.$$

Therefore by virtue of the choice (3.11) of the levels k we have

$$\iint_{Q^t} |b(x,\tau,u,Du)\psi\psi'\zeta^p| dx d\tau \le \frac{C_o}{2} \iint_{Q^t} |Du|^p (1+\psi)\psi'^2\zeta^p dx d\tau$$
$$+ \frac{1}{c}\ln\left(\frac{H_k^{\pm}}{c}\right) \iint_{Q^t} |\varphi_2| \chi\left[(u-k)_{\pm} > 0\right] dx d\tau.$$

Collecting these estimates we arrive at

$$\int_{K_\rho \times \{t\}} \Psi^2\left(H_k^{\pm}, (u-k)_{\pm}, c\right)\zeta^p dx$$
$$\le \int_{K_\rho \times \{-\theta\}} \Psi^2\left(H_k^{\pm}, (u-k)_{\pm}, c\right)\zeta^p dx + \gamma \iint_{Q(\theta,\rho)} \psi|\psi'|^{2-p}|D\zeta|^p dx d\tau$$
$$+ \frac{\gamma}{c^2}\left(1 + \ln\frac{H_k^{\pm}}{c}\right) \iint_{Q(\theta,\rho)} \left(\varphi_o + \varphi_1^{\frac{p}{p-1}} + \varphi_2\right)\chi\left[(u-k)_{\pm} > 0\right] dx d\tau$$

where we have used the fact that $c<1$. Treating the last integral as before proves the Proposition.

Remark 3.4. If the constant C_2 in $(\mathbf{A}_3)$ is zero, then we may take $\delta=\infty$ and there is no restriction on the levels k. Also if $\varphi_i \equiv 0$, $i=0,1,2$, then the last term on the right hand side of (3.14) can be discarded.

Remark 3.5. In any case, whence the constant δ_o has been chosen according to (3.11), the constant γ on the right hand side of either (3.8) or (3.14), is *independent of* u. It is only the levels k that might depend upon the solution u via (3.11).

4. Energy estimates near the boundary

We assume u is a weak solution of either the Dirichlet problem (2.1) or the Neumann problem (2.7), satisfying in addition

$$u \in L^{\infty}(\Omega_T) \quad \text{and} \quad u \in L^2\left(0,T;W^{1,p}(\Omega_T)\right). \tag{4.1}$$

The assumptions $(\mathbf{D}), (\mathbf{U_o}), (\mathbf{N}), (\mathbf{N}-(\mathbf{i}))$ on the boundary data will be retained. We will derive energy and logarithmic estimates, similar to those of Propositions 3.1 and 3.2, near the lateral boundary S_T as well as at $t=0$.

Fix a point (x_o,t_o) on S_T, and construct the box $[(x_o,t_o)+Q(\theta,\rho)]$, where θ is so small that $t_o-\theta>0$. In $[(x_o,t_o)+Q(\theta,\rho)]$ introduce a piecewise smooth cutoff function $(x,t)\to\zeta(x,t)$ satisfying (3.1). We observe that for all $t\in(t_o-\theta,t_o)$, $x\to\zeta(x,t)$ vanishes on the boundary of $[x_o+K_\rho]$ and *not* on the boundary of $[x_o+K_\rho]\cap\Omega$.

Here the *interior quantities* introduced in the previous section are modified as follows

$$\operatorname*{ess\,sup}_{[(x_o,t_o)+Q(\theta,\rho)]\cap\Omega_T} \left|(u-k)_\pm\right| \equiv D_k^\pm \le \delta, \tag{4.2}$$

where $\delta\le\delta_o$ and δ_o is a parameter chosen according to (3.11). Analogously we define the logarithmic function

$$\Psi\left(D_k^\pm,(u-k)_\pm,c\right) \equiv \ln^+\left\{\frac{D_k^\pm}{D_k^\pm-(u-k)_\pm+c}\right\}, \quad c<D_k^\pm, \tag{4.3}$$

and introduce the sets

$$B_{k,\rho}^\pm(\tau) \equiv \left\{x\in[x_o+K_\rho]\cap\Omega \,\middle|\, (u(x,\tau)-k)_\pm>0\right\}. \tag{4.4}$$

4-(i). Variational boundary data

Let u be a weak solution of (2.7) satisfying (4.1) and assume in addition that

$$\partial\Omega \text{ is of class } C^{1+\lambda} \text{ for some } \lambda\in(0,1). \tag{4.5}$$

PROPOSITION 4.1. *There exist constants γ and δ_o that can be determined a priori only in terms of the data and the quantities $\|u\|_{\infty,\Omega_T}$ and $\||\partial\Omega|\|_{1+\lambda}$, such that for every $(x_o,t_o)\in S_T$, for every cylinder $[(x_o,t_o)+Q(\theta,\rho)]$ such that $t_o-\theta>0$ and for every level k satisfying (4.2) for $\delta\le\min\{\delta_o\,;\,1\}$*

$$\begin{aligned}
&\sup_{t_o-\theta<t<t_o} \int_{[x_o+K_\rho]\cap\Omega} (u-k)_\pm^2\zeta^p(x,t)\,dx + \gamma^{-1}\iint_{[(x_o,t_o)+Q(\theta,\rho)]\cap\Omega_T} |D(u-k)_\pm\zeta|^p dxd\tau \\
&\quad\le \int_{[x_o+K_\rho]\cap\Omega} (u-k)_\pm^2\zeta^p(x,t_o-\theta)\,dx + \gamma\iint_{[(x_o,t_o)+Q(\theta,\rho)]\cap\Omega_T} (u-k)_\pm^{p}|D\zeta|^p dxd\tau \\
&\quad+\gamma\iint_{[(x_o,t_o)+Q(\theta,\rho)]\cap\Omega_T} (u-k)_\pm^2\zeta^{p-1}\zeta_t dxd\tau + \gamma\left\{\int_{t_o-\theta}^{t_o} |B_{k,\rho}^\pm(\tau)|^{\frac{r}{q}}d\tau\right\}^{\frac{p(1+\kappa)}{r}}.
\end{aligned} \tag{4.6}$$

Moreover if ζ is independent of $t\in(t_o-\theta,t_o)$,

$$\tag{4.7} \begin{aligned} \sup_{t_o-\theta<t<t_o} &\int_{[x_o+K_\rho]\cap\Omega} \Psi^2\left(D_k^\pm,(u-k)_\pm,c\right)(x,t)\zeta^p(x)\,dx \\ &\le \int_{[x_o+K_\rho]\cap\Omega} \Psi^2\left(D_k^\pm,(u-k)_\pm,c\right)(x,t_o-\theta)\zeta^p(x)\,dx \\ &+\gamma \iint_{[(x_o,t_o)+Q(\theta,\rho)]\cap\Omega_T} \Psi\left|\Psi_u\left(D_k^\pm,(u-k)_\pm,c\right)\right|^{2-p}|D\zeta|^p\,dxd\tau \\ &+\frac{\gamma}{c^2}\left(1+\ln\frac{D_k^\pm}{c}\right)\left\{\int_{t_o-\theta}^{t_o}\left|B_{k,\rho}^\pm(\tau)\right|^{\frac{r}{q}}d\tau\right\}^{\frac{p(1+\kappa)}{r}} \end{aligned}$$

where the numbers q,r,κ satisfy (3.6)-(3.7).

Remark 4.1. If the Neumann data are homogeneous, i.e., $\psi_i\equiv 0$, $i=0,1$, and $\varphi_i\equiv 0$, $i=0,1,2$, and $b(x,t,u,Du)\equiv 0$, then we may take $\delta=\infty$ and the levels k are not restricted.

Remark 4.2. The proof below is local in nature and it shows that (4.6)-(4.7) hold true for weak solutions that satisfy the Neumann data on a portion of S_T. Accordingly, only such a portion is required to be of class $C^{1+\lambda}$. Also, no reference to initial data is necessary.

PROOF OF PROPOSITION 4.1: Fix $(x_o,t_o)\in S_T$, assume that (x_o,t_o) coincides with the origin and work with cubes K_ρ and cylinders $Q(\theta,\rho)$. Since $\partial\Omega$ is of class $C^{1+\lambda}$, for sufficently small ρ, the portion of $\partial\Omega$ within the cube K_ρ, can be represented in a local system of coordinates as a portion of the hyperplane $x_N=0$ and $K_\rho\cap\Omega\subset\{x_N>0\}$. Set

$$\tag{4.8} K_\rho^+\equiv K_\rho\cap\{x_N>0\}\ \text{ and }\ Q^+(\theta,\rho)\equiv Q(\theta,\rho)\cap\{x_N>0\}.$$

Without loss of generality we may assume that (2.11) is written in such a coordinate system. To derive (4.6), in (2.11) we take the testing functions

$$\varphi=\pm(u_h-k)_\pm\zeta^p,$$

and let $h\to 0$. All the terms in (2.11) are treated as in the proof of Proposition 3.1, except for the boundary integral. We arrive at

(4.9)
$$\sup_{-\theta<t<0}\int_{K_\rho^+}(u-k)_\pm^2\zeta^p(x,t)dx+\frac{C_o}{2}\iint_{Q^+(\theta,\rho)}|D(u-k)_\pm\zeta|^p dxd\tau$$
$$\le\int_{K_\rho}(u-k)_\pm^2\zeta^p(x,-\theta)dx$$
$$+\gamma\iint_{Q^+(\theta,\rho)}(u-k)_\pm^p|D\zeta|^p dxd\tau+\gamma\iint_{Q^+(\theta,\rho)}(u-k)_\pm^2\zeta^{p-1}\zeta_t dxd\tau$$
$$+\gamma\iint_{Q^+(\theta,\rho)}\left(\varphi_o+\varphi_1^{\frac{p}{p-1}}+\varphi_2\right)\chi\left[(u-k)_\pm>0\right]dxd\tau$$
$$+\gamma\int_{-\theta}^{0}\int_{\tilde K_\rho}\psi(x,\tau,u)(u-k)_\pm\zeta^p d\bar x d\tau.$$

where $\tilde K_\rho\equiv K_\rho\cap\partial\Omega$ is the $(N-1)$-dimensional cube

$$\tilde K_\rho\equiv\left\{\bar x\equiv(x_1,x_2,\dots,x_{N-1})\in\mathbf{R}^{N-1}\,\middle|\,\max_{1\le i\le(N-1)}|x_i|<\rho\right\}.$$

We estimate such a boundary integral by transforming it into an interior integral as follows

$$\left|\int_{-\theta}^{0}\int_{\tilde K_\rho}\psi(x,\tau,u)(u-k)_\pm\zeta^p d\bar x d\tau\right|$$
$$=\left|\int_{-\theta}^{0}\int_{\tilde K_\rho}\left(\int_0^\rho\frac{\partial}{\partial x_N}\left(\hat\psi(x,\tau,u)(u-k)_\pm\zeta^p\right)dx_N\right)d\bar x d\tau\right|$$
$$\le\gamma\iint_{Q^+(\theta,\rho)}\left(|\hat\psi_{x_N}|(u-k)_\pm\zeta^p+|\hat\psi||D(u-k)_\pm\zeta|\right)dxd\tau$$
$$+\gamma\iint_{Q^+(\theta,\rho)}\left(|\hat\psi|(u-k)_\pm\zeta^{p-1}|D\zeta|+|\hat\psi_u||D(u-k)_\pm|(u-k)_\pm\zeta^p\right)dxd\tau.$$

By virtue of assumption (**N**),

$$\iint_{Q^+(\theta,\rho)}|\hat\psi_{x_N}|(u-k)_\pm\zeta^p dxd\tau\le D_k^\pm\iint_{Q^+(\theta,\rho)}\psi_1\chi\left[(u-k)_\pm>0\right]dxd\tau,$$

where $D_k^\pm$ is defined in (4.2). Also by Young's inequality

$$
\begin{aligned}
&\iint_{Q^+(\theta,\rho)} |\hat{\psi}| \left\{ |D(u-k)_\pm \zeta| + (u-k)_\pm \zeta^{p-1} |D\zeta| \right\} dx d\tau \\
&\quad \le \frac{C_o}{8} \iint_{Q^+(\theta,\rho)} |D(u-k)_\pm \zeta|^p dx d\tau + \gamma(p) \iint_{Q^+(\theta,\rho)} (u-k)_\pm^p |D\zeta|^p dx d\tau \\
&\qquad + \gamma\left(p, \|u\|_{\infty,\Omega_T}\right) \iint_{Q^+(\theta,\rho)} (\psi_o + \psi_1)^{\frac{p}{p-1}} \chi\left[(u-k)_\pm > 0\right] dx d\tau.
\end{aligned}
$$

Finally

$$
\begin{aligned}
&\iint_{Q^+(\theta,\rho)} |\hat{\psi}_u| |D(u-k)_\pm| (u-k)_\pm \zeta^p dx d\tau \\
&\quad \le \gamma \iint_{Q^+(\theta,\rho)} |\hat{\psi}_u| \left\{ |D(u-k)_\pm \zeta| (u-k)_\pm \zeta^{p-1} + (u-k)_\pm^2 \zeta^{p-1} |D\zeta| \right\} dx d\tau \\
&\quad \le \frac{C_o}{8} \iint_{Q^+(\theta,\rho)} |D(u-k)_\pm \zeta|^p dx d\tau + \gamma(p, \delta_o) \iint_{Q^+(\theta,\rho)} (u-k)_\pm^p |D\zeta|^p dx d\tau \\
&\qquad + \gamma\left(p, \|u\|_{\infty,\Omega_T}\right) \iint_{Q^+(\theta,\rho)} \psi_o^{\frac{p}{p-1}} \chi\left[(u-k)_\pm > 0\right] dx d\tau.
\end{aligned}
$$

Combining these estimates implies that the boundary integral on the right hand side of (4.9) can be estimated by

$$
\begin{aligned}
&\frac{C_o}{8} \iint_{Q^+(\theta,\rho)} |D(u-k)_\pm \zeta|^p dx d\tau + \gamma\left(p, \|u\|_{\infty,\Omega_T}\right) \iint_{Q^+(\theta,\rho)} (u-k)_\pm^p |D\zeta|^p dx d\tau \\
&\qquad + \gamma(p, \delta_o) \iint_{Q^+(\theta,\rho)} (1 + \psi_o + \psi_1)^{\frac{p}{p-1}} \chi\left[(u-k)_\pm > 0\right] dx d\tau.
\end{aligned}
$$

We put this in (4.9) and, to conclude the proof, estimate the integral involving the functions φ_i, $i=0,1,2$, and ψ_i, $i=0,1$, as in the proof of Proposition 3.1.

The proof of the logarithmic estimate (4.7) near the lateral boundary S_T is similar to the proof of the interior logarithmic estimate (3.14), modulo the modifications indicated above and we omit the details.

4-(ii). Dirichlet boundary data

Let u be a weak solution of the Dirichlet problem (2.1), which in addition satisfies (4.1). The assumption $(\mathbf{D})$ on the boundary datum g is retained.

Fix $(x_o, t_o) \in S_T$ and consider the cylinder $[(x_o, t_o) + Q(\theta, \rho)]$, where θ is so small that $t_o - \theta > 0$. Local energy estimates for u near (x_o, t_o) are obtained by taking, in the weak formulation (2.5), the testing functions

$$\varphi_h^{\pm} = \pm(u_h - k)_{\pm}\zeta^p,$$

integrating over $[(x_o, t_o) + Q(\theta, \rho)]$ and letting $h \to 0$. Such a choice of testing functions is admissible if for a.e. $t \in (t_o - \theta, t_o]$,

$$(4.10) \qquad (u(\cdot, t) - k)_{\pm}\, \zeta^p(x, t) \in W_o^{1,p}([x_o + K_\rho] \cap \Omega).$$

Since $x \to \zeta(x, t)$ vanishes on the boundary of $[x_o + K_\rho]$ and *not* on the boundary of $[x_o + K_\rho] \cap \Omega$, condition (4.10) will be verified if for a.e. $t \in (t_o - \theta, t_o]$

$$(u - k)_{\pm} = 0 \ \text{ in the sense of the traces on } \partial\,[x_o + K_\rho] \cap \Omega.$$

In view of Lemma 3.1 of Chap. I, this can be realised for the function $(u - k)_+$ if k is chosen to satisfy

$$(4.11) \qquad k \geq \sup_{[(x_o, t_o) + Q(\theta, \rho)] \cap S_T} g.$$

Analogously the functions $-(u_h - k)_- \zeta^p$ can be taken as testing functions in (2.5) if

$$(4.12) \qquad k \leq \inf_{[(x_o, t_o) + Q(\theta, \rho)] \cap S_T} g.$$

With these choices of k we may repeat calculations in all analogous to those of Proposition 3.1 and derive energy inequalities for u near S_T. Analogous considerations hold for a version of the logarithmic estimates along the lines of Proposition 3.2. We summarise

PROPOSITION 4.2. *There exist constants γ and δ_o that can be determined a priori only in terms of the data and such that for every $(x_o, t_o) \in S_T$, for every cylinder $[(x_o, t_o) + Q(\theta, \rho)]$ such that $t_o - \theta > 0$ and for every level k satisfying (4.2) for $\delta \leq \delta_o$ and in addition (4.11) for the functions $(u - k)_+$ and (4.12) for $(u - k)_-$, the following inequalities hold:*

$$(4.13) \quad \sup_{t_o - \theta < t < t_o} \int_{[x_o + K_\rho] \cap \Omega} (u - k)_{\pm}^2 \zeta^p(x, t)\,dx + \gamma^{-1} \iint_{[(x_o, t_o) + Q(\theta, \rho)] \cap \Omega_T} |D(u - k)_{\pm}\zeta|^p\,dx d\tau$$
$$\leq \int_{[x_o + K_\rho] \cap \Omega} (u - k)_{\pm}^2 \zeta^p(x, t_o - \theta)\,dx + \gamma \iint_{[(x_o, t_o) + Q(\theta, \rho)] \cap \Omega_T} (u - k)_{\pm}^p |D\zeta|^p\,dx d\tau$$
$$+\gamma \iint_{[(x_o, t_o) + Q(\theta, \rho)] \cap \Omega_T} (u - k)_{\pm}^2 \zeta^{p-1}\zeta_t\,dx d\tau + \gamma \left\{ \int_{t_o - \theta}^{t_o} |B_{k,\rho}^{\pm}(\tau)|^{\frac{r}{q}}\,d\tau \right\}^{\frac{p(1+\kappa)}{r}}.$$

Moreover if the cutoff function ζ is independent of $t \in (t_o - \theta, t_o)$,

$$\sup_{t_o-\theta<t<t_o} \int_{[x_o+K_\rho]\cap\Omega} \Psi^2\left(D_k^\pm, (u-k)_\pm, c\right)(x,t)\zeta^p(x)dx \tag{4.14}$$
$$\leq \int_{[x_o+K_\rho]\cap\Omega} \Psi^2\left(D_k^\pm, (u-k)_\pm, c\right)(x, t_o-\theta)\zeta^p(x)dx$$
$$+\gamma \iint_{[(x_o,t_o)+Q(\theta,\rho)]\cap\Omega_T} \Psi\left|\Psi_u\left(D_k^\pm, (u-k)_\pm, c\right)\right|^{2-p}|D\zeta|^p dxd\tau$$
$$+\frac{\gamma}{c^2}\left(1+\ln\frac{D_k^\pm}{c}\right)\left\{\int_{t_o-\theta}^{t_o} |B_{k,\rho}^\pm(\tau)|^{\frac{r}{q}}d\tau\right\}^{\frac{p(1+\kappa)}{r}}$$

where the numbers q, r, κ satisfy (3.6)-(3.7).

Local considerations as those in §3 apply to the present case. In particular the Proposition continues to hold for weak solutions that satisfy the Dirichlet data on a portion of S_T.

4-(iii). Initial data

Consider a weak solution of (1.1) that takes the initial datum u_o in the sense that

$$\frac{1}{h}\int_0^h u(\cdot,\tau)d\tau \longrightarrow u_o \quad in\ L^2_{loc}(\Omega)\ as\ h\to 0. \tag{4.15}$$

Thus u could be a solution of either the Dirichlet problem (2.1) or the Neumann problem (2.7). In either case the assumption $(\mathbf{U_o})$ is in force.

Fix $(x_o, t_o) \in \Omega_T$ and consider the cylinder $[(x_o,t_o)+Q(\theta,\rho)]$ where θ is such that $t_o-\theta=0$. Therefore $[(x_o,t_o)+Q(\theta,\rho)]$ lies on the bottom of the cylindrical domain Ω_T. Consider a cutoff function ζ satisfying (3.1) and in addition

$$\zeta \quad \text{is independent of} \ \in (0, t_o).$$

Local energy estimates for u near $t=0$ are derived by taking in the weak formulation (1.5) testing functions

$$\varphi_h^\pm = \pm(u_h-k)_\pm\zeta^p,$$

integrating over $(0,t), t\in(0,t_o)$, and letting $h\to 0$. The first term in (1.5) gives

$$\frac{1}{2}\int_{[x_o+K_\rho]} (u_h-k)_\pm^2(x,t)\zeta^p dx - \frac{1}{2}\int_{[x_o+K_\rho]} (u_h-k)_\pm^2(x,0)\zeta^p dx.$$

If k is chosen so that $k\geq \sup_{[x_o+K_\rho]} u_o$, then in view of (4.15) we have

$$\int_{[x_o+K_\rho]} (u_h(x,0)-k)_+^2\,\zeta^p dx \longrightarrow 0 \text{ as } h\to 0.$$

Also from the definition (3.12) of the function $\Psi(\cdot)$, it follows that

$$\Psi\left(D_k^{\pm},(u-k)_{\pm},c\right)=0 \quad \text{whenever } (u-k)_{\pm}=0.$$

Thus if $k\geq \sup_{[x_o+K_\rho]} u_o$, then

$$\int_{[x_o+K_\rho]} \Psi^2\left(D_k^{+},(u_h-k)_+,c\right)(x,0)\zeta^p(x)dx \longrightarrow 0 \quad \text{as } h\to 0.$$

Analogous considerations hold for $(u_h-k)_-\,\zeta^p$. We summarise

PROPOSITION 4.3. *There exist constants γ and δ_o that can be determined a priori only in terms of the data such that for every $(x_o,t_o)\in\Omega_T$, for every cylinder $[(x_o,t_o)+Q(\theta,\rho)]$ such that $t_o-\theta=0$ and for every level k satisfying (4.2) for $\delta\leq\delta_o$ and in addition*

$$\text{(4.16)}\qquad \begin{cases} k\geq \sup_{[x_o+K_\rho]} u_o & \textit{for the function } (u-k)_+ \\ k\leq \inf_{[x_o+K_\rho]} u_o & \textit{for the function } (u-k)_-, \end{cases}$$

the following inequalities hold:

$$\text{(4.17)}\qquad \sup_{t_o-\theta<t<t_o} \int_{[x_o+K_\rho]} (u-k)_{\pm}^2(x,t)\zeta^p(x)\,dx + \iint_{[(x_o,t_o)+Q(\theta,\rho)]} |D(u-k)_{\pm}\zeta|^p dxd\tau$$
$$\leq \gamma \iint_{[(x_o,t_o)+Q(\theta,\rho)]\cap\Omega_T} (u-k)_{\pm}^p|D\zeta|^p dxd\tau + \gamma\left\{\int_0^{t_o} |B_{k,\rho}^{\pm}(\tau)|^{\frac{r}{q}}d\tau\right\}^{\frac{p(1+\kappa)}{r}}.$$

Moreover

$$\text{(4.18)}\qquad \sup_{t_o-\theta<t<t_o} \int_{[x_o+K_\rho]\cap\Omega} \Psi^2\left(D_k^{\pm},(u-k)_{\pm},c\right)(x,t)\zeta^p(x)dx$$
$$+\gamma\iint_{[(x_o,t_o)+Q(\theta,\rho)]} \Psi\left|\Psi_u\left(D_k^{\pm},(u-k)_{\pm},c\right)\right|^{2-p}|D\zeta|^p dxd\tau$$
$$\leq \frac{\gamma}{c^2}\left(1+\ln\frac{D_k^{\pm}}{c}\right)\left\{\int_{t_o-\theta}^{t_o} |B_{k,\rho}^{\pm}(\tau)|^{\frac{r}{q}}d\tau\right\}^{\frac{p(1+\kappa)}{r}}$$

where the numbers q,r,κ satisfy (3.6)-(3.7).

Remark 4.3. Local considerations apply to this case along the lines of similar remarks in the previous sections.

Remark 4.4. The constant γ on the right hand sides of either (4.13)-(4.14) or (4.17)-(4.18) is *independent of* u. It is only the levels k that might depend upon the solution u via (3.11). Moreover if $\varphi_i \equiv 0$, $i=0,1,2$, and $C_2=0$, the levels k are independent of u.

Remark 4.5. We conclude this section by observing that all the energy estimates as well as logarithmic estimates for $(u-k)_+$ hold true if merely u is a subsolution of (1.1) and for $(u-k)_-$ if u is a supersolution of (1.1).

5. Restricted structures: the levels k and the constant γ

We will make a few remarks on the dependence of the constant γ in the energy and logarithmic estimates and on the restrictions to be placed on the levels k.

5-(i). About the constant γ

For the interior estimates of Propositions 3.1 and 3.2, the constant γ depends only upon the data and it is independent of the apriori knowledge of $\|u\|_{\infty,\Omega_T}$. It can be calculated apriori only in terms of the numbers N, p, r, κ, the constants C_i, $i=0,1,2$, and the norms

$$\|\varphi_o, \varphi_1^{\frac{p}{p-1}}, \varphi_2\|_{\hat{q},\hat{r}\,;\,\Omega_T}.$$

The same dependence holds for estimates near the parabolic boundary of Ω_T in the case of Dirichlet data (see §4-(ii) and §4-(iii)).

In the case of variational data, γ depends also upon the structure of $\partial\Omega$ (see §1, Chap. I), and the norms

$$\|\psi_1, \psi_o^{\frac{p}{p-1}}\|_{\hat{q},\hat{r}\,;\,\Omega_T}.$$

5-(ii). Restricted structures

The choices (3.11) and (4.2) of δ_o impose a restriction on the levels k. Such a restriction is needed to handle the lower order terms $b(x,t,u,Du)$ in (1.1). It follows from (3.10) and (3.10)$'$ that the choice (3.11) of δ_o permits the absorption of the term

$$C_2 \iint\limits_{Q^t} |D(u-k)_\pm|^p (u-k)_\pm \zeta^p dx d\tau \le \delta_o C_2 \iint\limits_{Q^t} |D(u-k)_\pm|^p \zeta^p dx d\tau$$

into the terms generated by the principal part of the operator in (1.1). Also, the coefficient of the integral involving φ_2 depends only upon the data (i.e., C_o, C_2), if the levels k are chosen according to (3.11).

Such a choice of δ_o imposes on k to be *close* to either the supremum or the infimum of u in $Q(\theta,\rho)$. Thus, in particular, the apriori knowledge of $\|u\|_{\infty,Q(\theta,\rho)}$, is required.

We will introduce structure conditions on (1.1) that yield energy and logarithmic estimates analogous to (3.8) and (3.14) for the truncated functions $(u-k)_\pm$, with *no restriction* on the levels k. We will limit ourselves to the interior estimates of Propositions 3.1 and 3.2.

First, it is obvious from the remarks above that Propositions 3.1 and 3.2 continue to hold for *all* the levels k if $b(x,t,u,Du)\equiv 0$. A more general condition is

$$(\mathbf{A}_3') \qquad |b(x,t,u,Du)| \le C_2|Du|^{p-1} + \varphi_2,$$

where

$$(\mathbf{A}_4') \qquad \varphi_2^{\frac{p}{p-1}} \in L^{\hat{q},\hat{r}}(\Omega_T),$$

and $\hat{q},\hat{r}$ satisfy $(\mathbf{A}_5)$ and $(\mathbf{A}_5\text{–}\mathbf{i})$–$(\mathbf{A}_5\text{–}\mathbf{iii})$. The structure condition $(\mathbf{A}_3')$ implies $(\mathbf{A}_3)$. The *source term* φ_2 is required to be *more integrable* than the corresponding source term in $(\mathbf{A}_3)$.

Let us consider local weak solutions u of (1.1) with the structure conditions $(\mathbf{A}_1),(\mathbf{A}_2),(\mathbf{A}_3'),(\mathbf{A}_4),(\mathbf{A}_4'),(\mathbf{A}_5)$ and $(\mathbf{A}_5\text{–}\mathbf{i})$–$(\mathbf{A}_5\text{–}\mathbf{iii})$. We do *not* require that u be locally bounded. To derive local energy and logarithmic estimates for u we proceed as in the proof of Propositions 3.1 and 3.2. The lower order terms in (3.10) are now estimated by repeated use of the Young's inequality as follows.

$$\begin{aligned}
&\iint_{Q^t} |b(x,\tau,u,Du)(u-k)_\pm\zeta^p|\,dxd\tau \\
&\le \frac{C_o}{4}\iint_{Q^t}|D(u-k)_\pm\zeta|^p dxd\tau + \gamma\iint_{Q^t}\varphi_2(u-k)_\pm\zeta^p dxd\tau \\
&+\gamma\iint_{Q^t}(u-k)_\pm^{p}\max\{|D\zeta|^p\,;\,\zeta^p\}\,dxd\tau.
\end{aligned}$$

We conclude that, for such solutions, inequalities (3.8)-(3.14) hold true for *every level* k, with a constant γ independent of u, provided the integral

$$\iint_{[(x_o,t_o)+Q(\theta,\rho)]}(u-k)_\pm^{p}|D\zeta|^p dxd\tau$$

in (3.8) is replaced by

$$\iint_{[(x_o,t_o)+Q(\theta,\rho)]}(u-k)_\pm^{p}\max\{|D\zeta|^p;\zeta^p\}\,dxd\tau$$

and the integral

$$\iint\limits_{[(x_o,t_o)+Q(\theta,\rho)]} \Psi \left| \Psi_u \left(H_k^{\pm}, (u-k)_{\pm}, c \right) \right|^{2-p} |D\zeta|^p dx d\tau$$

in (3.14) is replaced by

$$\iint\limits_{[(x_o,t_o)+Q(\theta,\rho)]} \Psi \left| \Psi_u \left(H_k^{\pm}, (u-k)_{\pm}, c \right) \right|^{2-p} \max \left\{ |D\zeta|^p; \zeta^p \right\} dx d\tau.$$

6. Bibliographical notes

When $p=2$, assumptions $(\mathbf{A}_1)$–$(\mathbf{A}_5)$ are optimal to obtain a Hölder modulus of continuity for the solutions (see [67]).

The weak formulation of local and global weak sub(super)-solutions is standard and we refer for example to [67,73].

When $1 < p < 2$, it seems more suitable to work with cubes of the type of K_ρ rather than balls. For this reason we have introduced a unified geometry. The notation $[x_o + K_\rho]$ to denote a cube about x_o is introduced in Krylov-Safonov [64].

The idea of deriving energy inequalities for the truncated functions $(u-k)_\pm$ seems to appear first in Bernstein [12], in a *global* way, i.e., with the integrals extended to the whole Ω_T. A *local* version of such estimates by use of local cutoff functions was introduced in the celebrated paper of DeGiorgi [33]. Since then they have been widely used especially in the russian literature (see for example [67] and references therein).

Logarithmic estimates seem to be crucial in the study of the local behaviour of solutions of elliptic and parabolic equations in divergence form. For this we refer to Kruzkov [60,61,62], Moser [81,82,83] and Serrin [92]. The logarithmic function in (3.12) has been introduced in [35] and is now a standard tool in studying the local behaviour of degenerate and singular p.d.e.'s.

III

Hölder continuity of solutions of degenerate parabolic equations

1. The regularity theorem

Consider solutions u of (1.1) or of Chap. II for the case $p > 2$. The equation is degenerate since the modulus of ellipticity vanishes when $|Du| = 0$. We will prove that if $u \in L^{\infty}_{loc}(\Omega_T)$, then it is Hölder continuous within its domain of definition. It will shown in Chap.V that local weak solutions of such degenerate equations are indeed locally bounded. To simplify the presentation we will assume that $u \in L^{\infty}(\Omega_T)$. If u is only locally bounded, it will suffice to work within a fixed compact subset of Ω_T. In the theorems below, the statement that a constant γ depends upon the data means that it can be determined a priori only in terms of the norm $\|u\|_{\infty,\Omega_T}$, the constants C_i, $i=0,1,2$, and the norms $\|\varphi_o, \varphi_1^{\frac{p}{p-1}}, \varphi_2\|_{\hat{q},\hat{r};\Omega_T}$ appearing in the structure conditions $(\mathbf{A}_1)$–$(\mathbf{A}_3)$. We let $\mathcal{K}$ denote a compact subset of Ω_T and let $p-\text{dist}\,(\mathcal{K};\Gamma)$ be the *intrinsic* parabolic distance from $\mathcal{K}$ to the parabolic boundary of Ω_T, i.e.,

$$(1.1) \qquad p - \text{dist}\,(\mathcal{K};\Gamma;p) \equiv \inf_{\substack{(x,t)\in\mathcal{K}\\(y,s)\in\Gamma}} \left(|x-y| + \|u\|_{\infty,\Omega_T}^{\frac{p-2}{p}} |t-s|^{1/p} \right).$$

1-(i). Interior Hölder continuity

THEOREM 1.1. *Let u be a bounded local weak solution of (1.1) of Chap. II in Ω_T. Then $(x,t) \to u(x,t)$ is locally Hölder continuous in Ω_T. Moreover there* *exists constants $\gamma > 1$ and $\alpha \in (0,1)$ depending only upon the data, such that $\forall \mathcal{K} \subset \Omega_T$,*

$$\left|u(x_1,t_1) - u(x_2,t_2)\right| \leq \gamma \|u\|_{\infty,\Omega_T} \left(\frac{|x_1 - x_2| + \|u\|_{\infty,\Omega_T}^{\frac{p-2}{p}} |t_1 - t_2|^{1/p}}{p - \operatorname{dist}(\mathcal{K};\Gamma;p)} \right)^{\alpha},$$

for every pair of points $(x_1,t_1),(x_2,t_2) \in \mathcal{K}$. If the lower order terms $b(x,t,u,Du)$ satisfy $(\mathbf{A}_3')$ of §5 of Chap. II, then γ and α are independent of $\|u\|_{\infty,\Omega_T}$.

1-(ii). Boundary regularity (Dirichlet data)

THEOREM 1.2. *Let u be a bounded weak solution of the Dirichlet problem (2.1) of Chap. II and let $(\mathbf{D})$ and $(\mathbf{U_o})$ hold. The boundary $\partial\Omega$ is assumed to satisfy the property of positive geometric density (1.1) of Chap. I. Then $u \in C\left(\overline{\Omega}_T\right)$, and there exists a continuous positive non-decreasing function $s \to \omega(s) : \mathbf{R}^+ \to \mathbf{R}^+$, such that*

$$\left|u(x_1,t_1) - u(x_2,t_2)\right| \leq \omega\left(|x_1 - x_2| + |t_1 - t_2|^{\frac{1}{p}}\right),$$

for every pair of points $(x_1,t_1),(x_2,t_2) \in \overline{\Omega}_T$. In particular if the boundary datum g is Hölder continuous in S_T with exponent say α_g, and if the initial datum u_o is Hölder continuous in $\overline{\Omega}$ with exponent say α_{u_o}, then u is Hölder continuous in $\overline{\Omega}_T$ and there exist constants $\gamma > 1$ and $\alpha \in (0,1)$ such that

$$\left|u(x_1,t_1) - u(x_2,t_2)\right| \leq \gamma \|u\|_{\infty,\Omega_T} \left(|x_1 - x_2| + \|u\|_{\infty,\Omega_T}^{\frac{p-2}{p}} |t_1 - t_2|^{1/p}\right)^{\alpha},$$

for every pair of points $(x_1,t_1),(x_2,t_2) \in \overline{\Omega}_T$.

The constants γ and α depend only upon the data. Moreover the constant α depends also upon the Hölder exponents α_g, α_{u_o} of g and u_o respectively.

If the lower order terms $b(x,t,u,Du)$ satisfy $(\mathbf{A}_3')$ of §5 of Chap. II, then γ and α are independent of $\|u\|_{\infty,\Omega_T}$.

Even though we have stated the Theorem in a *global way* the proof has a *local* thrust. For example the boundary datum g could be continuous or Hölder continuous only on a open portion of S_T (open in the relative topology of S_T), say Σ. Then the solution u of the Dirichlet problem would be continuous (respectively Hölder continuous) up to every compact subset of Σ.

Analogous remarks hold in the case u_o is only locally continuous or locally Hölder continuous. In particular to establish the continuity (Hölder continuity respectively) of u up to $\Omega \times \{0\}$, no reference is needed to the Dirichlet problem or any boundary value problem.

1-(iii). Boundary regularity (Variational data)

To stress such a *locality* we state our next theorem as if no information were available on the initial datum u_o.

THEOREM 1.3. *Let u be a weak solution of the Neumann problem (2.7) of Chap. II, satisfying*

$$u \in L^\infty \left(\overline{\Omega} \times [\varepsilon, T] \right), \qquad \varepsilon \in (0, T).$$

Assume that $\partial\Omega$ is of class $C^{1,\lambda}$ and let $(\mathbf{N})$ and $(\mathbf{N}-\mathbf{i})$ hold. Then u is Hölder continuous in $\overline{\Omega} \times [\varepsilon_1, T]$, for all $\varepsilon < \varepsilon_1 < T$, and there exist constants γ and α such that

$$\begin{aligned} &\left|u(x_1,t_1) - u(x_2,t_2)\right| \\ &\qquad \le \gamma \|u\|_{\infty,\overline{\Omega}\times[\varepsilon,T]} \left(|x_1 - x_2| + \|u\|_{\infty,\Omega\times(\varepsilon,T)}^{\frac{p-2}{p}} |t_1 - t_2|^{1/p} \right)^\alpha , \end{aligned}$$

for every pair of points $(x_1,t_1), (x_2,t_2) \in \overline{\Omega} \times [\varepsilon_1, T]$. The constants $\gamma > 1$ and α depend only upon ε, $\|u\|_{\infty,\overline{\Omega}\times[\varepsilon,T]}$ and the data, including the structure of $\partial\Omega$ and the norms $\|\psi_1, \psi_o^{\frac{p}{p-1}}\|_{\hat{q},\hat{r};\Omega_T}$ appearing in $(\mathbf{N}-\mathbf{i})$. In addition the constant γ depends upon the distance $(\varepsilon_1 - \varepsilon)$.

If the Neumann data are homogeneous, i.e., if $\psi_o \equiv \psi_1 \equiv 0$, and if in addition the lower order terms $b(x,t,u,Du)$ satisfy $(\mathbf{A}_3')$ of §5 of Chap. II, then γ and α are independent of $\|u\|_{\infty,\bar{\Omega}\times[\varepsilon,T]}$.

Remark 1.1. The continuity of u can be claimed up to $t = 0$ provided $(\mathbf{U_o})$ of Chap. II holds. Also, if u_o is Hölder continuous in $\overline{\Omega}$ then u is Hölder continuous in $\overline{\Omega}_T$.

2. Preliminaries

The Hölder continuity of u, either in the interior of Ω_T or at the parabolic boundary, will be, heuristically, a consequence of the following fact. The function $(x,t) \to u(x,t)$ can be modified in a set of measure zero to yield a continuous representative out of the equivalence class $u \in V_{loc}^{2,p}(\Omega_T)$, if for every $(x_o,t_o) \in \Omega_T$ there exist a family of nested and shrinking cylinders $[(x_o,t_o) + Q(\theta_n, \rho_n)]$ with the same *vertex* such that the essential oscillation ω_n of u in $[(x_o,t_o) + Q(\theta_n, \rho_n)]$ tends to zero as $n \to \infty$ in a way *quantitatively* determined by the structure conditions $(\mathbf{A}_1')!-(\mathbf{A}_6)$.

The key idea of the proof is to work with cylinders whose dimensions are suitably rescaled to reflect the degeneracy exhibited by the equation. To make this precise, fix $(x_o,t_o) \in \Omega_T$ and construct the cylinder

$$\left[(x_o,t_o) + Q\left(R^{p-\varepsilon}, 2R\right)\right] \subset \Omega_T$$

where ε is a small positive number to be determined later. After a translation we may assume that $(x_o, t_o) \equiv (0,0)$. Set

$$\mu^+ = \operatorname*{ess\,sup}_{Q(R^{p-\varepsilon},2R)} u, \quad \mu^- = \operatorname*{ess\,inf}_{Q(R^{p-\varepsilon},2R)} u, \quad \omega = \operatorname*{ess\,osc}_{Q(R^{p-\varepsilon},2R)} \equiv \mu^+ - \mu^-$$

and construct the cylinder

$$Q\left(a_o R^p, R\right), \qquad \frac{1}{a_o} = \left(\frac{\omega}{A}\right)^{p-2} \tag{2.1}$$

where A is a constant to be determined later only in terms of the data. We will assume that

$$\left(\frac{\omega}{A}\right)^{p-2} > R^{\varepsilon}. \tag{2.2}$$

This implies the inclusion

$$Q\left(a_o R^p, R\right) \subset Q\left(R^{p-\varepsilon}, 2R\right) \tag{2.3}$$

and the inequality

$$\operatorname*{ess\,osc}_{Q(a_o R^p,R)} u \le \omega.$$

By *cylinders rescaled to reflect the degeneracy*, we mean boxes of the type (2.1) where the length has been suitably stretched to accommodate the degeneracy. If $p=2$, these are the standard parabolic boxes reflecting the natural homogeneity of the space and time variables.

3. The main proposition

PROPOSITION 3.1. *There exist constants $\varepsilon_o, \eta \in (0,1)$ and $C, A > 1$, that can be determined a priori depending only upon the data, satisfying the following. Construct the sequences*

$$R_o = R, \qquad \omega_o = \omega$$

and for $n=1,2,\ldots$,

$$R_n = C^{-n} R, \qquad \omega_{n+1} = \max\left\{\eta\omega_n;\ C R_n^{\varepsilon_o}\right\}.$$

Construct also the family of cylinders

$$Q^{(n)} \equiv Q\left(a_n R_n^p, R_n\right), \qquad \frac{1}{a_n} = \left(\frac{\omega_n}{A}\right)^{p-2}, \qquad n = 0,1,2,\ldots.$$

Then for all $n=0,1,2,\ldots$

$$Q^{(n+1)} \subset Q^{(n)} \qquad \textit{and} \qquad \operatorname*{ess\,osc}_{Q^{(n)}} u \le \omega_n.$$

A consequence of this Proposition is:

LEMMA 3.1. *There exist constants $\gamma > 1$ and $\alpha \in (0,1)$ that can be determined a priori only in terms of the data, such that for all the cylinders*

$$0 < \rho \leq R, \quad Q\left(a_o \rho^p, \rho\right), \quad \frac{1}{a_o} = \left(\frac{\omega}{A}\right)^{p-2},$$

$$\operatorname*{ess\,osc}_{Q(a_o\rho^p,\rho)} u \leq \gamma\left(\omega + R^{\varepsilon_o}\right)\left(\frac{\rho}{R}\right)^{\alpha}.$$

PROOF: From the iterative construction of ω_n it follows that $\omega_{n+1} \leq \eta\omega_n + CR_n^{\varepsilon_o}$ and by iteration

$$\omega_n \leq \eta^n \omega + C\left(\sum_{i=o}^{n-1} \eta^i C^{-\varepsilon_o(n-i)}\right) R^{\varepsilon_o}.$$

We may assume without loss of generality that ε_o is so small that $\eta \leq C^{-\varepsilon_o}$. Then

$$\omega_n \leq \eta^n \omega + Cn\left(\frac{R}{C^n}\right)^{\varepsilon_o}.$$

Let now $0 < \rho \leq R$ be fixed. There exists a non-negative integer n such that

$$C^{-(n+1)}R \leq \rho \leq C^{-n}R.$$

This implies the inequalities

$$\begin{aligned}
(n+1) &\geq \ln\left(\frac{\rho}{R}\right)^{-\frac{1}{\ln C}}; \\
\eta^n &\leq \eta^{-1}\left(\frac{\rho}{R}\right)^{\alpha_1}, \qquad \alpha_1 = \frac{|\ln \eta|}{\ln C}; \\
Cn\left(\frac{R}{C^n}\right)^{\varepsilon_o} &\leq C^{1+\varepsilon_o} \ln\left(\frac{\rho}{R}\right)^{-\frac{1}{\ln C}} \rho^{\varepsilon_o} \\
&\leq C(\varepsilon_o) R^{\frac{\varepsilon_o}{2}} \rho^{\frac{\varepsilon_o}{2}}.
\end{aligned}$$

Therefore

$$\omega_n \leq \gamma\left(\omega + R^{\varepsilon_o}\right)\left(\frac{\rho}{R}\right)^{\alpha}, \qquad \alpha = \min\left\{\alpha_1; \frac{\varepsilon_o}{2}\right\}.$$

To conclude the proof we observe that since $\omega_n \leq \omega$, the cylinder $Q\left(a_o\rho^p, \rho\right)$ is included in $Q^{(n)} \equiv Q\left(a_n R_n^p, R_n\right)$, so that

$$\operatorname*{ess\,osc}_{Q(a_o\rho^p,\rho)} u \leq \omega_n.$$

Statements of Hölder continuity over a compact set now follow by a standard covering argument.

Remark 3.1. The proof of Proposition 3.1 will show that indeed it is sufficient to work with the number ω and the cylinder $Q(a_o R^p, R)$ linked by

$$\operatorname*{ess\,osc}_{Q(a_o R^p, R)} u \leq \omega. \tag{3.1}$$

This fact is in general not verifiable, for a given box, since its dimensions would have to be *intrinsically* defined in terms of the essential oscillation of u within it.

Therefore the role of having introduced the cylinder $Q(R^{p-\varepsilon}, 2R)$ and having assumed (2.2) is that (3.1) holds true for the *constructed* box $Q(a_o R^p, R)$. It will be part of the proof of Proposition 3.1 to show that at each step the cylinders $Q^{(n)}$ and the essential oscillation of u within them satisfy the intrinsic geometry dictated by (3.1).

To begin the proof, inside $Q(a_o R^p, R)$ consider subcylinders of *smaller size* constructed as follows. The number ω being fixed, let s_o be the smallest positive integer such that

$$\frac{\omega}{2^{s_o}} \leq \delta_o, \tag{3.2}$$

where the number δ_o is introduced in (3.11) of Chap. II in the derivation of the local energy estimates. Then construct cylinders

$$[(0, \bar{t}) + Q(dR^p, R)], \qquad \frac{1}{d} = \left(\frac{\omega}{2^{s_o}}\right)^{p-2}. \tag{3.3}$$

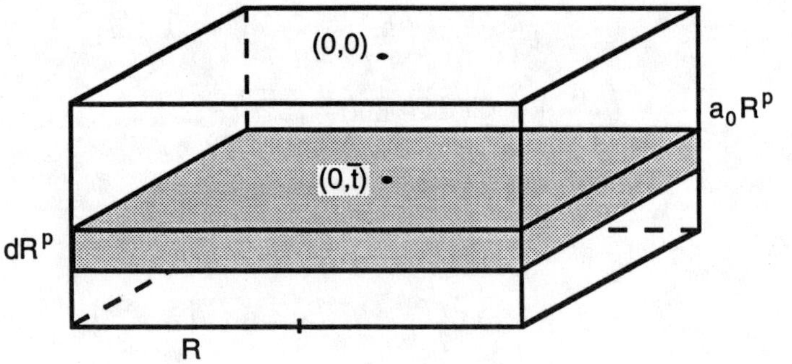

Figure 3.1

These are contained inside $Q\left(a_o R^p, R\right)$ if the number A is chosen larger that 2^{s_o} and if $\bar{t}$ ranges over

$$-\left\{A^{p-2}-\left(2^{s_o}\right)^{p-2}\right\}\frac{R^p}{\omega^{p-2}}<t<0.$$

The structure of the proof is based on studying separately two cases. Either we can find a cylinder of the type of $[(0,\bar{t})+Q\left(dR^p,R\right)]$ where u is *mostly* large, or such a cylinder cannot be found. In either case the conclusion is that the essential oscillation of u in a smaller cylinder about (x_o,t_o) decreases in a way that can be quantitatively measured. In the arguments to follow we assume (2.2) is in force and determine later the numbers A,ε and ε_o.

Remark 3.2. For later use we estimate the quantity

$$G(\omega,R)\equiv\gamma R^{N\kappa}\left(\frac{\omega}{2^{s_o}}\right)^{-2}d^{\frac{p(1+\kappa)}{r}},$$

where γ is a constant depending only upon the data and κ is defined in (3.5) of Chap. II. From the definition of d in (3.3) it follows that

$$G(\omega,R)\le A_1R^{N\kappa}\omega^{-b},\quad\text{where}\quad b=2+(p-2)\,\frac{p(1+\kappa)}{r}\tag{3.4}$$

and

$$A_1=A^{2+(p-2)\frac{p(1+\kappa)}{p}}.$$

Along the proof we will encounter quantities of the type $A_iR^{N\kappa}\omega^{-b}, i=1,2,\ldots,\ell$, where A_i are constants that can be determined a priori only in terms of the data and are independent of ω and R. We may assume without loss of generality that they satisfy

$$A_iR^{N\kappa}\omega^{-b}\le 1.\tag{3.5}$$

Indeed if not, we would have $\omega\le CR^{\varepsilon_o}$ for the choices

$$C=\max_{1\le i\le l}A_i^{1/b}\qquad\text{and}\qquad\varepsilon_o=\frac{N\kappa}{b},$$

and the first iterative step of the Proposition would be trivial.

Remark 3.3. The proof below and (2.2) show that the numbers ε and ε_o can be taken as

$$\varepsilon_o=\frac{N\kappa}{b},\qquad\qquad\varepsilon=(p-2)\varepsilon_o.$$

In the estimates to follow we denote with γ a generic positive constant that can be calculated a priori depending only upon the data and that may be different in different contexts.

3-(i). About the dependence on $\|u\|_{\infty,\Omega_T}$

We will use the energy and logarithmic estimates of Propositions 3.1 and 3.2 of Chap. II for the truncated functions $(u-k)_\pm$ over cylinders contained in $Q(a_oR^p,R)$. When working with $(u-k)_-$ we will use the levels

$$k=\mu^-+\frac{\omega}{2^{s_o+i}}\qquad \text{for some } i\geq 0.$$

These levels are admissible since

$$\|(u-k)_-\|_{\infty,Q(a_oR^p,R)}\leq\delta_o.$$

When working with $(u-k)_+$ we will take levels

$$k=\mu^+-\frac{\omega}{2^{s_o+i}}\qquad \text{for some } i\geq 0.$$

These are also admissible since

$$\|(u-k)_+\|_{\infty,Q(a_oR^p,R)}\leq\delta_o.$$

Let us fix δ_o as in (3.11) of Chap. II. Then, since $\omega\leq 2\|u\|_{\infty,\Omega_T}$, (3.2) holds true if we choose s_o so large that

$$2^{s_o}=\frac{8C_2}{C_o}\|u\|_{\infty,\Omega_T}.$$

Having chosen s_o this way, (3.2) is verified when working within *any* subdomain of Ω_T. The a priori knowledge of the norm $\|u\|_{\infty,\Omega_T}$ is required through the number s_o. If the lower order terms $b(x,t,u,Du)$ in (1.1) satisfy $(\mathbf{A}_3')$ of §5 of Chap. II, then, as remarked there, the energy and logarithmic inequalities hold true for the truncated functions $(u-k)_\pm$ with *no restriction* on the levels k. Thus in such a case s_o can be taken to be one and no a priori knowledge of $\|u\|_{\infty,\Omega_T}$ is needed.

The numbers A and A_i introduced in (3.5) will be chosen to be larger than 2^{s_o}. In the proof below we will choose them of the type

$$A\equiv A_o\qquad\text{and}\qquad A_i=2^{s_o+h_i},\qquad i=0,1,2,\ldots,$$

where $h_i\geq 0$ will be *independent of* $\|u\|_{\infty,\Omega_T}$. We have just remarked that if the lower order terms $b(x,t,u,Du)$ satisfy $(\mathbf{A}_3')$ of §5 of Chap. II, then s_o can be taken to be one. We conclude that for equations with such a structure, the numbers A_i can be determined a priori only in terms of the data and independent of the norm $\|u\|_{\infty,\Omega_T}$.

4. The first alternative

LEMMA 4.1. *There exists a number* $\nu_o \in (0,1)$ *independent of* ω, R, A *such that if for some cylinder of the type* $[(0,\bar{t}) + Q(dR^p, R)]$

$$\left|(x,t) \in [(0,\bar{t}) + Q(dR^p, R)] \,\middle|\, u(x,t) < \mu^- + \frac{\omega}{2^{s_o}}\right| \le \nu_o |Q(dR^p, R)|,$$

then

$$u(x,t) > \mu^- + \frac{\omega}{2^{s_o+1}} \qquad a.e.\ (x,t) \in \left[(0,\bar{t}) + Q\left(d(\tfrac{R}{2})^p, \tfrac{R}{2}\right)\right]. \tag{4.1}$$

PROOF: Fix a cylinder for which the assumption of the lemma holds. Up to a translation we may assume that $(0,\bar{t}) \equiv (0,0)$, and we may work within cylinders $Q(d\rho^p, \rho)$, $0 < \rho \le R$. Let

$$R_n = \frac{R}{2} + \frac{R}{2^{n+1}}, \qquad n = 0, 1, 2, \ldots,$$

construct the family of nested cylinders $Q(dR_n^p, R_n)$ and let ζ_n be a piecewise smooth cutoff function in $Q(dR_n^p, R_n)$ such that

$$\begin{cases} 0 < \zeta_n(x,t) \le 1, \quad \forall (x,t) \in Q(dR_n^p, R_n), \quad \text{and} \\ \zeta_n \equiv 1 \quad \text{in} \quad Q\left(dR_{n+1}^p, R_{n+1}\right); \\ \zeta_n = 0 \quad \text{on the parabolic boundary of} Q(dR_n^p, R_n); \\ |D\zeta_n| \le \frac{2^{n+1}}{R}, \quad 0 \le \frac{\partial}{\partial t}\zeta_n \le \frac{2^{p(n+1)}}{dR^p}, \quad \frac{1}{d} = \left(\frac{\omega}{2^{s_o}}\right)^{p-2}. \end{cases} \tag{4.2}$$

We will use the energy inequalities of Proposition 3.1 of Chap II. written over the cylinders $Q(dR_n^p, R_n)$, for the functions $(u - k_n)_-$, where for $n = 0, 1, 2, \ldots$,

$$k_n = \mu^- + \frac{\omega}{2^{s_o+1}} + \frac{\omega}{2^{s_o+1+n}}.$$

In this setting, (3.8) of Chap. II takes the form

$$\begin{aligned} &\operatorname*{ess\,sup}_{-dR_n^p < t < 0} \int_{K_{R_n}} (u - k_n)_-^2\, \zeta_n^p(x,t)dx + \iint_{Q(dR_n^p, R_n)} |D(u - k_n)_-\, \zeta_n|^p dx d\tau \\ &\le \gamma \frac{2^{np}}{R^p} \left\{ \iint_{Q(dR_n^p, R_n)} (u - k_n)_-^p\, dx d\tau + \frac{1}{d} \iint_{Q(dR_n^p, R_n)} (u - k_n)_-^2\, dx d\tau \right\} \\ &+ \gamma \left\{ \int_{-dR_n^p}^{0} |A_{k_n, R_n}^-(\tau)|^{\frac{r}{q}} d\tau \right\}^{\frac{p(1+\kappa)}{r}}. \end{aligned} \tag{4.3}$$

We will show that as $n \to \infty$

$$\iint_{Q(dR_n^p, R_n)} \chi\left[(u-k_n)_- > 0\right] dx d\tau \longrightarrow 0.$$

Since $k_n \searrow k_\infty = \mu^- + \frac{\omega}{2^{s_o+1}}$, this would imply that

$$\left|\left[u < \mu^- + \frac{\omega}{2^{s_o+1}}\right] \cap Q\left(d\left(\tfrac{R}{2}\right)^p, \tfrac{R}{2}\right)\right| = 0,$$

thereby proving the lemma. We observe that

$$\int_{K_{R_n}} (u-k_n)_-^2 \, \zeta_n^p(x,t) dx \geq \left(\frac{2^{s_o}}{\omega}\right)^{p-2} \int_{K_{R_n}} (u-k_n)_-^p \, \zeta_n^p(x,t) dx$$

and estimate above the first two terms on the right hand side of (4.3) by

$$\begin{aligned}
&\gamma \frac{2^{np}}{R^p} \left(\frac{\omega}{2^{s_o}}\right)^2 \left(\frac{\omega}{2^{s_o}}\right)^{p-2} \iint_{Q(dR_n^p, R_n)} \chi\left[(u-k_n)_- > 0\right] dx d\tau \\
&\quad + \gamma \frac{2^{np}}{R^p} \left(\frac{\omega}{2^{s_o}}\right)^2 \frac{1}{d} \iint_{Q(dR_n^p, R_n)} \chi\left[(u-k_n)_- > 0\right] dx d\tau \\
&\quad \leq \gamma \frac{2^{np}}{R^p} \left(\frac{\omega}{2^{s_o}}\right)^p \iint_{Q(dR_n^p, R_n)} \chi\left[(u-k_n)_- > 0\right] dx d\tau,
\end{aligned}$$

where we have used the definition (3.3) of d. Combining these remarks in (4.3) and dividing through by d, we arrive at

$$\begin{aligned}
\text{(4.4)} \quad &\operatorname*{ess\,sup}_{-dR_n^p < t < 0} \int_{K_{R_n}} (u-k_n)_-^p \, \zeta_n^p(x,t)\, dx + \frac{1}{d} \iint_{Q(dR_n^p, R_n)} |D(u-k_n)_- \zeta_n|^p dx d\tau \\
&\leq \gamma \frac{2^{np}}{R^p} \left(\frac{\omega}{2^{s_o}}\right)^p \frac{1}{d} \iint_{Q(dR_n^p, R_n)} \chi\left[(u-k_n)_- > 0\right] dx d\tau \\
&\quad + \gamma \left(\frac{\omega}{2^{s_o}}\right)^{p-2} d^{\frac{p(1+\kappa)}{r}} \left\{ \frac{1}{d} \int_{-dR_n^p}^{0} |A_{k_n,R_n}^-(\tau)|^{\frac{r}{q}} d\tau \right\}^{\frac{p(1+\kappa)}{r}}.
\end{aligned}$$

In (4.4) we introduce the change of time-variable $z = t/d$ which transforms $Q(dR_n^p, R_n)$ into

$$Q_n \equiv Q(R_n^p, R_n) \equiv K_{R_n} \times \{-R_n^p, 0\}.$$

Setting also

$$v(\cdot, z) = u(\cdot, zd) \quad \text{and} \quad \hat{\zeta}_n(\cdot, z) = \zeta_n(\cdot, zd),$$

the inequality (4.4) can be written more concisely as

$$\|(v-k_n)_-\hat{\zeta}_n\|^p_{V^p(Q_n)} \le \gamma \frac{2^{np}}{R^p}\left(\frac{\omega}{2^{s_o}}\right)^p |A_n|$$
$$+\gamma\left(\frac{\omega}{2^{s_o}}\right)^{p-2} d^{\frac{p(1+\kappa)}{r}} \left\{\int_{-R_n^p}^{0} |A_n(z)|^{\frac{r}{q}} dz\right\}^{\frac{p(1+\kappa)}{r}}$$

where we have set

$$A_n(z) \equiv \left\{x \in K_{R_n} \middle| v(x,z) < k_n\right\} \quad \text{and} \quad |A_n| = \int_{-R_n^p}^{0} |A_n(z)|dz.$$

Since $(v-k_n)_-\hat{\zeta}_n$ vanishes on the lateral boundary of Q_n, by Corollary 3.1 of Chap. I we have

$$(4.5)\qquad \|(v-k_n)_-\|^p_{p,Q_{n+1}} \le \|(v-k_n)_-\hat{\zeta}_n\|^p_{p,Q_n}$$
$$\le \|(v-k_n)_-\hat{\zeta}_n\|^p_{p\frac{N+p}{p},Q_n} |A_n|^{\frac{p}{N+p}}$$
$$\le \|(v-k_n)_-\hat{\zeta}_n\|^p_{V^p(Q_n)} |A_n|^{\frac{p}{N+p}}.$$

The left hand side of (4.5) is estimated below by

$$\|(v-k_n)_-\|^p_{p,Q_{n+1}} \ge |k_n-k_{n+1}|^p |A_{n+1}| \ge \frac{1}{2^{p(n+2)}}\left(\frac{\omega}{2^{s_o}}\right)^p |A_{n+1}|.$$

Combining these estimates gives

$$(4.6)\qquad |A_{n+1}| \le \gamma 4^{np} \frac{|A_n|^{1+\frac{p}{N+p}}}{R^p}$$
$$+\gamma 4^{np}\left(\frac{\omega}{2^{s_o}}\right)^{-2} d^{\frac{p(1+\kappa)}{r}} |A_n|^{\frac{p}{N+p}} \left\{\int_{-R_n^p}^{0} |A_n(z)|^{\frac{r}{q}} dz\right\}^{\frac{p(1+\kappa)}{r}}.$$

Divide by $|Q_{n+1}|$ and introduce the quantities

$$Y_n = \frac{|A_n|}{|Q_n|}, \qquad Z_n = \frac{1}{|K_{R_n}|}\left(\int_{-R_n^p}^{0} |A_n(z)|^{\frac{r}{q}} dz\right)^{\frac{p}{r}}.$$

Using also the fact that, by virtue of Remark 3.2 and (3.5)

$$R^{N\kappa}\left(\frac{\omega}{2^{s_o}}\right)^{-2} d^{\frac{p(1+\kappa)}{r}} \le 1,$$

we obtain from (4.6) in dimensionless form

$$Y_{n+1} \leq \gamma 4^{np} \left\{ Y_n^{1+\frac{p}{N+p}} + Y_n^{\frac{p}{N+p}} Z_n^{1+\kappa} \right\}, \qquad \forall n = 0, 1, 2, \ldots.$$

Next by the embedding of Proposition 3.3 of Chap. I

$$\begin{aligned} Z_{n+1} \left(k_n - k_{n+1}\right)^p &\leq \left|K_{R_{n+1}}\right|^{-1} \| \left(v - k_n\right)_- \|_{q,r;Q_{n+1}}^p \\ &\leq \left|K_{R_{n+1}}\right|^{-1} \| \left(v - k_n\right)_- \hat{\zeta}_n \|_{q,r;Q_n}^p \\ &\leq \gamma R^{-N} \| \left(v - k_n\right)_- \hat{\zeta}_n \|_{V^p(Q_n)}^p. \end{aligned}$$

Therefore

$$Z_{n+1} \leq \gamma 4^{np} \left\{ Y_n + Z_n^{1+\kappa} \right\}, \qquad \forall n = 0, 1, 2, \ldots.$$

From Lemma 4.2 of Chap. I it follows that Y_n and Z_n tend to zero as $n \to \infty$, provided

$$Y_o + Z_o^{1+\kappa} \leq (4\gamma)^{-(1+\kappa)/\theta_o} 4^{-p(1+\kappa)/\theta_o^2} \equiv \nu_o,$$

where $\theta_o = \min\left\{\frac{p}{N+p}; \kappa\right\}$.

5. The first alternative continued

Suppose the assumptions of Lemma 4.1 are verified for some box of the type $[(0, \bar{t}) + Q\,(dR^p, R)]$. We will exploit the fact that at the time level

$$-\theta = \bar{t} - \left(\frac{\omega}{2^{s_o}}\right)^{2-p} \left(\frac{R}{2}\right)^p,$$

the function $x \to u(x, -\theta)$ is strictly above the level $\mu^- + \frac{\omega}{2^{s_o+1}}$, in the cube $K_{R/2}$.

To simplify the symbolism let us set $\rho = \frac{R}{2}$ and construct the cylinder

$$Q\,(\theta, \rho) \equiv K_\rho \times (-\theta, 0), \qquad \rho = R/2.$$

The length θ of such a cylinder satisfies

$$\left(\frac{\omega}{2^{s_o}}\right)^{2-p} \leq \frac{\theta}{\rho^p} \leq \left(\frac{\omega}{A}\right)^{2-p} 2^p. \tag{5.1}$$

The next lemma asserts that, owing to (4.1), the set where $u(\cdot, t)$ is close to μ^-, within the smaller cube $K_{R/4}$, can be made arbitrarily small for all time levels $-\theta \leq t \leq 0$.

LEMMA 5.1. *For every number $\nu_1 \in (0,1)$, there exists a positive integer s_1, depending only upon the data and independent of ω, R, such that*

$$\left| x \in K_{R/4} \middle| u(x,t) < \mu^- + \frac{\omega}{2^{s_1}} \right| \le \nu_1 |K_{R/4}|, \qquad \forall t \in (-\theta, 0).$$

PROOF: Consider the logarithmic estimates of Proposition 3.2 of Chap. II, written over the cylinder $Q(\theta,\rho)$, for

$$(u-k)_-, \qquad k = \mu^- + \frac{\omega}{2^{s_o+1}}.$$

As a number c in the definition of Ψ, we take

$$c = \frac{\omega}{2^{s_o+1+n}}, \qquad n > 1,$$

where n is to be chosen. Thus we take

$$\Psi \equiv \ln^+ \left\{ \frac{H_k^-}{H_k^- - \left(u - \left(\mu^- + \frac{\omega}{2^{s_o+1}}\right)\right)_- + \frac{\omega}{2^{s_o+1+n}}} \right\},$$

where

$$H_k^- = \operatorname*{ess\,sup}_{Q(\theta,\rho)} \left(u - \left(\mu^- + \frac{\omega}{2^{s_o+1}} \right) \right)_- \le \frac{\omega}{2^{s_o+1}}.$$

For $t=-\theta$, by virtue of (4.1) we have $\left(u - \left(\mu^- + \frac{\omega}{2^{s_o+1}}\right)\right)_- = 0$, and therefore

$$\Psi(x,-\theta) = 0, \qquad \forall x \in K_\rho.$$

These remarks in (3.14) of Chap. II yield

$$\int_{K_\rho} \Psi^2(x,t)\zeta^p(x)dx \le \frac{\gamma}{\rho^p} \iint_{Q(\theta,\rho)} \Psi |\Psi_u|^{2-p} dx d\tau \tag{5.2}$$

$$+ \gamma \left(\frac{\omega}{2^{s_o+1+n}} \right)^{-2} \left(1 + \ln H_k^- \left(\frac{\omega}{2^{s_o+1+n}} \right)^{-1} \right) \left(\int_{-\theta}^{0} |A_{k,\rho}^-(\tau)|^{\frac{r}{q}} d\tau \right)^{\frac{p(1+\kappa)}{r}},$$

where $A_{k,\rho}^-(\cdot)$ is defined in (3.4) of Chap. II and $x \to \zeta(x)$ is a piecewise smooth cutoff function in K_ρ that equals one on $K_{\rho/2}$ and such that $|D\zeta| \le 4/\rho$. Next

$$\Psi \le \ln \left(\frac{\frac{\omega}{2^{s_o+1}}}{\frac{\omega}{2^{s_o+1+n}}} \right) = n \ln 2$$

and

$$|\Psi_u|^{2-p} = \left| H_k^- - (u-k)_- + \frac{\omega}{2^{s_o+1+n}} \right|^{p-2} \le \left(\frac{\omega}{2^{s_o}} \right)^{p-2}.$$

Therefore, in also view of (5.1), the first term on the right hand side of (5.2) is estimated above by

$$\frac{\gamma}{\rho^p}\iint\limits_{Q(\theta,\rho)} \Psi|\Psi_u|^{2-p}dxd\tau \leq \gamma n\frac{\theta}{\rho^p}\left(\frac{\omega}{2^{s_o}}\right)^{p-2}|K_{\rho/2}| \leq \gamma n A^{p-2}|K_{\rho/2}|,$$

where γ and A are constants depending only upon the data, and A has to be determined later. The second term is estimated by using the conditions (3.6), (3.7) of Chap. II, linking the parameters r, q, κ. This gives

$$\gamma\Big(\frac{\omega}{2^{s_o+1+n}}\Big)^{-2}\Big(1+\ln H_k^-\Big(\frac{\omega}{2^{s_o+1+n}}\Big)^{-1}\Big)\Bigg(\int_{-\theta}^{0}|A_{k,\rho}^-(\tau)|^{\frac{r}{q}}d\tau\Bigg)^{\frac{p(1+\kappa)}{r}}$$

$$\leq \gamma n\left(\frac{\omega}{2^{s_o+1+n}}\right)^{-2}\left(\frac{\omega}{A}\right)^{-\frac{p(1+\kappa)}{r}(p-2)}R^{N\kappa}|K_{\rho/2}|.$$

The number n will be determined shortly, depending only upon the data and independent of ω, ρ. Therefore by virtue of Remark 3.2 and (3.5) we may estimate

$$n\left(\frac{\omega}{2^{s_o+1+n}}\right)^{-2}\left(\frac{\omega}{A}\right)^{-\frac{p(1+\kappa)}{r}(p-2)}R^{N\kappa} \leq 1.$$

Combining these remarks into (5.2) yields

$$\int_{K_{\rho/2}} \Psi^2(x,t)\,dx \leq \gamma n A^{p-2}|K_{\rho/2}|, \tag{5.3}$$

where we have used the fact that $\zeta \equiv 1$ on $K_{\rho/2}$. The integral in (5.3) is estimated below by extending the integration to the smaller set

$$\left\{x \in K_{\rho/2}\,\middle|\, u(x,t) < \mu^- + \frac{\omega}{2^{s_o+1+n}}\right\}, \qquad t \in (-\theta, 0).$$

On such a set

$$\Psi^2 \geq \ln^2\left\{\frac{H_k^-}{H_k^- - \frac{\omega}{2^{s_o+1}} + \frac{\omega}{2^{s_o+n}}}\right\}$$

and since the right hand side of this inequality is a decreasing function of H_k^-, we have

$$\Psi^2 \geq \ln^2\left(\frac{\frac{\omega}{2^{s_o+1}}}{\frac{\omega}{2^{s_o+n}}}\right) = (n-1)^2\ln^2 2.$$

Putting this into (5.3) gives that for all $t \in (-\theta, 0)$

$$\left|x \in K_{\rho/2}\,\middle|\, u(x,t) < \mu^- + \frac{\omega}{2^{s_o+1+n}}\right| \leq \gamma A^{p-2}\frac{n}{(n-1)^2}|K_{\rho/2}|. \tag{5.4}$$

To prove the lemma we have only to choose n sufficiently large.

6. The first alternative concluded

The information in Lemma 5.1 will imply that u is strictly bounded away from μ^- in a smaller cylinder. To make this precise, consider the box

$$Q\left(\theta, \frac{\rho}{2}\right) \equiv K_{\rho/2} \times (-\theta, 0), \qquad \rho = R/2,$$

where θ satisfies the bounds in (5.1).

LEMMA 6.1. *The numbers $\nu_1 \in (0,1)$ and $s_1 \gg 1$ can be chosen a priori dependent only upon the data and independent of ω and R, so that*

$$u(x,t) > \mu^- + \frac{\omega}{2^{s_1+1}}, \qquad \text{a.e. } (x,t) \in Q\left(\theta, \frac{\rho}{4}\right).$$

PROOF: We will use the local energy estimates of Proposition 3.1 of Chap. II in the following setting. Let

$$\rho_n = \frac{\rho}{4} + \frac{\rho}{2^{n+2}}, \qquad n = 0, 1, 2, \ldots,$$

construct the cylinders $Q(\theta, \rho_n)$ and let $x \to \zeta_n(x)$ be a piecewise smooth cutoff function in K_{ρ_n} that equals one on $K_{\rho_{n+1}}$ and such that $|D\zeta_n| \le 2^{n+3}/\rho$. Write the inequalities (3.8) of Chap. II over $Q(\theta, \rho_n)$ for the functions $(u-k_n)_-$, where

$$k_n = \mu^- + \frac{\omega}{2^{s_1+1}} + \frac{\omega}{2^{s_1+1+n}},$$

and observe that, owing to the conclusion of Lemma 4.1,

$$(u-k_n)_-(x,-\theta) = 0, \quad \forall x \in K_{\rho_n}, \quad n = 0, 1, 2, \ldots.$$

With these choices, the energy inequalities yield

$$\text{(6.1)} \qquad \sup_{-\theta<t<0} \int_{K_{\rho_n}} (u-k_n)_-^2 \zeta_n^p dx + \gamma^{-1} \iint_{Q(\theta,\rho_n)} |D(u-k_n)_- \zeta_n|^p dx d\tau$$

$$\le \frac{\gamma 2^{np}}{\rho^p} \iint_{Q(\theta,\rho_n)} (u-k_n)_-^p \, dx d\tau + \gamma \left\{ \int_{-\theta}^{0} |A^-_{k_n,\rho_n}(\tau)|^{\frac{r}{q}} d\tau \right\}^{\frac{p(1+\kappa)}{r}}.$$

The first term on the left hand side is estimated below, for all $t \in (-\theta, 0)$, by

$$\int_{K_{\rho_n}} (u-k_n)_-^2 \zeta_n^p dx \ge \left(\frac{\omega}{2^{s_1}}\right)^{2-p} \int_{K_{\rho_n}} (u-k_n)_-^p \zeta_n^p dx$$

$$\ge 2^{-p} \left(\frac{2^{s_1}}{A}\right)^{p-2} \frac{\theta}{\rho^p} \int_{K_{\rho_n}} (u-k_n)_-^p \zeta_n^p dx \ge \frac{\theta}{\rho^p} \int_{K_{\rho_n}} (u-k_n)_-^p \zeta_n^p dx,$$

if s_1 is chosen so large as to satisfy the conclusion of Lemma 5.1 and the inequality $2^{-p}(2^{s_1}/A)^{p-2} \geq 1$. We put this in (6.1), divide through by θ/ρ^p and introduce in the cylinders $Q(\theta,\rho_n)$, the change of variable $z = t\rho^p/\theta$. This maps $Q(\theta,\rho_n)$ into the boxes

$$Q_n \equiv K_{\rho_n} \times (-\rho^p, 0).$$

Let us also set $v(x,z) = u(x, z\theta/\rho^p)$ and

$$A_n(z) \equiv \{x \in K_{\rho_n} \big| v(x,z) < k_n\}, \qquad |A_n| = \int_{-\rho^p}^{0} |A_n(z)| dz.$$

Then (6.1) yields in a more concise way

$$\|(v-k_n)_-\zeta_n\|^p_{V^p(Q_n)} \leq \frac{\gamma 2^{np}}{\rho^p}\left(\frac{\omega}{2^{s_1}}\right)^p |A_n| + \gamma\left(\frac{\rho^p}{\theta}\right)^{1-\frac{p(1+\kappa)}{r}} \left\{\int_{-\rho^p}^{0} |A_n(z)|^{\frac{r}{q}} dz\right\}^{\frac{p(1+\kappa)}{r}}.$$

By the embedding of Corollary 3.1 of Chap. I,

$$\begin{aligned}
2^{-(n+2)p}\left(\frac{\omega}{2^{s_1}}\right)^p |A_{n+1}| &\leq \iint_{Q_{n+1}\cap[v<k_{n+1}]} (v-k_n)_-^p \, dxdz \\
&\leq \iint_{Q_n} (v-k_n)_-^p \, \zeta_n^p dxdz \\
&\leq \gamma \|(v-k_n)_-\zeta_n\|^p_{V^p(Q_n)} |A_n|^{\frac{p}{N+p}} \\
&\leq \frac{\gamma 2^{np}}{\rho^p}\left(\frac{\omega}{2^{s_1}}\right)^p |A_n|^{1+\frac{p}{N+p}} \\
&\quad + \gamma\left(\frac{\rho^p}{\theta}\right)^{1-\frac{p(1+\kappa)}{r}} |A_n|^{\frac{p}{N+p}} \left\{\int_{-\rho^p}^{0} |A_n(z)|^{\frac{r}{q}} dz\right\}^{\frac{p(1+\kappa)}{r}}.
\end{aligned}$$

Divide through by the coefficient of $|A_{n+1}|$ and set

$$Y_n = \frac{|A_n|}{|Q_n|}, \qquad Z_n = \frac{1}{|K_{\rho_n}|}\left\{\int_{-\rho^p}^{0} |A_n(z)|^{\frac{r}{q}} dz\right\}^{\frac{p(1+\kappa)}{r}}.$$

Using also (5.1) and (3.5) to estimate

$$\gamma\left(\frac{\omega}{2^{s_1}}\right)^{-p}\left(\frac{\rho^p}{\theta}\right)^{1-\frac{p(1+\kappa)}{r}} R^{N\kappa} \leq 1,$$

we arrive at

$$Y_{n+1} \le \gamma 4^{np} \left\{ Y_n^{1+\frac{p}{N+p}} + Y_n^{\frac{p}{N+p}} Z_n^{1+\kappa} \right\}.$$

Proceeding as in the proof of Lemma 4.1, we have

$$Z_{n+1} \le \gamma 4^{np} \left\{ Y_n + Z_n^{1+\kappa} \right\}.$$

By Lemma 4.2 of Chap. I it follows that Y_n and Z_n tend to zero as $n \to \infty$, provided

$$Y_o + Z_o^{1+\kappa} \le (4\gamma)^{-(1+\kappa)/\theta_o} 4^{-p(1+\kappa)/\theta_o^2} \equiv \nu_1, \tag{6.2}$$

where $\theta_o = \min\left\{ \frac{p}{N+p} \, ; \, \kappa \right\}$. To prove the lemma, we fix ν_1 as in (6.2) and pick s_1 according to Lemma 5.2.

We summarise the results obtained so far.

PROPOSITION 6.1. *There exists numbers $\nu_o, \eta_o \in (0,1)$ and $A_1 \gg 1$ depending only upon the data and independent of ω, R, such that if for some cylinder of the type $[(0,\bar{t}) + Q\,(dR^p, R)]$,*

$$\left| (x,t) \in [(0,\bar{t}) + Q\,(dR^p, R)] \,\middle|\, u(x,t) < \mu^- + \frac{\omega}{2^{s_o}} \right| \le \nu_o \left| Q\,(dR^p, R) \right|, \tag{6.3}$$

then either

$$\omega \le A_1 R^{\frac{N\kappa}{b}} \tag{6.4}$$

or

$$\operatorname*{ess\,osc}_{Q\left(d\left(\frac{R}{8}\right)^p, \frac{R}{8}\right)} u \le \eta_o\, \omega \tag{6.5}$$

where b is introduced in (3.4).

PROOF: Assume (6.4) is violated. By Lemma 6.1, we can determine a positive number s_1 such that

$$\operatorname*{ess\,inf}_{Q\left(\theta, \frac{R}{4}\right)} u \ge \mu^- + \frac{\omega}{2^{s_1+1}}$$

where θ satisfies (5.1) with $\rho = R/2$. Change the sign of this inequality and add the quantity $\operatorname{ess\,sup}_{Q\left(\theta,\frac{R}{2}\right)} u$ to the left hand side and μ^+ to the right hand side. This gives

$$\operatorname*{ess\,osc}_{Q\left(\theta, \frac{R}{8}\right)} u \le \left(1 - \frac{1}{2^{s_1+1}} \right) \omega.$$

Therefore the proposition follows with $\eta_o = \left(1 - \frac{1}{2^{s_1+1}}\right)$, since

$$Q\left(d \left(\frac{R}{8} \right)^p, \frac{R}{8} \right) \subset Q\left(\theta, \frac{R}{8} \right).$$

Remark 6.1. Let us trace the dependence of η_o and A_1 upon $\|u\|_{\infty,\Omega_T}$. The numbers ν_o and ν_1 depend only upon the data and are independent of u. The number s_1 is given by $s_1 = s_o + n$ where n is chosen from (5.4). Thus

$$\eta_o = 1 - \frac{1}{2^{s_o+n+1}}$$

depends upon $\|u\|_{\infty,\Omega_T}$ via s_o through (3.2). Also A and A_1 are of the type $A_i = 2^{s_o+h_i}$, where h_i, $i=0,1,2,\ldots$, can be determined a priori only in terms of the data and are independent of $\|u\|_{\infty,\Omega_T}$. We conclude that if the lower order terms $b(x,t,u,Du)$ satisfy the structure condition $(\mathbf{A}_3')$ of §5 of Chap. II, we have $s_o = 1$ and therefore η_o, A, A_1 can be determined a priori only in terms of the data and are independent of $\|u\|_{\infty,\Omega_T}$.

7. The second alternative

We assume in this section that the assumptions of Lemma 4.1 are violated, i.e. for *every* subcylinder $[(0,\bar{t}) + Q(dR^p, R)] \subset Q(a_o R^p, R)$

$$\left| (x,t) \in [(0,\bar{t}) + Q(dR^p, R)] \,\middle|\, u(x,t) < \mu^- + \frac{\omega}{2^{s_o}} \right| > \nu_o \left| Q(dR^p, R) \right|.$$

Since

$$\mu^+ - \frac{\omega}{2^{s_o}} \geq \mu^- + \frac{\omega}{2^{s_o}}, \qquad \forall s_o \geq 2,$$

we rewrite this as

$$\left| (x,t) \in [(0,\bar{t}) + Q(dR^p, R)] \,\middle|\, u(x,t) > \mu^+ - \frac{\omega}{2^{s_o}} \right| \leq (1-\nu_o) \left| Q(dR^p, R) \right|, \tag{7.1}$$

valid *for all* cylinders

$$[(0,\bar{t}) + Q(dR^p, R)] \subset Q(a_o R^p, R) \qquad \frac{1}{a_o} = \left(\frac{\omega}{A}\right)^{p-2}.$$

In view of (7.1) we will study the behaviour of u near its supremum μ^+ and will be working with the truncated functions $(u-k)_+$ for the levels

$$k = \mu^+ - \frac{\omega}{2^{s_o+i}}, \qquad i \geq 0.$$

LEMMA 7.1. *Let* $[(0,\bar{t}) + Q(dR^p, R)] \subset Q(a_o R^p, R)$ *be fixed and let (7.1) hold. There exists a time level*

$$t^* \in \left[\bar{t} - dR^p, \bar{t} - \frac{\nu_o}{2} dR^p\right],$$

such that

$$\left|x \in K_R \mid u(x,t^*) > \mu^+ - \frac{\omega}{2^{s_o}}\right| \le \left(\frac{1-\nu_o}{1-\nu_o/2}\right)|K_R|.$$

PROOF: If not, for all $t \in \left[\bar{t} - dR^p, \bar{t} - \frac{\nu_o}{2} dR^p\right]$,

$$\left|x \in K_R \mid u(x,t) > \mu^+ - \frac{\omega}{2^{s_o}}\right| > \left(\frac{1-\nu_o}{1-\nu_o/2}\right)|K_R|$$

and

$$\begin{aligned}
&\left|(x,t) \in [(0,\bar{t}) + Q(dR^p, R)] \mid u(x,t) > \mu^+ - \frac{\omega}{2^{s_o}}\right| \\
&\qquad \ge \int_{\bar{t}-dR^p}^{\bar{t}-\frac{\nu_o}{2}dR^p} \left|x \in K_R \mid u(x,\tau) > \mu^+ - \frac{\omega}{2^{s_o}}\right| d\tau \\
&\qquad > (1-\nu_o)\left|Q(dR^p, R)\right|,
\end{aligned}$$

contradicting (7.1).

The lemma asserts that at some time level t^* the set where u is close to its supremum occupies only a portion of the cube K_R. The next lemma claims that this indeed occurs for all time levels near the top of the cylinder $[(0,\bar{t}) + Q(dR^p, R)]$.

LEMMA 7.2. *There exists a positive integer $s_2 > s_o$ such that*

$$\left|x \in K_R \mid u(x,t) > \mu^+ - \frac{\omega}{2^{s_2}}\right| \le \left(1 - \left(\frac{\nu_o}{2}\right)^2\right)|K_R|,$$

for all $t \in \left[\bar{t} - \frac{\nu_o}{2} dR^p, \bar{t}\right]$.

PROOF: Consider the logarithmic inequalities (3.14) of Chap. II written over the box $K_R \times (t^*, \bar{t})$ for the function $(u-k)_+$ for the levels $k = \mu^+ - \frac{\omega}{2^{s_o}}$. As for the number c in the definition of the function Ψ, we take

$$c = \frac{\omega}{2^{s_o+n}}, \qquad n > 0 \text{ to be chosen.}$$

Thus we take

$$\Psi = \ln^+\left\{\frac{H_k^+}{H_k^+ - \left(u - \left(\mu^+ - \frac{\omega}{2^{s_o}}\right)\right)_+ + \frac{\omega}{2^{s_o+n}}}\right\}, \tag{7.2}$$

where

$$H_k^+ \equiv \operatorname*{ess\,sup}_{[(0,\bar{t})+Q(dR^p,R)]} \left(u - \left(\mu^+ - \frac{\omega}{2^{s_o}}\right)\right)_+.$$

The cutoff function $x \to \zeta(x)$ is taken so that $\zeta = 1$ in the cube $K_{(1-\sigma)R}$, $\sigma \in (0,1)$, and $|D\zeta| \le (\sigma R)^{-1}$. With these choices, inequality (3.14) of Chap. II yields for all $t \in (t^*, \bar{t})$

(7.3)
$$\int_{K_{(1-\sigma)R}} \Psi^2(x,t)dx \le \int_{K_R} \Psi^2(x,t^*)dx + \frac{\gamma}{(\sigma R)^p} \int_{t^*}^{\bar t}\int_{K_R} \Psi|\Psi_u|^{2-p} dx d\tau$$
$$+\gamma\left(\frac{\omega}{2^{s_o+n}}\right)^{-2}\left[1+\ln H_k^+\left(\frac{\omega}{2^{s_o+n}}\right)^{-1}\right]\left\{\int_{t^*}^{\bar t} |A_k^+(\tau)|^{\frac{r}{q}} d\tau\right\}^{\frac{p(1+\kappa)}{r}}.$$

The various terms in (7.3) are estimated as follows. First

$$\Psi \le n\ln 2;\ \ |\Psi_u|^{2-p} \le 2^p\left(\frac{\omega}{2^{s_o}}\right)^{p-2};\ \ \left[1+\ln H_k^+\left(\frac{\omega}{2^{s_o+n}}\right)^{-1}\right] \le \gamma n \ln 2.$$

Next, from (7.2) it follows that Ψ vanishes on the set $\left[u<\mu^+ - \frac{\omega}{2^{s_o}}\right]$. Therefore, using Lemma 7.1, the first integral on the right hand side of (7.3) is estimated above by

$$\int_{K_R} \Psi^2(x,t^*)dx \le n^2 \ln^2 2 \left(\frac{1-\nu_o}{1-\nu_o/2}\right)|K_R|.$$

The second integral is estimated by

$$\frac{\gamma}{(\sigma R)^p}\int_{t^*}^{\bar t}\int_{K_R} \Psi|\Psi_u|^{2-p} dx d\tau \le \frac{\gamma}{\sigma^p} n|K_R|,$$

since $\bar t - t^* \le dR^p$, and d is given by (3.3). Finally for the last term, we have

$$\gamma\left(\frac{\omega}{2^{s_o+n}}\right)^{-2}\left[1+\ln H_k^+\left(\frac{\omega}{2^{s_o+n}}\right)^{-1}\right]\left\{\int_{t^*}^{\bar t} |A_k^+(\tau)|^{\frac{r}{q}} d\tau\right\}^{\frac{p(1+\kappa)}{r}}$$
$$\le \gamma n A_2 \omega^{-b} R^{N\kappa}|K_R|,$$

where $A_2 = 2^{(s_o+n)b}$ and b is defined in (3.4). By Remark 3.2 and (3.5) we may assume that $nA_2\omega^{-b}R^{N\kappa} \le 1$. Combining these remarks in (7.3) we conclude that for all $t \in (t^*, \bar t)$

(7.4)
$$\int_{K_{(1-\sigma)R}} \Psi^2(x,t)\,dx \le n^2 \ln^2 2\left(\frac{1-\nu_o}{1-\nu_o/2}\right)|K_R| + \frac{\gamma}{\sigma^p} n|K_R|.$$

The left hand side of (7.4) is estimated below by integrating over the smaller set

$$\left\{x \in K_{(1-\sigma)R} \middle| u(x,t) > \mu^+ - \frac{\omega}{2^{s_o+n}}\right\}.$$

On such a set, since the function Ψ in (7.2) is a decreasing function of H_k^+, we estimate

$$\Psi^2 \geq \ln^2 \left(\frac{\frac{\omega}{2^{s_o}}}{\frac{\omega}{2^{s_o+n-1}}} \right) = (n-1)^2 \ln^2 2.$$

After carrying this in (7.4) and dividing through by $(n-1)^2 \ln^2 2$ we obtain

$$\left| x \in K_{(1-\sigma)R} \mid u(x,t) > \mu^+ - \frac{\omega}{2^{s_o+n}} \right|$$
$$\leq \left(\frac{n}{n-1} \right)^2 \left(\frac{1-\nu_o}{1-\nu_o/2} \right) |K_R| + \frac{\gamma}{\sigma^p n} |K_R|.$$

On the other hand

$$\left| x \in K_R \mid u(x,t) > \mu^+ - \frac{\omega}{2^{s_o+n}} \right|$$
$$\leq \left| x \in K_{(1-\sigma)R} \mid u(x,t) > \mu^+ - \frac{\omega}{2^{s_o+n}} \right| + \left| K_R \backslash K_{(1-\sigma)R} \right|$$
$$\leq \left| x \in K_{(1-\sigma)R} \mid u(x,t) > \mu^+ - \frac{\omega}{2^{s_o+n}} \right| + N\sigma |K_R|.$$

Therefore

$$\left| x \in K_R \mid u(x,t) > \mu^+ - \frac{\omega}{2^{s_o+n}} \right|$$
$$\leq \left[\left(\frac{n}{n-1} \right)^2 \left(\frac{1-\nu_o}{1-\nu_o/2} \right) + \frac{\gamma}{\sigma^p n} + N\sigma \right] |K_R|,$$

for all $t \in (t^*, \bar{t})$. Choose σ so small that $\sigma N \leq \frac{3}{8}\nu_o^2$ and then n so large that

$$\left(\frac{n}{n-2} \right)^2 \leq \left(1 - \frac{\nu_o}{2} \right) (1+\nu_o) \quad \text{and} \quad \frac{\gamma}{\sigma^p n} \leq \frac{3}{8} \nu_o^2.$$

Then for such a choice of n the lemma follows with $s_2 = s_o + n$.

Remark 7.1. Since the number ν_o is independent of ω and R, also s_2 is independent of these parameters. The number A that determines the length of $Q\,(a_o R^p, R)$ is still to be chosen. We will determine it later independent of ω and R and subject to the condition $A > 2^{s_2}$.

Since (7.1) holds for all cylinders of the type $[(0, \bar{t}) + Q\,(dR^p, R)]$, the conclusion of Lemma 7.2 holds true for all time levels satisfying

$$t \geq -(a_o - d)\, R^p = -\left(1 - \left(\frac{2^{s_o}}{A} \right)^{p-2} \right) a_o R^p,$$

where a_o and d are defined in (2.1) and (3.3) respectively. If the number A is chosen sufficiently large, we deduce

COROLLARY 7.1. *For all* $t \in \left(-\frac{a_o}{2}R^p, 0\right)$,

$$\left| x \in K_R \mid u(x,t) > \mu^+ - \frac{\omega}{2^{s_2}} \right| \le \left(1 - \left(\frac{\nu_o}{2}\right)^2\right) |K_R|.$$

From now on we will focus on the cylinder $Q\left(\frac{a_o}{2}R^p, R\right)$ and to simplify the symbolism we set

$$A_s(t) = \left\{ x \in K_R \middle| u(x,t) > \mu^+ - \frac{\omega}{2^s} \right\},$$
$$A_s = \left\{ (x,t) \in Q\left(\frac{a_o}{2}R^p, R\right) \middle| u(x,t) > \mu^+ - \frac{\omega}{2^s} \right\}.$$

8. The second alternative continued

The information of Corollary 7.1 will be employed to deduce that the set where u is close to its supremum μ^+, within the cylinder $Q\left(\frac{a_o}{2}R^p, R\right)$, can be made arbitrarily small. In this section we will also determine the length of the cylinder $Q\left(a_o R^p, R\right)$ by determining the number A.

LEMMA 8.1. *For every* $\nu_* \in (0,1)$ *there exists a number* $s_* > s_2$ *independent of* ω *and* R, *such that*

$$|A_{s_*}| \le \nu_* \left| Q\left(\frac{a_o}{2}R^p, R\right) \right|.$$

Remark 8.1. Assume for the moment that the number s_* has been chosen. Then we determine the length of the cylinder $Q\left(a_o R^p, R\right)$ by choosing

$$A \equiv 2^{s_*}. \tag{8.1}$$

PROOF OF LEMMA 8.1: Consider the local energy estimates (3.8) of Chap. II written over the box $Q\left(a_o R^p, 2R\right)$, for the functions $(u-k)_+$. The levels k are given by

$$k = \mu^+ - \frac{\omega}{2^s},$$

where $s_2 \le s \le s_*$ and s_* is to be chosen. We take a cutoff function ζ that equals one on $Q\left(\frac{a_o}{2}R^p, R\right)$, vanishes on the parabolic boundary of $Q\left(a_o R^p, 2R\right)$ and such that

$$|D\zeta| \le \frac{1}{R}, \qquad 0 \le \zeta_t \le \frac{2}{a_o R^p}.$$

Neglecting the first term on the left hand side of these energy estimates, and using the indicated choices, we obtain

(8.2)
$$\iint\limits_{Q(\frac{a_o}{2}R^p,R)} |D(u-k)_+|^p dxd\tau \le \frac{\gamma}{R^p} \iint\limits_{Q(a_oR^p,2R)} (u-k)_+^p \, dxd\tau$$
$$+ \frac{\gamma}{a_oR^p} \iint\limits_{Q(a_oR^p,2R)} (u-k)_+^2 \, dxd\tau + \gamma \left\{ \int_{-a_oR^p}^{0} |A_{k,2R}^+(\tau)|^{\frac{r}{q}} d\tau \right\}^{\frac{p(1+\kappa)}{r}}.$$

The various term on the right hand side of (8.2) are estimated as follows. First

(i)
$$\frac{\gamma}{R^p} \iint\limits_{Q(a_oR^p,2R)} (u-k)_+^p \, dxd\tau \le \frac{\gamma}{R^p} \left(\frac{\omega}{2^s}\right)^p \left| Q\left(\frac{a_o}{2}R^p, R\right) \right|.$$

Next by virtue of the choice (8.1) of the parameter A, and the definition (2.1) of a_o,

(ii)
$$\frac{\gamma}{a_oR^p} \iint\limits_{Q(a_oR^p,2R)} (u-k)_+^2 \, dxd\tau \le \frac{\gamma}{R^p} \left(\frac{\omega}{2^s}\right)^p \left| Q\left(\frac{a_o}{2}R^p, R\right) \right|.$$

Finally making use of Remark 3.2 and (3.5)

(iii)
$$\gamma \left\{ \int_{-a_oR^p}^{0} |A_{k,2R}^+(\tau)|^{\frac{r}{q}} d\tau \right\}^{\frac{p(1+\kappa)}{r}}$$
$$\le \frac{\gamma}{R^p} \left(\frac{\omega}{2^s}\right)^p \left| Q\left(\frac{a_o}{2}R^p, R\right) \right| (A_3\omega^{-b}R^{N\kappa})$$
$$\le \frac{\gamma}{R^p} \left(\frac{\omega}{2^s}\right)^p \left| Q\left(\frac{a_o}{2}R^p, R\right) \right|,$$

where $A_3 = 2^{bs_*}$ and b is defined in (3.4). These estimates in (8.2) give

(8.3)
$$\iint\limits_{A_s} |Du|^p dxd\tau \le \frac{\gamma}{R^p} \left(\frac{\omega}{2^s}\right)^p \left| Q\left(\frac{a_o}{2}R^p, R\right) \right|.$$

Next we use Lemma 2.2 of Chap. I applied to the function $u(\cdot,t)$ for all times $-\frac{a_o}{2}R^p \le t \le 0$, and for the levels

$$k = \mu^+ - \frac{\omega}{2^s}, \qquad l = \mu^+ - \frac{\omega}{2^{s+1}}; \qquad (l-k) = \frac{\omega}{2^{s+1}}.$$

Notice that by virtue of Corollary 7.1 we have

$$\left| x \in K_R \mid u(x,t) < \mu^+ - \frac{\omega}{2^s} \right| \equiv |K_R| - |A_s(t)| \ge \left(\frac{\nu_o}{2}\right)^2 |K_R|.$$

Applying Lemma 2.2 of Chap. I in this setting, gives

$$\frac{\omega}{2^{s+1}}\left|A_{s+1}(t)\right| \leq \frac{4\gamma}{\nu_o^2}\frac{R^{N+1}}{\left|K_R\right|}\int_{A_s(t)\backslash A_{s+1}(t)}|Du|\,dx,$$

for all $t\in\left(-\frac{a_o}{2}R^p, 0\right)$. From this, integrating over such a time interval we get

$$\frac{\omega}{2^{s+1}}\left|A_{s+1}\right| \leq \frac{\gamma}{\nu_o^2}R\iint_{A_s\backslash A_{s+1}}|Du|\,dxd\tau$$

$$\leq \frac{\gamma}{\nu_o^2}R\left(\iint_{A_s}|Du|^p dxd\tau\right)^{\frac{1}{p}}\left|A_s\backslash A_{s+1}\right|^{\frac{p-1}{p}}.$$

Take the $\frac{p}{p-1}$ power, estimate the integral on the right hand side by (8.3) and divide through by $\left(\frac{\omega}{2^{s+1}}\right)^{\frac{p}{p-1}}$. This gives

$$\left|A_{s+1}\right|^{\frac{p}{p-1}} \leq \gamma\left(\nu_o\right)^{-\frac{2p}{p-1}}\left|Q\left(\frac{a_o}{2}R^p, R\right)\right|^{\frac{1}{p-1}}\left|A_s\backslash A_{s+1}\right|.$$

These inequalities are valid for all $s_2\leq s\leq s_*$. We add them for

$$s = s_2, s_2+1, s_2+2, \ldots\ s_*-1.$$

The right hand side can be majorized by a convergent series bounded above by $\left|Q\left(\frac{a_o}{2}R^p, R\right)\right|$. Therefore

$$\left(s_*-s_2\right)\left|A_{s_*}\right|^{\frac{p}{p-1}} \leq \gamma\left(\nu_o\right)^{-\frac{2p}{p-1}}\left|Q\left(\frac{a_o}{2}R^p, R\right)\right|^{\frac{p}{p-1}}.$$

To prove the lemma we divide by (s_*-s_2) and take s_* so large that

$$\frac{\gamma}{\nu_o^2\left(s_*-s_2\right)^{\frac{p-1}{p}}} \leq \nu_*.$$

Remark 8.2. If ν_* is independent of ω and R, also s_* and hence A are independent of these quantities.

Remark 8.3. The process described in Lemma 8.1 has a double scope. Given ν_*, it determines a *level* $\mu^+ - \frac{\omega}{2^{s_*}}$ *and a cylinder* so that the measure of the set where u is above such a level can be made smaller than ν_*, on that particular cylinder.

9. The second alternative concluded

Next we show that indeed u is strictly below its supremum μ^+ in a smaller box coaxial with $Q\left(\frac{a_o}{2}R^p, R\right)$ and with the same *vertex*. To simplify the symbolism let us set

$$a_* = \frac{1}{2}a_o = \frac{1}{2}\left(\frac{A}{\omega}\right)^{p-2},$$

and write accordingly $Q\left(\frac{a_o}{2}R^p, R\right)\equiv Q\left(a_*R^p, R\right)$.

LEMMA 9.1. *The number ν_* (and hence s_* and A) can be chosen so that*

$$u(x,t) \le \mu^+ - \frac{\omega}{2^{s_*+1}}, \qquad a.e.\ Q\left(a_*\left(\tfrac{R}{2}\right)^p, \tfrac{R}{2}\right).$$

PROOF: We will apply the local energy estimates of Proposition 3.1 of Chap. II over the boxes $Q(a_* R_n^p, R_n)$ to the function $(u-k_n)_+$, where for all $n = 0, 1, 2, \ldots$,

$$R_n = \frac{R}{2} + \frac{R}{2^{n+1}} \qquad \text{and} \qquad k_n = \mu^+ - \frac{\omega}{2^{s_*+1}} - \frac{\omega}{2^{s_*+1+n}}.$$

The cutoff functions ζ_n are taken to satisfy

$$\begin{cases} 0 < \zeta_n(x,t), \quad \forall (x,t) \in Q(a_* R_n^p, R_n), \quad \text{and} \\ \zeta_n \equiv 1 \quad \text{in } Q\left(a_* R_{n+1}^p, R_{n+1}\right); \\ \zeta_n = 0 \quad \text{on the parabolic boundary of} \quad Q(a_* R_n^p, R_n), \\ |D\zeta_n| \le \frac{2^{n+1}}{R}, \quad 0 \le \frac{\partial}{\partial \tau}\zeta_n \le \frac{2^{p(n+1)}}{a_* R^p}, \\ \frac{1}{a_*} = 2\left(\frac{\omega}{A}\right)^{p-2}, \quad A = 2^{s_*}. \end{cases}$$

With these choices, inequalities (3.8) of Chap. II take the form

$$\text{(9.1)} \quad \operatorname*{ess\,sup}_{-a_* R_n^p \le t \le 0} \int_{K_{R_n}} (u-k_n)_+^2 \zeta_n^p(x,t)\,dx + \iint_{Q(a_* R_n^p, R_n)} |D(u-k_n)_+ \zeta_n|^p dx d\tau$$
$$\le \frac{\gamma 2^{np}}{R^p} \iint_{Q(a_* R_n^p, R_n)} (u-k_n)_+^p\, dx d\tau + \frac{\gamma 2^{np}}{a_* R^p} \iint_{Q(a_* R_n^p, R_n)} (u-k_n)_+^2\, dx d\tau$$
$$+ \gamma \left\{ \int_{-a_* R_n^p}^{0} |A_{k_n,R_n}^+(\tau)|^{\frac{r}{q}} d\tau \right\}^{\frac{p(1+\kappa)}{r}}.$$

First by the definition of a_*

$$\int_{K_{R_n}} (u-k_n)_+^2 \zeta_n^p(x,t)\,dx \ge \left(\frac{\omega}{2^{s_*}}\right)^{2-p} \int_{K_{R_n}} (u-k_n)_+^p \zeta_n^p(x,t)\,dx$$
$$\ge 2a_* \|(u-k_n)_+ \zeta_n\|_{p,K_{R_n}}^p(t).$$

Next, using again the definition of a_*, the first two terms on the right hand side of (9.1) are estimated above by

$$\frac{\gamma 2^{np}}{R^p}\left(\frac{\omega}{2^{s_*}}\right)^p \iint_{Q(a_* R_n^p, R_n)} \chi\left[(u-k_n)_+ > 0\right] dx d\tau.$$

Substituting this in (9.1) and dividing through by a_* gives

$$(9.2)\quad \operatorname*{ess\,sup}_{-a_* R_n^p \le t \le 0} \int_{K_{R_n}} (u-k_n)_+^p \zeta_n^p(x,t)\,dx + \frac{1}{a_*} \iint_{Q(a_* R_n^p, R_n)} |D(u-k_n)_+ \zeta_n|^p dx d\tau$$

$$\le \frac{\gamma 2^{np}}{R^p}\left(\frac{\omega}{2^{s_*}}\right)^p \frac{1}{a_*} \iint_{Q(a_* R_n^p, R_n)} \chi\left[(u-k_n)_+ > 0\right] dx d\tau$$

$$+ \gamma a_*^{\frac{p(1+\kappa)}{r}-1} \left\{ \frac{1}{a_*} \int_{-a_* R_n^p}^{0} |A_{k_n,R_n}^+(\tau)|^{\frac{r}{q}} d\tau \right\}^{\frac{p(1+\kappa)}{r}}.$$

By (3.5) and Remark 3.2 we may estimate

$$a_*^{\frac{p(1+\kappa)}{r}-1} \le \left(\frac{\omega}{2^{s_*}}\right)^p A_4 \omega^{-b} \le \left(\frac{\omega}{2^{s_*}}\right)^p R^{-N\kappa},$$

where $A_4 = A^b$. Next, in the cylinders $Q(a_* R_n^p, R_n)$ we introduce the change of variable $z = t/a_*$ which maps $Q(a_* R_n^p, R_n)$ into $Q_n \equiv K_{R_n} \times (-R_n^p, 0)$. Setting

$$v(\cdot, z) \equiv u(\cdot, a_* z), \qquad \hat{\zeta}_n \equiv \zeta_n(\cdot, a_* z)$$

and

$$A_n(z) \equiv \left\{x \in K_{R_n} \middle| v(x,z) > k_n\right\}, \qquad |A_n| \equiv \int_{-R_n^p}^{0} |A_n(z)| dz,$$

inequality (9.2) can be rewritten more concisely as

$$\|(v-k_n)_+ \hat{\zeta}_n\|_{V^p(Q_n)}^p \le \gamma \frac{2^{np}}{R^p}\left(\frac{\omega}{2^{s_*}}\right)^p |A_n|$$

$$+ \gamma \left(\frac{\omega}{2^{s_*}}\right)^p R^{-N\kappa} \left\{ \int_{-R_n^p}^{0} |A_n(z)|^{\frac{r}{q}} dz \right\}^{\frac{p(1+\kappa)}{r}}.$$

This inequality and Corollary 3.1 of Chap. I give

$$
\begin{aligned}
&2^{-(n+2)p}\left(\frac{\omega}{2^{s_*}}\right)^p |A_{n+1}| \\
&= (k_{n+1}-k_n)^p \left|(x,z)\in Q_{n+1} \mid v(x,z)>k_{n+1}\right| \\
&\leq \|\,(v-k_n)_+\,\|^p_{Q_{n+1}} \\
&\leq \|\,(v-k_n)_+\,\hat{\zeta}_n\|^p_{Q_n} \\
&\leq \gamma |A_n|^{\frac{p}{N+p}} \|\,(v-k_n)_+\,\hat{\zeta}_n\|^p_{V^p(Q_n)} \\
&\leq \gamma\left(\frac{\omega}{2^{s_*}}\right)^p \frac{2^{np}}{R^p}|A_n|^{1+\frac{p}{N+p}} \\
&+\gamma\left(\frac{\omega}{2^{s_*}}\right)^p R^{-N\kappa}|A_n|^{\frac{p}{N+p}}\left\{\int_{-R_n^p}^{0}|A_n(z)|^{\frac{r}{q}}dz\right\}^{\frac{p(1+\kappa)}{r}}.
\end{aligned}
$$

Thus setting

$$
Y_n=\frac{|A_n|}{|Q_n|}, \qquad Z_n=\frac{1}{|K_{R_n}|}\left(\int_{-R_n^p}^{0}|A_n(z)|^{\frac{r}{q}}dz\right)^{\frac{p}{r}},
$$

we have the recursive inequalities

$$
\begin{aligned}
Y_{n+1} &\leq \gamma 4^{np}\left\{Y_n^{1+\frac{p}{N+p}}+Y_n^{\frac{p}{N+p}}Z_n^{1+\kappa}\right\}, \\
Z_{n+1} &\leq \gamma 4^{np}\left\{Y_n+Z_n^{1+\kappa}\right\}.
\end{aligned}
$$

It follows from these with the aid of Lemma 4.2 of Chap. I that Y_n and Z_n tend to zero as $n\to\infty$ provided

$$
Y_o+Z_o^{1+\kappa}\leq \gamma^{-\frac{1+\kappa}{\theta_o^2}}4^{-p\frac{1+\kappa}{\theta_o^2}}\equiv \nu_*,
$$

where $\theta_o=\min\left\{\frac{p}{N+p};\kappa\right\}$.

The following proposition summarises the results of the second alternative and it is proved arguing as in the proof of Proposition 6.1

PROPOSITION 9.1. *There exists numbers $\nu_o,\eta_1\in(0,1)$ and $A_2\gg 1$ depending only upon the data and independent of ω and R, such that if for all cylinders of the type $[(0,\bar{t})+Q\,(dR^p,R)]$*

$$
\left|(x,t)\in[(0,\bar{t})+Q\,(dR^p,R)]\,\middle|\,u(x,t)>\mu^+-\frac{\omega}{2^{s_o}}\right| \leq (1-\nu_o)\,|Q\,(dR^p,R)\,|,
$$

then either

$$
\omega\leq A_2\,R^{\frac{N\kappa}{b}} \tag{9.3}
$$

or

(9.4)
$$\operatorname*{ess\,osc}_{\left[Q\left(a_*\left(\frac{R}{2}\right)^p,\frac{R}{2}\right)\right]} u \le \eta_1\,\omega$$

where b is introduced in (3.4).

Remark 9.1. The constants $\eta_1 \in (0,1)$ and A_2 depend only upon the data and, in general, also upon the norm $\|u\|_{\infty,\Omega_T}$ via the number s_o. If the lower order term $b(x,t,u,Du)$ satisfies the structure condition $(\mathbf{A}_3')$ of §5 of Chap. II, we have $s_o = 1$ and therefore η_1, A, A_2 can be determined a priori only in terms of the data and are independent of $\|u\|_{\infty,\Omega_T}$.

10. Proof of Proposition 3.1

The two alternatives just discussed can be combined to prove the main Proposition 3.1. Let us recall that

$$\frac{1}{d} = \left(\frac{\omega}{2^{s_o}}\right)^{p-2}, \qquad \frac{1}{a_o} = \left(\frac{\omega}{A}\right)^{p-2}.$$

The concluding statement of the first alternative is that, starting from the cylinder

$$Q\left(R^{p-\varepsilon}, 2R\right)$$

and going down to the smaller cylinder

$$Q\left(d\left(\tfrac{R}{8}\right)^p, \tfrac{R}{8}\right),$$

the essential oscillation ω decreases by a factor $\eta_o \in (0,1)$, unless $\omega \le A_1 R^{\frac{N\kappa}{b}}$, where A_1 is a large constant that can be computed a priori only in terms of the data and the number s_o is introduced in (3.2). Analogously, the conclusion of the second alternative is that starting from the same cylinder and going down to the smaller box

$$Q\left(\tfrac{a_o}{2}\left(\tfrac{R}{2}\right)^p, \tfrac{R}{2}\right),$$

the number ω decreases by a factor $\eta_1 \in (0,1)$, unless $\omega \le A_2 R^{\frac{N\kappa}{b}}$, where A_2 is a constant that can be computed a priori in terms of the data. We combine these two facts into

LEMMA 10.1. *There exist constants*

$$\eta = \max\{\eta_o; \eta_1\} \qquad \textit{and} \qquad A = \max\{A_1; A_2\}$$

that can be determined a priori only in terms of the data, such that either

$$\omega \le A R^{\frac{N\kappa}{b}} \qquad \textit{or} \qquad \operatorname*{ess\,osc}_{Q\left(d\left(\frac{R}{8}\right)^p, \frac{R}{8}\right)} u \le \eta\,\omega.$$

We comment further on the content of Remark 3.1. The arguments presented do not require that the starting cylinder be $Q\left(R^{p-\varepsilon}, 2R\right)$. It would have been sufficient to have started from the box

$$Q\left(a_o R^p, R\right),$$

if we had known a priori that

$$\operatorname*{ess\,osc}_{Q(a_o R^p, R)} u \le \omega. \tag{10.1}$$

Next we will construct a box for which information of the type of (10.1) can be derived. Set

$$\omega_1 \equiv \max\left\{\eta\omega; AR^{\frac{N\kappa}{b}}\right\} \quad \text{and} \quad \frac{1}{a_1} = \left(\frac{\omega_1}{A}\right)^{p-2},$$

and let us estimate from below the length of the cylinder $Q\left(d\left(\frac{R}{8}\right)^p, \frac{R}{8}\right)$ for which the conclusion of Lemma 10.1 holds. We have

$$\begin{aligned} d\left(\frac{R}{8}\right)^p &= \left(\frac{2^{s_o}}{\omega}\right)^{p-2}\frac{R^p}{8^p} \\ &\ge 2^{-3p}\gamma^{-1}\eta^{p-2}\left(\frac{2^{s_o}}{A}\right)^{p-2}\left(\frac{A}{\omega_1}\right)^{p-2} R^p \\ &= a_1 R_1^p, \end{aligned}$$

where

$$R_1 = C^{-1}R \qquad \text{and} \qquad C = 8\left(\frac{1}{\eta}\right)^{\frac{p-2}{p}}\left(\frac{A}{2^{s_o}}\right)^{\frac{p-2}{2}}.$$

It follows that, for the cylinder $Q\left(a_1 R_1^p, R_1\right)$, inequality (10.1) is verified and the process can now be repeated starting from such a box, thereby proving Proposition 3.1.

As indicated in §3 this implies the interior Hölder continuity stated in Theorem 1.1. The constant dependence indicated in the statement of the theorem follows from the arguments of §3-(I) and Remarks 6.1 and 9.1.

11. Regularity up to $t=0$

Let u be a weak solution of (1.1) of Chap. II that takes initial data u_o in Ω. We assume u_o is continuous with modulus of continuity, say $\omega_o(\cdot)$. The regularity of u up to $t=0$ will follow from a proposition analogous to Proposition 3.1. Fix $(x_o, 0)\in\Omega\times\{0\}$, and $R>0$ so that $[x_o + K_{2R}]\subset\Omega$. After a translation we may assume $x_o=0$ and construct the cylinder

$$Q_o\left(R^{p-\varepsilon}, 2R\right) \equiv K_{2R}\times\left\{0, R^{p-\varepsilon}\right\},$$

where ε is a positive number to be chosen. As before, set

$$\mu^+ = \operatorname*{ess\,sup}_{Q_o(R^{p-\varepsilon},2R)} u, \ \mu^- = \operatorname*{ess\,inf}_{Q_o(R^{p-\varepsilon},2R)} u, \ \omega = \operatorname*{ess\,osc}_{Q_o(R^{p-\varepsilon},2R)} u.$$

Let s_o be the smallest positive integer satisfying (3.2) and construct the box

$$Q_o\left(dR^p, R\right) \equiv K_R \times \{0, dR^p\}, \qquad \frac{1}{d} = \left(\frac{\omega}{2^{s_o}}\right)^{p-2}.$$

For all $R>0$, these boxes are lying on the bottom of Ω_T.

PROPOSITION 11.1. *There exist constants $\varepsilon_o, \bar{\eta} \in (0,1)$ and $C, A > 1$ that can be determined a priori depending only upon the data, satisfying the following. Construct the sequences $R_o = R, \omega_o = \omega$ and*

$$R_n = C^{-n} R, \quad \omega_{n+1} = \max\left\{\bar{\eta}\omega_n; C R_n^{\varepsilon_o}\right\}, \quad n=1,2,\dots,$$

and the family of boxes

$$Q_o^{(n)} \equiv Q_o\left(a_n R_n^p, R_n\right), \qquad \frac{1}{a_n} = \left(\frac{\omega_n}{A}\right)^{p-2}, \qquad n = 0,1,2,\dots.$$

Then for all $n=0,1,2,\dots$

$$Q_o^{(n+1)} \subset Q_o^{(n)} \qquad \text{and} \qquad \operatorname*{ess\,osc}_{Q_o^{(n)}} u \le \max\left\{\omega_n \,;\, 2 \operatorname*{ess\,osc}_{K_{R_n}} u_o\right\}.$$

The proof of the continuity (or the Hölder continuity) of u up to $t=0$ follows from a simple variant of Lemma 3.1. Statement and proof of such a variant goes along the lines of similar arguments in §3. Here we indicate how to prove Proposition 11.1. Assume without loss of generality that $\mu^+ \ge |\mu^-|$ and that $dR^p < R^{p-\varepsilon}$, i.e., $\left(\frac{\omega}{2^{s_o}}\right)^{p-2} > R^\varepsilon$. Indeed otherwise we would have

$$\omega \le 2^{s_o} R^{\varepsilon_o}, \qquad \varepsilon_o = \frac{\varepsilon}{p-2}.$$

This implies that $Q_o\left(dR^p, R\right)$ is all contained in the box $Q_o\left(R^{p-\varepsilon}, 2R\right)$, and we may work within $Q_o\left(dR^p, R\right)$. Also without loss of generality we may assume that $s_o \ge 2$. Set

$$\mu_o^+ = \operatorname*{ess\,sup}_{K_R} u_o, \ \mu_o^- = \operatorname*{ess\,inf}_{K_R} u_o, \ \omega_o(R) = \operatorname*{ess\,osc}_{K_R} u_o,$$

and consider the two inequalities

$$\text{(11.1)} \qquad \mu^+ - \frac{\omega}{2^{s_o}} < \mu_o^+, \qquad \mu^- + \frac{\omega}{2^{s_o}} > \mu_o^-.$$

If both hold, subtracting the second from the first we obtain

$$\operatorname*{ess\,osc}_{Q_o(dR^p,R)} u \le 2 \operatorname*{ess\,osc}_{K_R} u_o,$$

and there is nothing to prove. Let us assume, for example, that the second of (11.1) is violated. Then for all $s \geq s_o$, the levels

$$k = \mu^- + \frac{\omega}{2^s},$$

satisfy the second of (4.16) of Chap. II. Therefore we may derive energy and logarithmic estimates for the truncated functions $(u-k)_-$. These take the form

$$\sup_{0\leq t\leq dR^p} \int_{K_R} (u-k)_-^2(x,t)\zeta^p dx + \iint_{Q_o(dR^p,R)} |D(u-k)_-\zeta|^p dx d\tau \tag{11.2}$$

$$\leq \gamma \iint_{Q_o(dR^p,R)} (u-k)_-^p |D\zeta|^p dx d\tau + \gamma \left\{ \int_0^{dR^p} |B_{k,R}^-(\tau)|^{\frac{q}{r}} d\tau \right\}^{\frac{p(1+\kappa)}{r}},$$

$$\sup_{0\leq t\leq dR^p} \int_{K_R} \Psi^2\left(D_k^-, (u-k)_-, c\right)\zeta^p(x)\,dx \tag{11.3}$$

$$\leq \gamma \iint_{Q_o(dR^p,R)} \Psi|\Psi_u\left(D_k^-, (u-k)_-, c\right)|^{2-p}|D\zeta|^p dx d\tau$$

$$+\frac{\gamma}{c^2}\left(1+\ln\frac{D_k^-}{c}\right)\left\{\int_0^{dR^p} |B_{k,R}^-(\tau)|^{\frac{r}{q}} d\tau\right\}^{\frac{p(1+\kappa)}{r}},$$

where $D_k^{\pm}$ and $B_{k,R}^{\pm}$ are defined as in (4.2) and (4.4) of Chap. II. The proof can now be completed as follows. First by using the logarithmic estimates (11.3) and proceeding as in Lemma 5.1, given any $e_o \in (0,1)$ we can find positive numbers ℓ_o and $\mathcal{A}_o$, depending only upon the data, such that either

$$\omega \leq A_3 R^{\frac{N\kappa}{b}} \qquad (b \text{ defined in (3.4)}) \tag{11.4}$$

or, for all $t \in (0, dR^p)$,

$$\left| x \in K_{R/2} \,\middle|\, u(x,t) < \mu^- + \frac{\omega}{2^{\ell_o}} \right| < e_o |K_{R/2}|.$$

Second, using the energy inequalities (11.2) and the procedure of Lemma 6.1, we conclude that if (11.4) does not hold, then

$$\operatorname*{ess\,inf}_{Q_o\left(d\left(\frac{R}{8}\right)^p, \frac{R}{8}\right)} u > \mu^- + \frac{\omega}{2^{\ell_o+1}}. \tag{11.5}$$

Changing the sign of (11.5) and adding $\operatorname{ess\,sup}_{Q_o\left(d\left(\frac{R}{8}\right)^p, \frac{R}{8}\right)} u$ to the left hand side and μ^+ to the right hand side we obtain

$$\operatorname*{ess\,osc}_{Q_o\left(d\left(\frac{R}{8}\right)^p,\frac{R}{8}\right)} u \le \bar{\eta}\omega, \qquad \bar{\eta} = 1 - \frac{1}{2^{\ell_o+1}}.$$

If the first of (11.1) is violated, we write the energy and the logarithmic inequalities for $(u-k)_+$, $k=\mu^+ - \frac{\omega}{2^s}$ for $s \ge s_o$ and proceed as before.

To summarise, going down from $Q_o\,(R^{p-\varepsilon}, 2R)$ to the smaller box

$$Q_o\left(d\left(\frac{R}{8}\right)^p, \frac{R}{8}\right),$$

the essential oscillation decreases by a factor of $\bar{\eta}$, unless either

$$\omega \le 2 \operatorname*{ess\,osc}_{K_R} u_o \qquad \text{or} \qquad \omega < A_o R^{\frac{N\kappa}{b}}.$$

LEMMA 11.1. *There exist constants $A_o > 1$ and $\bar{\eta} \in (0,1)$, that can be computed a priori only in terms of the data, such that either*

$$\omega < A_o R^{\frac{N\kappa}{b}} \qquad \textit{or} \qquad \operatorname*{ess\,osc}_{Q_o\left(d\left(\frac{R}{8}\right)^p,\frac{R}{8}\right)} u \le \max\left\{\bar{\eta}\omega\,;\, 2\operatorname*{osc}_{K_R} u_o\right\}.$$

To prove Proposition 11.1 we iterate this process over a sequence of boxes *all lying on the bottom of Ω_T*. This is done by arguments similar to those in the previous sections.

12. Regularity up to S_T. Dirichlet data

Let (x_o, t_o) be fixed and consider the cylinder $[(x_o,t_o) + Q\,(R^{p-\varepsilon}, 2R)]$, where

$$\varepsilon = \varepsilon_o(p-2), \qquad \varepsilon_o = \frac{N\kappa}{b},$$

where the number b is defined in (3.5) and κ is introduced in (3.2) of Chap. II. We let $R>0$ be so small that $t_o - R^{p-\varepsilon} \ge 0$, and change variables so that $(x_o,t_o) \equiv (0,0)$. The function u solves (1.1) of Chap. II and takes boundary data g on S_T in the sense of the traces of functions in $V^{2,p}(\Omega_T)$. The Dirichlet datum $(x,t) \to g(x,t)$ is continuous in S_T with modulus of continuity $\omega_g(\cdot)$. Set

$$\mu^+ = \operatorname*{ess\,sup}_{Q(R^{p-\varepsilon},2R)\cap\Omega_T} u, \quad \mu^- = \operatorname*{ess\,inf}_{Q(R^{p-\varepsilon},2R)\cap\Omega_T} u, \quad \omega = \operatorname*{ess\,osc}_{Q(R^{p-\varepsilon},2R)\cap\Omega_T} u,$$

and construct the box

$$Q\,(dR^p, R), \qquad \frac{1}{d} = \left(\frac{\omega}{2^{s_o}}\right)^{p-2}$$

where the number s_o is introduced in (3.2). Let also

$$\mu_g^+ = \sup_{Q(dR^p,R)\cap S_T} g, \qquad \mu_g^- = \inf_{Q(dR^p,R)\cap S_T} g.$$

If the two inequalities

$$\mu^+ - \frac{\omega}{2^{s_o}} \le \mu_g^+, \qquad \mu^- + \frac{\omega}{2^{s_o}} \ge \mu_g^-, \tag{12.1}$$

are both true, subtracting the second from the first gives

$$\omega \le 2 \operatorname*{osc}_{Q(dR^p,R)\cap S_T} g,$$

and the oscillation of u over $Q(dR^p, R) \cap \Omega_T$ is comparable to the oscillation of g over $Q(dR^p, R) \cap S_T$. Let us assume, for example, that the first of (12.1) is violated. Then the levels

$$k = \mu^+ - \frac{\omega}{2^s}, \qquad \forall s \ge s_o,$$

satisfy (4.11) of Chap. II, and we may derive energy estimates for $(u-k)_+$. Since $(u-k)_+$ vanishes on $Q(dR^p, R)\cap S_T$, we may extend it to the whole $Q(dR^p, R)$ by setting it to be zero outside Ω_T within the box $Q(dR^p, R)$. Also, in (4.13) of Chap. II we take a cutoff function vanishing on the parabolic boundary of $Q(dR^p, R)$. Taking into account these remarks, we obtain the energy estimates

$$\begin{aligned}
&\sup_{-dR^p \le t \le 0} \int_{K_R} (u-k)_+^2 \zeta^p(x,t)\,dx + \iint_{Q(dR^p,R)} |D(u-k)_+ \zeta|^p\,dxd\tau \\
&\le \gamma \iint_{Q(dR^p,R)} (u-k)_+^p |D\zeta|^p dxd\tau + \gamma \iint_{Q(dR^p,R)} (u-k)_+^2 \zeta^{p-1}\zeta_t\,dxd\tau \\
&+\gamma \left\{ \int_{-dR^p}^0 |B_{k,R}^+(\tau)|^{\frac{r}{q}} d\tau \right\}^{\frac{p(1+\kappa)}{r}},
\end{aligned} \tag{12.2}$$

where $B_{k,R}^+(\tau)$ is defined in (4.4) of Chap. II.

We observe that the conclusion of Lemma 7.2 is automatically verified for $(u-k)_+$. Indeed the function $x \to (u(x,t)-k)_+$, vanishes outside $\Omega \cap K_R$, for all $t \in (-dR^p, 0)$ and $\partial\Omega$ satisfies the property of positive geometric density of Chap. I. Therefore we may use Lemma 8.1 and its proof to deduce that for all $e_1 \in (0,1)$, there exist positive numbers $\mathcal{A}_1$ and ℓ_1 that can be determined a priori only in terms of the data such that either $\omega < \mathcal{A}_1 R^{\frac{N\kappa}{b}}$ or

$$\left| (x,t) \in Q(dR^p, R) \mid u(x,t) > \mu^+ - \frac{\omega}{2^{\ell_1}} \right| < e_1 |Q(dR^p, R)|.$$

An application of Lemma 9.1 now gives

LEMMA 12.1. *There exist numbers* $\mathcal{A}_1 > 1$ *and* $\tilde{\eta} \in (0,1)$ *that can be computed a priori only in terms of the data such that either*

$$\omega < \mathcal{A}_1 R^{\frac{N\kappa}{b}} \quad \text{or} \quad \operatorname*{ess\,osc}_{Q\left(d\left(\frac{R}{4}\right)^p, \frac{R}{4}\right)} u \le \max\left\{\tilde{\eta}\omega; \operatorname*{osc}_{Q(dR^p, R)\cap S_T} g\right\}.$$

The proof of the theorem can now be completed by stating a proposition similar to Proposition 11.1.

13. Regularity at S_T. Variational data

First we remark that the proof of interior regularity is only based on the energy and logarithmic estimates of §3 of Chap. II. In particular if such estimates were available for some locally bounded function $u \in V^p_{loc}(\Omega_T)$, then the conclusion of Theorem 1.1 would hold for u, irrespective of the differential equation u might satisfy.

Keeping this in mind, one realises that the proof of Theorem 1.3 is the same as that of interior Hölder continuity, owing to the energy and logarithmic inequalities of Proposition 4.1 of Chap. II.

If $(x_o, t_o) \in S_T$ is fixed, after a translation to $(x_o, t_o) \equiv (0,0)$ and a local flattening of $\partial\Omega$, inequalities (4.6) and (4.7) of §4 of Chap. II, can be viewed as written over cylinders of the type $Q^+(\theta, \rho)$, defined in (4.8) of Chap. II.

The cutoff function $x \to \zeta(x,t)$ vanishes on the boundary of K_ρ and *not* on the boundary of K^+_ρ. This affects the proof only in the application of the embedding Corollary 3.1 of Chap. I. Such an embedding was applied, after rescaling, to functions $v \in V^p(Q_n)$, where $Q_n \equiv K_{R_n} \times \{-R^p_n, 0\}$ (see Lemmas 4.1, 6.1, 8.1). Now for these domains, the ratio $T/|\Omega|^{p/N}$ is a constant depending only upon the dimension N.

We also remark that the application of Lemma 2.2 of Chap. I, in the context of *half* cubes K^+_ρ, is possible since such a lemma holds for convex domains (see Remark 2.2 of Chap. I).

14. Remarks on stability

As $p \searrow 2$ the equation becomes *less degenerate*. The proof presented in the previous sections shows that γ and α are *stable* in the sense that

$$\lim_{p \searrow 2} \gamma(p) = \gamma(2) < \infty \quad \text{and} \quad \lim_{p \searrow 2} \alpha(p) = \alpha(2) \in (0,1).$$

Thus the classical results of Hölder continuity of weak solutions of quasilinear non-degenerate parabolic equations can be recovered from our results by letting $p \searrow 2$ in the structure conditions of §1 of Chap. II.

14-(i). Continuous dependence on the operator

A similar *stability* holds for the local behaviour of solutions of a family of equations of the type of (1.1) of Chap. II. To be specific, let us consider as an example the family of equations

$$\frac{\partial}{\partial \tau} u_\lambda - \operatorname{div} \mathbf{a}_\lambda (x, t, u_\lambda, Du_\lambda) = b_\lambda (x, t, u_\lambda, Du_\lambda) \qquad \text{in} \quad \Omega_T,$$

$$u_\lambda \in C_{loc}\left(0, T; L^2_{loc}(\Omega)\right) \cap L^p_{loc}\left(0, T; W^{1,p}_{loc}(\Omega)\right),$$

where λ ranges over some subset I of the real numbers. Assume that for all $\lambda \in I$, the functions

$$(x, t, u_\lambda, Du_\lambda) \longrightarrow \mathbf{a}_\lambda (x, t, u_\lambda, Du_\lambda) \quad \text{and} \quad b_\lambda (x, t, u_\lambda, Du_\lambda),$$

satisfy the structure conditions $(\mathbf{A}_1)$–$(\mathbf{A}_5)$ uniformly in λ, i.e., for constants C_i and functions φ_i, $i=0,1,2$, independent of λ. Assume moreover that

$$u_\lambda \in L^\infty_{loc}(\Omega_T),$$

uniformly in λ. Then

LEMMA 14.1. *$\{u_\lambda\}$ is a family of uniformly Hölder continuous functions over compact subsets of Ω_T.*

Results of this kind are referred to in the literature as continuous dependence of the solution on the operator. Stability results also hold for a family of equations where also the parameter p ranges over a compact subset of $[2, \infty)$.

15. Bibliographical notes

Questions regarding the local behaviour of solutions of equations of the type of the p–laplacian were raised by Ladyzenkaja–Solonnikov–Ural'tzeva [67], Aronson–Serrin [7] and Trudinger [97]. The first results for the *elliptic* case appear in Ural'tzeva [100] and Uhlenbeck [99] and, for the parabolic case, in [39]. These results hold also for systems and we will comment further on them in Chap. VIII. The proof presented here is taken from [36,37]. The structure conditions $(\mathbf{A}_1)$–$(\mathbf{A}_5)$ are optimal for Theorems 1.1-1.3 to hold, as pointed out in [67] in the non-degenerate case $p = 2$. The iteration technique of Lemma 4.1 is a *parabolic* version of a similar *elliptic* technique due to DeGiorgi [33]. The new input regards the space–time geometry intrinsically defined by the solution itself. A first version of this technique appears in [38] in a simpler situation. It turns out that the same idea can be used to establish the local Hölder continuity of solutions of the porous medium equations and its generalisations. Here we mention the contributions of [37] and [24]. It can also be used to prove the local Hölder continuity of doubly degenerate equations. To be specific consider the p.d.e. (1.1) of Chap. II satisfying the structure conditions

$$
\begin{array}{ll}
(\mathbf{A}_1) & \mathbf{a}(x,t,u,Du)\cdot Du \geq C_o\Phi(|u|)|Du|^p - \varphi_o(x,t), \\
(\mathbf{A}_2) & |\mathbf{a}(x,t,u,Du)| \leq C_1\Phi(|u|)|Du|^{p-1} + \Phi^{\frac{1}{p}}(u)\varphi_1(x,t), \\
(\mathbf{A}_3) & |b(x,t,u,Du)| \leq C_2\Phi(|u|)|Du|^p + \varphi_2(x,t).
\end{array}
$$

The non-negative functions φ_i, $i=0,1,2$, satisfy $(\mathbf{A}_4)$-$(\mathbf{A}_5)$ of §1 of Chap. II. The function $\Phi(\cdot)$ is degenerate near the origin in the sense that there exists a number $\sigma_o > 0$ such that

$$\gamma_1 s^{\beta_1} \leq \Phi(s) \leq \gamma_2 s^{\beta_2}, \qquad \forall 0 \leq s \leq \sigma_o,$$

for given positive constants $0<\gamma_1\leq\gamma_2$ and $0\leq\beta_2\leq\beta_1$. This behaviour has to hold only near the degeneracy, i.e., for s near zero. For $s>\sigma_o$ it will suffice that $\Phi(s)$ be bounded above and below by given positive constants, i.e., for example,

$$\Lambda_o \leq \Phi(s) \leq \Lambda_1, \qquad s \geq \sigma_o.$$

We require that

$$u \in C_{loc}\left(0,T;L^2_{loc}(\Omega)\right), \quad \Phi^{\frac{1}{p-1}}(u)|Du| \in L^p_{loc}(\Omega_T).$$

Let $F(\cdot)$ denote a primitive of $\Phi^{\frac{1}{p-1}}(\cdot)$. Then the p.d.e. can be interpreted weakly by requiring that

$$F(u) \in L^p_{loc}\left(0,T;W^{1,p}_{loc}(\Omega)\right).$$

If $\Phi(s)=1, \forall s>0$, then (1.1) is of the $p-$laplacian type. If $p=2$ and $\Phi(s)=s^{m-1}$ for some $m>0$, then (1.1) exhibits a degeneracy ($m>1$), or singularity ($0<m<1$) of the type of porous medium equation. In the latter case a weak solution is required to satisfy

$$|u|^m \in L^2_{loc}\left(0,T;W^{1,2}_{loc}(\Omega_T)\right).$$

The Hölder continuity of solutions of such doubly degenerate equations can be proved by methods similar to the ones presented here and has been established independently by Porzio–Vespri [88] and Ivanov [52]. The technique is also flexible enough to handle equations bearing a power-like degeneracy at two values of the solutions. These arise in the flow of immiscible fluids in a porous medium and have as a prototype

$$u_t = \Delta u(1-u) = 0, \qquad 0 \leq u \leq 1.$$

Results on continuous dependence appear in [9] in a different context.

IV

Hölder continuity of solutions of singular parabolic equations

1. Singular equations and the regularity theorems

Evolution equations of the type of (1.1) of Chap. II for $1<p<2$ are *singular* since their modulus of ellipticity becomes unbounded when $|Du|=0$. We will lay out a theory of local and global Hölder continuity of solutions u of such singular p.d.e.'s. We assume that $u \in L^\infty(\Omega_T)$. If u is only *locally* bounded it will suffice to work within compact subsets $\mathcal{K}$ of Ω_T. The *intrinsic* p–distance dist $(\mathcal{K};\Gamma;p)$ from $\mathcal{K}$ to the parabolic boundary of Ω_T is defined as in (1.1) of Chap. III. In the theorems below, the statement that a constant γ depends upon the data means that it can be determined a priori only in terms of $\|u\|_{\infty,\Omega_T}$, the constants C_i, $i=0,1,2$, and the norms $\|\varphi_o, \varphi_1^{\frac{p}{p-1}}, \varphi_2\|_{\hat{q},\hat{r};\Omega_T}$ appearing in the structure conditions $(\mathbf{A}_1)$-$(\mathbf{A}_5)$. For p in the singular range $1<p<2$, let , $p-$ dist $(\mathcal{K};\Gamma)$ denote the *intrinsic* parabolic distance from $\mathcal{K}$ to the parabolic boundary of Ω_T, i.e,

$$p - \operatorname{dist}(\mathcal{K};\Gamma) \equiv \inf_{\substack{(x,t)\in\mathcal{K}\\(y,s)\in\Gamma}} \left(\|u\|_{\infty,\Omega_T}^{\frac{2-p}{p}} |x-y| + |t-s|^{\frac{1}{p}} \right).$$

1-(i). Hölder continuity in the interior

THEOREM 1.1. *Let u be a bounded local weak solution of (1.1) of Chap. II and let $(\mathbf{A}_1)$-$(\mathbf{A}_5)$ hold. Then u is locally Hölder continuous in Ω_T, and there exists constants $\gamma>1$ and $\alpha\in(0,1)$ depending only upon the data, such that $\forall\mathcal{K}\subset\Omega_T$,*

$$\left|u(x_1,t_1)-u(x_2,t_2)\right|\le\gamma\|u\|_{\infty,\Omega_T}\left(\frac{\|u\|_{\infty,\Omega_T}^{\frac{2-p}{p}}|x_1-x_2|+|t_1-t_2|^{1/p}}{p-\operatorname{dist}(\mathcal{K};\Gamma)}\right)^{\alpha},$$

for every pair of points $(x_1,t_1),(x_2,t_2)\in\mathcal{K}$. *If the lower order terms* $b(x,t,u,Du)$ *satisfy* $(\mathbf{A}_3')$ *of §5 of Chap. II, then* γ *and* α *are independent of* $\|u\|_{\infty,\Omega_T}$.

1-(ii). Boundary regularity (Dirichlet data)

THEOREM 1.2. *Let* u *be a bounded weak solution of the Dirichlet problem (2.1) of Chap. II and let* $(\mathbf{D})$ *and* $(\mathbf{U_o})$ *hold. Assume also that the boundary* $\partial\Omega$ *has the property of positive geometric density (1.1) of Chap. I. Then* $u\in C\left(\overline{\Omega}_T\right)$ *and there exists a continuous non-decreasing function* $s\to\omega(s):\mathbf{R}^+\to\mathbf{R}^+$*, such that* $\omega(0)=0$ *and*

$$\left|u(x_1,t_1)-u(x_2,t_2)\right|\le\omega\left(|x_1-x_2|+|t_1-t_2|^{\frac{1}{p}}\right),$$

for every pair of points $(x_1,t_1),(x_2,t_2)\in\overline{\Omega}_T$. *In particular, if the boundary datum* g *is Hölder continuous in* S_T *with exponent say* α_g*, and if the initial datum* u_o *is Hölder continuous in* $\bar{\Omega}$ *with exponent say* α_{u_o}*, then* $(x,t)\to u(x,t)$ *is Hölder continuous in* $\overline{\Omega}_T$ *and there exist constants* $\gamma>1$ *and* $\alpha\in(0,1)$ *such that*

$$\left|u(x_1,t_1)-u(x_2,t_2)\right|\le\gamma\|u\|_{\infty,\Omega_T}\left(\|u\|_{\infty,\Omega_T}^{\frac{2-p}{p}}|x_1-x_2|+|t_1-t_2|^{\frac{1}{p}}\right)^{\alpha},$$

for every pair of points $(x_1,t_1),(x_2,t_2)\in\bar{\Omega}_T$.

The constants γ *and* α *depend only upon the data and the number* α^* *of (1.1) of Chap. I. Moreover the constant* α *also depends upon the Hölder exponents* α_g,α_{u_o} *of* g *and* u_o *respectively.*

If the lower order terms $b(x,t,u,Du)$ *satisfy* $(\mathbf{A}_3')$ *of §5 of Chap. II, then* γ *and* α *are independent of* $\|u\|_{\infty,\Omega_T}$.

1-(iii). Boundary regularity (variational data)

THEOREM 1.3. *Let* u *be a bounded weak solution of the Neumann problem (2.7) of Chap. II and let* $(\mathbf{N})$ *and* $(\mathbf{N}-\mathbf{i})$ *hold. Assume that the boundary* $\partial\Omega$ *is of class* $C^{1,\lambda}$. *Then* u *is Hölder continuous in* $\overline{\Omega}_T$ *and there exist constants* γ *and* α *such that*

$$\left|u(x_1,t_1)-u(x_2,t_2)\right|\le\gamma\|u\|_{\infty,\bar{\Omega}_T}\left(\|u\|_{\infty,\Omega_T}^{\frac{2-p}{p}}|x_1-x_2|+|t_1-t_2|^{\frac{1}{p}}\right)^{\alpha},$$

for every pair of points $(x_1,t_1),(x_2,t_2)\in\overline{\Omega}_T$.

The constants $\gamma > 1$ and α only depend upon $\|u\|_{\infty,\bar{\Omega}_T}$ and the data, including the structure of $\partial\Omega$ and the norms $\|\psi_1, \psi_o^{\frac{p}{p-1}}\|_{\hat{q},\hat{r};\Omega_T}$ appearing in the assumptions **(N) − i.**

If the Neumann data are homogeneous, i.e., if $\psi_o \equiv \psi_1 \equiv 0$, and if in addition the lower order terms $b(x,t,u,Du)$ satisfy $(\mathbf{A}_3')$ of §5 of Chap. II, then γ and α are independent of $\|u\|_{\infty,\bar{\Omega}_T}$.

1-(iv). Some comments

The last two Theorems have been stated in a *global way*. The proof however uses only local arguments so that they could be stated within any compact portion, say $\mathcal{K}$ of $\overline{\Omega}$. Accordingly, the hypotheses on the boundary data need only to hold within $\mathcal{K}$. For example, in the case of Dirichlet data, the boundary datum g could be continuous or Hölder continuous only on a open portion of S_T (open in the relative topology of S_T), say Σ. Then the solution u of the Dirichlet problem would be continuous (respectively Hölder continuous) up to every compact subset of Σ. Analogous considerations can be made for Neumann data satisfying **(N)-(N-i)** on relatively open portions of S_T.

Similar remarks hold if u_o is only locally continuous or locally Hölder continuous. In particular, to establish the continuity (Hölder continuity respectively) of u up to $\Omega \times \{0\}$, no reference is needed to any boundary data on S_T.

Finally we comment on the assumption that u be locally bounded. It will be shown in the next Chapter, that when $p > 2$, solutions of (1.1) are locally bounded. This is no longer true, in general, if $1 < p < 2$. A weak solutions of u of (1.1) is in $L^\infty_{loc}(\Omega_T)$, *only if*

$$u \in L^r_{loc}(\Omega_T) \quad \text{for some } r \geq 1 \text{ satisfying } N(p-2) + rp > 0$$

and such a condition is sharp. Thus, unlike the degenerate case, when p is near one, the local boundedness is not implicit into the notion of weak solution and must be obtained by other information such as boundary data. We refer to Chap. V for a systematic study of local and global boundedness.

2. The main proposition

The Hölder continuity of u, either in the interior of Ω_T or at the parabolic boundary, will be, heuristically, a consequence of the following fact. The function $(x,t) \to u(x,t)$ can be modified in a set of measure zero to yield a continuous representative out of the equivalence class $u \in V^{2,p}_{loc}(\Omega_T)$, if for every $(x_o, t_o) \in \Omega_T$ there exist a family of nested and shrinking cylinders $[(x_o, t_o) + Q(\theta_n, \rho_n)]$ with same *vertex* such that the essential oscillation ω_n of u in $[(x_o, t_o) + Q(\theta_n, \rho_n)]$ tends to zero as

$n \to \infty$ in a way that can be *quantitatively* determined by the structure conditions $(\mathbf{A}_1)$-$(\mathbf{A}_5)$.

To begin the proof of Theorem 1.1 we introduce a space–time configuration that reflects the singularity exhibited by the p.d.e. Fix $(x_o, t_o) \in \Omega_T$ and construct the cylinder

$$\left[(x_o, t_o) + Q\left(R^p, R^{1-\varepsilon}\right)\right] \subset \Omega_T,$$

where ε is a small positive number to be determined later. After a translation one may assume that $(x_o, t_o) \equiv (0, 0)$ and set

$$\mu^+ = \operatorname*{ess\,sup}_{Q(R^p, R^{1-\varepsilon})} u, \quad \mu^- = \operatorname*{ess\,inf}_{Q(R^p, R^{1-\varepsilon})} u, \quad \omega = \operatorname*{ess\,osc}_{Q(R^p, R^{1-\varepsilon})} u \equiv \mu^+ - \mu^-.$$

Consider the box

$$Q\left(R^p, c_o R\right), \quad \text{where} \quad c_o = \left(\frac{\omega}{A}\right)^{\frac{p-2}{p}} \tag{2.1}$$

and where A is a constant to be determined later only in terms of the data. If we assume that

$$\left(\frac{\omega}{A}\right)^{\frac{p-2}{p}} < R^{-\varepsilon}, \tag{2.2}$$

then we have

$$Q\left(R^p, c_o R\right) \subset Q\left(R^p, R^{1-\varepsilon}\right) \quad \text{and} \quad \operatorname*{ess\,osc}_{Q(R^p, c_o R)} u \le \omega.$$

Cylinders of the type of (2.1) have the space variables stretched by a factor $(\omega/A)^{\frac{p-2}{p}}$, which is intrisically determined by the solution. If $p = 2$ these are the standard parabolic cylinders with the natural homogeneity of the space and time variables.

PROPOSITION 2.1. *There exist constants $\varepsilon_o, \eta \in (0,1)$ and $C, A, \mathcal{A} > 1$, that can be determined a priori depending only upon the data, satisfying the following. Construct the sequences $R_o = R, \omega_o = \omega$*

$$R_n = C^{-n} R, \quad \omega_{n+1} = \max\left\{\eta\omega_n; \mathcal{A} R_n^{\varepsilon_o}\right\}, \quad n = 1, 2, \ldots,$$

and the boxes

$$Q_{(n)} \equiv Q\left(R_n, c_n R_n\right), \quad c_n = \left(\frac{\omega_n}{A}\right)^{\frac{p-2}{p}}, \quad n = 0, 1, 2, \ldots.$$

Then for all $n = 0, 1, 2, \ldots$

$$Q_{(n+1)} \subset Q_{(n)} \quad \textit{and} \quad \operatorname*{ess\,osc}_{Q_{(n)}} u \le \omega_n.$$

A consequence of this proposition is

LEMMA 2.1. *There exist constants* $\gamma > 1$ *and* $\alpha \in (0,1)$ *that can be determined a priori only in terms of the data, such that for all the cylinders*

$$0 < \rho \le R, \qquad Q\left(\rho^p, c_o\rho\right), \qquad c_o = \left(\frac{\omega}{A}\right)^{\frac{p-2}{p}},$$

$$\operatorname*{ess\,osc}_{Q(\rho^p, c_o\rho)} u \le \gamma\left(\omega + R^{\varepsilon_o}\right)\left(\frac{\rho}{R}\right)^{\alpha}.$$

This is the analog of Lemma 3.1 of Chap. III. The proof is the same and it implies the Hölder continuity of u over compact subsets of Ω_T via a covering argument.

Remark 2.1. The proof of Proposition 2.1 will show that it would suffice to work with the number ω and the cylinder $Q\left(R^p, c_o R\right)$ linked by

$$\operatorname*{ess\,osc}_{Q(R^p, c_o R)} u \le \omega. \tag{2.3}$$

This fact is in general not verifiable for a given box since its dimensions would have to be *intrinsically* defined in terms of the essential oscillation of u within it.

The reason for introducing the cylinder $Q\left(R^p, R^{1-\varepsilon}\right)$ and assuming (2.2) is that (2.3) holds true for the *constructed* box $Q\left(R^p, c_o R\right)$. It will be part of the proof of Proposition 2.1 to show that at each step the cylinders $Q_{(n)}$ and the essential oscillation of u within them satisfy the intrinsic geometry dictated by (2.3).

Remark 2.2. Such a geometry is not the only possible. For example, one could introduce a scaling with different parameters in the space and time variables. Examples of such *mixed scalings* will occur along the proof of Proposition 2.1. Here we mention that the proof could be structured by introducing the boxes $Q\left(R^{p-\varepsilon}, 2R\right)$ and $Q\left(a_o R^p, R\right)$ *formally* identical to those of §2 of Chap. III and rephrasing the Proposition 2.1 in terms of such a geometry.

3. Preliminaries

Inside $Q\left(R^p, c_o R\right)$ consider subcylinders of *smaller size* constructed as follows. The number ω being fixed, let s_o be the smallest positive integer such that

$$\frac{\omega}{2^{s_o}} \le \delta_o, \tag{3.1}$$

where δ_o is introduced in (3.11) of Chap. II. Then construct cylinders

$$\left[(\bar{x}, 0) + Q\left(R^p, d_o R\right)\right], \qquad d_o = \left(\frac{\omega}{2^{s_o}}\right)^{\frac{p-2}{p}}. \tag{3.2}$$

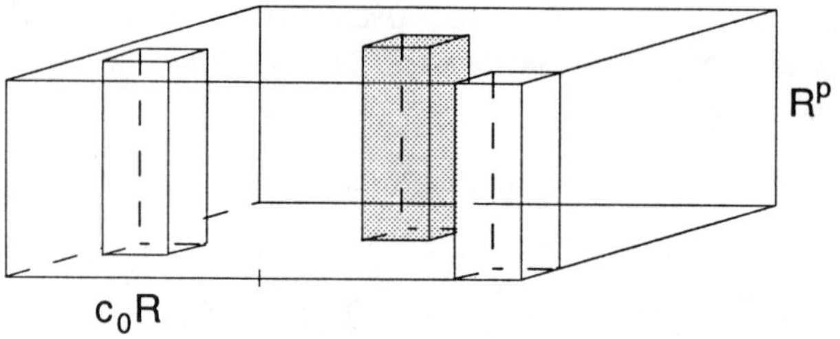

Figure 3.1

These are contained inside $Q(R^p, c_oR)$ if the number A is larger that 2^{s_o} and if $\bar{x}$ ranges over the cube $K_{\mathcal{R}(\omega)}$, where

$$
\begin{aligned}
\mathcal{R}(\omega) &\equiv \left\{A^{\frac{2-p}{p}} - (2^{s_o})^{\frac{2-p}{p}}\right\}\omega^{\frac{p-2}{p}} R \\
&= \left\{\left(\frac{A}{2^{s_o}}\right)^{\frac{2-p}{p}} - 1\right\}\left(\frac{\omega}{2^{s_o}}\right)^{\frac{p-2}{p}} R \qquad (3.3)\\
&= L_o(d_oR), \text{ where } L_o \equiv \left(\frac{A}{2^{s_o}}\right)^{\frac{2-p}{p}} - 1.
\end{aligned}
$$

One may view these as boxes moving inside $Q(R^p, c_oR)$ as the coordinates $\bar{x}$ of their *vertices* range over the cube $K_{\mathcal{R}(\omega)}$. The cylinders $[(\bar{x}, 0) + Q(R^p, d_oR)]$ can also be viewed as the blocks of a partition of $Q(R^p, c_oR)$. Indeed we may arrange that L_o be an integer and view the cube K_{c_oR} as the union, up to a set of measure zero, of L_o^N disjoint cubes each congruent to K_{d_oR}. Analogously $Q(R^p, c_oR)$ is the disjoint union, up to a set of measure zero of L_o^N open boxes each congruent to $Q(R^p, d_oR)$. The proof of Theorem 1.1, is based on studying the following two cases. Let ν_o be a small positive number. Then either

the first alternative

there exists a cylinder of the type of $[(\bar{x}, 0) + Q(R^p, d_oR)]$, making up the partition of $Q(R^p, c_oR)$, such that

$$
\begin{aligned}
\text{meas}&\left\{(x,t) \in [(\bar{x}, 0) + Q(R^p, d_oR)] \;\middle|\; u(x,t) < \mu^- + \frac{\omega}{2^{s_o}}\right\} \qquad (3.4)\\
&\qquad < \nu_o |Q(R^p, d_oR)|,
\end{aligned}
$$

or

the second alternative

for *all* cylinders $[(\bar{x}, 0) + Q(R^p, d_oR)]$ making up the partition of $Q(R^p, c_oR)$,

$$\text{(3.5)} \qquad \operatorname{meas}\left\{(x,t) \in [(\bar{x},0) + Q\left(R^p, d_o R\right)] \,\middle|\, u(x,t) < \mu^- + \frac{\omega}{2^{s_o}}\right\} \geq \nu_o \left|Q\left(R^p, d_o R\right)\right|.$$

In either case the conclusion is that the oscillation of u in a smaller cylinder with *vertex* at the origin, decreases in a way that can be quantitatively measured. In the arguments to follow we assume (2.2) holds. Indeed if not,

$$\omega \leq A\, R^{\varepsilon_o}, \qquad \varepsilon_o = \frac{p\varepsilon}{(2-p)}$$

and the first iterative step of Proposition 2.1 would be trivial.

Remark 3.1. Along the proof we will encounter quantities of the type

$$A_i R^{N\kappa} \omega^{-b_o}, \qquad i = 1, 2, \ldots, l \in \mathbf{N},$$

where A_i are constants that can be determined a priori only in terms of the data and

$$\text{(3.6)} \qquad b_o = 2 + N(p-2)\left(\frac{1}{p} - \frac{1+\kappa}{q}\right).$$

From the range of κ and q as defined in (3.5)-(3.7) of Chap. II, one checks that $b_o \geq p$. We may assume that

$$\text{(3.7)} \qquad A_i R^{N\kappa} \omega^{-b_o} \leq 1\,.$$

Indeed if not, we would have $\omega \leq \mathcal{A} R^{\varepsilon_o}$ for the choices

$$\mathcal{A} = \max_{1 \leq i \leq l} A_i^{\frac{1}{b_o}} \qquad \text{and} \qquad \varepsilon_o = \frac{N\kappa}{b_o}.$$

Remark 3.2. The proof shows that the numbers ε and ε_o can be taken as

$$\varepsilon_o = \frac{N\kappa}{b_o}, \qquad \varepsilon = \frac{\varepsilon_o(2-p)}{p}.$$

3-(i). About the dependence on $\|u\|_{\infty,\Omega_T}$

In the arguments below we will use the energy and logarithmic estimates of Propositions 3.1 and 3.2 of Chap. II, for the truncated functions $(u-k)_{\pm}$, over cylinders contained in $Q\left(R^p, c_o R\right)$. When working with $(u-k)_-$ we will use the levels

$$k = \mu^- + \frac{\omega}{2^{s_o+i}}, \qquad \text{for some } i \geq 0,$$

and when working with $(u-k)_+$ we will take

$$k = \mu^+ - \frac{\omega}{2^{s_o+i}}, \qquad \text{for some } i \geq 0.$$

These are admissible since

$$\|(u-k)_{\pm}\|_{\infty,Q(R^p,c_oR)} \le \frac{\omega}{2^{s_o+i}} \le \delta_o.$$

Let us fix δ_o as in (3.11) of Chap. II. Then, since $\omega \le 2\|u\|_{\infty,\Omega_T}$, (3.1) holds true, within *any* subdomain of Ω_T, if we choose s_o so large that

$$2^{s_o} = \frac{4^{p+8}\,C_2}{C_o}\|u\|_{\infty,\Omega_T}. \tag{3.8}$$

The a priori knowledge of the norm $\|u\|_{\infty,\Omega_T}$ is required through the number s_o. If the lower order terms $b(x,t,u,Du)$ in (1.3) satisfy $(\mathbf{A}_3')$ of §5 of Chap. II, then, as remarked there, the energy and logarithmic inequalities hold true for the truncated functions $(u-k)_{\pm}$ with *no restriction* on the levels k. Thus in such a case, s_o can be taken to be one and no a priori knowledge of $\|u\|_{\infty,\Omega_T}$ is needed.

The numbers A and A_i introduced in (3.7) will be chosen to be larger than 2^{s_o}. In the proof below we will choose them of the type

$$A \equiv A_o \qquad A_i = 2^{s_o+\ell_i}, \qquad i=1,2,\dots,$$

where $\ell_i \ge 0$ will be *independent of* $\|u\|_{\infty,\Omega_T}$. We have just remarked that if the lower order terms $b(x,t,u,Du)$ satisfy $(\mathbf{A}_3')$ of §5 of Chap. II, then s_o can be taken to be one. We conclude that for equations with such a structure the numbers A_i can be determined a priori only in terms of the data and independent of the norm $\|u\|_{\infty,\Omega_T}$.

4. Rescaled iterations

The following rescaled iteration technique applies to any subcylinder of Ω_T and it is crucial in both alternatives. Let $m>0$ be given by

$$m = m_1 + m_2, \qquad \text{where } m_1 \ge s_o, \text{ and } m_2 \ge 0$$

and consider the cube

$$K_{d_1R} \equiv \left\{ \max_{1\le i\le N} |x_i| < d_1 R \right\}, \quad d_1 = \left(\frac{\omega}{2^{m_1}}\right)^{\frac{p-2}{p}},$$

and the box

$$\mathcal{Q}_R(m_1,m_2) \equiv K_{d_1R} \times \left\{ -2^{m_2(p-2)}R^p, 0 \right\}.$$

Fix $(\bar{x},\bar{t}) \in \Omega_T$, and let $R>0$ be so small that

$$[(\bar{x},\bar{t}) + \mathcal{Q}_R(m_1,m_2)] \subset \Omega_T.$$

Remark 4.1. If $(\bar{x},\bar{t}) \equiv (0,0)$ and $2^{m_1}=A$, $m_2=0$, then the cylinder $[(\bar{x},\bar{t}) + \mathcal{Q}_R(m_1,m_2)]$ coincides with $Q(R^p,c_oR)$. Analogously, if $m_2=0$, $m_1=s_o$ and $\bar{t}=0$, then, for a suitable choice of $\bar{x}$ the cylinder $[(\bar{x},\bar{t}) + \mathcal{Q}_R(m_1,m_2)]$ coincides with one of the boxes making up the partition of $Q(R^p,c_oR)$.

LEMMA 4.1. *There exists a number ν_o that can be determined a priori only in terms of the data and independent of ω, R and m_1, m_2 such that:*

(I). If u is a super–solution of (1.3) in $[(\bar{x},\bar{t}) + \mathcal{Q}_R(m_1,m_2)]$ satisfying

$$\operatorname*{ess\,osc}_{[(\bar{x},\bar{t})+\mathcal{Q}_R(m_1,m_2)]} u \le \omega$$

and

$$\text{meas}\left\{(x,t) \in [(\bar{x},\bar{t}) + \mathcal{Q}_R(m_1,m_2)] \mid u(x,t) < \mu^- + \frac{\omega}{2^m}\right\} \le \nu_o \big|\mathcal{Q}_R(m_1,m_2)\big|,$$

then either

$$\omega^{b_o} \le \mathcal{A}_o R^{N\kappa},$$

or

$$u(x,t) \ge \mu^- + \frac{\omega}{2^{m+1}}, \qquad \forall (x,t) \in \mathcal{Q}_{\frac{R}{2}}(m_1,m_2),$$

where b_o is defined in (3.6) and $\mathcal{A}_o$ is a constant depending only upon the data and the numbers m_1, m_2. Analogously

(II). If u is a sub–solution of (1.3) in $[(\bar{x},\bar{t}) + \mathcal{Q}_R(m_1,m_2)]$ satisfying

$$\operatorname*{ess\,osc}_{[(\bar{x},\bar{t})+\mathcal{Q}_R(m_1,m_2)]} u \le \omega$$

and

$$\text{meas}\left\{(x,t) \in [(\bar{x},\bar{t}) + \mathcal{Q}_R(m_1,m_2)] \mid u(x,t) > \mu^+ - \frac{\omega}{2^m}\right\} \le \nu_o \big|\mathcal{Q}_R(m_1,m_2)\big|,$$

then either

$$\omega^{b_o} \le \mathcal{A}_o R^{N\kappa},$$

or

$$u(x,t) \le \mu^+ - \frac{\omega}{2^{m+1}}, \qquad \forall (x,t) \in \mathcal{Q}_{\frac{R}{2}}(m_1,m_2),$$

PROOF: We only prove the statement regarding super–solutions. Assume $(\bar{x},\bar{t}) \equiv (0,0)$ and construct the decreasing sequences of numbers

$$R_n = \frac{R}{2} + \frac{R}{2^{n+1}}, \quad k_n = \mu^- + \frac{\omega}{2^{m+1}} + \frac{\omega}{2^{m+1+n}}, \qquad n=0,1,2,\ldots,$$

and the families of nested cubes and cylinders

$$K_n \equiv K_{d_1 R_n}, \qquad d_1 = \left(\frac{\omega}{2^{m_1}}\right)^{\frac{p-2}{p}},$$

$$\mathcal{Q}_n \equiv \mathcal{Q}_{R_n}(m_1,m_2) = K_n \times \left\{-2^{(p-2)m_2} R_n^p, 0\right\}.$$

Consider (3.8) of Chap. II, written over the boxes $\mathcal{Q}_n$ for $(u-k_n)_-$ and with the choice of the cutoff functions ζ_n

$$\begin{cases} 0<\zeta_n(x,t)\le 1,\ \forall (x,t)\in\mathcal{Q}_n, \text{ and } \zeta_n\equiv 1 \text{ in } \mathcal{Q}_{n+1}; \\ \zeta_n=0 \quad \text{on the parabolic boundary of } \mathcal{Q}_n; \\ |D\zeta_n|\le \frac{2^{n+2}}{R}\left(\frac{\omega}{2^{m_1}}\right)^{\frac{2-p}{p}},\ 0\le \zeta_{n,t}\le 2^{(2-p)m_2}\frac{2^{p(n+2)}}{R^p}. \end{cases}$$

In this setting, (3.8) takes the form

$$\begin{aligned}(4.1)\qquad &\sup_{-2^{(p-2)m_2}R_n^p\le t\le 0}\int_{K_n}(u-k_n)_-^2\,\zeta_n^p(x,t)dx \\ &\quad+\iint_{\mathcal{Q}_n}\left|D(u-k_n)_-\,\zeta_n\right|^p dxd\tau \\ &\quad\le \gamma\frac{2^{pn}}{R^p}\left(\frac{\omega}{2^{m_1}}\right)^{2-p}\iint_{\mathcal{Q}_n}(u-k_n)_-^p\,dxd\tau \\ &\quad+\gamma\frac{2^{pn}}{R^p}2^{(2-p)m_2}\iint_{\mathcal{Q}_n}(u-k_n)_-^2\,dxd\tau \\ &\quad+\gamma\left\{\int_{-2^{(p-2)m_2}R_n^p}^{0}\left|A^-_{k_n,d_1R_n}(\tau)\right|^{\frac{r}{q}}d\tau\right\}^{\frac{p(1+\kappa)}{r}}.\end{aligned}$$

Since

$$\sup_{\mathcal{Q}_n}(u-k_n)_-\le\frac{\omega}{2^{m_1+m_2}},$$

the first two terms on the right hand side of (4.1) are estimated above by

$$\gamma\frac{2^{pn}}{R^p}\left(\frac{\omega}{2^m}\right)^2 2^{(2-p)m_2}\iint_{\mathcal{Q}_n}\chi\left[(u-k_n)_->0\right]dxd\tau.$$

To estimate below the two integrals on the left hand side, introduce the level

$$\bar{k}_n\equiv\frac{1}{2}(k_n+k_{n+1})<k_n.$$

Then for all $t\in\left(-2^{(p-2)m_2}R_n^p,0\right)$

$$\begin{aligned}\int_{K_n}(u-k_n)_-^2\,\zeta_n^p(x,t)dx &\ge \int_{K_n}(k_n-\bar{k}_n)^{2-p}(u-\bar{k}_n)_-^p\,\zeta_n^p(x,t)dx \\ &\ge\left(\frac{\omega}{2^m}\right)^{2-p}2^{(p-2)(n+3)}\int_{K_n}(u-\bar{k}_n)_-^p\,\zeta_n^p(x,t)dx.\end{aligned}$$

Also we have

$$\iint_{Q_n} \left|D\left(u-k_n\right)_-\zeta_n\right|^p dx d\tau \geq \iint_{Q_n} \left|D\left(u-\bar{k}_n\right)_-\zeta_n\right|^p dx d\tau$$

$$-\gamma \frac{2^{pn}}{R^p}\left(\frac{\omega}{2^m}\right)^2 2^{(2-p)m_2} \iint_{Q_n} \chi\left[\left(u-k_n\right)_->0\right] dx d\tau.$$

Put these estimates in (4.1), divide through by

$$\left(\frac{\omega}{2^m}\right)^{2-p} 2^{(p-2)(n+3)}$$

and in the resulting integrals introduce the change of variables

$$y=\left(\frac{\omega}{2^{m_1}}\right)^{\frac{2-p}{p}} x, \qquad z=2^{(2-p)m_2} t,$$

which maps $\mathcal{Q}_n$ into

$$Q_n \equiv K_{R_n} \times\{-R_n^p, 0\}.$$

Setting also

$$v(y, z)=u\left(d_1 y, 2^{(p-2)m_2} t\right), \quad \hat{\zeta}(y, z)=\hat{\zeta}\left(d_1 y, 2^{(p-2)m_2} t\right),$$

and

$$A_n(z) \equiv\left\{y \in K_{R_n} \middle| v(y, z)<k_n\right\}, \quad |A_n| \equiv \int_{-R_n^p}^0 \left|A_n(z)\right| dz,$$

we arrive at

$$\| \left(v-\bar{k}_n\right)_- \hat{\zeta}_n \|_{V^p(Q_n)}^p \leq \frac{\gamma 2^{2n}}{R^p}\left(\frac{\omega}{2^m}\right)^p |A_n|$$

$$+\gamma\left(\frac{\omega}{2^m}\right)^p A_o R^{N\kappa} \omega^{-b_o}\left\{\int_{-R_n^p}^0 \left|A_n(z)\right|^{\frac{r}{q}} dz\right\}^{\frac{p(1+\kappa)}{r}}.$$

By (3.7), $A_o R^{N\kappa} \omega^{-b_o} \leq 1$, and by Corollary 3.1 of Chap. I,

$$\begin{aligned}
&2^{-(n+3)p}\left(\frac{\omega}{2^m}\right)^p |A_{n+1}| \\
&\leq \left(\bar{k}_n-k_{n+1}\right)^p \left|(y, z) \in Q_{n+1} \,\middle|\, v(y, z)<k_{n+1}\right| \\
&\leq \| \left(v-k_n\right)_+ \|_{p, Q_{n+1}}^p \\
&\leq \| \left(v-k_n\right)_+ \hat{\zeta}_n \|_{p, Q_n}^p \\
&\leq \gamma |A_n|^{\frac{p}{N+p}} \| \left(v-k_n\right)_+ \hat{\zeta}_n \|_{V^p(Q_n)}^p
\end{aligned}$$

$$\leq \gamma\left(\frac{\omega}{2^m}\right)^p\left[\frac{2^{np}}{R^p}|A_n|^{1+\frac{p}{N+p}}+|A_n|^{\frac{p}{N+p}}\left\{\int_{-R_n^p}^0 \left|A_n(z)\right|^{\frac{r}{q}} dz\right\}^{\frac{p(1+\kappa)}{r}}\right].$$

Thus setting

$$Y_n = \frac{|A_n|}{|Q_n|} \quad \text{and} \quad Z_n = \frac{1}{|K_{R_n}|}\left(\int_{-R_n^p}^{0} |A_n(z)|^{\frac{r}{q}} dz\right)^{\frac{p}{r}},$$

we have the recursive inequalities

$$Y_{n+1} \le \gamma 4^{np}\left\{Y_n^{1+\frac{p}{N+p}} + Y_n^{\frac{p}{N+p}} Z_n^{1+\kappa}\right\},$$
$$Z_{n+1} \le \gamma 4^{np}\left\{Y_n + Z_n^{1+\kappa}\right\}.$$

It follows from these and Lemma 4.2 of Chap. I that Y_n and Z_n tend to zero as $n\to\infty$, provided

$$Y_o + Z_o^{1+\kappa} \le (2\gamma)^{-(1+\kappa)/\theta_o} 4^{-p(1+\kappa)/\theta_o^2} \equiv \nu_o,$$

where $\theta_o = \min\left\{\frac{p}{N+p}; \kappa\right\}$.

Remark 4.2. The proof shows that the number ν_o depends upon p but it is 'stable' as $p \nearrow 2$, i.e.,

$$\nu_o(p) \longrightarrow \nu_o(2), \qquad \text{as } p \longrightarrow 2.$$

Remark 4.3. The conclusion of Lemma 4.1 continues to hold for cylinders of the type

$$\text{(4.2)} \qquad \mathcal{Q}_R(m,\beta) \equiv K_r \times (-\beta R^p, 0), \quad r = \left(\frac{\omega}{2m}\right)^{\frac{p-2}{p}} R, \qquad \beta > 0,$$

provided β is independent of ω and R. In such a case we take $m = m_1$ and ν_o will depend also upon β.

5. The first alternative

Suppose that there exists a cylinder of the type of $[(\bar{x},0) + Q(R^p, d_o R)]$ making up the partition of $Q(R^p, c_o R)$ for which (3.4) holds. Then we apply Lemma 4.1 with $m_1 = s_o$ and $m_2 = 0$ to conclude that

$$\text{(5.1)} \qquad u(x,t) \ge \mu^- + \frac{\omega}{2^{s_o+1}}, \quad \forall (x,t) \in [(\bar{x},0) + Q((\tfrac{R}{2})^p, d_o \tfrac{R}{2})].$$

We view the box $[(\bar{x},0) + Q((\frac{R}{2})^p, d_o \frac{R}{2})]$ as a block inside $Q(R^p, c_o R)$. Let $\mathcal{R}(\omega)$ be the *'radius'* introduced in (3.3). The location of $\bar{x}$ within the cube $K_{\mathcal{R}(\omega)}$ is only known qualitatively. We will show that the *'positivity'* of (5.1) *'spreads'* over the full cube $K_{c_o R}$, for all times

$$(\tfrac{R}{8})^p \le t \le 0.$$

In a precise way we will prove

PROPOSITION 5.1. *Assume (5.1) holds for some $\bar{x} \in K_{\mathcal{R}(\omega)}$. There exists positive numbers $\mathcal{A}_1$ and l_1 that can be determined a priori only in terms of the data and the number A in the definition of $Q\left(R^p, c_o R\right)$, such that either*

$$\omega^{b_o} \leq \mathcal{A}_1 R^{N\kappa}, \tag{5.2}$$

or

$$u(x,t) \geq \mu^- + \frac{\omega}{2^{s_o+l_1}}, \quad \forall (x,t) \in Q\left((\tfrac{R}{8})^p, c_o R\right). \tag{5.3}$$

As a consequence we may rephrase the first alternative in the following form.

COROLLARY 5.1. *Assume that (3.4) holds for some cylinder of the type of $[(\bar{x},0) + Q\left(R^p, d_o R\right)]$ making up the partition of $Q\left(R^p, c_o R\right)$. There exists positive numbers $\mathcal{A}_1$ and $\bar{s}$ that can be determined a priori only in terms of the data and the number A in the definition of $Q\left(R^p, c_o R\right)$, such that either*

$$\omega^{b_o} \leq \mathcal{A}_1 R^{N\kappa},$$

or

$$\operatorname*{ess\,osc}_{Q(\rho^p, c_o\rho)} u \leq \eta_1 \omega, \qquad \forall \rho \in (0, R/8),$$

where

$$\eta_1 \equiv 1 - 2^{-(s_o+\bar{s})}.$$

We regard $\bar{x}$ as the centre of a large cube

$$[\bar{x} + K_{8c_o R}], \qquad c_o \equiv \left(\frac{\omega}{A}\right)^{\frac{p-2}{p}},$$

which we may assume is contained in the cube $K_{R^{1-\varepsilon}}$. Indeed if not, we would have

$$16 c_o > R^{-\varepsilon}, \ \text{ i.e., } \omega < 16^{\frac{p}{2-p}} A R^{\varepsilon_o}, \quad \varepsilon_o \equiv \frac{p\varepsilon}{2-p}.$$

We will be working within the box

$$\left[(\bar{x},0) + Q\left((\tfrac{R}{2})^p, 8 c_o R\right)\right]$$

and will show that the conclusion of Proposition 5.1 holds within the cylinder

$$\left[(\bar{x},0) + Q\left((\tfrac{R}{8})^p, 2 c_o R\right)\right].$$

This contains $Q\left((\frac{R}{8})^p, c_o R\right)$, regardless of the location of $\bar{x}$ in the cube $K_{\mathcal{R}(\omega)}$.

5-(i). The p.d.e. in dimensionless form

Introduce the change of variables

$$x \longrightarrow \frac{x - \bar{x}}{2 c_o R}, \qquad t \longrightarrow \frac{4^p\, t}{(R/2)^p},$$

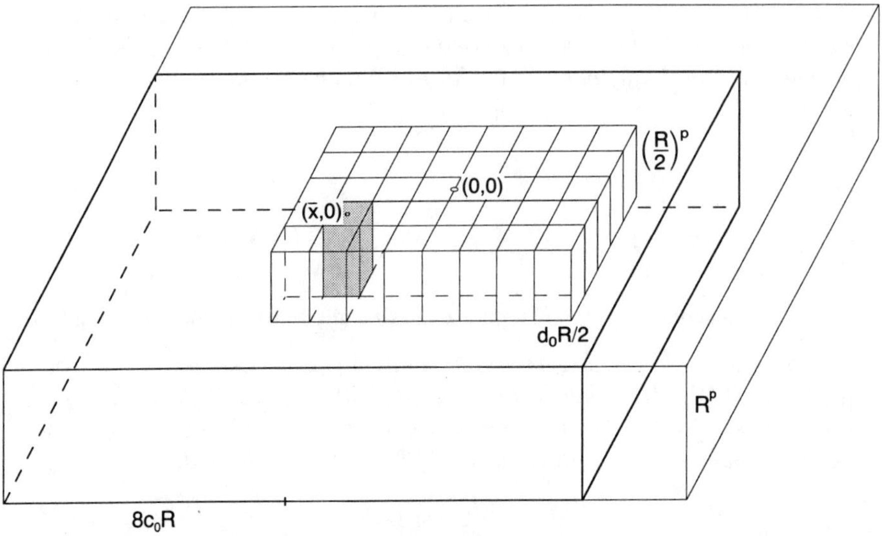

Figure 5.1

which maps $\left[(\bar{x},0)+Q\left(\left(\frac{R}{2}\right)^p,8c_oR\right)\right]$ into $Q_4\equiv K_4\times(-4^p,0)$. Also introduce the function

$$v=\left(u-\mu^-\right)\frac{2^{s_o}}{\omega}. \tag{5.4}$$

Denoting again with x and t the new variables, the function v satisfies the p.d.e.

$$v_t-\operatorname{div}\tilde{\mathbf{a}}(x,t,v,Dv)+\tilde{b}(x,t,v,Dv)=0,\quad \text{in } \mathcal{D}'(Q_4), \tag{5.5}$$

where $\tilde{\mathbf{a}}:Q_4\times\mathbf{R}^{N+1}\to\mathbf{R}^N$ and $\tilde{b}:Q_4\times\mathbf{R}^{N+1}\to\mathbf{R}$, satisfy the structure conditions

$$\tilde{\mathbf{a}}(x,t,v,Dv)\cdot Dv\geq\frac{C_o}{2^{4p}}\left(\frac{2^{s_o}}{A}\right)^{2-p}|Dv|^p-\tilde{\varphi}_o. \tag{5.6}$$

$$\left|\tilde{\mathbf{a}}(x,t,v,Dv)\right|\leq\frac{C_1}{2^{4p}}\left(\frac{2^{s_o}}{A}\right)^{2-p}|Dv|^{p-1}+\tilde{\varphi}_1. \tag{5.7}$$

$$\left|\tilde{b}(x,t,v,Dv)\right|\leq\frac{C_2}{2^{4p}}\left(\frac{2^{s_o}}{A}\right)^{2-p}\left(\frac{\omega}{2^{s_o}}\right)|Dv|^p+\tilde{\varphi}_2. \tag{5.8}$$

Here C_i, $i=0,1,2$, are the constants appearing in the structure conditions $(\mathbf{A}_1)$-$(\mathbf{A}_3)$ of Chap. II. Moreover, setting

(5.9) $$g \equiv \tilde{\varphi}_o + \tilde{\varphi}_1^{\frac{p}{p-1}} + \tilde{\varphi}_2,$$

the function g satisfies

(5.10) $$\|g\|_{\hat{q},\hat{r},Q_4} \leq \gamma R^{N\kappa} \omega^{-b_o}$$

where $\gamma = \gamma(N, p, A, s_o, \text{data})$ is a constant depending only upon the indicated quantities and b_o is defined in (3.6). The numbers κ and q and $\hat{q}, \hat{r}$ satisfy (3.5)–(3.7) of Chap. II. The information (5.1) translates into

(5.11) $$v(x,t) > \tfrac{1}{2} \quad \text{a.e. } (x,t) \in Q(h_o) \equiv \{|x| < h_o\} \times \{-4^p < t \leq 0\},$$

where

(5.12) $$h_o = \frac{d_o R}{8 c_o R} = \frac{1}{8}\left(\frac{2^{s_o}}{A}\right)^{\frac{2-p}{p}} < 1.$$

We regard $Q(h_o)$ as a thin cylinder sitting at the '*centre*' of Q_4. We will prove that the relative largeness of v in $Q(h_o)$, spreads sidewise[(1)] over Q_2.

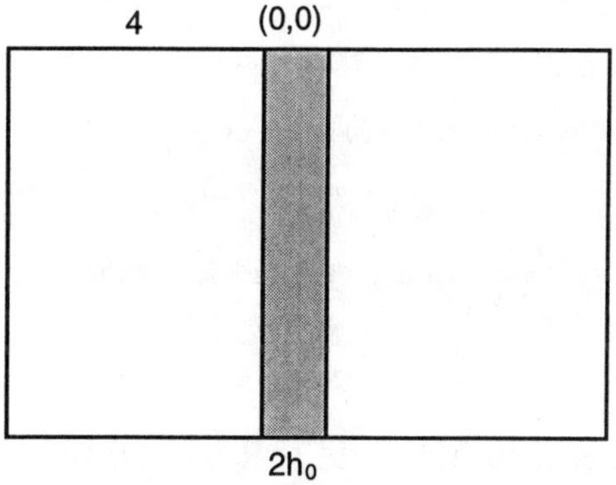

Figure 5.2

Proposition 5.1 will be a consequence of the following fact.

LEMMA 5.1. *For every $\nu \in (0,1)$ there exists positive numbers $A^* > 1$ and $\delta^* \in (0,1)$ that can be determined a priori only in terms of ν, N, p and the data, such that either*

(5.13) $$\omega^{b_o} \leq A^* R^{N\kappa},$$

or

(1) For further comments on this phenomenon we refer to §14-(i).

(5.14) $$\operatorname{meas}\left\{x \in K_2 \;\middle|\; v(x,t) \le \delta^*\right\} \le \nu |K_2|,$$

for all time levels $t \in [-2^p, 0]$.

Remark 5.1. The key feature of the lemma is that the set where v is small can be made arbitrarily small for *every* time level in $[-2^p, 0]$.

5-(ii). Proof of Proposition 5.1 assuming Lemma 5.1

In Lemma 5.1 we choose $\nu = \nu_o$ where ν_o is the number claimed by Lemma 4.1, and determine $\delta^* \equiv \delta^*(\nu_o)$ accordingly. We let m_2 be defined by

$$2^{-m_2} = \delta^*(\nu_o)$$

and apply Lemma 4.1 with $\omega = 1$, $\mu^- = 0$, $R = 2$, over the boxes

$$(0,\bar{t}) + K_2 \times \left\{-2^{m_2(p-2)} 2^p, 0\right\} \equiv (0,\bar{t}) + \mathcal{Q}_2(0, m_2)$$

as long as they are contained in Q_2, i.e., for $\bar{t}$ satisfying

(5.15) $$2^{m_2(p-2)} 2^p - 2^p \le \bar{t} \le 0.$$

Since (5.14) holds true for *all* time levels $-2^p \le t \le 0$, each such box satisfies

$$\operatorname{meas}\left\{(x,t) \in [(0,\bar{t}) + \mathcal{Q}_2(0,m_2)] \;\middle|\; v(x,t) \le 2^{-m_2}\right\} \le \nu_o \left|\mathcal{Q}_2(0,m_2)\right|.$$

Therefore by Lemma 4.1 either (5.2) holds or

$$v(x,t) \ge 2^{-(m_2+1)} \quad \forall (x,t) \in Q_1.$$

Returning to the original coordinates and redefining the various constants accordingly proves Proposition 5.1.

6. Proof of Lemma 5.1. Integral inequalities

First we prove the lemma under the additional assumptions

(6.1) $$\frac{d}{dt} v \in C\left(-4^p, 0; L^1(K_4)\right) \quad \text{and} \quad \|g(\cdot)\|_{\hat{q}, K_4} \in C\left(-4^p, 0\right).$$

These will simplify some of the calculations and will be removed later. The weak formulation of (5.5) is

(6.2) $$\int_{K_4} v_t \varphi(x,t) dx + \int_{K_4} \tilde{\mathbf{a}}(x,t,v,Dv) \cdot D\varphi \, dx = -\int_{K_4} \tilde{b}(x,t,v,Dv) \varphi dx$$

for all $-4^p < t < 0$ and all testing functions

$$\varphi \in C(Q_4) \cap C\left(-4^p, 0; W_o^{1,p}(K_4)\right).$$

Let

$$G(t) \equiv (\delta k)^{-(1+\frac{p}{\hat{q}-1})} \int_{-4^p}^{t} \|g(\tau)\|_{\hat{q},K_4}^{\frac{\hat{q}}{\hat{q}-1}} \, d\tau, \tag{6.3}$$

where k and δ are positive parameters to be chosen later, g is defined in (5.9) and $\hat{q}$ is the number entering in the structure conditions (5.10). We define the new unknow function

$$w \equiv v + G(t), \tag{6.4}$$

and rewrite (6.2) in terms of w. Next, by the parabolic structure[1] of (6.2), the truncation $(k-w)_+$ is a subsolution of (6.2), i.e., for all testing functions $\varphi \geq 0$

$$\begin{aligned} &\int_{K_4} \frac{\partial}{\partial \tau}(k-w)_+ \varphi(x,t)dx + \int_{K_4} \tilde{\mathbf{a}}\left(x,t,v,D(k-w)_+\right) \cdot D\varphi dx \\ &\leq -\int_{K_4} \tilde{b}\left(x,t,v,D(k-w)_+\right)\varphi dx - G'(t)\int_{K_4} \varphi(x,t)dx. \end{aligned} \tag{6.5}$$

In this formulation we take the testing function

$$\varphi \equiv \frac{\zeta^p}{\left[k-(k-w)_+ + \delta\, k\right]^{p-1}},$$

where $\zeta \equiv \zeta_1(x)\zeta_2(t)$ is a piecewise smooth cutoff function in Q_4, satisfying

$$\begin{cases} 0 \leq \zeta \leq 1 \quad \text{in } Q_4, \quad \text{and } \zeta \equiv 1 \text{ in } Q_2\,; \\ \zeta = 0 \quad \text{on the parabolic boundary of } Q_4; \\ |D\zeta_1| \leq 1, \quad 0 \leq \zeta_{2,t} \leq 1\,; \\ \text{the sets } \left\{x \in K_4 \mid \zeta_1(x) > k\right\} \text{ are convex } \forall k \in (0,1). \end{cases} \tag{6.6}$$

We use the structure conditions (5.6)-(5.8), with the symbolism

$$\tilde{C}_i \equiv \frac{C_i}{2^{4p}}\left(\frac{2^{s_o}}{A}\right)^{2-p}.$$

Set also

$$\Phi_k(w) = \int_0^{(k-w)_+} \frac{ds}{[k-s+\delta k]^{p-1}}, \tag{6.7}$$

$$\Psi_k(w) = \ln\left[\frac{k(1+\delta)}{k(1+\delta)-(k-w)_+}\right]. \tag{6.8}$$

---

(1) See §1-(i) of Chap. II.

Then we obtain

$$\begin{aligned}
&\frac{d}{dt}\int_{K_4}\Phi_k(w)\zeta^p dx+\tilde{C}_o\int_{K_4}\left|D\Psi_k(w)\right|^p\zeta^p dx\\
&\qquad\le \tilde{C}_1\int_{K_4}\left(\left|D\Psi_k(w)\right|\zeta\right)^{p-1}|D\zeta|dx\\
&\qquad+\tilde{C}_2 2^{2(p+1)}\left(\frac{\omega}{2^{s_o}}\right)\int_{K_4}\left|D\Psi_k(w)\right|^p\zeta^p dx\\
&\qquad+p\int_{K_4}\left(\Phi_k(w)\right)\zeta^{p-1}\zeta_t\,dx\\
&\qquad+\int_{K_4}\left[\tilde{\varphi}_o+\tilde{\varphi}_1^{\frac{p}{p-1}}+\tilde{\varphi}_2\right]\frac{\zeta^p}{\left[k-(k-w)_++\delta\,k\right]^p}\,dx\\
&\qquad+\frac{1}{p}\int_{K_4}|D\zeta|^p\,dx\;-G'(t)\int_{K_4}\varphi(x,t)\,dx.
\end{aligned}$$

By the choice (3.8) of the number s_o, the second term involving $\left|D\,\Psi_k(w)\right|^p$ is absorbed in the analogous term of the left hand side. The integral involving $\left|D\Psi_k(w)\right|^{p-1}$ is treated by means of Young's inequality and the resulting term involving $\left|D\Psi_k(w)\right|^p$ is absorbed in the analogous term on the left hand side. The remaining term is majorised by an absolute constant depending only upon $\tilde{C}_i$, $i=0,1$. Next, if we stipulate to take k in the interval $(0,1]$, the integral involving ζ_t is majorised by $\gamma/(2-p)$, where γ is an absolute constant depending only upon p. Finally the sum of the last two integrals can be majorised by an absolute constant. Indeed

$$\begin{aligned}
&\int_{K_4}\left[\tilde{\varphi}_o+\tilde{\varphi}_1^{\frac{p}{p-1}}+\tilde{\varphi}_2\right]\frac{\zeta^p}{\left[k-(k-w)_++\delta\,k\right]^p}\,dx\\
&\qquad\le\|g(\cdot,t)\|_{\hat{q},K_4}\left(\int_{K_4}\frac{\zeta^{p\frac{\hat{q}}{\hat{q}-1}}}{\left[k-(k-w)_++\delta\,k\right]^{p\frac{\hat{q}}{\hat{q}-1}}}\,dx\right)^{\frac{\hat{q}-1}{\hat{q}}}\\
&\qquad\le(\delta k)^{-\left(1+\frac{p}{\hat{q}-1}\right)\frac{\hat{q}-1}{\hat{q}}}\,\|g(\cdot,t)\|_{\hat{q},K_4}\left(\int_{K_4}\varphi dx\right)^{\frac{\hat{q}-1}{\hat{q}}}\\
&\qquad\le G'(t)\int_{K_4}\varphi dx\;+1.
\end{aligned}$$

We conclude that there exist constants $\tilde{\gamma}_o$ and $\tilde{\gamma}$ depending only upon N, p, A, s_o and the data, such that

(6.9)
$$\frac{d}{dt}\int_{K_4} \Phi_k(w)\zeta^p dx + \tilde{\gamma}_o \int_{K_4} \left|D\Psi_k(w)\right|^p \zeta^p dx \le \frac{\tilde{\gamma}}{2-p}.$$

Next, since $k \in (0, \frac{1}{2})$ the function $\Psi_k(w)$ vanishes for all $|x| \le h_o$. This follows from the definition (6.4) of w and (5.11). Therefore we may apply the Poincaré inequality (2.9) of Proposition 2.1 of Chap. I, to minorise the second term on the left hand side of (6.9). We summarise:

LEMMA 6.1. *There exists two constants γ_o and γ that can be determined a priori only in terms of N, p, A, s_o such that*

(6.10)
$$\frac{d}{dt}\int_{K_4} \Phi_k(w)\zeta^p dx + \gamma_o \int_{K_4} \Psi_k^p(w)\, \zeta^p dx \le \gamma,$$

where $\Phi_k(w)$ and $\Psi_k(w)$ are defined in (6.7)–(6.8).

Remark 6.1. The function $G(\cdot)$ introduced in (6.3) is defined through the numbers k and δ which are still to be chosen. By virtue of the structure conditions (5.10), we have

$$\begin{aligned} G(t) &\le \gamma\,(\delta\, k)^{-\left(1+\frac{p}{\hat{q}-1}\right)} \, \|g\|_{\hat{q},\hat{r};Q_4}^{\frac{\hat{q}}{\hat{q}-1}} \\ &\le \gamma\,(\delta\, k)^{-\left(1+\frac{p}{\hat{q}-1}\right)} \left[R^{N\kappa}\,\omega^{-b_o}\right]^{\frac{\hat{q}}{\hat{q}-1}}. \end{aligned}$$

If we choose $k\delta = \delta^* \in (0, \frac{1}{2})$ depending only upon the data, we may assume without loss of generality that

(6.11)
$$G(t) \le \gamma\,(\delta k)^{-\left(1+\frac{p}{\hat{q}-1}\right)} \left[R^{N\kappa}\,\omega^{-b_o}\right]^{\frac{\hat{q}}{\hat{q}-1}} \le \delta^{*2}.$$

Indeed, otherwise for such a selection of δ^*

$$\omega^{b_o} \le A^*\, R^{N\kappa}, \qquad A^* \equiv \gamma \delta_*^{\theta},$$

for some positive number θ depending only upon $\hat{q}$ and p and some γ depending only upon the data.

The number δ^* will be chosen shortly only in terms of the data. In view of (6.11) we may regard the function w introduced in (6.4) as independent of k and δ.

7. An auxiliary proposition

Introduce the quantities

(7.1)
$$Y_n \equiv \sup_{-4^p \le t \le 0} \int_{K_4 \cap [w(\cdot,t) < \delta^n]} \zeta^p\,(x,t)\, dx, \qquad n = 0, 1, 2, \ldots.$$

The proof of Lemma 5.1 is a consequence of the following:

PROPOSITION 7.1. *The number $\nu \in (0,1)$ being fixed, we may find numbers $\delta, \sigma \in (0,1)$ depending only upon N, p the data and ν, such that for $n=0,1,2,\dots$, either*

$$Y_n \le \nu \tag{7.2}$$

or

$$Y_{n+1} \le \max\{\nu\,;\,\sigma Y_n\}. \tag{7.3}$$

PROOF OF LEMMA 5.1: Iterating (7.2)-(7.3) gives

$$Y_n \le \max\{\nu\,;\,\sigma^{n-1} Y_o\}, \qquad n=1,2,\dots.$$

Since $Y_o \le |K_4|$, we have only to take $n = n_o$ so large that

$$\sigma^{n_o - 1} \le \nu 2^{-N}.$$

Then the lemma follows with $2\delta^* = \delta^{n_o}$. Indeed, $Y_{n_o} \le \sigma^{n_o-1} Y_o$ implies

$$\text{meas}\{x \in K_2 \mid w(x,t) < 2\delta^*\} < \nu |K_2|.$$

Recalling the definition (6.4) of w and the upper bound (6.11), this in turn yields (5.14) and concludes the proof of the lemma.

7-(i). *Proof of Proposition 7.1*

In (6.10) we take $k = \delta^n$, $n \in \mathbf{N}$, where $\delta \in (0,1)$ is to be chosen. From the definition (7.1) of Y_n, it follows that for every $\varepsilon \in (0,1)$ there exists $t_o \in (-4^p, 0)$, such that

$$\int_{K_4 \cap [w(\cdot, t_o) < \delta^{n+1}]} \zeta^p(x,t_o)\, dx \;\ge\; Y_{n+1} - \varepsilon. \tag{7.4}$$

The numbers $n \in \mathbf{N}$ and $t_o \in (-4^p, 0)$ being fixed, we consider the following two cases:

either

$$\frac{d}{dt} \int_{K_4} \zeta^p(x,t_o)\, \Phi_{\delta^n}(w(x,t_o))\, dx \ge 0 \tag{7.5}$$

or

$$\frac{d}{dt} \int_{K_4} \zeta^p(x,t_o)\, \Phi_{\delta^n}(w(x,t_o))\, dx < 0. \tag{7.6}$$

In either case we may assume that $Y_n > \nu$, otherwise the proposition becomes trivial. Also, in (7.4) we may take ε arbitrarily small within the range $(0, \nu/2)$.

7-(ii). The case (7.5)

If (7.5) holds, it follows from (6.10) with $k = \delta^n$, that

$$\int_{K_4} \zeta^p(x, t_o) \Psi^p_{\delta^n} (w(x, t_o))\, dx \le C \equiv \frac{\gamma}{\gamma_o}. \tag{7.7}$$

We minorise this integral by extending the integration over the smaller set

$$[w(\cdot, t_o) < \delta^{n+1}] \cap K_4.$$

On such a set

$$\Psi_{\delta^n}(w) \ge \ln \frac{\delta^n(1+\delta)}{2\delta^{n+1}}.$$

Therefore

$$\left(\ln \frac{1+\delta}{2\delta}\right)^p \int_{K_4 \cap [w(\cdot, t_o) < \delta^{n+1}]} \zeta^p(x, t_o)\, dx \le \int_{K_4} \zeta^p(x, t_o) \Psi^p_{\delta^n}(w(x, t_o))\, dx.$$

From this, (7.7) and (7.4)

$$Y_{n+1} \le \varepsilon + C\left(\ln \frac{1+\delta}{2\delta}\right)^{-p}.$$

To prove the proposition in such a case, we choose δ so small that

$$C\left(\ln \frac{1+\delta}{2\delta}\right)^{-p} \le \frac{\nu}{2}.$$

Such a choice depends only upon the constants γ, γ_o and ν and therefore it depends only upon the data.

8. Proof of Proposition 7.1 when (7.6) holds

If (7.6) holds true, define

$$t_* \equiv \sup \left\{ t \in (-4^p, t_o) \,\middle|\, \frac{d}{dt} \int_{K_4} \zeta^p(x, t)\, \Phi_{\delta^n}(w(x, t))\, dx \ge 0 \right\}.$$

By the definition of t_*,

$$\int_{K_4} \zeta^p(x, t_o)\, \Phi_{\delta^n}(w(x, t_o))\, dx \le \int_{K_4} \zeta^p(x, t_*)\, \Phi_{\delta^n}(w(x, t_*))\, dx. \tag{8.1}$$

By the arguments of the first alternative

$$\int_{K_4} \zeta^p(x,t_*)\Psi^p_{\delta^n}\left(w\left(x,t_*\right)\right)\,dx \le C$$

and $\forall s \in [0,1]$,

$$\int_{K_4\cap\left[(\delta^n-w)_+>s\delta^n\right]} \zeta^p\left(x,t_*\right)\,dx \le C\left[\ln\frac{1+\delta}{1+\delta-s}\right]^{-p},$$

where C is an absolute constant depending only upon N and p and the data. By the definition (7.1) of Y_n, we have for all $s \in [0,1]$

$$\tag{8.2} \int_{K_4\cap\left[(\delta^n-w)_+>s\delta^n\right]} \zeta^p\left(x,t_*\right)\,dx \le \min\left\{Y_n\,;\,C\left[\ln\frac{1+\delta}{1+\delta-s}\right]^{-p}\right\}$$
$$= \begin{cases} Y_n & \text{if } 0 \le s < s_* \\ C\left[\ln\frac{1+\delta}{1+\delta-s}\right]^{-p} & \text{if } s_* \le s < 1, \end{cases}$$

where s_* is the root of the equation

$$Y_n = C\left[\ln\frac{1+\delta}{1+\delta-s_*}\right]^{-p}.$$

Solving it we find

$$\tag{8.3} s_* = \frac{e^{(C/Y_n)^{1/p}}-1}{e^{(C/Y_n)^{1/p}}}\,(1+\delta).$$

Since $Y_n > \nu$, we have

$$\tag{8.4} s_* < \frac{e^{(C/\nu)^{1/p}}-1}{e^{(C/\nu)^{1/p}}}\,(1+\delta) \equiv \sigma_o(1+\delta).$$

Next we estimate the integral on the right hand side of (8.1). By the Fubini theorem

$$\int_{K_4} \zeta^p\left(x,t_*\right)\Phi_{\delta^n}\left(w\left(x,t_*\right)\right)dx$$
$$= \int_{K_4} \zeta^p(x,t_*)\left(\int_0^{\delta^n} \frac{\chi\left[s<(\delta^n-w)_+\right]}{\left[\delta^n(1+\delta)-s\right]^{p-1}}ds\right)dx$$
$$= \int_0^{\delta^n} \frac{1}{\left[\delta^n(1+\delta)-s\right]^{p-1}}\left(\int_{K_4} \zeta^p(x,t_*)\chi\left[s<(\delta^n-w)_+\right]dx\right)ds$$
$$= \int_0^1 \frac{\delta^{n(2-p)}}{\left[1+\delta-s\right]^{p-1}}\left(\int_{K_4} \zeta^p(x,t_*)\chi\left[(\delta^n-w)_+>s\delta^n\right]dx\right)ds.$$

Therefore (8.1) yields

$$\int_{K_4} \zeta^p(x,t_o)\,\Phi_{\delta^n}(w(x,t_o))\,dx \le \int_0^1 \frac{\delta^{n(2-p)}}{[1+\delta-s]^{p-1}} \left(\int_{K_4\cap[(\delta^n-w)_+>s\delta^n]} \zeta^p(x,t_*)\,dx \right) ds. \tag{8.5}$$

The last integral on the right hand side of (8.5) is estimated by means of (8.2). Taking into account the definition of s_* in (8.3), we have

$$\begin{aligned}
&\int_{K_4} \zeta^p(x,t_o)\,\Phi_{\delta^n}(w(x,t_o))\,dx \\
&\le \int_0^{s_*} \frac{\delta^{n(2-p)}}{[1+\delta-s]^{p-1}} \left(\int_{K_4\cap[(\delta^n-w)_+>s\delta^n]} \zeta^p(x,t_*)\,dx \right) ds \\
&+ \int_{s_*}^1 \frac{\delta^{n(2-p)}}{[1+\delta-s]^{p-1}} \left(\int_{K_4\cap[(\delta^n-w)_+>s\delta^n]} \zeta^p(x,t_*)\,dx \right) ds \\
&\le \int_0^{s_*} \frac{\delta^{n(2-p)}}{[1+\delta-s]^{p-1}} Y_n\,ds \\
&+ \int_{s_*}^1 \frac{\delta^{n(2-p)}}{[1+\delta-s]^{p-1}} C\left[\ln\frac{1+\delta}{1+\delta-s}\right]^{-p} ds \\
&= \int_0^1 \frac{\delta^{n(2-p)}}{[1+\delta-s]^{p-1}} Y_n\,ds \\
&- \int_{s_*}^1 \left\{ Y_n - C\left[\ln\frac{1+\delta}{1+\delta-s}\right]^{-p} \right\} \frac{\delta^{n(2-p)}}{[1+\delta-s]^{p-1}} ds \\
&= Y_n\,\delta^{n(2-p)}\;F(Y_n,\delta),
\end{aligned}$$

where

$$\begin{aligned}
F(Y_n,\delta) \equiv{}& \int_0^1 \frac{ds}{[1+\delta-s]^{p-1}} \\
&- \int_{s_*}^1 \left(1-\frac{C}{Y_n}\left[\ln\frac{1+\delta}{1+\delta-s}\right]^{-p}\right) \frac{ds}{[1+\delta-s]^{p-1}}.
\end{aligned}$$

From $Y_n \geq \nu$ and (8.4) we have

$$F(Y_n,\delta) \leq \int_0^1 \frac{ds}{[1+\delta-s]^{p-1}}$$
$$- \int_{\sigma_o(1+\delta)}^1 \left(1 - \frac{C}{\nu}\left[\ln\frac{1+\delta}{1+\delta-s}\right]^{-p}\right)\frac{ds}{[1+\delta-s]^{p-1}}.$$

These estimates in (8.5) give

$$\text{(8.6)}\qquad \int_{K_4} \zeta^p(x,t_o)\,\Phi_{\delta^n}(w(x,t_o))\,dx \leq Y_n\,[1-f(\delta)]\int_0^{1-\delta}\frac{\delta^{n(2-p)}}{[1+\delta-s]^{p-1}}ds,$$

where

$$\text{(8.7)}\qquad f(\delta)\int_0^{1-\delta}\frac{ds}{[1+\delta-s]^{p-1}} \equiv$$
$$\int_{\sigma_o(1+\delta)}^1 \left(1-\frac{C}{\nu}\left[\ln\frac{1+\delta}{1+\delta-s}\right]^{-p}\right)\frac{ds}{[1+\delta-s]^{p-1}} - \int_{1-\delta}^1 \frac{ds}{[1+\delta-s]^{p-1}}.$$

Estimating below the left hand side of (8.6) we find

$$\int_{K_4}\zeta^p(x,t_o)\,\Phi_{\delta^n}(w(x,t_o))\,dx \geq \int_{K_4\cap[w(\cdot,t_o)<\delta^{n+1}]}\zeta^p(x,t_o)\,dx\int_0^{1-\delta}\frac{\delta^{n(2-p)}}{[1+\delta-s]^{p-1}}\,ds$$
$$\geq (Y_{n+1}-\varepsilon)\int_0^{1-\delta}\frac{\delta^{n(2-p)}}{[1+\delta-s]^{p-1}}\,ds.$$

This and (8.6) yield

$$\text{(8.8)}\qquad Y_{n+1}-\varepsilon \leq Y_n\,(1-f(\delta))\,.$$

We estimate $f(\delta)$ below. For this let $\sigma_1 \geq \sigma_o$ be defined by

$$\sigma_1 = \frac{e^{(2C/\nu)^{1/p}}-1}{e^{(2C/\nu)^{1/p}}}\qquad\text{(see also (8.4))}.$$

Then integrating the first integral on the right hand side of (8.7) over the smaller interval $[\sigma_1(1+\delta),1]$, we derive the estimate

$$f(\delta) > \frac{1}{2}(1-\sigma_1)^{2-p} - \left(\frac{2\delta}{1+\delta}\right)^{2-p}.$$

We choose δ so small that

$$f(\delta) > \frac{1}{4}(1-\sigma_1)^{2-p}$$

and set

$$\sigma = 1 - \frac{1}{4}(1-\sigma_1)^{2-p}.$$

Since $\varepsilon \in (0, \nu/2)$ is arbitrary, we obtain from (8.8)

$$Y_{n+1} \le \sigma Y_n.$$

This proves the Proposition if (6.1) holds.

9. Removing the assumption (6.1)

Inequality (6.10) holds in any case in the integrated form

$$\int_{K_4(t)} \zeta^p \Phi_k dx - \int_{K_4(t-h)} \zeta^p \Phi_k dx + \gamma_o \int_{t-h}^{t} \int_{K_4} \zeta^p \Psi_k^p(v) dx \le \gamma h,$$
$$\text{for all } t \in [-4^p + h, 0], \quad h > 0.$$

We divide by h and let $h \to 0$ to obtain (6.10) where the term involving the $t-$derivative is replaced by

$$\left(\frac{d}{d\tau}\right)^{-} \int_{K_4} \zeta^p(x,t)\Phi_k(w)\, dx$$
$$\equiv \limsup_{h\to 0} \frac{1}{h} \left\{ \int_{K_4(t)} \zeta^p \Phi_k(w) dx - \int_{K_4(t-h)} \zeta^p \Phi_k(w)\, dx \right\} dx.$$

Define the set

$$S \equiv \left\{ t \in (-4^p, 0) | \left(\frac{d}{d\tau}\right)^{-} \int_{K_4(t)} \zeta^p \Phi_k(w)\, dx \ge 0 \right\},$$

and let t_o be defined as in (7.4). If $t_o \in S$, we have

$$\int_{K_4(t_o)} \zeta^p \Psi_k^p(v)\, dx \le \gamma. \tag{9.1}$$

If $t_o \notin S$ but

$$\sup \left\{ t < t_o \;\middle|\; t \in S \right\} = t_o,$$

by working with a sequence of time levels $t_n \in S$ and $\{t_n\} \to t_o$, we see that (9.1) continues to hold. If $t_o \notin S$ and

$$\tau \equiv \sup\left\{t < t_o \,\middle|\, t \in S\right\} < t_o,$$

we derive the two inequalities

$$\begin{cases} \displaystyle\int_{K_4(\tau)} \zeta^p \Psi_k^p(w)dx \le \gamma, \\ \displaystyle\int_{K_4(t_o)} \zeta^p \Phi_k(w)dx \le \int_{K_4(\tau)} \zeta^p \Phi_k(w)dx. \end{cases}$$

The remainder of the proof remains the same.

10. The second alternative

We assume here that (3.5) holds true for all cylinders $[(\bar{x},0)+Q\,(R^p,d_oR)]$ making up the partition of $Q\,(R^p,c_oR)$. Since $s_o \ge 1$ we have

$$\mu^+ - \frac{\omega}{2^{s_o}} \ge \mu^- + \frac{\omega}{2^{s_o}},$$

so that we may rephrase (3.5) as

$$\left|(x,t)\in[(\bar{x},0)+Q\,(R^p,d_oR)] \;\middle|\; u(x,t) > \mu^+ - \frac{\omega}{2^{s_o}}\right| < (1-\nu_o)|Q\,(R^p,d_oR)\,|, \tag{10.1}$$

for *all* boxes $[(\bar{x},0)+Q\,(R^p,d_oR)]$ making up the partition of $Q\,(R^p,c_oR)$.

Let n be a positive number to be selected and arrange that $2^{n\frac{2-p}{p}}$ is an integer. Then we combine $2^{\frac{n(2-p)}{p}N}$ of these cylinders to form boxes congruent to

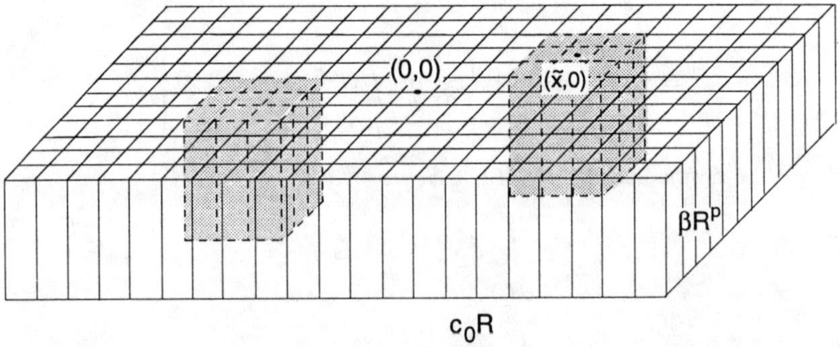

Figure 10.1

$$(10.2)\quad Q\left(R^p, d_* R\right) \equiv K_{d_* R} \times\left(-R^p, 0\right) \quad d_*=\left(\frac{\omega}{2^{s_o+n}}\right)^{\frac{p-2}{p}}=d_o\left(2^n\right)^{\frac{2-p}{p}} .$$

The cylinders obtained this way are contained in $Q\left(R^p, c_o R\right)$, if the abscissa $\tilde{x}$ of their 'vertices' ranges over the cube $K_{\mathcal{R}_1(\omega)}$, where

$$\begin{aligned}
\mathcal{R}_1(\omega) & =\left\{A^{2-p}-2^{\left(s_o+n\right) \frac{2-p}{p}}\right\} \omega^{\frac{p-2}{p}} R \\
& =\left\{\left(\frac{A}{2^{s_o+n}}\right)^{\frac{2-p}{p}}-1\right\}\left(\frac{\omega}{2^{s_o+n}}\right)^{\frac{p-2}{p}} R \\
& =L_1\, d_* R, \qquad \text{where } L_1 \equiv\left(\frac{A}{2^{s_o+n}}\right)^{\frac{2-p}{p}}-1 .
\end{aligned}$$

We will take A larger than 2^{s_o+n} and arrange that L_1 is an integer. Then we regard $Q\left(R^p, c_o R\right)$ as the union, up to a set of measure zero, of L_1^N pairwise disjoint boxes each congruent to $Q\left(R^p, d_* R\right)$. Each of the cylinders $\left[(\tilde{x}, 0)+Q\left(R^p, d_* R\right)\right]$ is the pairwise disjoint union of boxes $\left[(\bar{x}, 0)+Q\left(R^p, d_o R\right)\right]$ satisfying (10.1). Therefore we rephrase (3.5) as

$$\begin{aligned}
(10.3)\qquad \left|(x, t) \in\left[(\tilde{x}, 0)+Q\left(R^p, d_* R\right)\right] \mid u(x, t)>\mu^{+}-\frac{\omega}{2^{s_o}}\right| & \\
<\left(1-\nu_o\right)\left|Q\left(R^p, d_* R\right)\right|, &
\end{aligned}$$

for all boxes $\left[(\tilde{x}, 0)+Q\left(R^p, d_* R\right)\right]$ making up the partition of $Q\left(R^p, c_o R\right)$.

LEMMA 10.1. *Let* $\left[(\tilde{x}, 0)+Q\left(R^p, d_* R\right)\right]$ *be any box contained in* $Q\left(R^p, c_o R\right)$ *and satisfying (10.3). There exists a time level*

$$t^* \in\left(-R^p,-\frac{\nu_o}{2} R^p\right),$$

such that for all $s \geq s_o+1$,

$$(10.4)\qquad \left|x \in\left[\tilde{x}+K_{d_* R}\right] \mid u\left(x, t^*\right)>\mu^{+}-\frac{\omega}{2^s}\right|<\left(\frac{1-\nu_o}{1-\nu_o / 2}\right)\left|K_{d_* R}\right| .$$

PROOF: If (10.4) is violated for all $t \in\left(-R^p,-\frac{\nu_o}{2} R^p\right)$, then

$$\begin{aligned}
& \left|(x, t) \in\left[(\tilde{x}, 0)+Q\left(R^p, d_* R\right)\right] \mid u(x, t)>\mu^{+}-\frac{\omega}{2^{s_o}}\right| \\
& \quad \geq \int_{-R^p}^{-\frac{\nu_o}{2} R^p}\left|x \in\left[\bar{x}+K_{d_* R}\right] \mid u(x, t)>\mu^{+}-\frac{\omega}{2^s}\right| d t \\
& \quad \geq\left(1-\nu_o\right)\left|Q\left(R^p, d_* R\right)\right|,
\end{aligned}$$

contradicting (10.3).

The next lemma asserts that a property similar to (10.4) continues to hold for all time levels from t^* up to 0. The proof of the lemma will also determine the number n.

LEMMA 10.2. *There exists a positive integer n such that for all $t^*<t<0$,*

$$\left|x\in[\tilde{x}+K_{d_*R}]\mid u(x,t)>\mu^+-\frac{\omega}{2^{s_o+n}}\right|<\left(1-\left(\frac{\nu_o}{2}\right)^2\right)|K_{d_*R}|. \tag{10.5}$$

PROOF: Modulo a translation we may assume that $\tilde{x}\equiv 0$. Consider the logarithmic estimates (3.14) of Chap. II, written over the cylinder $K_{d_*R}\times(t^*,0)$, for the function $(u-k)_+$ and for the levels

$$k=\mu^+-\frac{\omega}{2^{s_o}}.$$

As for the number c in the definition (3.12) of the function Ψ we take

$$c=\frac{\omega}{2^{s_o+n}},$$

where n is a positive number to be chosen. Thus we take

$$\Psi=\ln\left\{\frac{H_k^+}{H_k^+-\left(u-\left(\mu^+-\frac{\omega}{2^{s_o}}\right)\right)_++\frac{\omega}{2^{s_o+n}}}\right\}, \tag{10.6}$$

where

$$H_k^+\equiv\operatorname*{ess\,sup}_{K_{d_*R}\times(t^*,0)}\left(u-\left(\mu^+-\frac{\omega}{2^{s_o}}\right)\right)_+.$$

The cutoff function $x\to\zeta(x)$ is taken so that

$$\begin{cases}\zeta\equiv 1, & \text{on the cube } K_{(1-\sigma)d_*R},\quad \sigma\in(0,1),\\ |D\zeta|\le(\sigma d_*R)^{-1}.\end{cases}$$

With these choices, the inequalities (3.14) of Chap. II yield for all $t^*<t<0$,

$$\begin{aligned}\int_{K_{(1-\sigma)d_*R}}\Psi^2(x,t)dx&\le\int_{K_{d_*R}}\Psi^2(x,t^*)dx+\frac{\gamma}{(\sigma d_*R)^p}\int_{t^*}^{0}\int_{K_{d_*R}}\Psi\Psi_u^{2-p}dxd\tau\\&+\gamma\left(\frac{\omega}{2^{s_o+n}}\right)^{-2}\left[1+\ln H_k^+\left(\frac{\omega}{2^{s_o+n}}\right)^{-1}\right]\left\{\int_{t^*}^{0}|A_{k,d_*R}^+(\tau)|^{\frac{r}{q}}d\tau\right\}^{\frac{2}{pt}\frac{p(1+\kappa)}{r}}.\end{aligned} \tag{10.7}$$

To estimate the various terms in (10.7) we first observe that

$$\Psi\le n\ln 2,\ \ \Psi_u^{2-p}\le\left(\frac{\omega}{2^{s_o+n}}\right)^{p-2},\ \ \left[1+\ln H_k^+\left(\frac{\omega}{2^{s_o+n}}\right)^{-1}\right]\le\gamma n\ln 2.$$

We estimate the first integral on the right hand side of (10.7). For this observe that Ψ vanishes on the set $\left[u<\mu^+-\frac{\omega}{2^{s_o}}\right]$. Therefore using Lemma 10.1,

$$\int_{K_{d_*R}}\Psi^2(x,t^*)\,dx\le n^2\ln^2 2\left(\frac{1-\nu_o}{1-\nu_o/2}\right)|K_{d_*R}|.$$

For the second integral we have

$$\frac{\gamma}{(\sigma d_* R)^p} \iint\limits_{t^* K_{d_* R}}^{0} \Psi |\Psi_u|^{2-p} dx d\tau \leq \frac{\gamma n}{(d_* \sigma)^p} \left(\frac{\omega}{2^{s_o+n}} \right)^{p-2} |K_{d_* R}|$$

$$\leq \frac{\gamma n}{\sigma^p} |K_{d_* R}|.$$

This estimation justifies the choice of the cylinders $[(\tilde{x}, 0) + Q(R^p, d_* R)]$ over the boxes $[(\bar{x}, 0) + Q(R^p, d_o R)]$. Indeed the integrand grows like $2^{n(2-p)}$ due to the singularity of the equation. This is balanced by taking a *parabolic geometry* where the space dimensions are stretched by a factor $2^{\frac{n(2-p)}{p}}$.

Finally the last term on the right hand side of (10.7) is estimated above by

$$\gamma n \left(\frac{\omega}{2^{s_o+n}} \right)^{-2} |K_{d_* R}|^{\frac{p(1+\kappa)}{q}} R^{p \frac{p(1+\kappa)}{r}}$$

$$\leq \gamma n \left(\frac{\omega}{2^{s_o+n}} \right)^{-2} d_*^{N \frac{p(1+\kappa)}{q} - N} R^{N\kappa} |K_{d_* R}|$$

$$= \gamma A_2 \omega^{-b_o} R^{N\kappa} |K_{d_* R}|$$

where $A_2 = n 2^{(s_o+n)b_o}$ and b_o is defined in (3.6). Combining these remarks in (10.7) and taking into account (3.7), we obtain for all $t^* < t < 0$,

$$\int\limits_{K_{(1-\sigma)d_* R}} \Psi^2(x,t)\, dx \leq n^2 \ln^2 2 \left(\frac{1-\nu_o}{1-\nu_o/2} \right) |K_{d_* R}| + \frac{\gamma n}{\sigma^p} |K_{d_* R}|. \tag{10.8}$$

The left hand side of (10.8) is estimated below by integrating over the smaller set

$$\left\{ x \in K_{(1-\sigma)d_* R} \;\middle|\; u(x,t) > \mu^+ - \frac{\omega}{2^{s_o+n}} \right\}.$$

On such a set, since Ψ is a decreasing function of H_k^+, we estimate

$$\Psi^2 \geq \ln^2 \left(\frac{\frac{\omega}{2^{s_o}}}{\frac{\omega}{2^{s_o+n-1}}} \right) = (n-1)^2 \ln^2 2.$$

We carry this in (10.8) and divide through by $(n-1)^2 \ln^2 2$, to obtain for all $t^* < t < 0$,

$$\left| x \in K_{(1-\sigma)d_* R} \;\middle|\; u(x,t) > \mu^+ - \frac{\omega}{2^{s_o+n}} \right|$$

$$\leq \left(\frac{n}{n-1} \right)^2 \left(\frac{1-\nu_o}{1-\nu_o/2} \right) |K_{d_* R}| + \frac{\gamma}{n\sigma^p} |K_{d_* R}|.$$

On the other hand

$$\left|x \in K_{d_* R} \mid u(x,t) > \mu^+ - \frac{\omega}{2^{s_o+n}}\right|$$
$$\leq \left|x \in K_{(1-\sigma)d_* R} \mid u(x,t) > \mu^+ - \frac{\omega}{2^{s_o+n}}\right| + \left|K_{d_* R} \backslash K_{(1-\sigma)d_* R}\right|$$
$$\leq \left|x \in K_{(1-\sigma)d_* R} \mid u(x,t) > \mu^+ - \frac{\omega}{2^{s_o+n}}\right| + N\sigma \left|K_{d_* R}\right|.$$

Therefore for all $t^* < t < 0$,

$$\left|x \in K_{(1-\sigma)d_* R} \mid u(x,t) > \mu^+ - \frac{\omega}{2^{s_o+n}}\right|$$
$$\leq \left[\left(\frac{n}{n-1}\right)^2 \left(\frac{1-\nu_o}{1-\nu_o/2}\right) + \frac{\gamma}{n\sigma^p} + N\sigma\right] \left|K_{d_* R}\right|.$$

To prove the lemma we choose σ so small that $\sigma N \leq \frac{3}{8}\nu_o^2$ and then n so large that

$$\left(\frac{n}{n-1}\right)^2 \leq \left(1 - \frac{\nu_o}{2}\right)(1+\nu_o) \quad \text{and} \quad \frac{\gamma}{n\sigma^p} \leq \frac{3}{8}\nu_o^2.$$

Remark 10.1. Since the number ν_o is independent of ω and R also n is independent of these parameters.

11. The second alternative concluded

The information of Lemma 10.2 will be exploited to show that in a small cylinder about $(0,0)$, the solution u is strictly bounded above by

$$\mu^+ - \frac{\omega}{2^m}, \qquad \text{for some } m > s_o + n.$$

In this process we also determine the number A introduced in §2 which defines the size of $Q(R^p, c_o R)$. To make this quantitative let us consider the box

$$Q(\beta R^p, c_o R), \qquad \beta = \frac{\nu_o}{2}, \qquad c_o = \left(\frac{\omega}{A}\right)^{\frac{p-2}{p}}.$$

We regard $Q(\beta R^p, c_o R)$ as partitioned into sub-boxes $[(\tilde{x}, 0) + Q(\beta R^p, d_* R)]$ where $\tilde{x}$ takes finitely many points within the cube $K_{\mathcal{R}_1(\omega)}$ introduced at the beginning of §10. For each of these subcylinders Lemma 10.2 holds.

LEMMA 11.1. *For every $\nu \in (0,1)$ there exist a number m dependent only upon the data and independent of ω and R such that for all cylinders $[(\tilde{x}, 0) + Q(\beta R^p, d_* R)]$ making up the partition of $Q(\beta R^p, c_o R)$,*

$$\text{meas}\left\{(x,t) \in [(\tilde{x}, 0) + Q(\beta R^p, d_* R)] \mid u(x,t) > \mu^+ - \frac{\omega}{2^m}\right\}$$
$$< \nu \left|Q(\beta R^p, d_* R)\right|.$$

PROOF: After a translation we may assume that $(\tilde{x},0)\equiv(0,0)$. Set $s_1=s_o+n$, and consider the energy inequality (3.8) of Chap. II written for $(u-k)_+$, where

$$k=\mu^+-\frac{\omega}{2^s}, \qquad s=s_1,s_1+1,s_1+2,\dots,m-1,$$

over the cylinder $Q\left(\beta(2R)^p,2d_*R\right)$. Over such a box

$$\left(u(x,t)-\left(\mu^+-\frac{\omega}{2^s}\right)\right)_+\le\frac{\omega}{2^s}\quad \text{a.e. } (x,t)\in Q\left(\beta(2R)^p,2d_*R\right).$$

The cutoff function $(x,t)\to\zeta(x,t)$ is taken to satisfy

$$\begin{cases}\zeta\equiv 1, & \text{on } Q\left(\beta R^p,d_*R\right),\\ \zeta=0, & \text{on the parabolic boundary of } Q\left(\beta(2R)^p,2d_*R\right),\\ |D\zeta|\le\frac{1}{d_*R^p}, & 0\le\zeta_t\le\frac{2}{\nu_oR^p}.\end{cases}$$

We put these estimates in (3.8) of Chap. II and discard the first non-negative term on the left hand side. This gives

$$\text{(11.1)}\qquad \iint_{Q(\beta R^p,d_*R)}\left|D\left(u-\left(\mu^+-\frac{\omega}{2^s}\right)\right)_+\right|^p dxd\tau\le\frac{\gamma}{(d_*R)^p}\left(\frac{\omega}{2^s}\right)^p\left|Q\left(\beta R^p,d_*R\right)\right|$$
$$+\frac{\gamma}{R^p}\left(\frac{\omega}{2^s}\right)^{2-p}\left(\frac{\omega}{2^s}\right)^p\left|Q\left(\beta R^p,d_*R\right)\right|+\gamma(d_*R)^{N\frac{p(1+\kappa)}{q}}R^{p\frac{p(1+\kappa)}{r}}.$$

We estimate above the various terms on the right hand side of (11.1) as follows. Since $s\ge s_1\equiv s_o+n$,

$$\left(\frac{\omega}{2^s}\right)^{2-p}\le\left(\frac{\omega}{2^{s_o+n}}\right)^{2-p}=\frac{1}{d_*^p}.$$

Therefore the sum of the first two terms is majorised by

$$\frac{\gamma}{(d_*R)^p}\left(\frac{\omega}{2^s}\right)^p\left|Q\left(\beta R^p,d_*R\right)\right|,$$

for a constant γ dependent only upon the data. The last term is majorised by making use of (3.6) of Chap. II and the definition (10.2) of d_*. This gives

$$\begin{aligned}\gamma\left(d_*R\right)^{N\frac{p(1+\kappa)}{q}}R^{p\frac{p(1+\kappa)}{r}}&\le\frac{\gamma}{R^p}d_*^{N\left(\frac{p(1+\kappa)}{q}-1\right)}R^{N\kappa}\left|Q\left(\beta R^p,d_*R\right)\right|\\&\le\frac{\gamma}{(d_*R)^p}A_3\omega^{-b_o}R^{N\kappa}\left|Q\left(\beta R^p,d_*R\right)\right|,\end{aligned}$$

where $A_3=2^{mb_o}$ and b_o is the number introduced in (3.6). Combining these remarks in (11.1) we deduce that there exists a constant γ depending only upon the data and independent of ω and R, such that

(11.2)

$$\iint_{Q(\beta R^p,d_*R)}\left|D\left(u-\left(\mu^+-\frac{\omega}{2^s}\right)\right)_+\right|^p dxd\tau\le\frac{\gamma}{(d_*R)^p}\left(\frac{\omega}{2^s}\right)^p\left|Q\left(\beta R^p,d_*R\right)\right|,$$

for all $s=s_1, s_1+1, s_1+2, \dots, m-1$. Next we apply Lemma 2.2 of Chap. I over the cube K_{d_*R} for the functions

$$v = u(\cdot,t), \qquad t \in \left(-\frac{\nu_o}{2}R^p, 0\right) \equiv (-\beta R^p, 0),$$

and the levels

$$l = \mu^+ - \frac{\omega}{2^{s+1}}, \qquad k = \mu^+ - \frac{\omega}{2^s}, \qquad s \geq s_1.$$

By virtue of Lemma 10.2

$$\left|\left[u(\cdot,t) < \mu^+ - \frac{\omega}{2^s}\right] \cap K_{d_*R}\right| \geq \left(\frac{\nu_o}{2}\right)^2 |K_{d_*R}|, \quad \forall t \in (-\beta R^p, 0).$$

To simplify the symbolism we set

$$A_s(t) \equiv \left\{x \in K_{d_*R} \mid u(x,t) > \mu^+ - \frac{\omega}{2^s}\right\}, \quad A_s = \int_{-\beta R^p}^{0} |A_s(\tau)| d\tau.$$

Then, with these specifications, (2.7) of Chap. I yields

$$\left(\frac{\omega}{2^s}\right)|A_{s+1}(t)| \leq \frac{\gamma}{\nu_o^2} d_* R \int_{A_s(t)\setminus A_{s+1}(t)} |Du(x,t)| dx, \qquad \forall t \in (-\beta R^p, 0).$$

First integrate both sides in $d\tau$ over $(-\beta R^p, 0)$, then take the $p-$power and majorise the right hand side by making use of the Hölder inequality and (11.2). We obtain

$$\begin{aligned}\left(\frac{\omega}{2^s}\right)^p |A_{s+1}|^p &\leq \gamma (d_* R)^p \left(\int_{-\beta R^p}^{0} \int_{A_s(\tau)} |Du|^p dx d\tau\right) |A_s \setminus A_{s+1}|^{p-1} \\ &\leq \gamma \left(\frac{\omega}{2^s}\right)^p |Q(\beta R^p, d_* R)| |A_s \setminus A_{s+1}|^{p-1}.\end{aligned}$$

From this,

$$|A_{s+1}|^{\frac{p}{p-1}} \leq \gamma |Q(\beta R^p, d_* R)|^{\frac{1}{p-1}} |A_s \setminus A_{s+1}|.$$

Adding these inequalities for $s=s_1, s_1+1, s_1+2, \dots, m-1$,

$$\begin{aligned}(m-s_1)|A_m|^{\frac{p}{p-1}} &\leq \gamma |Q(\beta R^p, d_* R)|^{\frac{1}{p-1}} \sum_{s=s_1}^{m} |A_s \setminus A_{s+1}| \\ &\leq \gamma |Q(\beta R^p, d_* R)|^{\frac{p}{p-1}}.\end{aligned}$$

To prove the lemma we have only to choose m so large that

$$\left(\frac{\gamma}{m-s_1}\right)^{\frac{p-1}{p}} \leq \nu.$$

Remark 11.1. This estimate deteriorates as $p \searrow 1$, i.e., $m \nearrow \infty$ as $p \searrow 1$. However the choice of m is 'stable' as $p \nearrow 2$.

To proceed we return to the box $Q(\beta R^p, c_o R)$ and recall that it is the finite union, up to a set of measure zero, of mutually disjoint boxes $[(\tilde{x}, 0) + Q(\beta R^p, d_* R)]$. Therefore Lemma 11.1 implies

COROLLARY 11.1. *For every $\nu \in (0,1)$ there exist a number m dependent only upon the data and independent of ω and R such that*

$$\text{meas}\left\{(x,t) \in Q(\beta R^p, c_o R) \mid u(x,t) > \mu^+ - \frac{\omega}{2^m}\right\} < \nu |Q(\beta R^p, c_o R)|. \tag{11.3}$$

We finally determine the size of the cylinder $Q(\beta R^p, c_o R)$ as follows. First in Corollary 11.1 select $\nu = \nu_o$ and determine m accordingly. Then let m_2 be given by

$$\beta = \frac{\nu_o}{2} = 2^{m_2(p-2)}, \tag{11.4}$$

and assume, by taking m even larger if necessary, that $m \geq m_2$. Then determine A from

$$A = 2^{m_1} \quad \text{and } m = m_1 + m_2. \tag{11.5}$$

With these choices the box $Q(\beta R^p, c_o R)$ coincides with the cylinder $\mathcal{Q}_R(m_1, m_2)$ introduced in §4. By Lemma 4.1, there exists a constant $\mathcal{A}_o$ dependent only upon the data and independent of ω and R such that either

$$\omega^{b_o} \leq \mathcal{A}_o R^{N\kappa},$$

or

$$u(x,t) \leq \mu^+ - \frac{\omega}{2^{m+1}} \qquad \forall (x,t) \in \mathcal{Q}_{\frac{R}{2}}(m_1, m_2),$$

where b_o is the number introduced in (3.6). We summarise:

PROPOSITION 11.1. *Suppose that (3.5) holds for all cylinders $[(\bar{x}, 0) + Q(R^p, d_o R)]$ making up the partition of $Q(R^p, c_o R)$. There exists a constant $\mathcal{A}_o$ dependent only upon the data and independent of ω and R such that either*

$$\omega^{b_o} \leq \mathcal{A}_o R^{N\kappa} \tag{11.6}$$

or for all $0 < \rho \leq R/2$,

$$\operatorname*{ess\,osc}_{Q(\beta\rho^p, c_o\rho)} u \leq \eta_o \omega, \quad \textit{where } \eta_o \equiv 1 - 2^{-(m+1)}. \tag{11.7}$$

12. Proof of the main proposition

The main Proposition 2.1 now follows by combining the two alternatives. Set

$$\mathcal{A} \equiv \max\{\mathcal{A}_o ; \mathcal{A}_1\}, \qquad \eta \equiv \min\{\eta_o ; \eta_1\},$$

and let C be defined by

$$\frac{1}{C} \equiv \left(\frac{\beta}{4^p}\right)^{1/p} = \left(\frac{\nu_o}{2^{2p+1}}\right)^{1/p}.$$

Then setting $R_1 \equiv R/C$ both alternatives can be combined into the following statement: either

$$\omega^{b_o} < A R^{N\kappa} \qquad \text{or} \qquad \operatorname*{ess\,osc}_{Q(\rho^p, c_o\rho)} u \le \eta\,\omega.$$

The process can now be repeated and continued as indicated in Proposition 2.1. Indeed, by Remark 2.1, the process can be continued as long as (2.3) holds.

12-(i). Stability for p near 2

The proof of the first alternative is based on the integral inequalities (6.9) and (6.10). Because of the right hand side of (6.9), Proposition 5.1 holds with constants that deteriorate as $p \nearrow 2$. We briefly indicate how to prove Proposition 5.1 with constants that are '*stable*' as $p \nearrow 2$. First, using the information (5.11)-(5.12) we can show that there exists a positive number l_* such that

$$\operatorname{meas}\left\{(x,t) \in Q_2 \;\middle|\; v(x,t) < 2^{-l_*}\right\} \le \nu_o \left|Q_2\right|.$$

This is accomplished by the same technique as for Lemma 12.1. This technique involves constants of the type $2^{l_*(2-p)}$. We may select p_* so close to 2 that

$$2^{l_*(2-p)} \le 2, \qquad \forall\, p_* \le p \le 2.$$

With such a choice the method can be carried out as if the p.d.e. was not singular. Next, an application of Lemma 4.1 implies that

$$v(x,t) \ge 2^{-(l_*+1)} \qquad \text{a.e. } (x,t) \in Q_1.$$

The application of Lemma 4.1 over the box Q_2 involves again terms of the type $2^{l_*(2-p)}$. As remarked before, these are majorised by an absolute constant for $p \in [p_*, 2]$.

13. Boundary regularity

The proof of Theorems 1.2 and 1.3 regarding the regularity up to the lateral boundary of Ω_T, is similar to the proof of the interior regularity. The few changes needed can be modelled after similar modifications presented in §11-13 of Chap. III for the degenerate case $p > 2$. However the proof of regularity up to $t = 0$ exhibits some differences.

13-(i). Regularity up to t=0

Assume that u_o is continuous with modulus of continuity say $\omega_o(\cdot)$. Fix $(x_o, 0) \in \Omega \times \{0\}$ and $R > 0$ so that $[x_o + K_{2R}] \subset \Omega$. After a translation we may assume $x_o = 0$ and construct the cylinder

$$Q_o(R^p, 2R) \equiv K_{2R} \times \{0, R^p\}.$$

Set

$$\mu^+ = \operatorname*{ess\,sup}_{Q_o(R^p,2R)} u, \quad \mu^- = \operatorname*{ess\,inf}_{Q_o(R^p,2R)} u, \quad \omega = \operatorname*{ess\,osc}_{Q_o(R^p,2R)} u.$$

Let s_o be the smallest positive integer satisfying (3.1) and construct the box

$$Q_o(dR^p, R) \equiv K_R \times \{0, dR^p\}, \quad d = \left(\frac{\omega}{2^m}\right)^{2-p}, \quad m > s_o,$$

where the number $m > 1$ is to be chosen. Notice that for all $R > 0$, these boxes are lying on the bottom of Ω_T. Also withous loss of generality we may assume that $\frac{\omega}{2^m} \leq 1$ so that there holds

$$Q_o(dR^p, R) \subset Q_o(R^p, 2R) \qquad \text{and} \qquad \operatorname*{osc}_{Q_o(dR^p,R)} u \leq \omega.$$

PROPOSITION 13.1. *There exist constants $\eta, \varepsilon_o \in (0,1)$ and $C, m > 1$ that can be determined a priori depending only upon the data satisfying the following. Construct the sequences $R_o = R, \omega_o = \omega$ and*

$$R_n = \frac{1}{C^n} R, \quad \omega_{n+1} = \max\{\eta\omega_n; C R_n^{\varepsilon_o}\}, \qquad n = 1, 2, \ldots,$$

and the family of boxes

$$Q_o^{(n)} \equiv Q_o(d_n R_n^p, R_n) \qquad d_n = \left(\frac{\omega_n}{2^m}\right)^{2-p}, \quad n = 0, 1, 2, \ldots.$$

Then for all $n = 0, 1, 2, \ldots$

$$Q_o^{(n+1)} \subset Q_o^{(n)} \quad \text{and} \quad \operatorname*{ess\,osc}_{Q_o^{(n)}} u \leq \max\left\{\omega_n; 2 \operatorname*{ess\,osc}_{K_{R_n}} u_o\right\}.$$

We indicate how to prove the first iterative step of the Proposition and show, in the process, how to determine the number m. Set

$$\mu_o^+ = \operatorname*{ess\,sup}_{K_R} u_o, \quad \mu_o^- = \operatorname*{ess\,inf}_{K_R} u_o, \quad \omega_o(R) = \operatorname*{ess\,osc}_{K_R} u_o,$$

and consider the two inequalities

$$\text{(13.1)} \qquad \mu^+ - \frac{\omega}{2^{s_o}} < \mu_o^+, \qquad \mu^- + \frac{\omega}{2^{s_o}} > \mu_o^-, \qquad s_o \geq 2.$$

If both hold, subtracting the second from the first gives

$$\operatorname*{ess\,osc}_{Q_o(dR^p,R)} u \le 2 \operatorname*{ess\,osc}_{K_R} u_o,$$

and there is nothing to prove. Let us assume for example that the second of (13.1) is violated. Then for all $s \ge s_o$, the levels

$$k = \mu^- + \frac{\omega}{2^s},$$

satisfy the second of (4.16) of Chap. II. Therefore we may derive energy and logarithmic estimates for the truncated functions $(u-k)_-$. These take the form

$$\sup_{0\le t\le dR^p} \int_{K_R} (u-k)_-^2(x,t)\zeta^p dx + \iint_{Q_o(dR^p,R)} |D(u-k)_-\zeta|^p dx d\tau \tag{13.2}$$

$$\le \gamma \iint_{Q_o(dR^p,R)} (u-k)_-^p |D\zeta|^p dx d\tau + \gamma \left\{ \int_0^{dR^p} |B_{k,R}^-(\tau)|^{\frac{q}{r}} d\tau \right\}^{\frac{p(1+\kappa)}{r}},$$

$$\sup_{0\le t\le dR^p} \int_{K_R} \Psi^2\left(D_k^-, (u-k)_-, c\right) \zeta^p(x)\, dx \tag{13.3}$$

$$\le \gamma \iint_{Q_o(dR^p,R)} \Psi |\Psi_u\left(D_k^-, (u-k)_-, c\right)|^{2-p} |D\zeta|^p dx d\tau$$

$$+ \frac{\gamma}{c^2}\left(1 + \ln \frac{D_k^-}{c}\right) \left\{ \int_0^{dR^p} |B_{k,R}^-(\tau)|^{\frac{r}{q}} d\tau \right\}^{\frac{p(1+\kappa)}{r}},$$

where $D_k^\pm$ and $B_{k,R}^\pm$ are defined as in (4.2) and (4.4) of Chap. II.

LEMMA 13.1. *For every $\nu \in (0,1)$, there exists a number $m > s_o \ge 2$ depending only upon the data and independent of ω and R such that either*

$$\omega^{b_o} \le 2^m R^{N\kappa} \qquad (b_o \textit{ defined in (3.6)}) \tag{13.4}$$

or

$$\left| \left\{ (x,t) \in Q_o(d\rho^p, \rho) \mid u(x,t) < \mu^- + \frac{\omega}{2^m} \right\} \right| < \nu |Q_o(d\rho^p, \rho)|, \tag{13.5}$$

where

$$\rho = \frac{R}{2} \qquad \textit{and} \qquad d = \left(\frac{\omega}{2^m}\right)^{2-p}.$$

PROOF: Consider inequalities (13.3) written for $k = \mu^- \frac{\omega}{2^{s_o}}$. As a constant c appearing in the definition of Ψ (see (3.12) of Chap. II), we take $c = \frac{\omega}{2^m}$. Thus we take

$$\Psi \equiv \ln^+ \left\{ \frac{D_k^-}{D_k^- - \left(u - \left(\mu^- + \frac{\omega}{2^{s_o}}\right)\right)_- + \frac{\omega}{2^m}} \right\},$$

where

$$D_k^- \equiv \|(u-k)_-\|_{\infty, Q_o(dR^p, R)}.$$

The cutoff function ζ is taken to satisfy

$$\begin{cases} \zeta \equiv 1, & \text{on } K_\rho, \\ \zeta & \text{vanishes for } |x| = R, \\ |D\zeta| \le 2/\rho. \end{cases}$$

By considerations analogous to those developed in §12 we have the estimates

$$\Psi \le (m - s_o)\ln 2, \qquad |\Psi_u|^{2-p} \le \left(\frac{\omega}{2^m}\right)^{p-2},$$

and

$$\gamma \iint_{Q_o(dR^p, R)} \Psi |\Psi_u|^{2-p} |D\zeta|^p dx d\tau \le \gamma m |K_\rho|.$$

Moreover the last term on the right hand side of (13.3) is estimated above by

$$\gamma m \left(\frac{\omega}{2^m}\right)^{-2} \left(\frac{\omega}{2^m}\right)^{(2-p)\frac{p(1+\kappa)}{r}} R^{N\kappa} |K_\rho| \le \gamma m 2^{m b_o} \omega^{-b_o} R^{N\kappa} |K_\rho|,$$

where b_o is the number introduced in (3.6). Combining these estimates in (13.3), we deduce that if (13.4) is violated, then

$$\text{(13.6)} \qquad \int_{K_\rho(t)} \Psi^2 \left(D_k^-, (u-k)_-, c\right) dx \le \gamma m |K_\rho|, \qquad \forall t \in (0, dR^p).$$

We minorise the left hand side of (13.6) by integrating over the smaller set

$$\left\{ x \in K_\rho \mid u(x,t) < \mu^- + \frac{\omega}{2^m} \right\}, \qquad \forall t \in (0, dR^p).$$

On such a set

$$\Psi^2 \ge \ln^2 \frac{\frac{\omega}{2^{s_o}}}{\frac{\omega}{2^{m+1}}} = (m - s_o - 1)^2 \ln^2 2.$$

These remarks in (13.6) give

$$\left| \left\{ x \in K_\rho \mid u(x,t) < \mu^- + \frac{\omega}{2^m} \right\} \right| \le \frac{\gamma m}{(m - s_o - 1)^2} |K_\rho|, \quad \forall t \in (0, dR^p),$$

for a constant γ depending only upon the data. To prove the lemma we have only to choose m so large that

$$\frac{\gamma m}{(m - s_o - 1)^2} < \nu.$$

Remark 13.1. The process described has a double meaning. It defines a level $\mu^- + \frac{\omega}{2^m}$ for the function u and the size of the box

$$Q_o\left(d\rho^p, \rho\right), \qquad d = \left(\frac{\omega}{2^m}\right)^{2-p},$$

within which the set where $u < \mu^- + \frac{\omega}{2^m}$ is small.

To conclude the proof of Proposition 13.1, choose $\nu = \nu_o$, where ν_o is the number claimed by Lemma 4.1. Then define m accordingly. By Lemma 4.1 and inequalities (13.2)

$$u(x,t) > \mu^- + \frac{\omega}{2^{m+1}}, \qquad \forall\, (x,t) \in K_{\frac{R}{4}} \times (0, dR^p).$$

Notice that no shrinking occurs in the t–direction. This is due to the fact that in (13.1) the cutoff function ζ can be taken independent of t.

14. Miscellaneous remarks

14-(i). Expansion of positivity

A crucial fact in the proof of Theorem 1.1 is the expansion of positivity of Proposition 5.1. To focus on this phenomenon let us consider homogeneous equations with measurable coefficients of the type

$$(14.1) \qquad \begin{cases} v \in L^\infty_{loc}\left(0,T; L^2_{loc}(\Omega)\right) \cap L^p_{loc}\left(0,T; W^{1,p}_{loc}(\Omega)\right), \; 1<p<2, \\ v_t - \left(|Dv|^{p-2} a_{ij}(x,t)\, u_{x_i}\right)_{x_j} = 0 \quad \text{in } \Omega_T, \end{cases}$$

where the entries $(x,t) \to a_{ij}(x,t)$ of the matrix (a_{ij}) are only measurable and satisfy the ellipticity condition

$$(14.2) \qquad \Lambda^{-1}|\xi|^2 \le a_{ij}(x,t)\xi_i\xi_j \le \Lambda|\xi|^2, \quad \text{a.e. } (x,t) \in \Omega_T, \quad \forall \xi \in \mathbf{R}^N,$$

for some $\Lambda > 0$. In such a case, the various costants A_i and $\mathcal{A}_i$ appearing in the proof of Proposition 5.1 are all zero and the function w introduced in (6.4) coincides with v. The information (5.3) has been translated into the dimensionless estimate (5.11)-(5.12) (see Fig. 5.2).

Let us think of (14.1) as defined weakly in the cylindrical domain $K_4 \times (-4^p, 0)$. The information (5.11)-(5.12) is that at the *'centre'* of $K_4 \times (-4^p, 0)$ there is a *thin* cylinder $K_{h_o} \times (-4^p, 0)$ where $v > 1$. The conclusion of the arguments of §5-9 is that there exists a small positive number γ_o that can be determined a priori only in terms of N, p and Λ, such that

$$v(x,t) \ge \gamma_o, \qquad \forall\, (x,t) \in K_1 \times (-1, 0).$$

Thus the '*positivity*' of v over K_{h_o} spreads over a full cube K_1. Actually the information (5.11)-(5.12) is only used to apply the Poincaré inequality of Proposition 2.1 of Chap. I to derive the integral inequality (6.10). Precisely

$$\int_{K_4\times\{t\}} \Psi_k^p(v)\zeta^p dx \le \frac{\gamma}{\text{meas}\,\{[\Psi_k[v(x,t)]=0]\cap[\zeta=1]\}} \int_{K_4\times\{t\}} |D\Psi_k(v)|^p\zeta^p dx,$$

for all $t\in(-4^p,0)$. Now to apply such an inequality it only suffices to have the information

$$\text{meas}\,\{[\Psi_k[v(x,t)]=0]\cap K_2\}\ge\alpha_o>0,\ \ \text{for some}\ \alpha_o>0,\quad \forall t\in(-4^p,0).$$

In particular it is not necessary to know that the set $[\Psi_k(v)=0]\equiv[v\ge1]$ is concentrated in a cylinder about the origin. We summarise:

THEOREM 14.1. *Let v be a non–negative weak solution of (14.1) in the cylindrical domain $Q_4(4\delta)\equiv K_4\times(0,4\delta)$ for some $\delta>0$. Assume moreover that*

$$\text{meas}\,\big\{x\in K_2\ \big|\ v(x,t)>k_o\big\}\ge\alpha_o,$$

for some positive numbers k_o and α_o and all $t\in(2\delta,4\delta)$. Then there exists a number $\gamma_o=\gamma_o\,(N,p,\Lambda,\delta,\alpha_o,k_o)$ that can be determined a priori only in terms of the indicated quantities, such that

$$v(x,t)\ge\gamma_o,\qquad\qquad \forall\,(x,t)\in K_1\times(3\delta,4\delta).$$

14-(ii). Extinction in finite time

Weak solutions of (14.1) may become extinct in finite time. We refer to §2-3 of Chap. VII for a precise description of this phenomenon. The extinction profile is the set $\partial\,[v=0]\cap\Omega_T$. Theorem 14.1 implies that the extinction profile is a portion of a hyperplane normal to the $t-$axis. Indeed if $u(x_o,t_o)>0$ for some $(x_o,t_o)\in\Omega_T$, by continuity we may construct a box about (x_o,t_o) where the assumptions of Theorem 14.1 are verified. It follows that the positivity of v at (x_o,t_o) expands at the same time t_o to the whole domain of definition of $v(\cdot,t_o)$.

14-(iii). Continuous dependence on the operator

We only remark that the comments on stability made in §14 of Chap. III, in the context of degenerate equations, carry over with no change to the case of singular equations.

15. Bibliographical notes

Theorems 1.1-1.3 were established in [26] for the case when the principal part of the operator is independent of t. This restriction has been removed in [27] and entails the new iteration technique presented in §6-10. This technique differs substantially from the classical iteration of Moser [81,82,83] or DeGiorgi [33]. It extracts the 'almost elliptic' nature of the singular p.d.e. as follows from the remarks in §14-(i). We will further discuss this point in Chap. VII in the context of Harnack estimates. The method is rather flexible and adapts to a variety of singular parabolic equations. For example it implies the Hölder continuity of solutions of singular equations of porous medium type. To be specific, consider the p.d.e.

$$u_t - \operatorname{div} \mathbf{a}(x,t,u,Du) + b(x,t,u,Du) = 0, \quad \text{in } \Omega_T,$$

with the structure conditions

$$\begin{aligned}
&(\mathbf{A}_1) \quad \mathbf{a}(x,t,u,Du)\cdot Du \geq C_o\,|u|^{m-1}\,|Du|^2 - \varphi_o(x,t), \quad m \in (0,1),\\
&(\mathbf{A}_2) \quad |\mathbf{a}(x,t,u,Du)| \leq C_1\,|u|^{m-1}\,|Du| + \varphi_1(x,t),\\
&(\mathbf{A}_3) \quad |b(x,t,u,Du)| \leq C_2|\,D\,|u|^m\,|^2 + \varphi_2(x,t).
\end{aligned}$$

We require that

$$u \in L^\infty_{loc}\left(0,T;L^2_{loc}(\Omega)\right) \quad \text{and} \quad |u|^m \in L^2_{loc}\left(0,T;W^{1,2}_{loc}(\Omega)\right).$$

The non-negative functions φ_i, $i=0,1,2$, satisfy $(\mathbf{A}_4)-(\mathbf{A}_5)$ of §1 of Chap. I with $p=2$. Further generalisations can be obtained by replacing s^{m-1}, $s>0$ with a function $\varphi(s)$ that blows up like a power when $s\to 0$ and is regular otherwise. Results concerning doubly non-linear equations bearing singularity and/or degeneracy are due to Ivanov [52,53,54] and Vespri [102]. A complete theory of doubly singular equations, however, is still lacking.

V

Boundedness of weak solutions

1. Introduction

Let u be a weak solution of equations of the type of (1.1) of Chap. II in Ω_T. We will establish local and global bounds for u in Ω_T. Global bounds depend on the data prescribed on the parabolic boundary of Ω_T. Local bounds are given in terms of local integral norms of u. Consider the cubes $K_\rho \subset K_{2\rho}$. After a translation we may assume they are contained in Ω provided ρ is sufficiently small. For $0 \le t_1 < t_o < t \le T$ consider the cylindrical domains

$$Q_o \equiv K_\rho \times (t_o, t), \quad Q_1 \equiv K_{2\rho} \times (t_1, t), \quad Q_o \subset Q_1 \subset \Omega_T.$$

The local estimates are of the type.

$$\|u\|_{\infty, Q_o} \le \gamma \left(1 + \iint_{Q_1} |u|^{q_o} \right)^{q_1}, \tag{1.1}$$

where the numbers q_i, $i=0,1$, are determined a priori in terms of p and N and the constant γ is determined a priori in terms of the structure conditions of the p.d.e. and Q_1.

Unlike the elliptic theory, the estimate (1.1) discriminates between the degenerate case $p > 2$ and the singular case $1 < p < 2$. To illustrate this point, consider local weak solutions of the elliptic equation

$$\begin{cases} u \in W^{1,p}_{loc}(\Omega), & p > 1 \\ \operatorname{div} |Du|^{p-2} Du = 0, & \text{in } \Omega. \end{cases}$$

These solutions satisfy the following estimate for any $p>1$. For every $\varepsilon>0$ there exists a constant $\gamma=\gamma(N,p,\varepsilon)$, such that

$$\|u\|_{\infty,K_\rho}\le\gamma\left(\fint_{K_{2\rho}}|u|^\varepsilon dx\right)^{\frac{1}{\varepsilon}}.$$

Consider now the corresponding parabolic equation

$$\text{(1.2)}\qquad\begin{cases}u\in C_{loc}\left(0,T;L^2_{loc}(\Omega)\right)\cap L^p_{loc}\left(0,T;W^{1,p}_{loc}(\Omega)\right), & p>1,\\ u_t-\operatorname{div}|Du|^{p-2}Du=0, & \text{in } \Omega_T,\end{cases}$$

and the cylindrical domain $Q^t_{2\rho}\equiv K_{2\rho}\times(0,t]$. Assume first that $p>2$. Then for all $\varepsilon\in(0,2]$ there exists a constant $\gamma=\gamma(N,p,\varepsilon)$ such that for all $\frac{1}{2}t\le s\le t$

$$\text{(1.3)}\qquad\|u(s)\|_{\infty,K_\rho}\le\gamma\left(\frac{t}{\rho^p}\iint_{Q^t_{2\rho}}|u|^{p-2+\varepsilon}dxd\tau\right)^{1/\varepsilon}\wedge\left(\frac{\rho^p}{t}\right)^{\frac{1}{p-2}}.$$

For the singular case $1<p<2$ a local sup-estimate can be derived only if u is *sufficiently* integrable. Introduce the numbers

$$\text{(1.4)}\qquad\lambda_r\equiv N(p-2)+rp,\qquad r\ge1,$$

and assume that $u\in L^r_{loc}(\Omega_T)$ for some $r\ge1$ such that $\lambda_r>0$. Then there exist a constant $\gamma=\gamma(N,p,r)$ such that for all $\frac{1}{2}t<s=t$ there holds

$$\text{(1.5)}\qquad\|u(s)\|_{\infty,K_\rho}\le\gamma\left(\frac{\rho^p}{t}\right)^{N/\lambda_r}\left(\iint_{Q^t_{2\rho}}|u|^r dxd\tau\right)^{p/\lambda_r}\wedge\left(\frac{t}{\rho^p}\right)^{\frac{1}{2-p}}.$$

When $1<p<2$ such an order of local integrability is not implicit in the notion of weak solution and it must be imposed. The counterexample of §12 of Chap. XII shows that it is sharp.

2. Quasilinear parabolic equations

Consider quasilinear evolution equations of the type

$$\text{(2.1)}\qquad u_t-\operatorname{div}\mathbf{a}(x,t,u,Du)=b(x,t,u,Du)\quad\text{in } \mathcal{D}'(\Omega_T).$$

The functions $\mathbf{a}:\Omega_T\times\mathbf{R}^{N+1}\to\mathbf{R}^N$ and $b:\Omega_T\times\mathbf{R}^{N+1}\to\mathbf{R}$, are measurable and satisfy

$$(\mathbf{B}_1) \qquad \mathbf{a}(x,t,u,Du)\cdot Du \ge C_o|Du|^p - c_o|u|^\delta - \varphi_o(x,t),$$

$$(\mathbf{B}_2) \qquad |\mathbf{a}(x,t,u,Du)| \le C_1|Du|^{p-1} + c_1|u|^{\delta\frac{p-1}{p}} + \varphi_1(x,t),$$

$$(\mathbf{B}_3) \qquad |b(x,t,u,Du)| \le C_2|Du|^{p\frac{\delta-1}{\delta}} + c_2|u|^{\delta-1} + \varphi_2(x,t)$$

for $p>1$ and a.e. $(x,t)\in\Omega_T$. Here C_i, c_i, $i=0,1,2$, are positive constants and δ is in the range

$$(\mathbf{B}_4) \qquad p \le \delta < p\frac{N+2}{N}.$$

The non-negative functions φ_i, $i=0,1,2$, are defined in Ω_T and satisfy

$$(\mathbf{B}_5) \qquad \varphi_o, \quad \varphi_1^{\frac{p}{p-1}}, \quad \varphi_2^{\frac{\delta}{\delta-1}} \in L^{\hat{q}}(\Omega_T)$$

where

$$(\mathbf{B}_6) \qquad \frac{1}{\hat{q}} = (1-\kappa_o)\frac{p}{N+p}, \qquad \kappa_o \in (0,1].$$

For $f\in L^1(\Omega_T)$ and $h\in(0,T)$ we let f_h denote the Steklov average of f. A function u is a *local* weak sub(super)–solution of (2.1) in Ω_T if

$$(2.2) \qquad u \in C_{loc}\left(0,T;L^2_{loc}(\Omega)\right)\cap L^p_{loc}\left(0,T;W^{1,p}_{loc}(\Omega)\right),$$

and for every compact subset $\mathcal{K}$ of Ω,
(2.3)

$$\int_{\mathcal{K}\times\{t\}}\left\{\frac{\partial}{\partial t}u_h\varphi + [\mathbf{a}(x,\tau,u,Du)]_h\cdot D\varphi - [b(x,\tau,u,Du)]_h\,\varphi\right\}dx \le (\ge) 0$$

for all $0<t\le T-h$ and all testing functions

$$(2.4) \qquad \varphi\in C_{loc}\left(0,T;L^2(\mathcal{K})\right)\cap L^p_{loc}\left(0,T;W^{1,p}_o(\mathcal{K})\right), \quad \varphi\ge 0.$$

The statement that a constant $\gamma=\gamma\,$(data) depends only upon the data means that it can be determined a priori only in terms of the numbers $N,p,\hat{q},\delta,\kappa_o$, the constants C_i, c_i, $i=0,1,2$, and the norms

$$\|\varphi_o, \varphi_1^{\frac{p}{p-1}}, \varphi_2^{\frac{\delta-1}{\delta}}\|_{\hat{q},\Omega_T}.$$

2-(i). *The Dirichlet problem*

Consider the boundary value problem

$$(2.5) \qquad \begin{cases} u_t - \operatorname{div}\mathbf{a}(x,t,u,Du) = b(x,t,u,Du), & \text{in } \Omega_T,\\ u(\cdot,t)\big|_{\partial\Omega} = g(\cdot,t), & \text{a.e. } t\in(0,T),\\ u(\cdot,0)=u_o, & \end{cases}$$

We retain the structure conditions $(\mathbf{B}_1)$-$(\mathbf{B}_6)$, and on the Dirichlet data g and u_o we assume

$$g \in L^{\infty}(S_T), \tag{2.6}$$

$$u_o \in L^2(\Omega). \tag{2.7}$$

The notion of weak solution is in (2.5) of Chap. II.

Remark 2.1. Unlike the assumption $(\mathbf{U_o})$ in Chap. II, we do not assume here that $u_o \in L^{\infty}(\Omega)$. Accordingly, our estimates of the norms $\|u(\cdot,t)\|_{\infty,\Omega}$ deteriorate as $t \searrow 0$.

2-(ii). *Homogeneous structures*

Local and global sup-bounds, take an elegant form for solutions of equations of the type

$$u_t - \operatorname{div} \mathbf{a}(x,t,u,Du) = 0, \quad \text{in } \Omega_T, \quad p > 1 \tag{2.8}$$

$$\begin{cases} \mathbf{a}(x,t,u,Du) \cdot Du \geq C_o |Du|^p, \\ |\mathbf{a}(x,t,u,Du)| \leq C_1 |Du|^{p-1}, \end{cases} \tag{2.9}$$

for two given constants $0 < C_o \leq C_1$. The lower order terms are zero and the principal part has the same structure as

$$u_t - \operatorname{div} |Du|^{p-2} Du = 0 \quad \text{or} \quad u_t - \left(|u_{xi}|^{p-2} u_{x_i} \right)_{x_i} = 0. \tag{2.10}$$

Because of the structural analogy with (2.10) we will refer to (2.8)-(2.9) as equations with homogeneous structure.

3. Sup-bounds

We let u be a non-negative weak subsolution of (2.1) and will state several upper bounds for it. The assumption that u is non-negative is not essential and is used here only to deduce that u is locally or globally bounded. If u is a subsolution, not necessarily bounded below, our results supply a priori bounds *above* for u. Analogous statements hold for non-positive local supersolutions and in particular for solutions.

The estimates of this section hold for p in the range

$$p > \max\left\{1\,;\ \frac{2N}{N+2}\right\}, \qquad \text{i.e., } \lambda_2 \equiv N(p-2) + 2p > 0 \tag{3.1}$$

The case $1 < p \leq \max\left\{1\,;\ \frac{2N}{N+2}\right\}$ will be discussed in §5. Let δ and κ_o be the numbers appearing in the structure conditions $(\mathbf{B}_1)$-$(\mathbf{B}_6)$ and set

$$q = p\frac{N+2}{N}, \qquad \kappa = \kappa_o \frac{p}{N}. \tag{3.2}$$

The range of δ in $(\mathbf{B}_4)$ is $p \le \delta < q$. We will assume that

$$\max\{p;\, 2\} \le \delta < q. \tag{3.3}$$

This is no loss of generality by possibly modifying the constants c_i and the functions φ_i, $i = 0, 1, 2$. We also observe that owing to (3.1), the range (3.3) of δ is non-empty. In the theorems below we will establish local or global bounds for solutions of (2.1). However precise quantitative estimates will be given only for the case

$$\varphi_i \in L^\infty(\Omega_T), \qquad i = 0, 1, 2. \tag{3.4}$$

In this case we may take $\kappa_o = 1$ in $(\mathbf{B}_6)$ and $\kappa = p/N$.

3-(i). Local estimates

THEOREM 3.1. *Let (3.1) hold. Every non-negative, local weak subsolution u of (2.1) in Ω_T is locally bounded in Ω_T. Moreover, if $\varphi_i \in L^\infty(\Omega_T)$, $i = 0, 1, 2$, there exists a constant $\gamma = \gamma\,(data)$ such that $\forall\, [(x_o, t_o) + Q\,(\rho^p, \rho)] \subset \Omega_T$ and $\forall \sigma \in (0, 1)$,*

$$\sup_{[(x_o,t_o)+Q(\sigma\rho^p,\sigma\rho)]} u \tag{3.5}$$

$$\le \gamma \left((1-\sigma)^{-(N+p)} + \left| Q\,(\rho^p, \rho) \right| \right)^{\frac{\kappa}{q-\delta}} \left(\;\; \iint\limits_{[(x_o,t_o)+Q(\rho^p,\rho)]} u^\delta \, dx d\tau \right)^{\frac{\kappa}{q-\delta}} \wedge 1.$$

3-(ii). Global estimates: Dirichlet data

THEOREM 3.2. *Let u be a non-negative weak subsolution of the Dirichlet problem (2.5) and let (2.6) hold. Then u is bounded in $\Omega \times (\varepsilon, T)$, $\forall \varepsilon \in (0, T)$. Moreover, if $\varphi_i \in L^\infty(\Omega_T)$, $i = 0, 1, 2$, there exists a constant $\gamma = \gamma\,(data)$, such that for all $0 < t \le T$,*

$$\sup_{\Omega} u(\cdot, t) \le \sup_{S_T} g + \gamma \left(t^{\frac{p}{N}} + \frac{1}{t} \right)^{\frac{1}{q-\delta}} \left(\int_0^t\!\!\int_\Omega u^\delta \, dx d\tau \right)^{\frac{\kappa}{q-\delta}} \wedge 1. \tag{3.6}$$

If in addition the initial datum u_o is bounded above, then

$$\sup_{\Omega} u(\cdot, t) \le \max\left\{ \sup_{S_T} g\,;\, \sup_{\Omega} u_o \right\} + \gamma \left(\iint\limits_{\Omega_t} u^\delta \, dx d\tau \right)^{\frac{\kappa}{q-\delta}} \wedge 1. \tag{3.6}$$

THEOREM 3.3 (THE WEAK MAXIMUM PRINCIPLE). *Let u be a non-negative weak subsolution of the Dirichlet problem (2.5) for equations with homogeneous structure as (2.8)-(2.9). Then*

$$\sup_{\Omega_T} u \le \max\left\{\operatorname*{ess\,sup}_{S_T} g\,;\ \operatorname*{ess\,sup}_{\Omega} u_o\right\}. \tag{3.7}$$

Remark 3.1. The weak maximum principle holds for equations with homogeneous structure for all $p>1$.

As a particular case, Theorems 3.1-3.3 give a priori sup-estimates for non-negative weak solutions of

$$u_t - \operatorname{div}|Du|^{p-2}Du = a_1 u^{\delta-1} + a_2\varphi, \quad a_i \in \mathbf{R}, \quad i=1,2, \tag{3.8}$$

where

$$1 \le \delta < p\frac{N+2}{N}$$

and

$$\varphi \in L^{\hat{q}}(\Omega_T), \qquad \frac{1}{\hat{q}} = \frac{p}{N+p}(1-\kappa_o), \quad \kappa_o \in (0,1].$$

These conditions on the lower order terms are optimal for a sup-bound to hold as it can be seen from the 'linear' case $p=2$. Set $p=2$ and $a_2=0$ in (3.8). For a local weak solution $u\in V^2_{loc}(\Omega_T)$ to be locally bounded, δ must not exceed $2\frac{N+2}{N}$. Likewise if $a_1=0$, the forcing term φ must satisfy

$$\varphi \in L^{\hat{q}}(\Omega_T), \qquad \text{where} \quad \hat{q} > \frac{N+2}{2}.$$

These are classical and optimal results for the *linear* case $p=2$ (see [67]).

4. Homogeneous structures. The degenerate case $p>2$

Here we consider non-negative local or global subsolutions of equations with the homogeneous structure (2.8)–(2.9). These structures reveal the basic difference between the degenerate case $p>2$ and the singular case $1<p<2$.

4-(i). Local estimates

THEOREM 4.1. *Every non-negative, local weak subsolution u of (2.8)-(2.9) in Ω_T is locally bounded in Ω_T. Moreover for all $\varepsilon \in (0,2]$ there exists a constant γ depending only upon the data and ε, such that $\forall\,[(x_o,t_o)+Q(\theta,\rho)]\subset\Omega_T$ and $\forall\sigma\in(0,1)$,*

$$\text{(4.1)}\qquad \sup_{[(x_o,t_o)+Q(\sigma\theta,\sigma\rho)]} u \le \frac{\gamma\,(\theta/\rho^p)^{1/\varepsilon}}{(1-\sigma)^{\frac{N+p}{\varepsilon}}} \left(\iint_{[(x_o,t_o)+Q(\theta,\rho)]} u^{p-2+\varepsilon}\,dx d\tau \right)^{1/\varepsilon} \wedge \left(\frac{\rho^p}{\theta}\right)^{\frac{1}{p-2}}.$$

Remark 4.1. If $\theta=\rho^p$, (4.1) is dimensionless but it is not homogeneous in u. In the linear case $p=2$, (4.1) holds for any positive number ε. In our case, ε is restricted in the range $(0,2]$.

It is of interest to have sup-estimates that involve 'low' integral norms of the solution. The next theorem is a result in this direction. Even though it is of local nature, it will be crucial in characterising the class of non-negative solutions in the strip $\mathbf{R}^N\times(0,T)$.(1)

THEOREM 4.2. *Let u be a non-negative, local subsolution u of (2.8)-(2.9) in Ω_T. There exists a constant $\gamma=\gamma\,(data)$, such that $\forall\,[(x_o,t_o)+Q\,(\theta,\rho)]\subset\Omega_T$ and $\forall\sigma\in(0,1)$,*

$$\text{(4.2)}\qquad \sup_{[(x_o,t_o)+Q(\sigma\theta,\sigma\rho)]} u \le \frac{\gamma\sqrt{\theta/\rho^p}}{(1-\sigma)^{\frac{N(p+1)+p}{2}}} \left(\sup_{t_o-\theta<\tau<t_o} \fint_{[x_o+K_\rho]} u(x,\tau)dx \right)^{p/2} \wedge \left(\frac{\rho^p}{\theta}\right)^{\frac{1}{p-2}}.$$

4-(ii). *Global estimates for solutions of the Dirichlet problem*

Consider a non-negative weak subsolution of the Dirichlet problem (2.5) for equations with homogeneous structure and let (2.6) hold. If the initial datum u_o is also bounded above, then the weak maximum principle estimate (3.7) holds true. If however u_o^+ is not bounded, it is of interest to investigate how the supremum of u behaves when $t\to 0$.

THEOREM 4.3. *Let u be a non-negative weak subsolution of the Dirichlet problem (2.5) and let (2.6) hold. There exists a constant $\gamma=\gamma\,(data)$, such that $\forall t\in(0,T)$,*

$$\text{(4.3)}\qquad \sup_{\Omega} u(\cdot,t) \le \sup_{S_T} g + \frac{\gamma}{t^{N/\lambda}} \left(\fint_0^t \int_\Omega u\,dx d\tau \right)^{p/\lambda}, \quad \lambda = N(p-2)+p.$$

(1) See §7 of Chap VI and §2 of Chap. XI.

Results of this kind could be used to construct solutions of the Dirichlet problem with initial data in $L^1(\Omega)$ or even finite measures. Indeed the regularity results of Chap. III supply the necessary compactness to pass to the limit in a sequence of approximating problems.

4-(iii). Estimates in $\Sigma_T \equiv \mathbf{R}^N \times (0,T)$

Consider a non-negative weak subsolution u of (2.8) in the whole strip Σ_T. By this we mean that u is a *local* weak subsolution of (2.8) in Ω_T for every *bounded* domain $\Omega \subset \mathbf{R}^N$. To derive global sup–estimates, we must impose some control on the behaviour of u as $|x| \to \infty$. We assume that the quantity

$$\|\!|u|\!\|_{\{r,t\}} \equiv \sup_{0<\tau<t} \sup_{\rho \geq r} \int_{K_\rho} \frac{u(x,\tau)}{\rho^{\lambda/(p-2)}}\, dx, \qquad \lambda = N(p-2)+p, \tag{4.4}$$

is finite for some $r>0$ and for all $t \in (0,T)$. The subsolution u at hand is not necessarily bounded. However it is locally bounded and and as $|x| \to \infty$ it grows no faster than $|x|^{\frac{p}{p-2}}$. This is the content of the next theorem.

THEOREM 4.4. *Let u be a non-negative subsolution of (2.8) in Σ_T, and assume (4.4) holds. There exist a constant $\gamma = \gamma\,(data)$, such that for all $t \in (0,T)$,*

$$\sup_{\rho \geq r} \frac{\|u(\cdot,t)\|_{\infty,K_\rho}}{\rho^{p/(p-2)}} \leq \gamma \sqrt{t}\, \|\!|u|\!\|_{\{r,t\}}^{p/2} \wedge t^{-\frac{1}{p-2}}. \tag{4.5}$$

PROOF: Apply Theorem 4.2 with the choices

$$(x_o,t_o) \equiv (0,t), \quad \theta = t, \quad \sigma = 1/2, \quad \rho \geq r,$$

and ρ replaced by 2ρ. It gives

$$\frac{\|u(\cdot,t)\|_{\infty,K_\rho}}{\rho^{p/(p-2)}} \leq \gamma\sqrt{t} \left(\sup_{0<\tau<t} \sup_{\rho \geq r} \int_{K_\rho} \frac{u(x,\tau)}{\rho^{\lambda/(p-2)}} \right)^{p/2} \wedge t^{-\frac{1}{p-2}}.$$

Remark 4.2. The assumption (4.4) is not restrictive. We will show in Chap. XI that it is necessary and sufficient for a non-negative solution of (2.8) to exist in Σ_T.

The right hand side of (4.5) blows up as $t \searrow 0$ at the rate of at least $t^{-\frac{1}{p-2}}$. Such a rate is not optimal. However the advantage of Theorem 4.4 is that it does hold for all $t \in (0,T)$. The purpose of the next theorem is two-fold. It gives an optimal estimate of how the local sup-bound for u may deteriorate as either $|x| \to \infty$ or $t \searrow 0$.

THEOREM 4.5. *Let u be a non-negative subsolution of (2.8) in Σ_T and assume (4.4) holds. There exists constants γ_* and γ depending only upon N, p and the constants C_i, $i=0,1$ in the structure condition (2.9), such that*

(4.6) *for all* $0 < t < \gamma_* |||u|||_{\{r,t\}}^{2-p}$ *and for all* $\rho \geq r$

$$\|u(\cdot,t)\|_{\infty,K_\rho} \leq \gamma \frac{\rho^{p/(p-2)}}{t^{N/\lambda}} |||u|||_{\{r,t\}}^{p/\lambda}, \quad \lambda = N(p-2)+p.$$

Information of this kind are of interest in investigating the behaviour of the solutions for t near zero and in studying the structure of the non-negative solutions in Σ_T.[1] The functional dependence in (4.6) is sharp as it can be verified from the explicit Barenblatt solution

$$\mathcal{B}(x,t) = t^{-\frac{N}{\lambda}} \left\{ 1 - \gamma_p \left(\frac{|x|}{t^{1/\lambda}} \right)^{\frac{p}{p-1}} \right\}_+^{\frac{p-1}{p-2}}, \quad t>0 \tag{4.7}$$

$$\gamma_p = \left(\frac{1}{\lambda} \right)^{\frac{1}{p-1}} \frac{p-2}{p}, \quad p>2.$$

The function $\mathcal{B}$ solves the Cauchy problem

$$\begin{cases} u_t - \operatorname{div} |Du|^{p-2} Du = 0, & \text{in } \mathbf{R}^N \times (0,\infty), \\ \mathcal{B}(\cdot,0) = M\delta_o, \end{cases} \tag{4.8}$$

where δ_o is the Dirac mass concentrated at the origin, and

$$M \equiv \|\mathcal{B}(\cdot,t)\|_{1,\mathbf{R}^N}, \quad \forall t>0.$$

The initial datum is taken in the sense of the measures, i.e., for every $\varphi \in C_o(\mathbf{R}^N)$

$$\int_{\mathbf{R}^N} \mathcal{B}(x,t)\varphi dx \longrightarrow M\varphi(0), \quad \text{as } t \searrow 0.$$

For $t>0$ and for every $\rho>0$ we have

$$\|\mathcal{B}(\cdot,t)\|_{\infty,K_\rho} = t^{-N/\lambda}.$$

5. Homogeneous structures. The singular case $1<p<2$

The estimates of §3 are valid for solutions $u \in L^2_{loc}(\Omega_T)$ as long as

$$p > \max\left\{1;\ \frac{2N}{N+2}\right\}.$$

---

(1) See Chap. XI.

In this section we will show that weak solutions $u \in L^r_{loc}(\Omega_T)$, $r \geq 1$, are bounded provided

$$p > \max\left\{1; \frac{2N}{N+r}\right\}.$$

Such integrability condition to insure boundedness is sharp. In §12 of Chap. XII we produce a solution of the homogeneous p.d.e. (1.2)

$$u \in L^1_{loc}(\Omega_T), \quad u \notin L^{1+\varepsilon}_{loc}(\Omega_T) \ \forall \varepsilon \in (0,1), \quad p = \frac{2N}{N+1},$$

that is unbounded.[1] Thus in the singular range $1 < p < \max\left\{1; \frac{2N}{N+2}\right\}$, the boundedness of a weak solutions is not a purely local fact and, if at all true, it must be deduced from some global information. One of them is the weak maximum principle of Theorem 3.3 and Remark 3.1. Another is a sufficiently high order of integrability.

5-(i). Local estimates

A sharp sufficient condition can be given in terms of the numbers

$$\lambda_r = N(p-2) + rp, \qquad r \geq 1. \tag{5.1}$$

We assume that u satisfies

$$u \in L^r_{loc}(\Omega_T), \quad \text{for some } r \geq 1 \text{ such that } \lambda_r > 0. \tag{5.2}$$

The global information needed here is

$$\begin{cases} u \text{ can be constructed as the weak limit in } L^r_{loc}(\Omega_T) \text{ of a} \\ \text{sequence of non-negative } \textit{bounded} \text{ subsolutions of (2.8).} \end{cases} \tag{5.3}$$

The notion of weak subsolution requires u to be in the class

$$u \in C_{loc}\left(0,T; L^2_{loc}(\Omega)\right) \cap L^p_{loc}\left(0,T; W^{1,p}_{loc}(\Omega)\right).$$

By the embedding of Proposition 3.1 of Chap. I, we have

$$u \in L^q_{loc}(\Omega_T), \qquad q = p\frac{N+2}{N}.$$

Therefore if p is so close to one that $\lambda_q \leq 0$, the order of integrability in (5.1)–(5.2) is not implicit in the notion of subsolution and must be imposed.

---

(1) The notion of solutions that are not in the function class (2.2) is discussed in Chap. XII

THEOREM 5.1. *Let u be a non-negative local weak subsolution of (2.8)-(2.9) in Ω_T and assume that (5.2) and (5.3) hold. There exists a constant $\gamma=\gamma(data, r)$, such that $\forall\,[(x_o,t_o)+Q(\theta,\rho)]\subset\Omega_T$ and $\forall\sigma\in(0,1)$,*

$$(5.4)\qquad \sup_{[(x_o,t_o)+Q(\sigma\theta,\sigma\rho)]} u \le \frac{\gamma\,(\rho^p/\theta)^{N/\lambda_r}}{(1-\sigma)^{\frac{p}{\lambda_r}(N+p)}}\left(\iint_{(x_o,t_o)+Q(\theta,\rho)} u^r\,dx d\tau\right)^{p/\lambda_r} \wedge \left(\frac{\theta}{\rho^p}\right)^{\frac{1}{2-p}}.$$

Remark 5.1. If $\theta=\rho^p$, (5.4) is dimensionless but it is not homogeneous in u.

5-(ii). Estimates near $t=0$

Fix $t\in(0,T)$ and let us rewrite (5.4) for the pair of boxes

$$K_{\sigma\rho}\times(\sigma t,t),\qquad K_\rho\times(0,t).$$

COROLLARY 5.1. *Let u be a non-negative local weak subsolution of (2.8)-(2.9) in Ω_T and let (5.2)-(5.3) hold. There exists a constant $\gamma=\gamma(data, r)$, such that for all $0<t\le T$ and for all $\sigma\in(0,1)$,*

$$(5.5)\qquad \sup_{K_{\sigma\rho}} u(\cdot,t) \le \frac{\gamma t^{-N/\lambda_r}}{(1-\sigma)^{\frac{N+p}{\lambda_r}}}\left(\int_0^t\!\!\int_{K_\rho} u^r\,dx d\tau\right)^{p/\lambda_r} \wedge \left(\frac{t}{\rho^p}\right)^{\frac{1}{2-p}}.$$

Remark 5.2. Assume that (5.2) holds with $r=1$, i.e.,

$$(5.6)\qquad p>\frac{2N}{N+1}.$$

Then the behaviour of the supremum of u as $t\searrow 0$ is *formally* the same as that of solutions of the Dirichlet problem for degenerate equations as in Theorem 4.3.

5-(iii). Global estimates: Dirichlet data

A peculiar phenomenon of these equations is that, unlike their degenerate counterparts, local and global estimates take essentially the same form. This appears, for example, by comparing (5.5) with the next global estimate.

THEOREM 5.2. *Let u be a non-negative weak subsolution of the Dirichlet problem (2.5) and let (2.6) and (5.2)-(5.3) hold. There exists a constant $\gamma = \gamma\,(data, r)$, such that for all $0<t\le T$,*

$$\sup_{\Omega} u(\cdot, t) \le \sup_{S_T} g + \frac{\gamma}{t^{N/\lambda_r}} \left(\int_0^t\!\!\int_{\Omega} u^r\, dx d\tau \right)^{p/\lambda_r}. \tag{5.7}$$

6. Energy estimates

The proof of the sup-bounds stated in the previous sections is based on local and global energy estimates similar to those of §3 of Chap. II.

6-(i). Local energy estimates

If $[(x_o, t_o) + Q\,(\theta, \rho)] \subset \Omega_T$ we let ζ denote a non-negative piecewise smooth cutoff function vanishing on the parabolic boundary of $[(x_o, t_o) + Q\,(\theta, \rho)]$.

PROPOSITION 6.1. *Let u be a non-negative local weak solution of (2.1) in Ω_T and let $(\mathbf{B}_1)$-$(\mathbf{B}_6)$ hold. There exist a constant $\gamma = \gamma\,(data)$, such that $\forall\,[(x_o, t_o) + Q\,(\theta, \rho)]\subset\Omega_T$ and for every level $k>0$*

$$\begin{aligned} &\sup_{t_o-\theta<t<t_o} \int_{[x_o+K_\rho]} (u-k)_+^2\, \zeta^p(x,t) dx + \gamma^{-1} \iint_{[(x_o,t_o)+Q(\theta,\rho)]} |D\,(u-k)_+\, \zeta|^p dx d\tau \\ &\le \gamma \iint_{[(x_o,t_o)+Q(\theta,\rho)]} (u-k)_+^p\, |D\zeta|^p dx d\tau + \gamma \iint_{[(x_o,t_o)+Q(\theta,\rho)]} (u-k)_+^2\, \zeta^{p-1} \zeta_t dx d\tau \\ &+\gamma \iint_{[(x_o,t_o)+Q(\theta,\rho)]} u^\delta \chi\,[u>0]\, dx d\tau + \gamma \left\{ \int_{t_o-\theta}^{t_o} \left|A_{k,\rho}^{\pm}(\tau)\right| d\tau \right\}^{\frac{N}{N+p}(1+\kappa)} \end{aligned} \tag{6.1}$$

where $\kappa = \kappa_o \frac{p}{N}$ and

$$\left|A_{k,\rho}(\tau)\right| \equiv \text{meas}\left\{x \in [x_o + K_\rho] \;\middle|\; u(x,\tau) > k\right\}.$$

In (6.1) the integral involving u^δ can be eliminated if

$$c_i = 0, \quad i = 0,1,2, \quad \text{and} \quad b(x,t,u,Du) \equiv 0.$$

Moreover the last term can be eliminated if $\varphi_i \equiv 0$, $i=0,1,2$.

PROOF: The proof is very similar to that of Proposition 3.1 of Chap. II. First we may assume that $(x_o, t_o) \equiv (0,0)$ modulo a translation. Then in (2.3) we take the testing functions

$$\varphi = (u_h - k)_+ \zeta^p,$$

where u_h is the Steklov average of u. All the terms are estimated as in §3 of Chap. II, with minor modifications, except the integrals involving the lower order terms $b(x,t,u,Du)$. For these, we let $h \to 0$ and use the structure condition $(\mathbf{B}_3)$ and Young's inequality to estimate

$$\begin{aligned}
(6.2)\qquad &\iint_{Q(\theta,\rho)} |b(x,\tau,u,Du)\,(u-k)_+\,\zeta^p| dx d\tau \\
&\le C_2 \iint_{Q(\theta,\rho)} |D\,(u-k)_+|^{p\frac{\delta-1}{\delta}}\,(u-k)_+\,\zeta^p dx d\tau \\
&+ c_2 \iint_{Q(\theta,\rho)} u^{\delta-1}\,(u-k)_+\,\zeta^p dx d\tau + \iint_{Q(\theta,\rho)} \varphi_2\,(u-k)_+\,\zeta^p dx d\tau \\
&\le \frac{C_o}{4} \iint_{Q(\theta,\rho)} |D\,(u-k)_+|^p \zeta^p dx d\tau + \gamma \iint_{Q(\theta,\rho)} u^\delta \chi\left[(u-k)_+ > 0\right] dx d\tau \\
&+ \gamma \iint_{Q(\theta,\rho)} \varphi_2^{\frac{\delta}{\delta-1}} \chi\left[(u-k)_+ > 0\right] dx d\tau\,.
\end{aligned}$$

Thus we arrive at

$$\begin{aligned}
&\sup_{-\theta<t<0} \int_{K_\rho} (u-k)_+^2\,\zeta^p(x,t) dx + \gamma^{-1} \iint_{Q(\theta,\rho)} |D\,(u-k)_+\,\zeta|^p dx d\tau \\
&\le \gamma \iint_{Q(\theta,\rho)} (u-k)_+^p\,|D\zeta|^p dx d\tau + \gamma \iint_{Q(\theta,\rho)} (u-k)_+^2\,\zeta^{p-1}\zeta_t dx d\tau \\
&+ \gamma \iint_{Q(\theta,\rho)} u^\delta \chi[(u-k)_+ > 0]\, dx d\tau + \gamma \iint_{Q(\theta,\rho)} \Phi \chi[(u-k)_+ > 0]\, dx d\tau,
\end{aligned}$$

where

$$\Phi = \varphi_o + \varphi_1^{\frac{p}{p-1}} + \varphi_2^{\frac{\delta}{\delta-1}}.$$

By the Hölder inequality and $(\mathbf{B}_5)$-$(\mathbf{B}_6)$,

$$\iint_{Q(\theta,\rho)} \Phi\, dx d\tau \le \|\Phi\|_{\hat{q},\Omega_T} \left\{ \int_{-\theta}^{0} |A_{k,\rho}(\tau)| d\tau \right\}^{\frac{N}{N+p}(1+\kappa)}.$$

Remark 6.1. Inequality (6.1) for the function $(u-k)_-$ holds true for local supersolutions of (2.1) and $k \le 0$.

Remark 6.2. Unlike inequalities (3.8) of Chap. II, the levels k here are not restricted.

6-(ii). Global energy estimates: Dirichlet data

PROPOSITION 6.2. *Let u be a non-negative weak sub-solution of the Dirichlet problem (2.5), let (2.6) hold and let k satisfy*

$$k \geq \sup_{S_T} g \wedge 0. \tag{6.2}$$

There exists a constant $\gamma = \gamma\,(data)$, such that for every non-negative function $t \to \zeta(t) \in C^1[0,T]$ and for every $0<t\leq T$,

$$\begin{aligned}\sup_{0<\tau<t} \int_\Omega (u-k)_+^2\, \zeta(\tau)\, dx + \iint_{\Omega_t} |D\,(u-k)_+|^p \zeta(\tau)\, dx d\tau \\ \leq \gamma \int_\Omega (u_o - k)_+^2\, \zeta(0)\, dx + \gamma \iint_{\Omega_t} (u-k)_+^2\, \zeta_t(\tau)\, dx d\tau \\ + \gamma \iint_{\Omega_t} u^\delta \chi\,[u>k]\,\zeta(\tau) dx d\tau + \gamma \left(\int_0^t |A_{k,\rho}(\tau)| \zeta(\tau) d\tau \right)^{\frac{N}{N+p}(1+\kappa)}\end{aligned} \tag{6.3}$$

where $\kappa = \kappa_o \frac{p}{N}$, and

$$|A_{k,\rho}(\tau)| \equiv \text{meas}\left\{ x \in \Omega \mid u(x,\tau) > k \right\}.$$

In (6.3) the integral involving u^δ can be eliminated if

$$c_i = 0, \quad i = 0,1,2, \quad \textit{and } b(x,t,u,Du) \equiv 0.$$

Moreover the last term can be eliminated if $\varphi_i \equiv 0$, $i = 0,1,2$.

A similar statement holds for the truncated functions $(u-k)_-$ provided

$$k \leq \inf_{S_T} g \vee 0. \tag{6.2$'$}$$

PROOF: If k satisfies (6.2), by Lemma 2.1 of Chap. I

$$(u_h(\cdot,t) - k)_+ \in W_o^{1,p}(\Omega), \qquad \forall 0 < t \leq T-h.$$

Therefore the testing functions

$$\varphi \equiv (u_h - k)_+ \zeta, \qquad \zeta \in C^1[0,T], \quad \zeta \geq 0,$$

are admissible in the weak formulation of the Dirichlet problem.

7. Local iterative inequalities

The common element in the proof of the sup-bounds stated in §3 and 4 is a set of iterative inequalities. We will derive them, starting from the energy inequalities of §6. Modulo a translation, we may assume that (x_o, t_o) coincides with the origin. Fix $\sigma \in (0,1)$ and consider the sequences

$$\rho_n = \sigma\rho + \frac{(1-\sigma)}{2^n}\rho, \quad \theta_n = \sigma\theta + \frac{(1-\sigma)}{2^n}\theta, \quad n = 0,1,2,\ldots,$$

and the corresponding cylinders $Q_n i \equiv Q(\theta_n, \rho_n)$. It follows from the definitions that

$$Q_o = Q(\theta, \rho), \quad \text{and} \quad Q_\infty = Q(\sigma\theta, \sigma\rho).$$

Consider also the family of boxes

$$\tilde{Q}_n = Q\left(\tilde{\theta}_n, \tilde{\rho}_n\right),$$

where for $n = 0,1,2,\ldots$

$$\tilde{\rho}_n = \frac{\rho_n + \rho_{n+1}}{2} = \sigma\rho + \frac{3(1-\sigma)}{2^{n+2}}\rho, \quad \tilde{\theta}_n = \frac{\theta_n + \theta_{n+1}}{2} = \sigma\theta + \frac{3(1-\sigma)}{2^{n+2}}\theta.$$

For these boxes we have the inclusion

$$Q_{n+1} \subset \tilde{Q}_n \subset Q_n, \qquad n = 0,1,2,\ldots.$$

Introduce the sequence of increasing levels

$$k_n = k - \frac{k}{2^n},$$

where k is a positive number to be chosen. We will work with the inequalities (6.1) written for the functions $(u - k_{n+1})_+$, over the boxes Q_n. The cutoff function ζ is taken to satisfy

$$\begin{cases} \zeta \text{ vanishes on the parabolic boundary of } Q_n, \\ \zeta \equiv 1 \text{ in } \tilde{Q}_n, \\ |D\zeta| \le \frac{2^{n+2}}{(1-\sigma)\rho}, \quad 0 \le \zeta_t \le \frac{2^{n+2}}{(1-\sigma)\theta}. \end{cases}$$

With these choices, (6.1) yields

$$\begin{aligned}
(7.1) \quad \sup_{-\theta_n < t < 0} \int_{K_{\rho_n}} (u - k_{n+1})_+^2 \zeta^p(x,t)dx + \iint_{Q_n} \left|D(u - k_{n+1})_+ \zeta\right|^p dxd\tau \\
\le \frac{\gamma 2^{np}}{(1-\sigma)^p \rho^p} \iint_{Q_n} (u - k_{n+1})_+^p \, dxd\tau \\
+ \frac{\gamma 2^n}{(1-\sigma)\theta} \iint_{Q_n} (u - k_{n+1})_+^2 \, dxd\tau \\
+ \gamma \iint_{Q_n} u^\delta \chi\left[(u - k_{n+1})_+ > 0\right] dxd\tau + \gamma \left|A_{n+1}\right|^{\frac{N}{N+p}(1+\kappa)},
\end{aligned}$$

where whe have set

$$|A_{n+1}| \equiv \text{meas}\left\{(x,t) \in Q_n \mid u(x,t) > k_{n+1}\right\}.$$

The last two terms can be eliminated for equations with homogeneous structure. First we observe that for all $s>0$

$$\begin{aligned}(7.2)\qquad \iint_{Q_n} (u-k_n)_+^s \,dxd\tau &\geq \iint_{Q_n} (u-k_n)_+^s \chi\left[u > k_{n+1}\right] dxd\tau \\ &\geq (k_{n+1}-k_n)^s |A_{n+1}| \\ &= \frac{k^s}{2^{(n+1)s}} |A_{n+1}|.\end{aligned}$$

Then we estimate

$$\begin{aligned}(7.3)\qquad \iint_{Q_n} (u-k_{n+1})_+^p \,dxd\tau &\leq \left(\iint_{Q_n} (u-k_{n+1})_+^\delta \,dxd\tau\right)^{\frac{p}{\delta}} |A_{n+1}|^{1-\frac{p}{\delta}} \\ &\leq \gamma \frac{2^{(\delta-p)n}}{k^{\delta-p}} \iint_{Q_n} (u-k_n)_+^\delta \,dxd\tau;\end{aligned}$$

$$\begin{aligned}(7.4)\qquad \iint_{Q_n} (u-k_{n+1})_+^2 \,dxd\tau &\leq \left(\iint_{Q_n} (u-k_{n+1})_+^\delta \,dxd\tau\right)^{\frac{2}{\delta}} |A_{n+1}|^{1-\frac{2}{\delta}} \\ &\leq \gamma \frac{2^{(\delta-2)n}}{k^{\delta-2}} \iint_{Q_n} (u-k_n)_+^\delta \,dxd\tau.\end{aligned}$$

To estimate the integral involving u^δ, first write

$$k_n = k_{n+1} \frac{2^{n+1}-2}{2^{n+1}-1}.$$

Then estimate below

$$\begin{aligned}(7.5)\qquad \iint_{Q_n} (u-k_n)_+^\delta \,dxd\tau &\geq \iint_{Q_n} (u-k_n)_+^\delta \chi\left[u > k_{n+1}\right] dxd\tau \\ &\geq \iint_{Q_n} u^\delta \left(1 - \frac{2^{n+1}-2}{2^{n+1}-1}\right)^\delta \chi\left[u > k_{n+1}\right] dxd\tau \\ &\geq \frac{1}{\gamma 2^{n\delta}} \iint_{Q_n} u^\delta \chi\left[u > k_{n+1}\right] dxd\tau.\end{aligned}$$

Finally

$$|A_{n+1}|^{\frac{N}{N+p}(1+\kappa)} \le \gamma\left(\frac{2^{n\delta}}{k^\delta}\iint_{Q_n}(u-k_n)_+^\delta\,dxd\tau\right)^{\frac{N}{N+p}(1+\kappa)}.$$

We combine these estimates into (7.1) to derive the following basic iterative inequalities

$$\text{(7.6)}\qquad \sup_{-\theta_n<t<0}\int_{K_{\tilde\rho_n}}(u-k_{n+1})_+^2(x,t)dx+\iint_{\tilde Q_n}|D(u-k_{n+1})_+|^p dxd\tau$$
$$\le \frac{\gamma 2^{n\delta}}{(1-\sigma)^p}\left(\frac{1}{\rho^p k^{\delta-p}}+\frac{1}{\theta k^{\delta-2}}\right)\iint_{Q_n}(u-k_n)_+^\delta\,dxd\tau$$
$$+\gamma 2^{n\delta}\iint_{Q_n}(u-k_n)_+^\delta\,dxd\tau$$
$$+\gamma 2^{n\delta}\left(\frac{1}{k^\delta}\iint_{Q_n}(u-k_n)_+^\delta\,dxd\tau\right)^{\frac{N}{N+p}(1+\kappa)}.$$

Moreover the last two terms can be eliminated for equations with homogeneous structure.

To proceed, construct a non-negative piecewise smooth cutoff function $\tilde\zeta_n$ in $\hat Q_n$, which equals one on Q_{n+1}, vanishes on the lateral boundary of $\tilde Q_n$ and such that

$$|D\tilde\zeta_n|\le 2^{n+2}/(1-\sigma)\rho.$$

Then the function $(u-k_{n+1})_+\tilde\zeta_n$ vanishes on the lateral boundary of $\tilde Q_n$ and by the multiplicative inequality of Proposition 3.1 of Chap. I,

$$\text{(7.7)}\qquad \iint_{\tilde Q_n}(u-k_{n+1})_+^q\,\tilde\zeta_n^q dxd\tau \le \gamma\left(\sup_{-\tilde\theta_n<t<0}\int_{K_{\tilde\rho_n}}(u-k_{n+1})_+^2\,dx\right)^{\frac{p}{N}}$$
$$\times\left(\iint_{\tilde Q_n}|D(u-k_{n+1})_+|^p dxd\tau+\iint_{\tilde Q_n}(u-k_{n+1})_+^p\,|D\tilde\zeta_n|^p dxd\tau\right).$$

Remark 7.1. The estimates in (7.2)–(7.5) and the inequalities (7.6), (7.7) are valid for any number

$$\delta\ge\max\{p;\,2\}.$$

The structural restriction $\delta=q$ does not play any role in the derivation of (7.6) and (7.7).

8. Local iterative inequalities $\left(p>\max\left\{1;\frac{2N}{N+2}\right\}\right)$

Introduce the dimensionless quantities

$$Y_n = \iint_{Q_n} (u-k_n)_+^{\delta}\,dxd\tau, \qquad n=0,1,2,\dots. \tag{8.1}$$

We will derive an iterative inequality for Y_n by estimating the right hand side of (7.7) by (7.6). We assume first

$$p>\max\left\{1;\,\frac{2N}{N+2}\right\}, \quad \max\{p;2\}\le\delta<q\equiv p\frac{N+2}{N} \tag{8.2}$$

and estimate

$$\begin{aligned}
Y_{n+1} &\le \gamma\iint_{\tilde Q_n}(u-k_{n+1})_+^{\delta}\,\tilde\zeta_n^{\delta}\,dxd\tau\\
&\le \gamma\left(\iint_{\tilde Q_n}(u-k_{n+1})_+^{q}\,\tilde\zeta_n^{q}\,dxd\tau\right)^{\frac{\delta}{q}}\left(\frac{|A_{n+1}|}{|Q_n|}\right)^{1-\frac{\delta}{q}}\\
&\le \gamma\left(\iint_{\tilde Q_n}(u-k_{n+1})_+^{q}\,\tilde\zeta_n^{q}\,dxd\tau\right)^{\frac{\delta}{q}}\left(\frac{\gamma 2^{n\delta}}{k^{\delta}}Y_n\right)^{1-\frac{\delta}{q}}.
\end{aligned}$$

We estimate the last integral by (7.7) and in turn estimate the right hand side of (7.7) by the inequalities (7.6) and (7.3). We arrive at the recursive inequalities

$$\begin{aligned}
Y_{n+1} &\le \frac{\gamma b^n}{k^{\frac{\delta}{q}(q-\delta)}(1-\sigma)^{p\frac{\delta}{q}\frac{N+p}{N}}}\,\mathcal{A}_k^{\frac{\delta}{q}\frac{p}{N}}\,Y_n^{1+\frac{\delta}{q}\frac{p}{N}}\\
&\quad+\frac{\gamma b^n\left(\rho^N\theta\right)^{\frac{\delta}{q}\frac{p}{N}}}{k^{\frac{\delta}{q}(q-\delta)}}\,Y_n^{1+\frac{\delta}{q}\frac{p}{N}}+\gamma b^n\left(\rho^N\theta\right)^{\frac{\delta}{q}\kappa}\left(\frac{1}{k^{\delta}}Y_n\right)^{1+\frac{\delta}{q}\kappa},
\end{aligned} \tag{8.3}$$

where

$$b=2^{\delta\left(1+\frac{\delta}{q}\frac{p}{N}\right)}$$

and

$$\mathcal{A}_k=\left(\left(\frac{\theta}{\rho^p}\right)k^{\frac{N+p}{p}(p-\delta)}+\left(\frac{\rho^p}{\theta}\right)^{\frac{N}{p}}k^{\frac{N+p}{p}(2-\delta)}\right). \tag{8.4}$$

The last two terms in (8.3) can be eliminated for solutions of equations with homogeneous structure.

9. Global iterative inequalities

We let u be any non−negative weak subsolution of the Dirichlet problem (2.5) and assume that (6.2) holds so that u satisfies the energy estimates (6.3). Fix $0<t\le T$ and introduce the sequence of increasing time levels

$$t_n = \sigma t\left(1-\frac{1}{2^n}\right), \quad \sigma\in(0,1), \quad n=0,1,2,\dots,$$

and the cutoff functions

$$\zeta_n(\tau) = \begin{cases} 1 & \text{if } t_{n+1}\le\tau\le t \\ \dfrac{\tau - t_n}{t_{n+1}-t_n} & \text{if } t_n<\tau<t_{n+1} \\ 0 & \text{if } 0\le\tau\le t_n. \end{cases}$$

Introduce also the sequence of increasing levels

$$k_n = \sup_{S_T} g + k - \frac{k}{2^n}, \quad n=0,1,2,\dots, \quad k>0 \text{ to be chosen}$$

and write (6.3) for the functions $(u-k_{n+1})_+$ and the cutoff functions ζ_n to obtain

$$\begin{aligned}(9.1)\quad &\sup_{t_{n+1}<\tau\le t}\int_\Omega (u-k_{n+1})_+^2\,dx + \int_{t_{n+1}}^t\int_\Omega |D(u-k_{n+1})_+|^p dx d\tau \\ &\le \frac{\gamma 2^n}{\sigma t}\int_{t_n}^t\int_\Omega (u-k_{n+1})_+^2\,dxd\tau + \gamma\int_{t_n}^t\int_\Omega u^\delta\chi[u>k_{n+1}]\,dxd\tau \\ &+\gamma\left(\int_{t_n}^t |A_{n+1}(\tau)|d\tau\right)^{\frac{N}{N+p}(1+\kappa)},\end{aligned}$$

where we have set

$$|A_n(\tau)| \equiv \text{meas}\left\{x\in\Omega \mid u(x,\tau)>k_n\right\}.$$

The last two terms of (9.1) can be eliminated for equations with homogeneous structure. Moreover the last term can be eliminated if in the structure conditions $(\mathbf{B}_1)$-$(\mathbf{B}_3)$, $\varphi_i\equiv 0$, $i=0,1,2$. If $\varphi_i\in L^\infty(\Omega_T)$, $i=0,1,2$, then $\kappa=\frac{p}{N}$.

Proceeding as in (7.2)−(7.5), we estimate

(i) $$\int_{t_n}^{t}\int_{\Omega}(u-k_{n+1})_+^2\,dxd\tau \le \frac{\gamma 2^{n(\delta-2)}}{k^{\delta-2}}\int_{t_n}^{t}\int_{\Omega}(u-k_n)_+^{\delta}\,dxd\tau,$$

(ii) $$\int_{t_n}^{t}\int_{\Omega}u^{\delta}\chi\left[u>k_{n+1}\right]dxd\tau \le \gamma 2^{n\delta}\int_{t_n}^{t}\int_{\Omega}(u-k_n)_+^{\delta}\,dxd\tau,$$

(iii) $$\int_{t_n}^{t}\left|A_{n+1}(\tau)\right|d\tau \le \frac{\gamma 2^{n\delta}}{k^{\delta}}\int_{t_n}^{t}\int_{\Omega}(u-k_n)_+^{\delta}\,dxd\tau.$$

Combining these remarks in (9.1) we arrive at the recursive inequalities

(9.2) $$\sup_{t_{n+1}<\tau\le t}\int_{\Omega}(u-k_{n+1})_+^2\,dx + \int_{t_{n+1}}^{t}\int_{\Omega}\left|D\left(u-k_{n+1}\right)_+\right|^p dxd\tau$$
$$\le \frac{\gamma 2^{\delta n}}{\sigma\, t\, k^{\delta-2}}\int_{t_n}^{t}\int_{\Omega}(u-k_n)_+^{\delta}\,dxd\tau + \gamma 2^{\delta n}\int_{t_n}^{t}\int_{\Omega}(u-k_n)_+^{\delta}\,dxd\tau$$
$$+\gamma 2^{\delta n}\left(\frac{1}{k^{\delta}}\int_{t_n}^{t}\int_{\Omega}(u-k_n)_+^{\delta}\,dxd\tau\right)^{\frac{N}{N+p}(1+\kappa)},$$

Remark 9.1. The structure restriction $\delta<q$ does not play any role in the derivation of (9.2). This inequality holds for all $\delta\ge\max\{p;2\}$.

9-(i). Global iterative inequalities. The case $p>\max\left\{1;\frac{2N}{N+2}\right\}$

Next we assume that the numbers p and δ are in the range (8.2). We apply the multiplicative embedding inequality of Proposition 3.1 of Chap. I, and proceed as in the case of the local inequalities. This process is indeed simpler, since $(u-k_n)_+(\cdot,t)\in W_o^{1,p}(\Omega)$ for a.e. $t\in(0,T)$. Setting

(9.3) $$Y_n \equiv \int_{t_n}^{t}\int_{\Omega}(u-k_n)_+^{\delta}\,dxd\tau,$$

we obtain

(9.4) $$Y_{n+1} \le \frac{\gamma b^n|\Omega_t|^{\frac{\delta}{q}\frac{p}{N}}}{k^{\frac{\delta}{q}(q-\delta)}}\left(\frac{1}{(\sigma t)^{\frac{N+p}{N}}\,k^{\frac{N+p}{N}(\delta-2)}}\right)^{\frac{\delta}{q}} Y_n^{1+\frac{\delta}{q}\frac{p}{N}}$$
$$+\frac{\gamma b^n|\Omega_t|^{\frac{\delta}{q}\frac{p}{N}}}{k^{\frac{\delta}{q}(q-\delta)}}Y_n^{1+\frac{\delta}{q}\frac{p}{N}} + \gamma b^n|\Omega_t|^{\kappa\frac{\delta}{q}}\left(\frac{1}{k^{\delta}}Y_n\right)^{1+\kappa\frac{\delta}{q}},$$

where $b=2^{\delta\left(1+\frac{\delta}{q}\frac{p}{N}\right)}$. In these, the last two terms can be eliminated for solutions of equations with homogeneous structure as in (2.8)-(2.9). Moreover the last term can be eliminated if, in the structure conditions $(\mathbf{B}_1)-(\mathbf{B}_3)$, $\varphi_i\equiv 0$, $i=0,1,2$. If $\varphi_i\in L^\infty(\Omega_T)$, $i=0,1,2$, then $\kappa=p/N$.

Suppose now that the initial datum u_o in (2.7) is bounded above and let us take in (6.3)

$$k_n=\max\left\{\sup_{S_T} g\,;\ \sup_{\Omega} u_o\right\}+k-\frac{k}{2^n},\qquad n=0,1,2,\dots,$$

where $k>0$ is to be chosen. Then the first integral on the right hand side of (6.3) is zero and we may take $\zeta\equiv 1$. In such a case, we arrive at an inequality analogous to (9.1), where the first integral on the right hand side is eliminated and where the integrals are all extended over the whole Ω_t. Proceeding as above we find that the quantities

$$Y_n\equiv\iint_{\Omega_t}(u-k_n)_+^\delta\,dxd\tau$$

satisfy the recursive inequalities

$$\text{(9.5)}\qquad \begin{aligned} Y_{n+1}&\le\frac{\gamma b^n|\Omega_t|^{\frac{p}{N}\frac{\delta}{q}}}{k^{\frac{\delta}{q}(q-\delta)}}Y_n^{1+\frac{p}{N}\frac{\delta}{q}}+\gamma b^n|\Omega_t|^{\kappa\frac{\delta}{q}}\left(\frac{1}{k^\delta}Y_n\right)^{1+\kappa\frac{\delta}{q}}\\ b&=2^{\delta\left(1+\frac{\delta}{q}\frac{N}{p}\right)}.\end{aligned}$$

For equations with homogeneous structure, all the terms on the right hand side of (9.5) are zero.

10. Homogeneous structures and $1<p\le\max\left\{1;\frac{2N}{N+2}\right\}$

Let u be a non-negative local weak subsolution of (2.8)–(2.9) in Ω_T. We assume that u satisfies

$$\text{(10.1)}\qquad u\in L^r_{loc}(\Omega_T),\qquad \text{for some } r\ge 1 \text{ such that } \lambda_r>0.$$

The numbers λ_r have been introduced in (5.1). We also assume that u can be constructed as the weak limit in $L^r_{loc}(\Omega_T)$ of a sequence of bounded subsolutions of (2.8). By possibly working with such approximations we may assume that u is *qualitatively* locally bounded. Below, we will derive iterative inequalities similar to (8.3) but involving the L^r_{loc}–norms as well as local sup-bounds of u.

If $1<p<\max\left\{1;\frac{2N}{N+2}\right\}$, we have $q<2$. If (10.1) holds for some $r\in[1,2]$, then $\lambda_2>0$ and $p>\max\left\{1;\frac{2N}{N+2}\right\}$. Therefore it suffices to assume that (10.1) holds for some $r>2$. In such a case we have

$$\text{(10.2)}\qquad r>q,\qquad q\equiv p\frac{N+2}{N}.$$

In (7.6) we discard the last two terms in view of the homogeneous structure of (2.8) and, owing to Remark 7.1, set also $\delta = r$. We obtain

$$\sup_{-\theta_n<t<0} \int_{K_{\tilde{\rho}_n}} (u-k_{n+1})_+^2 (x,t)dx + \iint_{\tilde{Q}_n} |D(u-k_{n+1})_+|^p dxd\tau \tag{10.3}$$
$$\leq \frac{\gamma 2^{nr}}{(1-\sigma)^p}\left(\frac{1}{\rho^p k^{r-p}} + \frac{1}{\theta k^{r-2}}\right) \iint_{Q_n} (u-k_n)_+^r \, dxd\tau.$$

Define

$$Y_n = \iint_{Q_n} (u-k_n)_+^r \, dxd\tau, \qquad n = 0,1,2,\dots,$$

and estimate

$$Y_{n+1} \leq \|u\|_{\infty,Q(\theta,\rho)}^{r-q} \iint_{\tilde{Q}_n} u^q dxd\tau.$$

We majorise the right hand side by means of (7.7) and in turn estimate the right hand side of (7.7) by (10.3). We arrive at the recursive inequalities

$$Y_{n+1} \leq \frac{\gamma b^n}{(1-\sigma)^{\frac{p}{N}(N+p)}} \|u\|_{\infty,Q(\theta,\rho)}^{r-q} \mathcal{B}_k^{\frac{p}{N}} Y_n^{1+\frac{p}{N}}, \tag{10.4}$$

where

$$\mathcal{B}_k \equiv \left\{ \left(\frac{\theta}{\rho^p}\right) k^{-(r-p)\frac{N+p}{p}} + \left(\frac{\rho^p}{\theta}\right)^{\frac{N}{p}} k^{-(r-2)\frac{N+p}{p}} \right\}. \tag{10.5}$$

The recursive estimates (10.4)-(10.5) have a global version. Let u be a non-negative weak subsolution of the Dirichlet problem (2.5) and assume that

$$u \in L^\infty(0,T;L^r(\Omega)), \quad \text{for some } r>2 \text{ satisfying } \lambda_r>0.$$

Then the quantities Y_n defined in (9.3) with $\delta = r$ satisfy

$$Y_{n+1} \leq \frac{\gamma b^n |\Omega_t|^{\frac{p}{N}}}{k^{(r-2)\frac{N+p}{N}}} \|u\|_{\infty,\Omega\times(t_n,t)}^{r-q} \frac{1}{t^{\frac{N+p}{N}}} Y_n^{1+\frac{p}{N}}. \tag{10.6}$$

11. Proof of Theorems 3.1 and 3.2

The starting point in the proof of Theorem 3.1 is the inequality (8.3). We take $\theta = \rho^p$ and stipulate to choose $k \geq 1$. Recalling that $\max\{p;2\} \leq \delta < q$, the quantity $\mathcal{A}_k$ in (8.4) is majorised by 2. To simplify the presentation consider first the case

$$\varphi_i \in L^\infty(\Omega_T), \qquad i = 0, 1, 2,$$

so that we may take $\kappa = \frac{p}{N}$. With these choices, (8.3) yields

$$Y_{n+1} \le \frac{\gamma b^n}{k^{\frac{\delta}{q}(q-\delta)}} \left((1-\sigma)^{-(N+p)} + \left| Q\left(\rho^p, \rho\right) \right| \right)^{\frac{\delta}{q}\frac{p}{N}} \; Y_n^{1+\frac{\delta}{q}\frac{p}{N}}.$$

It follows from Lemma 4.2 of Chap. I that $Y_n \to 0$ as $n \to \infty$, provided k is chosen to satisfy

$$k = \max\{k_o\,;\, 1\},$$

where

$$Y_o = \iint_{Q(\rho^p,\rho)} u^\delta dx d\tau = C k_o^{(q-\delta)\frac{N}{p}} \left((1-\sigma)^{-(N+p)} + \left| Q\left(\rho^p, \rho\right) \right| \right)^{-1},$$

for a constant C depending only upon the data. This in turn implies

$$\sup_{Q(\sigma\rho^p,\sigma\rho)} u \le \gamma \left((1-\sigma)^{-(N+p)} + \left| Q\left(\rho^p, \rho\right) \right| \right)^{\frac{p}{N}\frac{1}{q-\delta}} \left(\iint_{Q(\rho^p,\rho)} u^\delta dx d\tau \right)^{\frac{p}{N}\frac{1}{q-\delta}} \wedge 1.$$

The general case of $\kappa \in (0, \frac{p}{N})$ is proved by a minor modification of these arguments. It suffices to rewrite (8.3) as

$$Y_{n+1} \le \frac{\gamma b^n}{k^{\frac{\delta}{q}(q-\delta)}} \left((1-\sigma)^{-(N+p)} + \left| Q\left(\rho^p, \rho\right) \right| \right)^{\frac{\delta}{q}\kappa} \left\{ Y_n^{1+\frac{\delta}{q}\frac{p}{N}} + Y_n^{1+\frac{\delta}{q}\kappa} \right\}$$

and follow the iteration process of Lemma 4.1 of Chap. I. To prove Theorem 3.2 we refer to the global recursive inequalities (9.4). As before, we take $k \ge 1$ and consider first the case of $\varphi_i \in L^\infty(\Omega_T)$, so that $\kappa = p/N$,. Choosing $\sigma = \frac{1}{2}$ we arrive at

$$Y_{n+1} \le \frac{\gamma b^n |\Omega_t|^{\frac{\delta}{q}\frac{p}{N}}}{k^{\frac{\delta}{q}(q-\delta)}} \left(1 + t^{-\frac{N+p}{p}} \right)^{\frac{\delta}{q}\frac{p}{N}} Y_n^{1+\frac{\delta}{q}\frac{p}{N}}.$$

It follows from Lemma 4.1 of Chap. I that $Y_n \to 0$ as $n \to \infty$, provided k is chosen to satisfy

$$Y_o = \iint_{\Omega_t} u^\delta dx d\tau = \frac{C k^{(q-\delta)\frac{N}{p}}}{|\Omega_t|} \left(1 + t^{-\frac{N+p}{p}} \right)^{-1}$$

for a constant C depending only upon the data. This in turn implies that for all $0 < t \le T$

$$\|u(\cdot, t)\|_{\infty,\Omega} \le \gamma \left(t + \frac{1}{t^{\frac{N}{p}}} \right)^{\frac{p}{N}\frac{1}{q-\delta}} \left(\int_0^t \!\! \int_\Omega u^\delta dx d\tau \right)^{\frac{p}{N}\frac{1}{q-\delta}} \wedge 1.$$

The general case of $\kappa \in (0, \frac{p}{N})$ is proved by a minor modification of these arguments. The second part of Theorem 3.2 is proved exactly the same way, by starting from the recursive inequalities (9.5).

12. Proof of Theorem 4.1

We refer back to the iterative inequalities (8.3) and discard the last two terms because of the homogeneous structure of the p.d.e. in (2.8)–(2.9). Since the resulting inequalities hold for all $p \leq \delta < p\frac{N+2}{N}$, we take $\delta = p$ and rewrite them as

$$Y_{n+1} = \frac{\gamma\ b^n}{(1-\sigma)^{p\frac{N+p}{N+2}}\ k^{\frac{2p}{N+2}}} \mathcal{A}_k^{\frac{p}{N+2}} Y_n^{1+\frac{p}{N+2}}, \tag{12.1}$$

where Y_n are defined in (8.1) and $\mathcal{A}_k$ are defined in (8.4) with $\delta = p$, i.e.,

$$\mathcal{A}_k = \left(\left(\frac{\theta}{\rho^p}\right) + \left(\frac{\rho^p}{\theta}\right)^{\frac{N}{p}} \frac{1}{k^{(p-2)\frac{N+p}{p}}} \right).$$

We stipulate to take k so large that of the two terms making up $\mathcal{A}_k$ the first dominates the second, i.e.,

$$k \geq \left(\frac{\rho^p}{\theta}\right)^{\frac{1}{p-2}} \tag{12.2}$$

so that

$$\mathcal{A}_k \leq 2\mathcal{A}, \qquad \mathcal{A} \equiv \frac{\theta}{\rho^p}.$$

It follows from Lemma 4.1 of Chap. I that $Y_n \to 0$ as $n \to \infty$ if we choose k from

$$Y_o \equiv \iint_{Q(\theta,\rho)} u^p dx d\tau = C\mathcal{A}^{-1}(1-\sigma)^{(N+p)} k^2,$$

where C is a constant depending only upon γ, b, N and p. For such a choice and (12.2),

$$\operatorname*{ess\,sup}_{Q(\sigma\theta,\sigma\rho)} u \leq \frac{\gamma\sqrt{\mathcal{A}}}{(1-\sigma)^{\frac{N+p}{2}}} \left(\iint_{Q(\theta,\rho)} u^p dx d\tau \right)^{\frac{1}{2}} \wedge \left(\frac{\rho^p}{\theta}\right)^{\frac{1}{p-2}}. \tag{12.3}$$

This estimate proves the theorem for $\varepsilon = 2$. Fix $\varepsilon \in (0, 2)$ and consider the increasing sequences

$$\rho_o \equiv \sigma\rho, \qquad \theta_o = \sigma\theta$$

and for $n = 1, 2, \ldots$

(12.4) $$\rho_n = \sigma\rho + (1-\sigma)\rho\sum_{i=1}^{n} 2^{-i}, \qquad \theta_n = \sigma\theta + (1-\sigma)\theta\sum_{i=1}^{n} 2^{-i},$$

and the corresponding cylinders $Q^{(n)} \equiv Q(\theta_n, \rho_n)$. By construction

(12.5) $$Q^{(o)} \equiv Q(\sigma\theta, \sigma\rho) \quad \text{and} \quad Q^{(\infty)} \equiv Q(\theta, \rho).$$

Set

(12.6) $$M_n = \operatorname*{ess\,sup}_{Q^{(n)}} u$$

and write (12.3) for the pair of boxes $Q^{(n)}$ and $Q^{(n+1)}$. This gives

$$M_n \le \frac{\gamma\, 2^{n\frac{N+p}{2}}\sqrt{\mathcal{A}}}{(1-\sigma)^{\frac{N+p}{2}}} \left(\iint_{Q^{(n+1)}} u^p dx d\tau \right)^{\frac{1}{2}} \wedge \mathcal{A}^{\frac{1}{2-p}}$$

$$\le M_{n+1}^{\frac{2-\varepsilon}{2}} \frac{\gamma\, 2^{n\frac{N+p}{2}}\sqrt{\mathcal{A}}}{(1-\sigma)^{\frac{N+p}{2}}} \left(\iint_{Q(\theta,\rho)} u^{p-2+\varepsilon} dx d\tau \right)^{\frac{1}{2}} \wedge \mathcal{A}^{\frac{1}{2-p}}.$$

If $\eta \in (0,1)$, the right hand side of this inequality is majorised by

$$\eta M_{n+1} + B d^n, \qquad d = 2^{\frac{N+p}{\varepsilon}},$$

where

$$B = \left(\frac{\gamma^2 \mathcal{A}}{\eta^{2-\varepsilon}(1-\sigma)^{(N+p)}} \iint_{Q(\theta,\rho)} u^{p-2+\varepsilon} dx d\tau \right)^{\frac{1}{\varepsilon}} \wedge \mathcal{A}^{\frac{1}{2-p}}.$$

Combining these estimates we arrive at the recursive inequalities

$$M_n \le \eta M_{n+1} + B d^n, \qquad n = 0, 1, 2, \ldots.$$

From these, by iteration

$$M_o \le \eta^n M_{n+1} + B d \sum_{i=0}^{n} (\eta d)^i, \qquad \forall n \in \mathbf{N}.$$

We choose $\eta = \frac{1}{2d}$ so that the sum on the right hand side can be majorised by a convergent series and let $n \to \infty$ to obtain

$$\sup_{Q(\sigma\theta,\sigma\rho)} u \le \frac{\gamma \mathcal{A}^{\frac{1}{\varepsilon}}}{(1-\sigma)^{\frac{N+p}{\varepsilon}}} \left(\iint_{Q(\theta,\rho)} u^{p-2+\varepsilon} dx d\tau \right)^{\frac{1}{\varepsilon}} \wedge \mathcal{A}^{\frac{1}{2-p}}.$$

13. Proof of Theorem 4.2

The proof of the theorem is a consequence of the following:

PROPOSITION 13.1. *Let u be a non-negative local sub-solution of (2.8)-(2.9) in Ω_T, and let $p>2$. There exists a constant $\gamma=\gamma\,(data)$, such that $\forall\,[(x_o,t_o)+Q\,(\theta,\rho)]\subset\Omega_T$ and $\forall\sigma\in(\frac{1}{2},1)$,*

$$\iint\limits_{[(x_o,t_o)+Q(\sigma\theta,\sigma\rho)]} u^p\,dxd\tau \le \frac{\gamma}{(1-\sigma)^{Np}}\left(\sup_{t_o-\theta<\tau<t_o}\ \fint\limits_{[x_o+K_\rho]} u(x,\tau)\,dx\right)^p \wedge \left(\frac{\rho^p}{\theta}\right)^{\frac{p}{p-2}}. \tag{13.1}$$

PROOF: We may assume that $(x_o,t_o)\equiv(0,0)$, and having fixed $\sigma\in(\frac{1}{2},1)$, consider the increasing sequences $\{\rho_n\}$ and $\{\theta\}_n$ introduced in (12.4) and the corresponding cylinders $Q^{(n)}$. Let $(x,t)\to\zeta_n$ be a non$-$negative piecewise smooth cutoff function in $Q^{(n+1)}$ that equals one on $Q^{(n)}$, vanishes on the parabolic boundary of $Q^{(n+1)}$ and such that

$$|D\zeta_n|\le\frac{2^{n+1}}{(1-\sigma)\rho},\qquad 0\le\zeta_{n,t}\le\frac{2^{n+1}}{(1-\sigma)\theta}.$$

The function $(u\zeta_n)(\cdot,t)$ vanishes on $\partial K_{\rho_{n+1}}$. Therefore by the embedding inequality (3.1) of Chap. I applied with $m=1$,

$$\iint\limits_{Q_{n+1}}(u\zeta_n)^{p\left(\frac{N+1}{N}\right)}dxd\tau \le\gamma\iint\limits_{Q_{n+1}}|Du\zeta_n|^p dxd\tau\left(\sup_{-\theta_{n+1}\le\tau<0}\ \int\limits_{K_{\rho_{n+1}}}u(x,\tau)\,dx\right)^{p/N}. \tag{13.2}$$

The constant γ depends only upon the data and it is independent of ρ, θ and n. The energy estimates for solutions of (2.8) give

$$\begin{aligned}\iint\limits_{Q_{n+1}}|Du\zeta_n|^p dxd\tau &\le\frac{\gamma 2^{np}}{(1-\sigma)^p\rho^p}\iint\limits_{Q_{n+1}}u^p dxd\tau+\frac{\gamma 2^n}{(1-\sigma)\theta}\iint\limits_{Q_{n+1}}u^2dxd\tau\\ &\le\gamma\frac{2^{np}}{(1-\sigma)^p\theta}\left[\left(\frac{\theta}{\rho^p}\right)\iint\limits_{Q_{n+1}}u^p dxd\tau\wedge\left(\frac{\rho^p}{\theta}\right)^{\frac{2}{p-2}}\right],\end{aligned}$$

where we have estimated the second integral by Hölder inequality. Without loss of generality we may assume that

$$\left(\frac{\theta}{\rho^p}\right) \iint\limits_{Q_{n+1}} u^p \, dx d\tau > \left(\frac{\rho^p}{\theta}\right)^{\frac{2}{p-2}}, \qquad \text{for all } n = 0, 1, 2, \ldots,$$

otherwise the Proposition becomes trivial. Combining these remarks with (13.2) and setting

$$X_n \equiv \iint\limits_{Q_n} u^p \, dx d\tau,$$

we obtain the recursive inequalities

$$X_n \equiv \iint\limits_{Q_n} u^p \, dx d\tau \leq \iint\limits_{Q_{n+1}} (u\zeta_n)^p \, dx d\tau \leq \left(\iint\limits_{Q_{n+1}} (u\zeta_n)^{p\frac{N}{N+1}} \, dx d\tau \right)^{\frac{N}{N+1}}$$

$$\leq \frac{\gamma 2^{np\frac{N}{N+1}}}{(1-\sigma)^{p\frac{N}{N+1}}} X_{n+1}^{\frac{N}{N+1}} \left(\sup_{-\theta<\tau<0} \fint_{K_\rho} u(x,\tau) \, dx \right)^{\frac{p}{N+1}}.$$

By the interpolation Lemma 4.3 of Chap. I, we conclude that there exists a constant γ, depending only upon the data such that

$$\iint\limits_{Q(\sigma\theta,\sigma\rho)} u^p \, dx d\tau \leq \frac{\gamma}{(1-\sigma)^{Np}} \left(\sup_{-\theta<\tau<0} \fint_{K_\rho} u(x,\tau) \, dx \right)^p.$$

14. Proof of Theorem 4.3

We may assume that the boundary datum is non-positive, by possibly replacing u with

$$w = u - \sup_{S_T} g.$$

We start from the global iterative inequalities (9.4) and discard the last two terms on the right hand side since the p.d.e. has the homogeneous structure (2.8)-(2.9). Taking $\delta = p$ we obtain

$$Y_{n+1} \leq \frac{\gamma b^n \left|\Omega_t\right|^{\frac{p}{N+2}}}{(\sigma t)^{\frac{N+p}{N+2}} k^{\frac{\mu}{N+2}}} Y_n^{1+\frac{p}{N+2}},$$

where Y_n are defined in (9.3) and

$$\mu = N(p-2) + p^2. \tag{14.1}$$

It follows from Lemma 4.1 of Chap. I that $Y_n \to 0$ as $n \to \infty$ if k is chosen from

$$Y_o = \iint_{\Omega_t} u^p dx d\tau \le C\,(\sigma t)^{\frac{N+p}{p}}\, |\Omega_t|^{-1} k^{\mu/p},$$

for a constant C depending only upon γ, b, p and N. Thus for all $\sigma t < \tau \le t$

$$\text{(14.2)} \qquad \|u(\cdot,\tau)\|_{\infty,\Omega} \le \frac{\gamma}{\sigma^{(N+p)/\mu} t^{N/\mu}} \left(\int_0^t\!\!\!\!\!\!-\int_\Omega u^p dx d\tau \right)^{p/\mu}.$$

Consider the decreasing sequence of time levels

$$t_n = \frac{t}{2} - \frac{t}{4} \sum_{i=0}^{n} 2^{-i},$$

and apply (14.2) over the expanding domains $\Omega \times \{t_{n+1}, t\}$, with σ taken from

$$\sigma(t - t_{n+1}) = t_n, \quad \text{i.e.,} \quad \sigma = \frac{1 - \sum_{i=0}^{n} 2^{-(i+1)}}{1 + \sum_{i=0}^{n+1} 2^{-(i+1)}} \ge 2^{-(n+1)}.$$

Setting also

$$M_n = \sup_{t_n < \tau < t} \|u(\cdot,\tau)\|_{\infty,\Omega},$$

we obtain from (14.2)

$$M_n \le \frac{\gamma d^n}{t^{N/\mu}} \left(\int_{t_{n+1}}^t\!\!\!\!\!\!-\int_\Omega u^p dx d\tau \right)^{p/\mu}$$

$$\le \frac{\gamma d^n}{t^{N/\mu}}\, M_{n+1}^{p(p-1)/\mu} \left(\int_0^t\!\!\!\!\!\!-\int_\Omega u\, dx d\tau \right)^{p/\mu},$$

where $d = 2^{(N+p)/\mu}$. By the interpolation Lemma 4.3 of Chap. I we conclude that

$$M_o = \sup_{\frac{t}{2} < \tau < t} \|u(\cdot,\tau)\|_{\infty,\Omega} \le \frac{\gamma}{t^{N/\lambda}} \left(\int_0^t\!\!\!\!\!\!-\int_\Omega u\, dx d\tau \right)^{p/\lambda}, \qquad \lambda = N(p-2) + p.$$

15. Proof of Theorem 4.5

Even though the theorem is of global nature, our starting point is the recursive inequality (12.1). We begin by observing that in the proof of Theorem 4.1 the choice of $k \geq 1$ was made to guarantee that $\mathcal{A}_k$ could be majorised by a quantity independent of k. Here we stipulate to choose k satisfying

$$k \geq \frac{1}{2} \sup_{K_{\sigma\rho}} u(\cdot, 0),$$

and in (12.1)[1] replace $\mathcal{A}_k$ with the larger quantity

$$2\mathcal{A}_* = \left(\frac{\theta}{\rho^p}\right) + \left(\frac{\rho^p}{\theta}\right)^{\frac{N}{p}} \left(\frac{2}{\sup_{K_{\sigma\rho}} u(\cdot, 0)}\right)^{(p-2)\frac{N+p}{p}}.$$

The numbers ρ and σ being fixed, we let θ be so small that, of the two terms making up $\mathcal{A}_*$, the second dominates the first, i.e.,

$$\theta \leq \rho^p \left(\frac{2}{\sup_{K_{\sigma\rho}} u(\cdot, 0)}\right)^{p-2}. \tag{15.1}$$

The knowledge of such a θ at this stage is only qualitative. It is part of the proof to give an upper estimate for all the positive numbers θ for which (15.1) is verified.

With these choices, the recursive inequalities (12.1) imply

$$Y_{n+1} = \frac{\gamma b^n \, k^{-\frac{2p}{N+2}}}{(1-\sigma)^{p\frac{N+p}{N+2}}} \mathcal{A}_*^{\frac{p}{N+2}} Y_n^{1+\frac{p}{N+2}}.$$

We proceed now as before and arrive at an analog of (12.3); namely, there exists a constant γ dependent only upon the data such that for all $\sigma \in (0,1)$

$$\sup_{Q(\sigma\theta, \sigma\rho)} u \leq \frac{\gamma\sqrt{\mathcal{A}_*}}{(1-\sigma)^{\frac{N+p}{2}}} \left(\iint_{Q(\theta,\rho)} u^p \, dx d\tau\right)^{1/2} \wedge \left(\frac{\rho}{\theta}\right)^{\frac{1}{p-2}}. \tag{15.2}$$

If θ and $\sup_{K_{\sigma\rho}} u(\cdot, 0)$ satisfy (15.1), it follows from (15.2) and the indicated choices,

$$\sup_{K_{\sigma\rho}} u(\cdot, 0) \leq \frac{\gamma \left(\rho^p/\theta\right)^{N/2p}}{(1-\sigma)^{\frac{N+p}{2}}} \left(\sup_{K_{\sigma\rho}} u(\cdot, 0)\right)^{(2-p)\frac{N+p}{2p}} \left(\iint_{Q(\theta,\rho)} u^p \, dx d\tau\right)^{1/2}.$$

Therefore for $\mu = N(p-2) + p^2$,

$$\sup_{K_{\sigma\rho}} u(\cdot, 0) \leq \frac{\gamma}{(1-\sigma)^{\frac{p}{\mu}(N+p)}} \left(\frac{\rho^p}{\theta}\right)^{N/\mu} \left(\iint_{Q(\theta,\rho)} u^p \, dx d\tau\right)^{p/\mu}. \tag{15.3}$$

---

[1] The inequalities (12.1) are written over the cylinders $Q(\theta_n, \rho_n)$ introduced at the beginning of §7.

This inequality holds for all θ, ρ, σ for which (15.1) is verified. It also holds for any pair of boxes

$$[(x_o, t_o) + Q(\theta, \rho)] \quad \text{and} \quad [(x_o, t_o) + Q(\sigma\theta, \sigma\rho)],$$

with arbitrary 'vertices' provided they are contained in Σ_T. Fix any $t \in (0, T)$ and introduce the boxes

$$K_\rho \times \{\tfrac{1}{2}t, t\} \qquad \text{and} \qquad K_{\rho/2} \times \{\tfrac{3}{4}t, t\}.$$

We rewrite (15.3) and (15.1) in terms of these cylinders, for which $\sigma = \frac{1}{2}$.

LEMMA 15.1. *For all $t \in (0, T)$ and $\rho > 0$ for which*

$$t \le 2^{p-1}\rho^p \left(\sup_{K_{\rho/2}} u(x,t)\right)^{-(p-2)}, \tag{15.4}$$

there holds

$$\sup_{K_{\rho/2}} u(x,t) \le \gamma \left(\frac{\rho^p}{t}\right)^{N/\mu} \left(\fint_{t/2}^{t}\fint_{K_\rho} u^p \, dx d\tau\right)^{p/\mu}. \tag{15.5}$$

For $r > 0$ introduce the quantity

$$f(t) = \sup_{0<\tau<t} \left\{\tau^{N/\lambda} \sup_{\rho \ge r} \frac{\|u(\cdot,\tau)\|_{\infty,K_\rho}}{\rho^{\frac{p}{p-2}}}\right\}, \quad \lambda = N(p-2) + p. \tag{15.6}$$

By possibly working within the time interval (ε, T) and then letting $\varepsilon \searrow 0$, we may assume that $f(t)$ is finite. This follows from Theorem 4.4. Let $t^* \in (0, T)$ be the largest time level for which

$$t^{p/\lambda} \le 2^p [f(t)]^{-(p-2)}, \qquad \forall 0 < t \le t^*. \tag{15.7}$$

The knowledge of t^* is only qualitative. Shortly we will find a quantitative upper bound for t^*. Here we remark that owing to the definition of $f(t)$ the condition (15.4) holds for all $\rho > r$ and all $t \in (0, t^*]$. Consequently (15.5) holds for all $t \in (0, t^*]$. We estimate the integral on the right hand side of (15.5) as follows

$$\begin{aligned}
\fint_{t/2}^{t}\fint_{K_\rho} u^p \, dx d\tau &\le \rho^{\frac{p^2}{p-2}} \fint_{t/2}^{t} \left(\frac{\|u(\cdot,\tau)\|_{\infty,K_\rho}}{\rho^{\frac{p}{p-2}}}\right)^{p-1} \int_{K_\rho} \frac{u(x,\tau)}{\rho^{N+\frac{p}{p-2}}} dx d\tau \\
&\le \left(\frac{2}{t}\right)^{\frac{N(p-1)}{\lambda}} \rho^{\frac{p^2}{p-2}} \left\{\sup_{0<\tau<t} \tau^{N/\lambda} \sup_{\rho \ge r} \frac{\|u(\cdot,\tau\|_{\infty,K_\rho}}{\rho^{\frac{p}{p-2}}}\right\}^{p-1} \\
&\qquad \times \left\{\sup_{0<\tau<t} \sup_{\rho\ge r} \int_{K_\rho} \frac{u(x,\tau)}{\rho^{N+\frac{p}{p-2}}} dx d\tau\right\} \\
&= \left(\frac{2}{t}\right)^{\frac{N(p-1)}{\lambda}} \rho^{\frac{p^2}{p-2}} f^{p-1}(t) \, |||u|||_{\{r,t\}}
\end{aligned}$$

where the norm $\|\!|\cdot|\!\|_{\{r,t\}}$ is defined in (4.4). Putting this estimate in (15.5) gives

$$\sup_{K_{\rho/2}} u(x,t) \leq \gamma \frac{\rho^{\frac{p}{p-2}}}{t^{N/\lambda}} f(t)^{p(p-1)/\mu} \|\!|u|\!\|_{\{r,t\}}^{p/\mu}.$$

We divide by $(\rho/2)^{p/(p-2)}$ and multiply by $t^{N/\lambda}$. Then take the supremum for $\rho > r$ and use the fact that $t \in (0, t^*)$ is arbitrary to deduce

$$f(t) \leq \gamma f(t)^{p(p-1)/\mu} \|\!|u|\!\|_{\{r,t\}}^{p/\mu}, \qquad \forall 0 < t \leq t^*,$$

i.e.,

$$f(t) \leq \gamma \|\!|u|\!\|_{\{r,t\}}^{p/\lambda}, \qquad \forall 0 < t \leq t^*. \tag{15.8}$$

Thus it follows from (15.7) that (15.8) continues to hold for all $0 < t \leq t_*$, where

$$t_* = \gamma_* \|\!|u|\!\|_{\{r,t^*\}}^{-(p-2)}.$$

16. Proof of Theorems 5.1 and 5.2

We first prove Theorem 5.1 for the case when the assumptions (5.1)-(5.2) hold for some $1 \leq r \leq 2$. In such a case we have $p > \max\left\{1\,;\,\frac{2N}{N+2}\right\}$, and we may use the iterative estimates (8.3). In these we discard the last two terms and take $\delta = 2$. We also stipulate to take

$$k \geq \frac{1}{2} \sup_{Q(\sigma\theta,\sigma\rho)} u$$

and arrive at

$$Y_{n+1} \leq \frac{\gamma b^n}{(1-\sigma)^{p\frac{\delta}{q}\frac{N+p}{N}} k^{\frac{\delta}{q}(q-\delta)}} \mathcal{A}_\sigma^{\frac{\delta}{q}\frac{p}{N}} Y_n^{1+\frac{\delta}{q}\frac{p}{N}}, \qquad \delta = 2,$$

where

$$\mathcal{A}_\sigma = \left\{ \left(\frac{\theta}{\rho^p}\right) \left[\sup_{Q(\sigma\theta,\sigma\rho)} u \right]^{(p-\delta)\frac{N+p}{p}} + \left(\frac{\rho^p}{\theta}\right)^{\frac{N}{p}} \right\}, \qquad \delta = 2,$$

and Y_n are defined in (8.1). By Lemma 4.1 of Chap. I, $Y_n \to 0$ as $n \to \infty$, provided we choose k from

$$Y_o \equiv \iint_{Q(\theta,\rho)} u^2 dx d\tau = C\,(1-\sigma)^{N+p} \mathcal{A}_\sigma^{-1} k^{\frac{N}{p}(q-\delta)},$$

for a constant C depending only upon the data. This implies

$$(16.1)\qquad \sup_{Q(\sigma\theta,\sigma\rho)} u \le \frac{\gamma\mathcal{A}_\sigma^{\frac{p}{N(q-\delta)}}}{(1-\sigma)^{\frac{p(N+p)}{N(q-\delta)}}}\left(\iint_{Q(\theta,\rho)} u^2 dxd\tau\right)^{\frac{p}{N(q-\delta)}}, \qquad \delta=2.$$

We conclude the proof for the case $r\in[1,2]$ by means of an interpolation process similar to that of Lemma 4.3 of Chap. I; namely, consider the sequences ρ_n, θ_n and the corresponding cylinders $Q^{(n)}\equiv Q(\theta_n,\rho_n)$, introduced in (12.4)-(12.5). Define also the numbers M_n as in (12.6), and write (16.1) for the pair of boxes $Q^{(n)}$ and $Q^{(n+1)}$. This gives

$$(16.2)\qquad M_n \le \frac{\gamma 2^{np\frac{N+p}{N(q-\delta)}}\mathcal{A}_n^{\frac{p}{N(q-\delta)}}}{(1-\sigma)^{p\frac{N+p}{N(q-\delta)}}}\left(\iint_{Q^{(n+1)}} u^2 dxd\tau\right)^{\frac{p}{N(q-\delta)}}$$

where

$$\mathcal{A}_n = \left\{\left(\frac{\theta}{\rho^p}\right)M_n^{(p-\delta)\frac{N+p}{p}} + \left(\frac{\rho^p}{\theta}\right)^{\frac{N}{p}}\right\}.$$

Consider the two terms making up $\mathcal{A}_n$. If for some $n=0,1,2,\dots$ the first term dominates the second, we have

$$(16.3)\qquad M_n \le \left(\frac{\theta}{\rho^p}\right)^{1/(2-p)}$$

and there is nothing to prove. Otherwise, (16.3) fails for all $n=0,1,2,\dots$ and

$$\mathcal{A}_n \le 2\left(\frac{\rho^p}{\theta}\right)^{\frac{N}{p}}, \qquad n=0,1,2,\dots.$$

We deduce from (16.2)

$$M_n \le \frac{\gamma\, 2^{np\frac{N+p}{N(q-\delta)}}}{(1-\sigma)^{p\frac{N+p}{N(q-\delta)}}} M_{n+1}^{\frac{p(2-r)}{N(q-2)}}\left(\frac{\rho^p}{\theta}\right)^{\frac{1}{q-2}}\left(\iint_{Q^{(n+1)}} u^r dxd\tau\right)^{\frac{p}{N(q-2)}}.$$

The proof is now concluded as in Lemma 4.3 of Chap. I.

The proof of Theorem 5.1 for the case $r>2$ is based on the recursive inequalities (10.4). As before, we stipulate to take

$$k \ge \frac{1}{2}\sup_{Q(\sigma\theta,\sigma\rho)} u.$$

and majorise $\mathcal{B}_k$ by

$$\mathcal{B}_k \le k^{(2-r)\frac{N+p}{p}}\mathcal{B}_\sigma,$$

where

$$\mathcal{B}_\sigma = \left\{ \left(\frac{\theta}{\rho^p}\right) \left[\sup_{Q(\sigma\theta,\sigma\rho)} u \right]^{(p-2)\frac{N+p}{p}} + \left(\frac{\rho^p}{\theta}\right)^{\frac{N}{p}} \right\}.$$

With these choices, we obtain from (10.4) the recursive inequalities

$$Y_{n+1} \le \frac{\gamma b^n \mathcal{B}_\sigma^{\frac{p}{N}}}{(1-\sigma)^{\frac{p}{N}(N+p)} k^{(r-2)\frac{N+p}{N}}} \|u\|_{\infty,Q(\theta,\rho)}^{r-q} Y_n^{1+\frac{p}{N}}.$$

By Lemma 4.1 of Chap. I, $Y_n \to 0$ as $n \to \infty$, provided

$$Y_o \equiv \iint_{Q(\theta,\rho)} u^r \, dx d\tau = C(1-\sigma)^{N+p} \mathcal{B}_\sigma^{-1} k^{(r-2)\frac{N+p}{p}} \|u\|_{\infty,Q(\theta,\rho)}^{(q-r)\frac{N}{p}},$$

for a constant C depending only upon the data. Thus

$$(16.4) \qquad \sup_{Q(\sigma\theta,\sigma\rho)} u$$

$$\le \frac{\gamma}{(1-\sigma)^{\frac{p}{r-2}}} \mathcal{B}_\sigma^{\frac{p}{(r-2)(N+p)}} \|u\|_{\infty,Q(\theta,\rho)}^{\frac{r-q}{r-2}\frac{N}{N+p}} \left(\iint_{Q(\theta,\rho)} u^r \, dx d\tau \right)^{\frac{p}{(r-2)(N+p)}}.$$

Let $Q^{(n)} \equiv Q(\theta_n, \rho_n)$ and M_n be defined as in (12.4)-(12.6). Then from (16.4)

$$M_n \le \frac{\gamma}{(1-\sigma)^{\frac{p}{r-2}}} \mathcal{B}_n^{\frac{p}{(r-2)(N+p)}} M_{n+1}^{\frac{r-q}{r-2}\frac{N}{N+p}} \left(\iint_{Q(\theta,\rho)} u^r \, dx d\tau \right)^{\frac{p}{(r-2)(N+p)}},$$

where

$$\mathcal{B}_n = \left\{ \left(\frac{\theta}{\rho^p}\right) M_n^{(p-2)\frac{N+p}{p}} + \left(\frac{\rho^p}{\theta}\right)^{\frac{N}{p}} \right\}.$$

The proof is now concluded as in the case $r \in [1,2]$.

The proof of Theorem 5.2 is essentially the same. If $r \in [1,2]$, it follows from the recursive inequalities (9.4). If $r > 2$, we start from the global inequalities (10.6).

17. Natural growth conditions

Consider the Dirichlet problem

$$(17.1) \qquad \begin{cases} u_t - \operatorname{div} |Du|^{p-2} Du = |Du|^p, & \text{in } \Omega_T, \quad p>1, \\ u\big|_\Gamma = f \in L^\infty(\Gamma), \end{cases}$$

where Γ denotes the parabolic boundary of Ω_T. The lower order term has the 'natural' or Hadamard growth condition with respect to $|Du|$ (see [48]). The notion of weak solution is that of §2 of Chap. II. Here we stress that if we merely require that $|Du| \in L^p(\Omega_T)$, the testing functions must be bounded to account for the growth of the right hand side.

The problem we address here is that of finding a sup-bound for a solution u. It is known that weak solutions of (17.1) in general are not bounded, not even in the elliptic case (see [15]). This is due to the fast growth of the right hand side with respect to $|Du|$. On the other hand the existence theory is based on constructing solutions as limits, in some appropriate topology, of bounded solutions of some sequence of approximating problems. The limiting process is possible if one can find a uniform upper bound on the approximating solutions. Therefore the main problem regarding sup–estimates for solutions of (17.1) can be formulated as follows. Assuming that a weak solution u of (17.1) is *qualitatively* bounded, find a *quantitative* $L^\infty(\Omega_T)$ estimate, depending only upon the data. In such a form, the problem was first formulated by Stampacchia [93] in the context of elliptic equations.

THEOREM 17.1. *Let u be a bounded weak solution of (17.1). Then*

$$\|u\|_{\infty,\Omega_T} \leq F \equiv \|f\|_{\infty,\Gamma}. \tag{17.2}$$

PROOF: By working with u_+ and u_- separately, we may assume that u is non-negative. Set

$$M = \operatorname*{ess\,sup}_{\Omega_T} u.$$

If $M > F$, in the weak formulation of (17.1) we take the testing functions

$$(u-k)_+, \qquad \text{where} \quad k = M - \varepsilon \geq F, \qquad \text{for some } \varepsilon > 0,$$

modulo a Steklov averaging process. These are admissible since they vanish on the parabolic boundary of Ω_T and are bounded. We obtain

$$\begin{aligned} \operatorname*{ess\,sup}_{0<\tau<T} \int_{\Omega\times\{\tau\}} (u-k)_+^2\, dx + \iint_{\Omega_T} |D(u-k)_+|^p dx d\tau &\leq \iint_{\Omega_T} |D(u-k)_+|^p (u-k)_+\, dx d\tau \\ &\leq \varepsilon \iint_{\Omega_T} |D(u-k)_+|^p dx d\tau. \end{aligned}$$

Thus if $\varepsilon \in (0,1)$, we have $(u-k)_+ \equiv 0$ in Ω_T and

$$\operatorname*{ess\,sup}_{\Omega_T} u \leq M - \varepsilon.$$

This contradicts the definition of M and proves the theorem.

COROLLARY 17.1. *Assume $f \equiv 0$. Then (17.1) does not have any non–trivial bounded weak solution.*

17-(i). General structures

More generally we may consider the Dirichlet problem

$$
(17.3)\qquad \begin{cases} u \in C\left(0,T;L^2(\Omega)\right)\cap L^p\left(0,T;W^{1,p}(\Omega)\right), \\ u_t - \operatorname{div}\mathbf{a}(x,t,u,Du) = b(x,t,u,Du) \text{ in } \Omega_T, \\ u\big|_\Gamma = f \in L^\infty(\Gamma), \end{cases}
$$

where the p.d.e. satisfies the structure conditions

$$
(\mathbf{B}_1^*)\qquad \mathbf{a}(x,t,u,Du)\cdot Du \geq C_o|Du|^p - \varphi_o(x,t),
$$

$$
(\mathbf{B}_3^*)\qquad |b(x,t,u,Du)| \leq C_2|Du|^p + \varphi_2(x,t).
$$

The lower order terms have the Hadamard 'natural' growth condition. Here C_i, $i=0,2$, are given positive constants and the non-negative functions φ_i, $i=0,2$, satisfy

$$
(\mathbf{B}_5^*)\qquad \Phi_o = \varphi_o + \varphi_2 \in L^{\hat{q}}(\Omega_T),
$$

where

$$
(\mathbf{B}_6)\qquad \frac{1}{\hat{q}} = (1-\kappa_o)\frac{p}{N+p}, \qquad \kappa_o \in (0,1].
$$

THEOREM 17.2. *Let u be a qualitatively bounded weak solution of (17.3) in Ω_T. There exists a constant C that can be determined quantitatively a priori only in terms of the data, such that*

$$
\|u\|_{\infty,\Omega_T} \leq \max\left\{2\|f\|_{\infty,\Gamma};\ C\|u\|_{p,\Omega_T}\right\}.
$$

PROOF: As before we may assume that u is non-negative. If M is the essential supremum of u in Ω_T, we may assume that $M > 2\|f\|_{\infty,\Gamma}$; otherwise there is nothing to prove. In the weak formulation of (17.3), we take the testing function $(u-k)_+$, where

$$
\|f\|_{\infty,\Gamma} \leq k < M.
$$

This is admissible, modulo a Steklov averaging process, since it is bounded and it vanishes in the sense of the traces on the parabolic boundary of Ω_T. Calculations in all analogous to those in §4-(ii) of Chap. II give

$$
(17.4)\qquad \sup_{0<t<T}\int_{\Omega\times\{t\}} (u-k)_+^2\,dx + \frac{C_o}{2}\iint_{\Omega_T} |D(u-k)_+|^p dxd\tau
$$

$$
\leq C_2\iint_{\Omega_T} |D(u-k)_+|^p (u-k)_+\,dxd\tau
$$

$$
+\gamma\iint_{\Omega_T}\left\{\varphi_o\chi\left[u>k\right] + \varphi_2(u-k)_+\right\} dxd\tau.
$$

Here and in what follows we denote with γ a generic positive constant that can be determined a priori only in terms of the data. Next choose $k = M - 2\varepsilon$ where $\varepsilon \in (0,1)$ is so small that $M - 2\varepsilon \geq \|f\|_{\infty,\Gamma}$, and

$$C_2 \iint_{\Omega_T} |D(u-k)_+|^p (u-k)_+ \, dx d\tau \leq 2C_2\varepsilon \iint_{\Omega_T} |D(u-k)_+|^p dx d\tau \leq \frac{C_o}{4} \iint_{\Omega_T} |D(u-k)_+|^p dx d\tau.$$

Thus we may take

$$2\varepsilon = \min\left\{ \|f\|_{\infty,\Gamma};\ \frac{1}{4}\frac{C_o}{C_2} \right\}.$$

Combining these calculations in (17.4), we arrive at

$$\sup_{0<t<T} \int_{\Omega \times \{t\}} (u-k)_+^2 \, dx + \iint_{\Omega_T} |D(u-k)_+|^p dx d\tau \leq \gamma \iint_{\Omega_T} \Phi_o \chi\,[u>k]\, dx d\tau.$$

By Hölder inequality and $(\mathbf{B}_5^*)$-$(\mathbf{B}_6)$ the last term is majorised by

$$\gamma \|\Phi_o\|_{\hat{q},\Omega_T} |\tilde{A}_k|^{\frac{N}{N+p}(1+\kappa)}, \qquad \kappa = \kappa_o \frac{p}{N},$$

where γ is a constant depending only upon the data and

$$\tilde{A}_k \equiv \{(x,t) \in \Omega_T \mid u(x,t) > k\}.$$

Consider the sequence of increasing levels

$$k_n = M - \varepsilon - \frac{\varepsilon}{2^n}, \qquad n = 0,1,2,\ldots,$$

and the corresponding family of sets

$$A_n \equiv \{(x,t) \in \Omega_T \mid u(x,t) > k_n\}.$$

These remarks imply that for all $n \in \mathbf{N}$

$$\sup_{0<t<T} \int_{\Omega \times \{t\}} (u-k_n)_+^2 \, dx + \iint_{\Omega_T} |D(u-k_n)_+|^p dx d\tau \leq \gamma |A_n|^{\frac{N}{N+p}(1+\kappa)},$$

for a constant γ depending only upon the data. From this and the multiplicative inequality of Proposition 3.1 of Chap. I,

$$\left(\frac{\varepsilon}{2^{n+1}}\right)^{p\frac{N+2}{N}}\left|A_{n+1}\right| \leq \iint\limits_{[u>k_{n+1}]} (u-k_n)_+^{p\frac{N+2}{N}}\,dxd\tau$$

$$\leq \gamma\left(\sup_{0<t<T}\int\limits_{\Omega}(u-k_n)_+^2\,dx\right)^{\frac{p}{N}}\iint\limits_{\Omega_T}|D(u-k_n)_+|^p dxd\tau$$

$$\leq \gamma\left|A_n\right|^{1+\kappa},$$

i.e., for all $n=0,1,2,\ldots$,

$$\left|A_{n+1}\right| \leq \gamma b^n \varepsilon^{-p\frac{N+p}{N}}\left|A_n\right|^{1+\kappa}, \qquad b=2^{p\frac{N+2}{N}}.$$

It follows from Lemma 4.1 of Chap. I that $\left|A_n\right| \to 0$ as $n\to\infty$ if

$$\left|A_o\right| \leq \gamma^* \equiv \left(\frac{\varepsilon^{p\frac{N+p}{N}}}{\gamma}\right)^{1/\kappa} b^{1/\kappa^2}.$$

In this case we would have

$$u \leq M-\varepsilon \quad \text{a.e. } \Omega_T$$

which contradicts the definition of M. Now

$$\left(\frac{M}{2}\right)^p\left|A_o\right| \leq \left(\frac{M}{2}\right)^p\left|A_{M/2}\right| \leq \iint\limits_{\Omega_T}|u|^p dxd\tau$$

i.e.,

$$\left|A_o\right| \leq \left(\frac{2}{M}\right)^p\int\limits_{\Omega}|u|^p dxd\tau.$$

If the right hand side is less than γ^* we have a contradiction. Thus

$$\sup_{\Omega_T} u \leq 2(\gamma^*)^{1/p}\|u\|_{p,\Omega_T}.$$

To prove that $\|u\|_{p,\Omega_T}$ is bounded above only in terms of the data, we may assume, modulo a shift that involves the supremum of the boundary data, that u is a bounded non-negative weak solution of (17.3) vanishing on Γ in the sense of the traces. In the weak formulation of (17.3), take the testing function

$$\varphi=(e^{\alpha u}-1)$$

where α is a positicve parameter to be chosen. We may also assume without loss of generality that $u_t \in L^2(\Omega_T)$. We obtain

$$\int_o^t \frac{\partial}{\partial t} \int_\Omega \left(\int_o^u (e^{\alpha s} - 1)\, ds \right) dx d\tau + \alpha \iint_{\Omega_t} |Du|^p e^{\alpha u} dx d\tau$$
$$\leq C_2 \iint_{\Omega_t} |Du|^p e^{\alpha u} dx d\tau + \gamma \iint_{\Omega_t} (1 + \Phi_o)\, e^{\alpha u} dx d\tau,$$

for a constant $\gamma = \gamma\,(\text{data})$ and for all $t \in (0, T]$. We choose $\alpha = 2C_2$ and set

$$w = e^{\frac{\alpha}{p} u} - 1,$$

to obtain

$$\|w\|^p_{V_p(\Omega_T)} \leq \gamma_o + \gamma_1 \iint_{\Omega_T} (1 + \Phi_o)\, w^p dx d\tau,$$

for two constants $\gamma_i = \gamma_i(\text{data})$, $i = 0, 1$. Next by $(B_5^*) - (B_6)$,

$$\iint_{\Omega_T} (1 + \Phi_o)\, w^p dx d\tau$$
$$\leq \left(\iint_{\Omega_T} (1 + \Phi_o)^{\hat{q}}\, dx d\tau \right)^{\frac{1}{\hat{q}}} \left(\iint_{\Omega_T} w^{\frac{p(N+p)}{N(1+\kappa)}} dx d\tau \right)^{\frac{N(1+\kappa)}{N+p}}$$
$$\leq \gamma = \gamma\,(\text{data}) \left| \Omega_T \right|^{\frac{N\kappa}{N+p}} \|w\|^p_{p\frac{N+p}{N}, \Omega_T}.$$

Moreover since $w(\cdot, t)$ vanishes om $\partial\Omega$ for a.e. $t \in (0, T)$, by the embedding of Proposition 3.1 Chap. I,

$$\|w\|^p_{p\frac{N+p}{N}, \Omega_T} \leq \gamma \|w\|^p_{V_p(\Omega_T)}, \quad \gamma = \gamma\,(\text{data}).$$

Combining these remarks in (17.5) we conclude that there exist two constants $\mathcal{C}_o, \mathcal{C}_1$, depending only upon the data such that

$$\|w\|^p_{p\frac{N+p}{N}, \Omega_T} \leq \mathcal{C}_o + \mathcal{C}_1 \left| \Omega_T \right|^{\frac{N\kappa}{N+p}} \|w\|^p_{p\frac{N+p}{N}, \Omega_T}.$$

If T is so small that, say

$$\mathcal{C}_1 \left| \Omega_T \right|^{\frac{N\kappa}{N+p}} \leq 1/2$$

then

$$\|w\|^p_{p\frac{N+p}{N}, \Omega_T} \leq 2\mathcal{C}_o.$$

For arbitrary $T > 0$, the argument can be repeated up to covering the whole Ω_T in a finite number of steps.

18. Bibliographical notes

The sup-bounds of §3 are essentially due to Porzio [87]. They follow a parabolic version of DeGiorgi iteration technique (see [67]) and remain valid even in the 'linear' case $p=2$. An effort has been made to trace the dependence of the various constants upon the size of the domains where the estimates are derived. We have also computed how the various estimates deteriorate when $t \to 0$. In the case of homogeneous structures for degenerate equations (see §4), the interpolation estimate (4.1) is of particular interest. It reveals a behaviour dramatically different from the linear case $p=2$. An estimate of this kind (i.e., for small ε) had been proved by Moser [83] for solutions of linear parabolic equations with measurable coefficients. Generalizations to quasilinear equations with 'linear' growth $p=2$ are in [7,97]. The global estimates in Σ_T of §4-(IV) are taken from [41]. We have given a different and simpler proof. For the porous medium equation with power-type non-linearities, estimates of the same nature have been proved by Bénilan–Crandall–Pierre [10]. Analogous estimates for general non-linearities appear in [4]. Still in the context of the porous medium equation, rather precise local sup-estimates have been recently obtained by Andreucci [3]. For equation with singular structure ($1<p<2$), the theory of local and global boundedness has started only recently in [42] and [43]. Improvements to equations with general structures are in [87]. The results of Theorems 5.1-5.3 are sharp. They will play a central role in the Harnack estimates of Chap. VII. The integrability condition (5.2) is sharp as shown by the counterexample in §13 of Chap. XII. The arguments of §17 appear in [101].

VI

Harnack estimates: the case $p>2$

1. Introduction

We will establish a Harnack-type estimate for non-negative weak solutions of degenerate parabolic equations of the type

$$(1.1)\qquad \begin{cases} u \in C_{loc}\left(0,T;L^2_{loc}(\Omega)\right) \cap L^p_{loc}\left(0,T;W^{1,p}_{loc}(\Omega)\right), \; p>2, \\ u_t - \operatorname{div}|Du|^{p-2}Du = 0, \quad \text{in } \Omega_T. \end{cases}$$

Since the equation is invariant by the scaling $x \to hx$, $t \to h^p t$, $h>0$, it may seem plausible that the Harnack estimate of Hadamard [50] and Pini [86],(1) would hold in the geometry of the cylinders

$$(1.2)\qquad Q_\rho(x_o,t_o) \equiv B_\rho(x_o)\times(t_o-\rho^p,\, t_o)\,.$$

This is not the case, as one can verify for the explicit solution $(x,t)\to\mathcal{B}(x,t)$ introduced in (4.7) of Chap. V. Let (x_o,t_o) be a point of the *free* boundary $\{t=|x|^\lambda\}$, and let $\rho>1$. Then if t_o is sufficiently large, the ball $B_\rho(x_o)$ taken at the time level $t_o-\rho^p$ intersects the support of $x\to\mathcal{B}(x,t_o-\rho^p)$ in a open set. Therefore

$$\sup_{B_\rho(x_o)} \mathcal{B}(x,t_o-\rho^p)>0 \qquad \text{and} \qquad \mathcal{B}(x_o,t_o)=0.$$

This reveals a gap between the elliptic theory and the corresponding parabolic theory. Indeed non-negative weak solutions of

(1) See (2.2) in the Preface.

$$\operatorname{div}|Du|^{p-2}Du = 0, \quad u \in W^{1,p}_{loc}(\Omega), \quad p > 1,$$

satisfy the Harnack inequality,[(2)] whereas solutions of the corresponding parabolic equation (1.1) in general do not.

Let u be a non-negative local solution of the heat equation in Ω_T. Then for all $\varepsilon > 0$ there exists a constant γ depending only upon N and ε, such that for every cylinder $Q_\rho(x_o, t_o) \subset \Omega_T$ and for every $\sigma \in (0,1)$,

$$\sup_{Q_{\sigma\rho}(x_o,t_o)} u \le \frac{\gamma}{(1-\sigma)^{\frac{N+2}{2\varepsilon}}} \left(\iint_{Q_\rho(x_o,t_o)} u^\varepsilon \, dx d\tau \right)^{\frac{1}{\varepsilon}}, \tag{1.3}$$

where $Q_\rho(x_o, t_o)$ is defined by (1.2) with $p=2$. This local sup-bound of the solution in terms of the integral average of a *small* power of u, is a key fact in Moser's proof of the Harnack estimate. An estimate of this kind does not hold for solutions of (1.1) and it is replaced by the more structured inequality (4.1) of Chap. V. A study of [83] however reveals that (1.3) continues to hold for sufficiently smooth solutions of

$$\frac{\partial}{\partial t} u^{p-1} - \operatorname{div}\left(|Du|^{p-2} Du\right) = 0, \quad \text{in } \Omega_T. \tag{1.4}$$

With this in mind one may heuristically regard (1.1) as it were (1.4) written in a time scale intrinsic to the solution itself and, loosely speaking, of the order of $t\,[u(x,t)]^{2-p}$. Next we observe that (2.2) in the Preface is equivalent to

$$u(x_o, t_o) \le \gamma \inf_{B_\rho(x_o)} u\left(\cdot, t_o + \rho^2\right), \quad \gamma = \gamma(N). \tag{1.5}$$

The Harnack estimate of Krylov and Safonov [64] for non-divergence parabolic equations is given precisely in this form.

This suggests that the number $[u(x_o, t_o)]^{2-p}$ is the intrinsic scaling factor and leads to conjecture that non-negative solutions of (1.1) will satisfy the Harnack inequality with respect to such an intrinsic time scale.

2. The intrinsic Harnack inequality

The following theorem makes rigorous the heuristic remarks of the previous section.

THEOREM 2.1. *Let u be a non-negative weak solution of (1.1). Fix any $(x_o, t_o) \in \Omega_T$ and assume that $u(x_o, t_o) > 0$. There exist constants $\gamma > 1$ and $C > 1$, depending only upon N and p, such that*

$$u(x_o, t_o) \le \gamma \inf_{B_\rho(x_o)} u\left(\cdot, t_o + \theta\right), \tag{2.1}$$

---

(2) See [82,92,96].

where

$$\theta = \frac{C\rho^p}{[u(x_o,t_o)]^{p-2}}, \tag{2.2}$$

provided the cylinder

$$Q_{4\rho}(\theta) \equiv \{|x-x_o|<4\rho\} \times \{t_o-4\theta, t_o+4\theta\} \tag{2.3}$$

is contained in Ω_T.

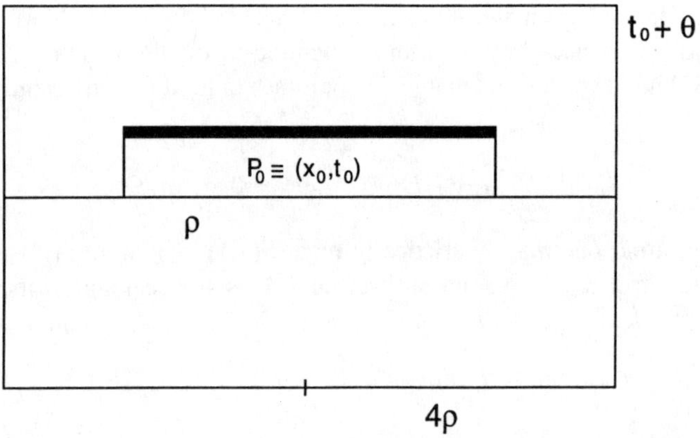

Figure 2.1

Remark 2.1. The values $u(x_o,t_o)$ are well defined since u is locally Hölder continuous in Ω_T.

Remark 2.2. The constants γ and C tend to infinity as $p\to\infty$. However they are 'stable' as $p\searrow 2$, i.e.,

$$\lim_{p\searrow 2} \gamma(N,p), C(N,p) = \gamma(N,2), C(N,2) < \infty.$$

Therefore by letting $p\to 2$ in (2.1) we recover, at least formally, the classical Harnack inequality for non-negative solutions of the heat equation. Such a limiting process can be made rigorous by the $C^{1,\alpha}_{loc}(\Omega_T)$ estimates of Chap. IX.

In Theorem 2.1 the level θ is connected to $u(x_o,t_o)$ via (2.2). It is convenient to have an estimate where the geometry can be prescribed a priori independent of the solution. This is the thrust of the next result which holds for all $\theta>0$.

THEOREM 2.2. *There exists a constant $B>1$ depending only upon N and p, such that*

(2.4) $\forall\,(x_o,t_o)\in\Omega_T,\ \forall\rho,\theta>0$ *such that* $Q_{4\rho}(\theta)\subset\Omega_T$,

$$u(x_o,t_o)\le B\left\{\left(\frac{\rho^p}{\theta}\right)^{\frac{1}{p-2}}+\left(\frac{\theta}{\rho^p}\right)^{N/p}\left[\inf_{B_\rho(x_o)}u(\cdot,t_o+\theta)\right]^{\lambda/p}\right\},$$

where

(2.5)
$$\lambda=N(p-2)+p.$$

Remark 2.3. Inequality (2.4) holds for all $p\in(2,\infty)$, but the constant B is not 'stable' as $p\searrow 2$, i.e.,

$$\lim_{p\searrow 2}B(N,p)=\infty.$$

In (2.4) the positivity of $u(x_o,t_o)$ is not required and $\theta>0$ is arbitrary so that Theorems 2.1 and 2.2 may seem markedly different. In fact they are equivalent, i.e.,

PROPOSITION 2.1. *Theorem 2.1* $\Longleftrightarrow$ *Theorem 2.2.*

In view of Remark 2.3, the equivalence is meant in the sense that (2.1) implies (2.4) in any case and (2.4) implies (2.1) with a constant $\gamma=\gamma(N,p)$ which may not be 'stable' as $p\searrow 2$. A consequence of Theorem 2.2 is

COROLLARY 2.1. *There exists a constant* $B>1$ *depending only upon* N *and* p, *such that*

(2.6) $\forall(x_o,t_o)\in\Omega_T,\ \forall\,\rho,\theta>0$ *such that* $Q_{4\rho}(\theta)\subset\Omega_T$,

$$\fint_{B_\rho(x_o)}u(x,t_o)dx\le B\left\{\left(\frac{\rho^p}{\theta}\right)^{\frac{1}{p-2}}+\left(\frac{\theta}{\rho^p}\right)^{N/p}[u(x_o,t_o+\theta)]^{\lambda/p}\right\}$$

2-(i). *Generalisations*

All the stated results remain valid if the right hand side of (1.1) contains a forcing term f, provided

(2.7)
$$f\in L^q_{loc}(\Omega_T),\qquad q>(N+p)/p$$

and f is non-negative. We will indicate later how to modify the proofs to include such a case.

3. Local comparison functions

Let $\rho>0$ and $k>0$ be fixed and consider the following *'fundamental solution'* of (1.1) with pole at $(\bar{x},\bar{t})$:

$$(3.1) \qquad \mathcal{B}_{k,\rho}\left(x,t;\bar{x},\bar{t}\right) \equiv \frac{k\rho^N}{S^{N/\lambda}(t)}\left\{1-\left(\frac{|x-\bar{x}|}{S^{1/\lambda}(t)}\right)^{\frac{p}{p-1}}\right\}_+^{\frac{p-1}{p-2}},$$

where λ is defined in (2.5) and

$$(3.2) \qquad S(t)=b(N,p)k^{p-2}\rho^{N(p-2)}(t-\bar{t})+\rho^\lambda, \qquad t\geq\bar{t},$$

$$b(N,p)=\lambda\left(\frac{p}{p-2}\right)^{p-1}.$$

By calculation, one verifies that $\mathcal{B}_{k,\rho}\left(x,t;\bar{x},\bar{t}\right)$ is a weak solution of (1.1) in $\mathbf{R}^N\times\{t>\bar{t}\}$. Moreover for $t=\bar{t}$ it vanishes outside the ball $B_\rho(\bar{x})$ and for $t>\bar{t}$ the function $x\to\mathcal{B}_{k,\rho}\left(x,t;\bar{x},\bar{t}\right)$ vanishes, in a C^1 fashion, across the boundary of the ball $\left\{|x-\bar{x}|<S^{1/\lambda}(t)\right\}$. One also verifies that

$$\mathcal{B}_{k,\rho}\left(x,t;\bar{x},\bar{t}\right)\leq k, \qquad |x|\in\mathbf{R}^N,$$

and that for $\bar{t}\leq t\leq t^*$, the support of $\mathcal{B}_{k,\rho}\left(x,t;\bar{x},\bar{t}\right)$

$$D^*\equiv\left\{|x-\bar{x}|\leq S^{1/\lambda}(t)\right\}\times[\bar{t}\,,\,t^*],$$

is contained in the cylindrical domain

$$Q^*\equiv B_{S^{1/\lambda}(t^*)}(\bar{x})\times[\bar{t}\,,\,t^*].$$

If u is a non-negative weak solution of (1.1) in Q^* satisfying

$$u(x,\bar{t})\geq k \quad \text{for } |x-\bar{x}|<\rho,$$

then

$$u(x,t)\geq\mathcal{B}_{k,\rho}\left(x,t;\bar{x},\bar{t}\right), \qquad \forall\,(x,t)\in Q^*.$$

This is a consequence of the following comparison principle.

LEMMA 3.1. *Let u and v be two solutions of (1.1) in Ω_T satisfying*

$$\begin{cases} u,v\in C\left(0,T;L^2(\Omega)\right)\cap L^p\left(0,T;W^{1,p}(\Omega)\right)\cap C\left(\bar{\Omega}_T\right)\\ u\geq v \ \text{ on the parabolic boundary of } \Omega_T. \end{cases}$$

Then $u\geq v$ in Ω_T.

PROOF: We write the weak form of (1.1) for u and v in terms of the Steklov-averages, as in (1.5) of Chap. II, against the testing function

$$[(v-u)_h]_+(x,t)=\left[\frac{1}{h}\int_t^{t+h}(v-u)(x,\tau)d\tau\right]_+,$$

$$h\in(0,T),\ t\in[0,T-h).$$

Differencing the two equations and integrating over $(0,t)$ gives

$$\int_{\Omega} [(v-u)_h]_+^2 (x,t)dx - \int_{\Omega} [(v-u)_h]_+^2 (x,0)dx$$

$$= -2 \iint_{\Omega_t} \left[|Dv|^{p-2}Dv - |Du|^{p-2}Du\right]_h \cdot D\left[(v-u)_h\right]_+ dxd\tau.$$

As $h \to 0$ the second term on the left hand side tends to zero since $(v-u)_+ \in C\left(\bar{\Omega}_T\right)$. Applying also Lemmas 3.2 and 4.4 of Chap. I we arrive at

$$\int_{\Omega} (v-u)_+^2 (x,t)dx$$

$$= -2 \iint_{\Omega_t \cap [v>u]} \left(|Dv|^{p-2}Dv - |Du|^{p-2}Du\right) \cdot D(v-u)dxd\tau \le 0.$$

3-(i). Local comparison functions: the case p near 2

The next comparison function is a subsolution of (1.1) for $p>2$ and for $p<2$ provided p is close enough to 2. For definiteness let us assume $p \in [2, 5/2]$ and consider the function

(3.3) $$\mathcal{G}_{k,\rho}\left(x,t;\bar{x},\bar{t}\right) \equiv \frac{k\rho^{\frac{\nu}{\lambda(\nu)}}}{\Sigma^{\nu}(t)} \left\{1 - \left(\frac{|x-\bar{x}|}{\Sigma^{\lambda(\nu)}(t)}\right)^{\frac{p}{p-1}}\right\}_+^{\frac{p}{p-1}},$$

(3.4) $$\Sigma(t) = k^{p-2}\rho^{(p-2)\frac{\nu}{\lambda(\nu)}}(t-\bar{t}) + \rho^{\frac{1}{\lambda(\nu)}}, \qquad t \ge \bar{t},$$

where the positive numbers ν and $\lambda(\nu)$ are linked by

(3.5) $$\lambda(\nu) = \frac{1-\nu(p-2)}{p}.$$

Introduce the number

(3.6) $$p(\nu) = 4(1+2\nu)/(1+4\nu),$$

and observe that

(3.7) $$\frac{1}{4} \le \lambda(\nu) \le \frac{1}{2}, \qquad \text{for} \quad p \in [2, p(\nu)].$$

LEMMA 3.2. *The number $\nu > 1$ can be determined a priori only in terms of N and independent of $p \in [2, 5/2]$, such that $\mathcal{G}_{k,\rho}$ is a classical subsolution of*

$$\frac{\partial}{\partial t}\mathcal{G}_{k,\rho} - \operatorname{div}\left(|D\mathcal{G}_{k,\rho}|^{p-2} D\mathcal{G}_{k,\rho}\right) \le 0 \quad \textit{in } \mathbf{R}^N \times \{t > \bar{t}\}.$$

PROOF: For $(x,t) \in \mathbf{R}^N \times \{t > \bar{t}\}$, set

$$\mathcal{L}^*\left(\mathcal{G}_{k,\rho}\right) \equiv \frac{\Sigma^{\nu+1}(t)}{\left(k\rho^{\frac{\nu}{\lambda(\nu)}}\right)^{p-1}}\left\{\frac{\partial}{\partial t}\mathcal{G}_{k,\rho} - \operatorname{div}\left(\left|D\mathcal{G}_{k,\rho}\right|^{p-2}D\mathcal{G}_{k,\rho}\right)\right\},$$

$$\|z\| \equiv \frac{|x-\bar{x}|}{\Sigma^{\lambda(\nu)}(t)}, \quad \mathcal{F} \equiv \left(1-\|z\|^{\frac{p}{p-1}}\right)_+, \quad a \equiv \left(\frac{p}{p-1}\right)^2.$$

Then, by calculation,

$$\begin{aligned}(3.8)\qquad \mathcal{L}^*\left(\mathcal{G}_{k,\rho}\right) &= -\nu\mathcal{F}^{\frac{p}{p-1}} + Na^{p-1}\mathcal{F}\\ &\quad - \frac{p}{p-1}a^{p-1}\|z\|^{\frac{p}{p-1}} + \lambda(\nu)a\mathcal{F}^{\frac{1}{p-1}}\|z\|^{\frac{p}{p-1}}.\end{aligned}$$

Introducing the set

$$\mathcal{E}_1 \equiv \left[\|z\|^{\frac{p}{p-1}} \ge \frac{1}{2}\left(1+\frac{N(p-1)}{N(p-1)+p}\right)\right],$$

we have

$$\mathcal{F} \le \frac{p}{2[N(p-1)+p]} \qquad \text{in } \mathcal{E}_1,$$

and therefore by (3.5)

$$\begin{aligned}\mathcal{L}^*\left(\mathcal{G}_{k,\rho}\right) &\le -\nu\mathcal{F}^{\frac{p}{p-1}} + Na^{p-1} + \lambda(\nu)a\mathcal{F}^{\frac{1}{p-1}}\\ &\quad - a^{p-1}\left(N+\frac{p}{p-1}\right)\|z\|^{\frac{p}{p-1}}\\ &\le a^{p-1}\left[\lambda(\nu)\left(\frac{p}{N(p-1)+p}\right)^{\frac{1}{p-1}} - \frac{p}{2(p-1)}\right]\\ &\le a^{p-1}\left[\frac{1}{2}\left(\frac{p}{N(p-1)+p}\right)^{\frac{1}{p-1}} - \frac{p}{2(p-1)}\right]\\ &\le a^{p-1}\left(\frac{1}{2}-\frac{p}{2(p-1)}\right) < 0.\end{aligned}$$

Within the set

$$\mathcal{E}_2 \equiv \left[\|z\|^{\frac{p}{p-1}} < \frac{1}{2}\left(1+\frac{N(p-1)}{N(p-1)+p}\right)\right],$$

we have

$$\mathcal{F} \ge \frac{p}{2[N(p-1)+p]} \qquad \text{in } \mathcal{E}_2.$$

It follows from (3.5) and (3.7) that

$$\begin{aligned}\mathcal{L}^*\left(\mathcal{G}_{k,\rho}\right) &\le -\nu\left(\frac{p}{2[N(p-1)+p]}\right)^{\frac{p}{p-1}} + a^{p-1}N + \lambda(\nu)a\\ &\le -\nu\left(\frac{p}{2[N(p-1)+p]}\right)^{\frac{p}{p-1}} + \frac{a}{p}\left[Npa^{p-2}+1\right].\end{aligned}$$

Choosing

$$\nu \equiv \max_{p\in[2,5/2]} \frac{a}{p}\left[Npa^{p-2}+1\right]\left(\frac{2[N(p-1)+p]}{p}\right)^{\frac{p}{p-1}}, \tag{3.9}$$

we have in either case

$$\mathcal{L}^*\left(\mathcal{G}_{k,\rho}\right) \le 0 \qquad \text{in } \mathbf{R}^N \times \{t > \bar{t}\}.$$

One verifies that for $t=\bar{t}$

$$\mathcal{G}_{k,\rho}\left(x,\bar{t};\bar{x},\bar{t}\right) \le k, \qquad x\in\mathbf{R}^N,$$

and that for $\bar{t}\le t\le t^*$, the support of $\mathcal{G}_{k,\rho}$,

$$\mathcal{R}^* \equiv \left\{|x-\bar{x}| < \Sigma^{\lambda(\nu)}(t)\right\}\times[\bar{t}\,,\,t^*],$$

is contained in the cylindrical domain

$$C^* \equiv \left\{|x-\bar{x}| < \Sigma^{\lambda(\nu)}(t^*)\right\}\times[\bar{t}\,,\,t^*].$$

Therefore if u is a solution of (1.1) in C^* such that

$$u(x,\bar{t}) \ge k \qquad \text{in } B_{\Sigma^{\lambda(\nu)}(\bar{t})}(\bar{x}),$$

then

$$u(x,t) \ge \mathcal{G}_{k,\rho}\left(x,t;\bar{x},\bar{t}\right) \qquad \text{in } C^*.$$

Remark 3.1. The same proof shows that gkr is a sub–solution of (1.1) also for $p<2$, provided p is *close* to 2. Precisely if $p\in(4-p(\nu),2)$.

4. Proof of Theorem 2.1

Let $(x_o,t_o)\in\Omega_T$ and $\rho>0$ be fixed, assume that $u(x_o,t_o)>0$ and consider the box

$$Q_{4\rho} \equiv \{|x-x_o|<4\rho\}\times\left\{t_o-\frac{4C\rho^p}{[u(x_o,t_o)]^{p-2}},\, t_o+\frac{4C\rho^p}{[u(x_o,t_o)]^{p-2}}\right\},$$

where C is a constant to be determined later. The change of variables

$$x \longrightarrow \frac{x-x_o}{\rho}, \qquad t\longrightarrow \frac{(t-t_o)\,[u(x_o,t_o)]^{p-2}}{\rho^p}$$

maps $Q_{4\rho}$ into the box $Q\equiv Q^+\cup Q^-$, where

$$Q^+\equiv B_4\times[0,4C), \qquad Q^-\equiv B_4\times(-4C,0].$$

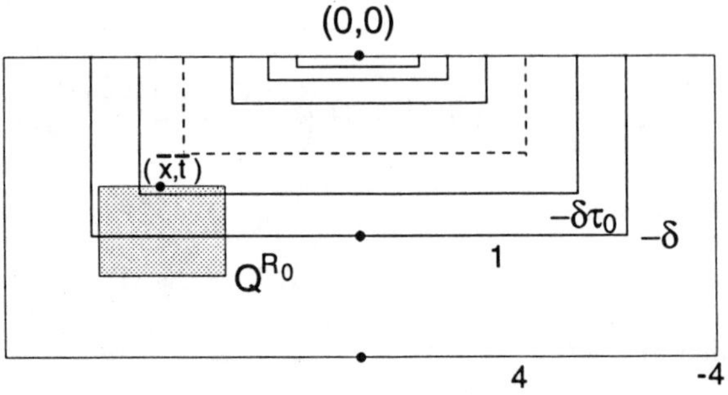

Figure 4.1

We denote again with x and t the new variables, and observe that the rescaled function

$$v(x,t) = \frac{1}{u(x_o,t_o)}\, u\left(x_o+\rho x,\, t_o+\frac{t\rho^p}{[u(x_o,t_o)]^{p-2}}\right)$$

is a *bounded* non-negative weak solution of

$$\begin{cases} v_t - \operatorname{div}\left(|Dv|^{p-2}Dv\right) = 0 & \text{in } Q \\ v(0,0) = 1. \end{cases}$$

To prove the Theorem it suffices to find constants $\gamma_o \in (0,1]$ and $C>1$ depending only upon N and p such that

$$\inf_{B_1} v(x,C) \ge \gamma_o.$$

Construct the family of nested and expanding boxes

$$Q_\tau \equiv \{|x|<\tau\}\times(-\tau^p,0], \qquad \tau\in(0,1],$$

and the numbers

$$M_\tau \equiv \sup_{Q_\tau} v, \quad N_\tau \equiv (1-\tau)^{-\beta}, \qquad \tau\in[0,1),$$

where $\beta>1$ will be chosen later. Let τ_o be the largest root of the equation $M_\tau = N_\tau$. Such a root is well defined since $M_o = N_o$, and as $\tau \nearrow 1$, the numbers M_τ remain bounded and $N_\tau \nearrow \infty$. By construction

$$\sup_{Q_\tau} v \le N_\tau, \qquad \forall\tau>\tau_o.$$

Since v is continuous in Q there exists within $\bar{Q}_{\tau_o}$ at least one point, say $(\bar{x},\bar{t})$, such that

$$v(\bar{x},\bar{t}) = N_{\tau_o} = (1-\tau_o)^{-\beta}.$$

The next arguments are intended to establish that within a small ball about $\bar{x}$ and at the same time-level $\bar{t}$ the function v is of the same order of $(1-\tau_o)^{-\beta}$. For this we make use of the Hölder continuity of v and more specifically of Lemma 3.1 of Chap. III.

Set

$$R = \frac{1-\tau_o}{2},$$

and consider the cylinder with *'vertex'* at $(\bar{x}, \bar{t})$

$$[(\bar{x},\bar{t}) + Q(R^p, R)] \equiv \left\{|x - \bar{x}| < \frac{1-\tau_o}{2}\right\} \times \left\{\bar{t} - \left(\frac{1-\tau_o}{2}\right)^p, \bar{t}\right\}.$$

By construction $[(\bar{x},\bar{t}) + Q(R^p, R)] \subset Q_{\frac{1+\tau_o}{2}}$ and therefore

$$\sup_{[(\bar{x},\bar{t})+Q(R^p,R)]} v \le N_{\frac{1+\tau_o}{2}} = 2^\beta (1-\tau_o)^{-\beta} \equiv \omega.$$

If A is the number determined by Proposition 3.1 of Chap. III, we may choose $\beta > 1$ so large that $(2^\beta/A) > 1$. Therefore the cylinder

$$[(\bar{x},\bar{t}) + Q(a_o R^p, R)], \qquad \frac{1}{a_o} \equiv \left(\frac{\omega}{A}\right)^{p-2} = \left[\frac{2^\beta (1-\tau_o)^{-\beta}}{A}\right]^{p-2} > 1$$

is contained in $[(\bar{x},\bar{t}) + Q(R^p, R)]$, and

$$\operatorname*{osc}_{[(\bar{x},\bar{t})+Q(a_o R^p,R)]} v \le \omega.$$

It follows that $[(\bar{x},\bar{t}) + Q(a_o R^p, R)]$ can be taken as the starting box in Lemma 3.1 of Chap. III. We conclude that there exist constants $\gamma > 1$ and $\alpha, \varepsilon_o \in (0,1)$ such that for all $r \in (0, R]$

$$\begin{aligned}\operatorname*{osc}_{\{|x-\bar{x}|<r\}} v(\cdot,\bar{t}) &\le \gamma(\omega + R^{\varepsilon_o})\left(\frac{r}{R}\right)^\alpha \\ &\le 2^{\beta+1}\gamma(1-\tau_o)^{-\beta}\left(\frac{r}{R}\right)^\alpha.\end{aligned}$$

We let $r = \sigma R$ and then choose σ so small that for all $\{|x - \bar{x}| < \sigma R\}$,

(4.1)
$$\begin{aligned} v(x,\bar{t}) &\ge v(\bar{x},\bar{t}) - 2^{\beta+1}\gamma(1-\tau_o)^{-\beta}\sigma^\alpha \\ &= \left(1 - 2^{\beta+1}\gamma\sigma^\alpha\right)(1-\tau_o)^{-\beta} \\ &= \frac{1}{2}(1-\tau_o)^{-\beta}.\end{aligned}$$

The various constants appearing in Proposition 3.1 and Lemma 3.1 of Chap. III, in our context, depend only upon N and p and are independent of v,(1) therefore the number σ can be determined a priori only in terms of N, p and β. We summarise:

(1) See §3-(I) of Chap. III.

LEMMA 4.1. *There exist a number $\sigma\in(0,1)$ depending only upon N, p and β such that*

$$v(x,\bar{t})\geq\frac{1}{2}(1-\tau_o)^{-\beta},\quad\forall\{|x-\bar{x}|<\sigma R\},\quad R=\frac{1}{2}(1-\tau_o).\tag{4.2}$$

Remark 4.1. The location of $(\bar{x},\bar{t})$ and the number τ_o (and hence R) are determined only *qualitatively*. However in view of (4.1) the number σ is *quantitatively* determined as soon as $\beta>1$ is quantitatively chosen.

4-(i). Expanding the positivity set

We will choose the constants $\beta>1$ and $C>1$ so that the qualitative *largeness* of $v(\cdot,\bar{t})$ in the *small* ball $B_{\sigma R}(\bar{x})$ turns into a quantitative bound below over the full sphere B_1 at some further time level C. This is achieved by means of the comparison functions of §3. Assume first that $p\in[2,\,p(\nu)]$, where ν is the number determined in Lemma 3.2, and consider the function $\mathcal{G}_{k,\rho}$ introduced in (3.3), with the choices

$$k=\frac{1}{2}(1-\tau_o)^{-\beta},\qquad \rho=\sigma R.\tag{4.3}$$

At the time level $t=C$ the support of $x\to\mathcal{G}_{k,\rho}\,(x,C;\bar{x},\bar{t})$ is the ball

$$\begin{aligned}|x-\bar{x}|&<\left\{\left[\frac{1}{2}(1-\tau_o)^{-\beta}\right]^{p-2}(\sigma R)^{(p-2)\frac{\nu}{\lambda(\nu)}}\,(C-\bar{t})+(\sigma R)^{1/\lambda(\nu)}\right\}^{\lambda(\nu)}\\&=\left\{\gamma^{p-2}\,(1-\tau_o)^{\left(\frac{\nu}{\lambda(\nu)}-\beta\right)(p-2)}\,(C-\bar{t})+(\sigma R)^{1/\lambda(\nu)}\right\}^{\lambda(\nu)},\end{aligned}$$

where

$$\gamma=\gamma(\sigma,\nu)=\frac{1}{2}\left(\frac{\sigma}{2}\right)^{\frac{\nu}{\lambda(\nu)}}.$$

Choose

$$\beta=\frac{\nu}{\lambda(\nu)}\qquad\text{and}\qquad C=\frac{3^{1/\lambda(\nu)}}{\gamma^{p-2}}.$$

Since $|\bar{x}|<1$ and $\bar{t}\in(-1,0]$, these choices imply that the support of $x\to\mathcal{G}_{k,\rho}\,(x,C;\bar{x},\bar{t})$ contains B_2, and by the comparison principle

$$\begin{aligned}\inf_{x\in B_1}v(x,C)&\geq\inf_{x\in B_1}\mathcal{G}_{k,\rho}\,(x,C;\bar{x},\bar{t})\\&\geq 2^{-(1+2\nu)}\left(\frac{\sigma}{2}\right)^{\frac{\nu}{\lambda(\nu)}}\left\{1-\left(\frac{2}{3}\right)^{\frac{p}{p-1}}\right\}^{\frac{p}{p-1}}\\&\equiv\gamma_o.\end{aligned}$$

The various constants depend only upon N and p and are *'stable'* as $p\searrow 2$.

Turning to the case $p \geq p(\nu)$, we consider the comparison function $\mathcal{B}_{k,\rho}(x,t;\bar{x},\bar{t})$ introduced in (3.1)-(3.2), with the choice of the parameters k and ρ as in (4.3). At $t=C$ the support of $\mathcal{B}_{k,\rho}(\cdot,C;\bar{x},\bar{t})$ is the ball

$$
\begin{aligned}
|x-\bar{x}|^{\lambda} &< \left\{ b\left[\frac{1}{2}(1-\tau_o)^{-\beta}\right]^{p-2} (\sigma R)^{N(p-2)} (C-\bar{t}) + (\sigma R)^{\lambda} \right\} \\
&= \left\{ b\,\gamma^{p-2} (1-\tau_o)^{(N-\beta)(p-2)} (C-\bar{t}) + (\sigma R)^{\lambda} \right\},
\end{aligned}
$$

where

$$
\gamma(N,\beta) = \frac{1}{2}\left(\frac{\sigma}{2}\right)^{N} \qquad \text{and} \qquad b = \lambda\left(\frac{p}{p-2}\right)^{p-1}.
$$

Choosing

$$
\beta = N \qquad \text{and} \qquad C = \frac{3^{\lambda}}{b\,\gamma^{p-2}}, \tag{4.4}
$$

we see that the support of $\mathcal{B}_{k,\rho}(\cdot,C;\bar{x},\bar{t})$ contains B_2, and by the comparison principle,

$$
\begin{aligned}
\inf_{x\in B_1} v(x,C) &\geq \inf_{x\in B_1} \mathcal{B}_{k,\rho}(x,C;\bar{x},\bar{t}) \\
&\geq (2)^{-(1+2\frac{N}{\lambda})} \left(\frac{\sigma}{2}\right)^{N} \left\{ 1 - \left(\frac{2}{3}\right)^{\frac{p}{p-1}} \right\}^{\frac{p}{p-2}} \\
&\equiv \gamma_o.
\end{aligned} \tag{4.5}
$$

Remark 4.2. These estimates involving the comparison function $\mathcal{B}_{k,\rho}$ hold for all $p>2$. However as $p \searrow 2$, the constant γ_o in (4.5) tends to zero. The purpose of introducing an auxiliary comparison function $\mathcal{G}_{k,\rho}$ for p near 2 is to have the constants under control as p approaches the non-degenerate case $p=2$. We also remark that $\mathcal{G}_{k,\rho}$ is a subsolution of (1.1) only for p close enough to 2.

5. Proof of Theorem 2.2

Let $(x_o,t_o) \in \Omega_T$, $\rho>0$ and $\theta>0$ be fixed so that the box $Q_{4\rho}(\theta)$ is contained in Ω_T. We may assume that (x_o,t_o) coincides with the origin and set $u_* \equiv u(0,0)$. If C and γ are the constants determined in Theorem 2.1, we may assume that

$$
t^* \equiv \frac{C\rho^p}{u_*^{p-2}} \leq \frac{\theta}{2}. \tag{5.1}
$$

Indeed otherwise

$$
u_* \leq B\left(\frac{\rho^p}{\theta}\right)^{\frac{1}{p-2}}, \qquad B \equiv (2C)^{\frac{1}{p-2}},
$$

and there is nothing to prove. By Theorem 2.1 and (5.1)

$$u_* \le \gamma u(x,t^*), \qquad \forall x \in B_\rho.$$

Consider the *'fundamental solution'* $\mathcal{B}_{k,\rho}$ with pole at $(0,t^*)$ and with $k=\gamma^{-1}u_*$. By the comparison principle, at the level $t=\theta$, we have

$$(5.2)\qquad u(x,\theta) \ge \frac{u_*\,\rho^N}{S^{\frac{N}{\lambda}}(t)}\left\{1-\left(\frac{|x|}{S^{1/\lambda}(t)}\right)^{\frac{p}{p-1}}\right\}_+^{\frac{p-1}{p-2}} \qquad \forall\,|x|<\rho,$$

where

$$\begin{aligned} S(t) &= \left[b\left(\gamma^{-1}u_*\right)^{p-2}\rho^{N(p-2)}(\theta-t^*)+\rho^\lambda\right] \\ &= \left[\frac{C\,b}{\gamma^{p-2}}\frac{u_*^{p-2}}{C\rho^p}\,(\theta-t^*)+1\right]\rho^\lambda. \end{aligned}$$

Here λ and b are defined in (2.5) and (3.2) respectively. It follows from (5.1) that

$$\left(\frac{C\,b}{\gamma^{p-2}}+1\right)\rho^\lambda \le S(t) \le \left(\frac{b}{\gamma^{p-2}}+\frac{1}{2C}\right)u_*^{p-2}\left(\frac{\theta}{\rho^p}\right)\rho^\lambda.$$

Therefore (5.2) gives

$$u(x,\theta) \ge u_*^{p/\lambda}\left(\frac{\rho^p}{\theta}\right)^{N/\lambda}\gamma_1, \qquad \gamma_1 \equiv \gamma_1(N,p),$$

and the theorem follows with

$$B = \max\left\{\gamma_1^{-\lambda/p};\,(2C)^{\frac{1}{p-2}}\right\}.$$

We have shown that Theorem 2.1 implies the estimate of Theorem 2.2. To prove the equivalence of Proposition 2.1, assume that (2.4) holds true for all $\theta>0$ such that $Q_{4\rho}(\theta)\subset\Omega_T$. Choose

$$\theta = \frac{(2B)^{p-2}\rho^p}{[u(x_o,t_o)]^{p-2}}.$$

Then if $Q_{4\rho}(\theta)\subset\Omega_T$, (2.4) gives

$$u(x_o,t_o) \le 2B^{N(p-2)/\lambda}\inf_{B_\rho(x_o)} u(\cdot,t_o+\theta).$$

5-(i). About the generalisations

The only tools we have used in the proof are the Hölder continuity of the solutions of (1.1) and the comparison principle. The integrability indicated in (2.7) guarantees the local Hölder continuity.(1) Moreover the comparison principle remains applicable since $f\ge 0$.

6. Global versus local estimates

The assumption that the cylinder $Q_{4\rho}(\theta)$ be contained in the domain of definition of the solution is essential for the Harnack estimates of Theorems 2.1 and 2.2 to hold. Indeed the function $(x,t)\rightarrow \mathcal{B}(x,t)$ introduced in (4.11) of Chap. V does not satisfy (2.4) for $x_o=0$ and t_o arbitrarily close to zero. This is not due to the pointwise nature of (2.1) and (2.4). A Harnack inequality, with t_o arbitrarily close to zero, fails to hold even in the *averaged* form (2.6). To see this let u be the unique weak solution of the boundary value problem

$$(\mathcal{P})\qquad \begin{cases} u_t-\left(|u_x|^{p-2}u_x\right)_x=0 & \text{in } Q\equiv(0,1)\times(0,\infty),\\ u(0,t)=u(1,t)=0 & \text{for all } t\ge 0,\\ u(\cdot,0)=u_o\in C_o^\infty(0,1) & \\ u_o(x)\in[0,1],\ \forall x\in(0,1) & \text{and } u_o(x)=1 \text{ for } x\in\left(\frac14,\frac34\right). \end{cases}$$

We claim that

$$u_t\ge\frac{-1}{p-2}\frac{u}{t}\qquad \text{in } \mathcal{D}'(Q). \tag{6.1}$$

Let us assume (6.1) for the moment. Since $0\le u\le 1$, by the comparison principle (6.1) implies that

$$-\left(|u_x|^{p-2}u_x\right)_x\le\frac{1}{(p-2)t},\qquad t>0.$$

At any fixed level t, the function $x\rightarrow u(x,t)$ is majorised by

$$v(x,t)=\frac{\gamma x^\delta}{t^{\frac{1}{p-1}}},\quad \delta\in\left(\frac{p-1}{p},1\right),\quad (\gamma\delta)^{p-1}(1-\delta)(p-1)\ge\frac{1}{p-2}.$$

Indeed

$$-\left(|v_x|^{p-2}v_x\right)_x\ge\frac{1}{(p-2)t}\quad\text{and}\quad v(0,t)=0,\quad v(1,t)>0.$$

Therefore for every $\delta\in\left(\frac{p-1}{p},1\right)$ there exists a constant $C=C(\delta)$, such that

$$u\left(\tfrac12,t\right)\le\frac{C(\delta)}{t^{1/(p-1)}}.$$

(1) See the structure conditions in §1 of Chap. II, Theorem 1.1 of Chap. III and Theorem 3.1 of Chap. V.

Now assume that (2.6) holds for $t_o=0$, $x_o=\frac{1}{2}$, $\theta=t$ and $\rho=\frac{1}{4}$. Then for $t>1$

$$1 \le \text{const}\left(t^{-\frac{1}{p-2}} + t^{-\frac{1}{p}}\right) \longrightarrow 0 \quad \text{as } t \longrightarrow \infty.$$

The proof of (6.1) is a particular case of the following

6-(i). *Regularising effects*

PROPOSITION 6.1. *Let $u \in L^p\left(0,T;W_o^{1,p}(\Omega)\right)$ be the unique non-negative weak solution of*

$$\begin{cases} u_t - \operatorname{div}|Du|^{p-2}Du = 0, & \text{in } \Omega_T,\ p>1, \\ u(\cdot,0) = u_o \in L^2(\Omega), & u_o \ge 0. \end{cases} \tag{6.2}$$

Then if $p>2$,

$$u_t \ge \frac{-1}{p-2}\frac{u}{t} \quad \text{in } \mathcal{D}'(\Omega) \ \text{a.e. } t>0, \tag{6.3}$$

and if $1<p<2$,

$$u_t \le \frac{1}{2-p}\frac{u}{t} \quad \text{in } \mathcal{D}'(\Omega) \ \text{a.e. } t>0. \tag{6.4}$$

PROOF: We only prove (6.4). By the homogeneity of the p.d.e., the unique solution v of (6.2) with initial datum

$$v(\cdot,0) = k^{\frac{1}{p-2}}u_o, \qquad k>0,$$

is given by

$$(x,t) \longrightarrow v(x,t) \ = \ k^{\frac{1}{p-2}}u(x,kt).$$

If $k \ge 1$, $v(\cdot,0) \le u_o$ and $v(\cdot,t) \le u(\cdot,t)$ in Ω, $\forall t \in (0,T)$. Fix $t \in (0,T)$ and let $k=\left(1+\frac{h}{t}\right)$ for a small positive number h. Then

$$\begin{aligned} u(x,t+h) - u(x,t) &= u(x,kt) - u(x,t) \\ &= k^{\frac{1}{2-p}}k^{\frac{1}{p-2}}u(x,kt) - u(x,t) \\ &= k^{\frac{1}{2-p}}v(x,t) - u(x,t) \\ &\le (k^{\frac{1}{2-p}} - 1)u(x,t). \end{aligned}$$

By the mean value theorem applied to $\left(k^{\frac{1}{2-p}} - 1\right)$,

$$u(x,t+h) - u(x,t) \le \frac{h}{2-p}(1+\xi)^{\frac{p-1}{2-p}}\frac{u(x,t)}{t} \tag{6.5}$$

for some $\xi \in \left(0,\frac{h}{t}\right)$. If $h<0$, and $|h| \ll 1$, we have $k<1$, $v(\cdot,0) \ge u_o$ and (6.5) holds with the inequality sign reversed. Divide by h and in (6.5) take the limit in $\mathcal{D}'(\Omega)$ as $h \to 0$.

Remark 6.1. In the proof of the proposition, the homogeneity of the operator and the positivity of the initial datum, are essential.

7. Global Harnack estimates

The *averaged* Harnack estimate (2.6) holds with t_o arbitrarily close to zero for non-negative local solutions of (1.1) in the strip $\Sigma_T \equiv \mathbf{R}^N \times (0, T]$, i.e.,

$$\begin{cases} u \in C_{loc}\left(0, T; L^2_{loc}(\mathbf{R}^N)\right) \cap L^p_{loc}\left(0, T; W^{1,p}_{loc}(\mathbf{R}^N)\right), \\ u_t - \operatorname{div}\left(|Du|^{p-2} Du\right) = 0 \quad \text{in } \Sigma_T. \end{cases} \tag{7.1}$$

THEOREM 7.1. *Let u be a non-negative solution of (7.1) in Σ_T. There exist a constant $B > 1$ depending only upon N and p, such that*

(7.2) $\forall\,(x_o, t_o) \in \Sigma_T, \ \ \forall\,\rho, \theta > 0$ *such that* $t_o + \theta < T$,

$$\fint_{B_\rho(x_o)} u(x, t_o)\,dx \le B \left\{ \left(\frac{\rho^p}{\theta}\right)^{\frac{1}{p-2}} + \left(\frac{\theta}{\rho^p}\right)^{N/p} \left[\inf_{B_\rho(x_o)} u(\cdot, t_o + \theta) \right]^{\lambda/p} \right\}.$$

Inequality (7.2) is more general than (2.6) in that the value $u(x_o, t_o + \theta)$ is replaced by the infimum of u over the ball $B_\rho(x_o)$ at the time level $t_o + \theta$.

In (7.1) no conditions are imposed on $x \to u(x, t)$ as $|x| \to \infty$ and no reference is made to possible *initial data*. The only global information is that the p.d.e. is solved in the whole strip Σ_T. Nevertheless (7.2) gives some control on the solution u as $|x| \to \infty$, namely,

COROLLARY 7.1. *Every non-negative solution of (7.1) in Σ_T satisfies*

(7.3) $\forall\, x_o \in \mathbf{R}^N, \quad \forall\, r > 0, \quad \forall\, \varepsilon \in (0, T)$

$$\sup_{0 < \tau \le T - \varepsilon} \ \sup_{\rho \ge r} \int_{B_\rho(x_o)} \frac{u(x, \tau)}{\rho^{\lambda/p}}\,dx \le \frac{B}{\varepsilon^{\frac{1}{p-2}}} \left[1 + \left(\frac{T}{r^p}\right)^{\frac{1}{p-2}} u(x_o, T - \varepsilon) \right]^{\lambda/p}.$$

PROOF: Apply (7.2) with $t_o = \tau \in (0, T - \varepsilon)$, divide by $\rho^{\frac{p}{p-2}}$ and take the supremum of both sides for $\rho \ge r$ and $\tau \in (0, T - \varepsilon)$.

8. Compactly supported initial data

The proof of (7.2) will be a consequence of the following:

PROPOSITION 8.1. *Let v be a non-negative solution of the Cauchy problem*

$$(8.1)\qquad \begin{cases} v \in C\left(\mathbf{R}^+; L^2(\mathbf{R}^N)\right) \cap L^p\left(\mathbf{R}^+; W^{1,p}(\mathbf{R}^N)\right), \\ v_t - \operatorname{div} |Dv|^{p-2} Dv = 0 \quad \text{in } \Sigma_\infty \equiv \mathbf{R}^N \times \mathbf{R}^+, \\ v(\cdot, 0) = v_o \geq 0 \text{ and } \begin{cases} \in C(\overline{B_r}) & \text{for some } r > 0, \\ \equiv 0 & \text{in } \mathbf{R}^N \backslash \overline{B_r}. \end{cases} \end{cases}$$

There exists a constant $B = B(N,p) > 1$, such that for all $\theta > 0$,

$$(8.2)\qquad \fint_{B_r} v_o(x) dx \leq B \left\{ \left(\frac{r^p}{\theta} \right)^{\frac{1}{p-2}} + \left(\frac{\theta}{r^p} \right)^{N/p} \left[\inf_{B_r} u(\cdot, \theta) \right]^{\lambda/p} \right\}.$$

Inequality (8.2) can be regarded as a special case of (7.2) when additional information are available on the initial datum.

Basic facts on the unique solvability of (8.1) are collected in §12. We assume the Proposition for the moment and proceed to gather a few facts about v.

LEMMA 8.1. *For each $t \in \mathbf{R}^+$, the function $x \to v(x,t)$ is compactly supported in $\mathbf{R}^N$, i.e.,*

$$(8.3)\qquad \begin{aligned} &\forall T \in \mathbf{R}^+, \ \exists R = R(T) > 0 \quad \text{such that} \\ &\operatorname{supp}\{v(\cdot, t)\} \subset B_{R(T)}, \quad \forall t \in (0, T]. \end{aligned}$$

Moreover the 'mass' is conserved, i.e.,

$$(8.4)\qquad \int_{\mathbf{R}^N} v(x,t)\, dx = \int_{B_r} v_o(x)\, dx, \qquad \forall t \geq 0.$$

PROOF: Consider the function $\mathcal{B}_{k,\rho}$ introduced in (3.1)-(3.2), with $\rho = 2r$ and $(\bar{x}, \bar{t}) \equiv (0,0)$. For $t = 0$ and $|x| < r$,

$$\begin{aligned} \mathcal{B}_{k,2r}(x, 0; 0, 0) &\geq k \left\{ 1 - \left(\frac{1}{2} \right)^{\frac{p}{p-1}} \right\}^{\frac{p-1}{p-2}} \\ &\geq \sup_{B_r} v_o, \end{aligned}$$

provided k is chosen sufficiently large. By the comparison principle, $v \leq \mathcal{B}_{k,\rho}$. The second statement follows from the first by integrating the p.d.e. over $\mathbf{R}^N \times (0,t)$.

Remark 8.1 (Existence of solutions). In view of (8.3), a solution of the Cauchy problem (8.1) can be determined by fixing any $T > 0$ and solving the p.d.e. in the *bounded* domain $\Omega_T \equiv B_{R(T)} \times (0,T)$ with homogeneous boundary data on $|x| = R(T)$ and the same initial conditions as in (8.1). It follows from the comparison principle of Lemma 3.1 that the solution is unique. This construction and Proposition 6.1 also give the regularising inequality

(8.5) $$v_t \geq \frac{-1}{p-2}\frac{v}{t}, \qquad \text{in } \mathcal{D}'(\Sigma_\infty),$$

and the estimate

(8.6) $$\sup_{\mathbf{R}^N} v(\cdot,t) \leq \sup_{\overline{B_r}} v_o.$$

COROLLARY 8.1. *The quantities*

(8.7) $$|||v|||_r \equiv \sup_{t\in\mathbf{R}^+}\sup_{\rho\geq r}\int_{B_\rho}\frac{v(x,t)}{\rho^{\lambda/(p-2)}}\,dx, \quad \lambda = N(p-2)+p,$$

(8.8) $$f(t) = \sup_{0<\tau<t}\left\{\tau^{N/\lambda}\sup_{\rho\geq r}\frac{\|v(\cdot,\tau)\|_{\infty,B_\rho}}{\rho^{p/(p-2)}}\right\}$$

are finite. Moreover there exists constants γ_* and γ depending only upon N, p, such that

(8.9) *for all* $0 < t < \gamma_* |||v|||_r^{2-p}$, *and for all* $\rho \geq r$,
$$\|v(\cdot,t)\|_{\infty,B_\rho} \leq \gamma\frac{\rho^{p/(p-2)}}{t^{N/\lambda}}|||v|||_r^{p/\lambda}, \quad \lambda = N(p-2)+p.$$

PROOF: The estimate (8.9) is the content of Theorem 4.5 of Chap. V.

8-(i). *Proof of Theorem 7.1 assuming (8.2)*

It suffices to prove (7.2) for $(x_o,t_o) \equiv (0,0)$ and $\theta \in (0,T)$. Fix $\rho = r > 0$ and consider the Cauchy problem (8.1) with initial datum

$$v_o \equiv \begin{cases} u(x,0) & \text{if } x \in \overline{B_r}, \\ 0 & \text{if } x \in \mathbf{R}^N\backslash\overline{B_r}. \end{cases}$$

By the results of Chaps. III and V, the solution u is locally bounded and locally Hölder continuous in Σ_T. Therefore up to the translation that maps (x_o,t_o) into the origin, $u(\cdot,0)$ is continuous in $\overline{B_r}$. By (8.3) the comparison principle of Lemma 3.1 can be applied over the *bounded* domain $B_{R(T)} \times (0,T)$ to yield $v \leq u$. Then (7.2) follows from (8.2).

9. Proof of Proposition 8.1

LEMMA 9.1. *There exists a constant* $\gamma=\gamma(N,p)$ *such that*

(9.1) *for all* $0<t<\gamma_* \|v\|_r^{2-p}$ *and for all* $\rho \geq r$,

$$\int_0^t \int_{B_\rho} |Dv|^{p-1} dx d\tau \leq \gamma t^{\frac{1}{\lambda}} \rho^{1+\frac{\lambda}{p-2}} \|v\|_r^{1+\frac{p-2}{\lambda}}.$$

PROOF: The calculations below are formal in that they require v to be strictly positive. They are made rigorous by replacing v with $v+\varepsilon$ and letting $\varepsilon \to 0$. Let $x \to \zeta(x)$ be a non-negative piecewise smooth cutoff function in $B_{2\rho}$ that equals one on B_ρ and such that $|D\zeta| \leq 1/\rho$. By the Hölder inequality

$$\begin{aligned}
&\int_0^t \int_{B_{2\rho}} |Dv|^{p-1} \zeta^{p-1} dx d\tau \\
&\quad \leq \int_0^t \int_{B_{2\rho}} \left(\tau^{\frac{p-1}{p^2}} |Dv|^{p-1} v^{-2\frac{p-1}{p^2}} \zeta^{p-1}\right) \left(\tau^{-\frac{p-1}{p^2}} v^{2\frac{p-1}{p^2}}\right) dx d\tau \\
&\quad \leq \left(\int_0^t \int_{B_{2\rho}} \tau^{1/p} |Dv|^p v^{-2/p} \zeta^p dx d\tau\right)^{\frac{p-1}{p}} \left(\int_0^t \int_{B_{2\rho}} \tau^{-\frac{p-1}{p}} v^{2\frac{p-1}{p}} dx d\tau\right)^{\frac{1}{p}} \\
&\quad \equiv [J_1(t)]^{\frac{p-1}{p}} [J_2(t)]^{\frac{1}{p}}.
\end{aligned}$$

To estimate $J_1(t)$, in the weak formulation of (8.1) we take the testing function $\varphi \equiv t^{1/p} v^{1-2/p} \zeta^p$ to obtain

$$\begin{aligned}
&\frac{p}{2(p-1)} \int_{B_{2\rho}} t^{\frac{1}{p}} v^{2\frac{p-1}{p}} \zeta^p dx + \frac{p-2}{p} \int_0^t \int_{B_{2\rho}} \tau^{\frac{1}{p}} |Dv|^p v^{-2/p} \zeta^p dx d\tau \\
&\qquad = p \int_0^t \int_{B_{2\rho}} \tau^{\frac{1}{p}} |Dv|^{p-2} v^{1-2/p} \zeta^{p-1} Dv \cdot D\zeta \, dx d\tau \\
&\qquad + \frac{1}{2(p-1)} \int_0^t \int_{B_{2\rho}} \tau^{\frac{1}{p}-1} v^{2\frac{p-1}{p}} \zeta^p dx d\tau.
\end{aligned}$$

In the estimates below γ denotes a generic positive constant that can be determined a priori only in terms of N and p and that might be different in different contexts. By Young's inequality and the structure of the cutoff function ζ

$$p \int_0^t \int_{B_{2\rho}} \tau^{\frac{1}{p}} |Dv|^{p-2} v^{1-2/p} \zeta^{p-1} Dv \cdot D\zeta \, dx d\tau$$

$$\leq \frac{p-1}{p} \frac{p}{2-p} \int_0^t \int_{B_{2\rho}} \tau^{\frac{1}{p}} |Dv|^p v^{-2/p} \zeta^p \, dx d\tau$$

$$+ \frac{\gamma(p)}{(p-2)^{p-1}} \int_0^t \int_{B_{2\rho}} \tau^{\frac{1}{p}} v^{p-\frac{2}{p}} |D\zeta|^p \, dx d\tau.$$

We conclude that there exists a constant $\gamma = \gamma(p)$ such that

$$J_1(t) \leq \frac{\gamma}{\rho^p} \int_0^t \int_{B_{2\rho}} \tau^{\frac{1}{p}} v^{p-\frac{2}{p}} \, dx d\tau + \gamma \int_0^t \int_{B_{2\rho}} \tau^{\frac{1}{p}-1} v^{2\frac{p-1}{p}} \, dx d\tau$$
$$\equiv L_1 + L_2.$$

Estimating L_i, $i = 1, 2$, separately we have

$$L_1 \leq \gamma \rho^{1+\frac{\lambda}{p-2}} \int_0^t \tau^{\frac{p+1}{\lambda}-1} \left(\tau^{N/\lambda} \frac{\|v(\cdot,\tau)\|_{\infty, B_{2\rho}}}{(2\rho)^{\frac{p}{p-2}}} \right)^{\frac{(p-2)(p+1)}{p}} \left(\int_{B_{2\rho}} \frac{v(x,\tau)}{\rho^{\frac{\lambda}{p-2}}} dx \right) d\tau$$

$$\leq \gamma \rho^{1+\frac{\lambda}{p-2}} \int_0^t \tau^{\frac{p+1}{\lambda}-1} \left[f(\tau) \right]^{\frac{(p-2)(p+1)}{p}} |||v|||_r \, d\tau$$

$$\leq \gamma \, \rho^{1+\frac{\lambda}{p-2}} \, t^{\frac{p+1}{\lambda}} \left[f(t) \right]^{\frac{(p-2)(p+1)}{p}} |||v|||_r.$$

By Corollary 8.1

$$f(t) \leq \gamma \, |||v|||_r^{p/\lambda}, \qquad \forall \, t \in \left(0, \gamma_* |||v|||_r^{2-p}\right).$$

Therefore for all such t,

$$L_1 \leq \gamma \, \rho^{1+\frac{\lambda}{p-2}} \, t^{\frac{1}{\lambda}} \left(t |||v|||_r^{p-2} \right)^{\frac{p}{\lambda}} |||v|||_r^{1+\frac{p-2}{\lambda}}$$
$$\leq \gamma \, \rho^{1+\frac{\lambda}{p-2}} \, t^{\frac{1}{\lambda}} \, |||v|||_r^{1+\frac{p-2}{\lambda}}.$$

Next

$$L_2 \leq \gamma \rho^{1+\frac{\lambda}{p-2}} \int_0^t \tau^{\frac{1}{\lambda}-1} \left(\tau^{N/\lambda} \frac{\|v(\cdot,\tau)\|_{\infty, B_{2\rho}}}{(2\rho)^{\frac{p}{p-2}}} \right)^{\frac{p-2}{p}} \left(\int_{B_{2\rho}} \frac{v(x,\tau)}{\rho^{\frac{\lambda}{p-2}}} \, dx \right) d\tau$$

$$\leq \gamma \, \rho^{1+\frac{\lambda}{p-2}} \, t^{\frac{1}{\lambda}} \, |||v|||_r^{1+\frac{p-2}{\lambda}}.$$

On the other hand $J_2(t) \equiv L_2$ and the Lemma follows.

Remark 9.1. The estimates above show that $\gamma=\gamma(N,p)\nearrow\infty$, as $p\searrow 2$.

Remark 9.2. The proof is independent of the fact that the initial datum is of compact support and that v is a solution in the whole Σ_∞. The lemma continues to hold for every non-negative solution in Σ_T for some $T>0$, provided the quantities

$$\sup_{0<\tau<T-\varepsilon}\ \sup_{\rho>r}\frac{\|u(\cdot,\tau)\|_{\infty,B_\rho}}{\rho^{p/(p-2)}}$$

are finite for all $\varepsilon\in(0,T)$. The conclusion will hold for all times

$$0<t\le\gamma_*|\!|\!|u|\!|\!|_{r,T-\varepsilon}. \tag{9.1$'$}$$

Remark 9.3. Lemma 9.1 is independent of the homogeneous structure of the p.d.e. Indeed it continues to hold, in the same form, for equations with homogeneous structure as in (2.8)-(2.9) of Chap. V, provided the analog of Corollary 8.1 is in force. A version of this Corollary can be proved, by essentially the same technique, for solutions of equations with general structure such as (2.1) of Chap. V.

Remark 9.4. The functional dependence upon t on the right hand side of (9.1) is optimal, as shown by the following example. The family

$$\mathcal{B}_r(x,t)=\left(t+r^\lambda\right)^{-N/\lambda}\left\{1-\gamma_p\left(\frac{|x|}{[t+r^\lambda]^{1/\lambda}}\right)^{\frac{p}{p-1}}\right\}_+^{\frac{p-1}{p-2}}, \tag{9.2}$$

$$\gamma_p=\left(\frac{1}{\lambda}\right)^{\frac{1}{p-1}}\frac{p-2}{p},\quad p>2;\ t,r>0,$$

solves (8.1) with initial data $\mathcal{B}_r(\cdot,0)$ supported in the ball B_r. By calculation we have, for all $\rho\ge r$

$$\int_0^t\int_{B_{2\rho}}|D\mathcal{B}_r|^{p-1}dx\,d\tau=\gamma(N,p)\left[t+r^\lambda\right]^{1/\lambda}-r,$$

where $\gamma(N,p)$ is an explicit constant independent of r and t. The assertion follows by letting $r\to 0$.

Let E_o denote the integral average of the initial datum, i.e.

$$E_o\equiv\fint_{B_r}v_o\,dx. \tag{9.3}$$

By the conservation of mass (8.4)

$$\begin{aligned}
|||v|||_r &\equiv \sup_{t\in\mathbf{R}^+}\sup_{\rho\geq r} \rho^{-\frac{\lambda}{p-2}} \int_{B_\rho} v(x,t)\,dx \\
&\leq \sup_{t\in\mathbf{R}^+}\sup_{\rho\geq r} \rho^{-\frac{\lambda}{p-2}} \int_{\mathbf{R}^N} v(x,t)\,dx \\
&= r^{-\frac{p}{p-2}} E_o.
\end{aligned}$$

Therefore Lemma 9.1 can be rephrased as

LEMMA 9.2. *There exists a constant* $\gamma=\gamma(N,p)$ *such that*

(9.4) $$\text{for all } 0<t<\gamma_* \frac{r^p}{E_o^{p-2}}, \text{ and for all } \rho\geq r$$

(9.5) $$\frac{1}{\rho}\int_0^t \fint_{B_\rho} |Dv|^{p-1}\,dx d\tau \leq \gamma \left(\frac{t}{r^p}\right)^{\frac{1}{\lambda}} \left(\frac{\rho}{r}\right)^{\frac{p}{p-2}} E_o^{1+\frac{p-2}{\lambda}}.$$

10. Proof of Proposition 8.1 continued

Let ζ be the standard cutoff function in B_{2r} that equals one on B_r. In the weak formulation of (8.1) take $x\rightarrow\zeta^p(x)$ as a testing function and integrate over $B_{2r}\times(0,t_o)$, where

(10.1) $$t_o = \varepsilon\,\gamma_* \frac{r^p}{E_o^{p-2}}$$

and ε is a small positive constant to be chosen. Making use of (9.5) we obtain

$$\begin{aligned}
\fint_{B_{2r}} v(x,t_o)\,dx &\geq 2^{-N} \fint_{B_r} v_o\,dx - \frac{p}{r}\int_0^{t_o}\fint_{B_{2r}} |Dv|^{p-1}\,dx \\
&\geq 2^{-N} E_o - \gamma\,(\varepsilon\gamma_*)^{1/\lambda} E_o \\
&= 2^{-(N+1)} E_o,
\end{aligned}$$

for the choice

$$\gamma\,(\varepsilon\gamma_*)^{1/\lambda} = 2^{-(N+1)}.$$

We summarise:

LEMMA 10.1. *There exist a constant c_* that can be determined a priori only in terms of N and p, such that*

$$\fint_{B_{2r}} v(x,t_o)\,dx \ge 2^{-(N+1)}\,E_o, \qquad t_o = c_*\frac{r^p}{E_o^{p-2}}.$$

Since v is continuous, there exists some $\bar{x}\in B_{2r}$ such that

(10.2) $$v(\bar{x},t_o) \ge 2^{-(N+1)}\,E_o.$$

Next we apply the Harnack estimate of Theorem 2.1. For this, construct the cylinder

$$Q \equiv B_{4\delta r}(\bar{x}) \times \left\{ t_o - \frac{4C\,(\delta r)^p}{[v(\bar{x},t_o)]^{p-2}}, t_o + \frac{4C\,(\delta r)^p}{[v(\bar{x},t_o)]^{p-2}} \right\},$$

where C is the constant determined in Theorem 2.1 and $\delta>0$ is to be chosen. Such a box is contained in Σ_∞ if

$$t_o \ge \frac{4C\,(\delta r)^p}{[v(\bar{x},t_o)]^{p-2}}.$$

Using (10.1) and (10.2) we see that this is the case if

$$\varepsilon\,\gamma_*\frac{r^p}{E_o^{p-2}} \ge \frac{4C(\delta r)^p 2^{(N+1)(p-2)}}{E_o^{p-2}}.$$

Therefore $Q\subset\Sigma_\infty$ for the choice

$$\delta^p = \frac{\varepsilon\,\gamma_*}{2^\lambda C}.$$

It follows from Theorem 2.1 that

$$\forall\,\{|x-\bar{x}|<\delta r\} \quad \text{at the time level} \quad \bar{t} = t_o + \frac{4C\,(\delta r)^p}{[v(\bar{x},t_o)]^{p-2}}$$

(10.3) $$\begin{aligned} v(x,\bar{t}) &\ge \gamma^{-1}\,v(\bar{x},t_o)\\ &\ge 2^{-(N+1)}\gamma^{-1}\,E_o\\ &\equiv c_o\,E_o. \end{aligned}$$

Therefore we have located a ball or radius δr about $\bar{x}$ and at the time level $\bar{t}$ where v is bounded below by c_oE_o. In view of (10.1) and (10.2) the time level $\bar{t}$ is bounded above by

(10.4) $$\begin{aligned} \bar{t} &\le \left(\varepsilon\,\gamma_* + C\delta^p 2^{(N+1)(p-2)}\right)\frac{r^p}{E_o^{p-2}}\\ &\equiv B_1\,\frac{r^p}{E_o^{p-2}}. \end{aligned}$$

11. Proof of Proposition 8.1 concluded

Let $\theta > 0$ be fixed. We may assume that

$$\theta \geq 2B_1 \frac{r^p}{E_o^{p-2}}. \tag{11.1}$$

Indeed otherwise

$$E_o \leq B \left(\frac{\theta}{r^p}\right)^{\frac{1}{p-2}}, \qquad B \equiv (2B_1)^{\frac{1}{p-2}},$$

and (8.2) becomes trivial. We will expand the bound below on v given by (10.3), up to the time level θ over the ball B_r. Consider the *'fundamental solution'* $\mathcal{B}_{k,\rho}$ introduced in (3.1)-(3.2), with pole at $(\bar{x}, \bar{t})$ and

$$\rho = \delta r, \qquad k = c_o E_o.$$

By the comparison principle,

$$v(x,t) \geq \mathcal{B}_{c_o E_o, \delta r}\left(x, t; \bar{x}, \bar{t}\right), \quad \forall x \in \mathbf{R}^N, \forall t \geq \bar{t}. \tag{11.2}$$

Let us estimate below the right hand side of (11.2) at the time level $t = \theta$. First by (10.4) and (11.1)

$$\theta - \bar{t} \geq \frac{1}{2}\theta,$$

therefore the support of $x \rightarrow \mathcal{B}_{c_o E_o, \delta r}\left(x, \theta; \bar{x}, \bar{t}\right)$ will cover the ball B_{4r} about the origin if

$$\begin{aligned} S(\theta) &= b\,(c_o E_o)^{p-2}\,(\delta r)^{N(p-2)}\,(\theta - \bar{t}) + (\delta r)^{\lambda} \\ &\geq \frac{b}{2}\,(c_o E_o)^{p-2}\,(\delta r)^{N(p-2)}\,\theta \\ &\geq (8r)^{\lambda}. \end{aligned}$$

This will occur if

$$E_o^{p-2} > B_2 \left(\frac{r^p}{\theta}\right) \qquad B_2 = \frac{2^{3\lambda+1}}{b\left[c_o \delta^N\right]^{p-2}}. \tag{11.3}$$

We may assume that (11.3) is in force and estimate above

$$\begin{aligned} S(\theta) &\leq b\left[c_o \delta^N\right]^{p-2} E_o^{p-2}\, r^{N(p-2)}\,\theta + \delta^{\lambda} r^{N(p-2)}\, r^p \\ &\leq \left\{ b\left[c_o \delta^N\right]^{p-2} + \frac{\delta^{\lambda}}{B_2} \right\} E_o^{p-2}\, r^{N(p-2)}\,\theta \\ &\equiv \delta_1\, E_o^{p-2}\, r^{N(p-2)}\,\theta. \end{aligned}$$

We return to (11.2). These estimates imply that for all $x \in B_r$ for $t = \theta$

$$
\begin{aligned}
v(x,\theta) &\geq \frac{c_o E_o\,(\delta r)^N}{\delta_1^{N/\lambda}\left[E_o^{p-2}\,r^{N(p-2)}\,\theta\right]^{N/\lambda}}\left\{1-\left(\frac{1}{2}\right)^{\frac{p}{p-1}}\right\}^{\frac{p-1}{p-2}}\\
&\equiv \delta_o\left(\frac{r^p}{\theta}\right)^{N/\lambda} E_o^{p/\lambda}.
\end{aligned}
$$

Therefore

$$
E_o \equiv \fint_{B_r} v_o\,dx \leq \delta_o^{-\lambda/p}\left(\frac{\theta}{r^p}\right)^{N/p}\left[\inf_{B_r} v(\cdot,\theta)\right]^{\lambda/p},
$$

and the Proposition follows with $B=\max\{(2B_1)^{\frac{1}{p-2}};B_2^{\frac{1}{p-2}};\delta_o^{-\lambda/p}\}$.

12. The Cauchy problem with compactly supported initial data

The proof of (7.2) is based on comparing u with the unique solution of the Cauchy problem

$$
(12.1)\qquad \begin{cases} v\in C\left(\mathbf{R}^+;L^2(\mathbf{R}^N)\right)\cap L^p\left(\mathbf{R}^+;W^{1,p}(\mathbf{R}^N)\right),\\ v_t-\operatorname{div}|Dv|^{p-2}Dv=0 \ \text{ in } \Sigma_\infty\equiv\mathbf{R}^N\times\mathbf{R}^+,\\ v(\cdot,0)=v_o\geq 0 \text{ and } \begin{cases}\in C(\overline{B}_r) & \text{for some } r>0,\\ \equiv 0 & \text{in } \mathbf{R}^N\backslash\overline{B}_r.\end{cases}\end{cases}
$$

Such a problem plays a role also in the theory of Harnack estimates for non-negative weak solutions of (1.1) in the *singular* case $1<p<2$. The Cauchy problem for general initial data in $L^1_{loc}\left(\mathbf{R}^N\right)$ and all $p>1$ will be studied in Chaps. XI and XII. To render the theory of Harnack inequalities self–contained we briefly discuss the unique solvability of (12.1) for all $p>1$. First, the notion of solution is:

(a) For every compact subset $\mathcal{K}\subset\mathbf{R}^N$ and for every $T>0$, u is a *local* solutions of the p.d.e. in $\mathcal{K}\times(0,T)$, in the sense of (1.2)-(1.4) of Chap. II.

(b) $v(\cdot,t)\longrightarrow v_o$ in $L^2(\mathbf{R}^N)$.

PROPOSITION 12.1. *There exists a unique solution to (12.1) for all $p>1$.*

PROOF: For $n=1,2,\dots$ let B_n be the ball of radius n about the origin and consider the boundary value problems

$$
(12.2)\qquad \begin{cases} v_n\in C\left(\mathbf{R}^+;L^2(B_n)\right)\cap L^p\left(\mathbf{R}^+;W_o^{1,p}(B_n)\right)\\ v_{n,t}-\operatorname{div}|Dv_n|^{p-2}Dv_n=0 \ \text{ in } B_n\times(0,n),\\ v_n(\cdot,0)=v_o\in L^2(B_n).\end{cases}
$$

The functions v_n vanish in the sense of the traces on $|x|=n$. We regard them as defined in the whole Σ_∞ by extending them to zero for $|x|>n$. The problems (12.2) can be uniquely solved by a Galerkin[1] procedure and give solutions v_n satisfying

$$\sup_{t\ge 0}\int_{\mathbf{R}^N}|v_n(t)|^2\,dx+2\iint_{\Sigma_\infty}|Dv_n|^p\,dxd\tau=\|v_o\|^2_{2,B_r},\qquad \forall n\in\mathbf{N}. \tag{12.3}$$

The sequence $\{v_n\}_{n\in\mathbf{N}}$ is equibounded[2] in Σ_∞, and uniformly Hölder continuous[3] in $\Sigma\times(\varepsilon,\infty)$ for all $\varepsilon>0$. In the weak formulation of (12.2), we take the testing function $(v_n+\varepsilon)^{p-2}v_n$, modulo a Steklov average. Letting $\varepsilon\to 0$ gives

$$\sup_{t\ge 0}\int_{\mathbf{R}^N}|v_n(t)|^p\,dx\le\|v_o\|^p_{p,B_r},\qquad \forall n\in\mathbf{N}. \tag{12.4}$$

Therefore

$$v_n\in L^p\left(\mathbf{R}^+;W^{1,p}(\mathbf{R}^N)\right)\quad\text{uniformly in } n.$$

A subsequence can be selected and relabelled with n such that $v_n\to v$ uniformly on compact subsets of Σ_∞ and weakly in $L^p\left(\mathbf{R}^+;W^{1,p}(\mathbf{R}^N)\right)$. The limit v is in the function space specified by (12.1), it is Hölder continuous in $\Sigma\times(\varepsilon,\infty)$ for all $\varepsilon>0$, and it satisfies the p.d.e. weakly in Σ_∞. To prove this we select a compact subset $\mathcal{K}\subset\mathbf{R}^N$ and some $T>0$. Then if n is so large that $\mathcal{K}\subset B_n$, we write (12.2) weakly against testing functions supported in $\mathcal{K}\times(0,T)$. The limiting process can be carried on the basis of the previous compactness and the non-linear term is identified by means of Minty's Lemma.[4] It remains to show that v takes the initial data v_o in the sense of $L^2(\mathbf{R}^N)$. Let $\eta\in(0,1)$ be arbitrary and let $v_{o,\eta}$ be a mollification of v_o such that

$$\|v_o-v_{o,\eta}\|_{2,\mathbf{R}^N}\longrightarrow 0\quad\text{as}\quad \eta\searrow 0.$$

In the weak formulation of (12.1), take the testing function $v_n-v_{o,\eta}$ modulo a Steklov average. If n is so large that $\operatorname{supp}[v_{o,\eta}]\subset B_n$, we obtain

$$\int_{\mathbf{R}^N}|v_n-v_{o,\eta}|^2(t)dx\le\|v_o-v_{o,\eta}\|^2_{2,\mathbf{R}^N}+\gamma\int_0^t\int_{\mathbf{R}^N}|Dv_{o,\eta}|^p dxd\tau,\quad \forall t>0,$$

for a constant γ depending only upon p. Letting $n\to\infty$,

$$\|v(\cdot,t)-v_o\|^2_{2,\mathcal{K}}\le 2\|v_o-v_{o,\eta}\|^2_{2,\mathbf{R}^N}+\gamma\int_0^t\int_{\mathbf{R}^N}|Dv_{o,\eta}|^p dxd\tau,\quad \forall t>0,$$

---

(1) See J.L.Lions [73] or Ladyzhenskaja–Solonnikov–Ural'tzeva [67].

(2) By the weak maximum principle of Theorem 3.3 of Chap. V.

(3) By the Hölder estimates of Theorem 1.2 of Chap. III and Theorem 1.2 of Chap. IV.

(4) See G. Minty [78].

for all compact subsets $\mathcal{K}\subset\mathbf{R}^N$. From this,

$$\lim_{t\searrow 0}\|v(\cdot,t)-v_o\|_{2,\mathcal{K}}=2\|v_o-v_{o,\eta}\|_{2,\mathbf{R}^N},\qquad \text{for all } \eta\in(0,1).$$

To prove uniqueness we first write the p.d.e. satisfied by the difference $w=v_1-v_2$ of two possibly distinct solutions originating from the same initial datum v_o, i.e.,

$$\text{(12.5)}\qquad \begin{cases} w\in C\left(\mathbf{R}^+;L^2(\mathbf{R}^N)\right)\cap L^p\left(\mathbf{R}^+;W^{1,p}(\mathbf{R}^N)\right),\\ w_t-\operatorname{div}\left(|Dv_1|^{p-2}Dv_1-|Dv_2|^{p-2}Dv_2\right)=0,\ \text{ in } \Sigma_\infty,\\ w(\cdot,0)=0,\ \text{ in } L^2_{loc}(\mathbf{R}^N).\end{cases}$$

In the weak formulation of (12.5), take the testing function $w\zeta$, modulo a Steklov average, where $x\rightarrow\zeta(x)$ is a non-negative piecewise smooth cutoff function in the ball B_{2R} that equals one on B_R and such that $|D\zeta|\le 1/R$. This gives, for all $t>0$,

$$\begin{aligned}\frac{1}{2}\int_{B_R}|w|^2(t)dx+\int_0^t\int_{B_{2R}}&\left\langle|Dv_1|^{p-2}Dv_1-|Dv_2|^{p-2}Dv_2,Dv_1-Dv_2\right\rangle\zeta\,dxd\tau\\ &=-\int_0^t\int_{B_{2R}}\left\langle|Dv_1|^{p-2}Dv_1-|Dv_2|^{p-2}Dv_2,D\zeta\right\rangle w\,dxd\tau.\end{aligned}$$

The second integral on the left hand side is non-negative[(1)] and it is discarded. Therefore

$$\int_{B_R}|w|^2(t)\,dx\le\frac{\|w\|_{p,\Sigma_\infty}\left(\|Dv_1\|_{p,\Sigma_\infty}+\|Dv_2\|_{p,\Sigma_\infty}\right)^{p-1}}{R}.$$

Uniqueness follows letting $R\rightarrow\infty$.

A similar argument proves the following weak comparison principle.

LEMMA 12.1. *Let v_i, $i=1,2$, be two weak solutions to (12.1) originating from bounded and compactly supported initial data $v_{o,i}$, $i=1,2$, satisfying $v_{o,1}\le v_{o,2}$. Then $v_1\le v_2$ in Σ_∞.*

PROOF: In the weak formulation of the difference $w=v_1-v_2$, take the testing function $w^+\zeta$ modulo a Steklov average.

---

(1) See Lemma 4.4 of Chap. I.

13. Bibliographical notes

In the classical work of Moser [81,82,83], the Hölder continuity is implied by the Harnack estimate. Conversely we use the Hölder estimates of Chaps. III and IV to establish a Harnack inequality. This point of view, even though not explicitly stated, is already present in the work of Krylov and Safonov [64]. The results of §2 have been established in [40]. A version of these holds for non-negative weak solutions of the porous medium equations

$$\left\{\begin{array}{l} u \in C_{loc}\left(0,T;L^2_{loc}(\Omega)\right),\ u^m \in L^2_{loc}\left(0,T;W^{1,2}_{loc}(\Omega)\right), \\ \dfrac{\partial}{\partial t}u - \Delta u^m = 0 \ \text{ in } \Omega_T,\ m>1. \end{array}\right. \tag{13.1}$$

In particular the intrinsic Harnack estimate takes the form

THEOREM 13.1. *Let u be a non-negative weak solution of (13.1). Fix any $(x_o,t_o)\in\Omega_T$ and assume that $u(x_o,t_o)>0$. There exist constants $\gamma>1$ and $C>1$, depending only upon N and m, such that*

$$u(x_o,t_o) \le \gamma \inf_{B_\rho(x_o)} u(\cdot,t_o+\theta), \qquad \theta = \frac{C\rho^2}{[u(x_o,t_o)]^{m-1}} \tag{13.2}$$

provided the cylinder

$$Q_{4\rho}(\theta) \equiv \{|x-x_o|<4\rho\}\times\{t_o-4\theta,t_o+4\theta\}$$

is contained in Ω_T.

A version of Corollary 2.1 for (13.1) appears in [6]. For the remaining results we refer to [40]. The 'fundamental solutions' $\mathcal{B}_{k,\rho}$ are due to Barenblatt [8]. The comparsion function $\mathcal{G}_{k,\rho}$ for p close to 2, is introduced in [40]. The technical device of the family of expanding cylinders Q_τ in §4 appears in Krylov–Safonov [64]. The regularising effects of Proposition 6.1 are due to Bénilan and Crandall [9].

VII

Harnack estimates and extinction profile for singular equations

1. The Harnack inequality

We will investigate the local behaviour of non-negative solutions of the singular p.d.e.,

$$(1.1)\qquad \begin{cases} u \in C_{loc}\left(0,T;L^2_{loc}(\Omega)\right) \cap L^p_{loc}\left(0,T;W^{1,p}_{loc}(\Omega)\right), & 1<p<2 \\ u_t - \operatorname{div}|Du|^{p-2}Du = 0, & \text{in } \Omega_T. \end{cases}$$

Weak solutions of (1.1) exhibit an intriguing behaviour. Even though in general they are not locally bounded,[1] they might become extinct after a finite time. It turns out however that the Harnack inequality of Theorem 2.1 of Chap. VI continues to hold provided p satisfies the further restriction

$$(1.2)\qquad \frac{2N}{N+1} < p < 2.$$

We will show that such a range of p is optimal for a Harnack estimate to hold. The extinction in finite time, the Harnack inequality and the L^∞–estimates are linked by the range (1.2) of the parameter p.

THEOREM 1.1. *Let u be a non-negative weak solution of (1.1) and let (1.2) hold. Fix any $(x_o,t_o)\in\Omega_T$ and assume that $u(x_o,t_o)>0$. There exist constants $\gamma>1$ and $c\in(0,1)$, depending only upon N and p, such that*

$$(1.3)\qquad u(x_o,t_o) \le \gamma \inf_{B_\rho(x_o)} u(\cdot, t_o+\theta),$$

[1] See §5-(IV) of Chap. V.

where

(1.4) $$\theta = c\left[u(x_o,t_o)\right]^{2-p}\rho^p,$$

provided the cylinder

(1.5) $$Q_{4\rho}(\theta) \equiv \{|x-x_o|<4\rho\}\times\{t_o-4\theta, t_o+4\theta\}$$

is contained in Ω_T.

Remark 1.1. The statement of Theorem 1.1 is the same as that of Theorem 2.1 of Chap. VI except that now the constant c is *'relatively small'*; that is, the positivity of $u(x_o,t_o)$ spreads over the ball $B_\rho(x_o)$ but is preserved only for the *'relatively small'* time $c\left[u(x_o,t_o)\right]^{2-p}\rho^p$.

Remark 1.2. As $p \searrow \frac{2N}{N+1}$, the constant γ tends to infinity and c tends to zero. However these constants are *'stable'* as $p \nearrow 2$, i.e.,

$$\lim_{p\nearrow 2}\gamma(N,p),\ c^{-1}(N,p) = \gamma(N,2),\ c^{-1}(N,2) < \infty.$$

Therefore the classical Harnack inequality for non-negative solutions of the heat equation can be recovered by letting $p \nearrow 2$ in (1.3). The limiting process can be made rigorous by the $C^{1,\alpha}_{loc}(\Omega_T)$ estimates of Chap. IX.

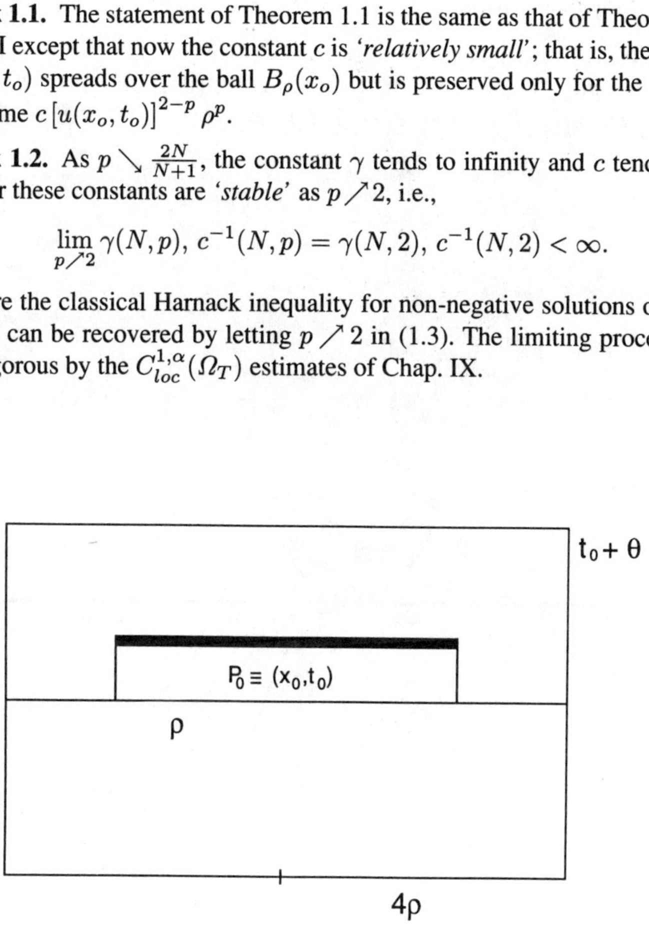

Figure 1.1

Fix $(x_o,t_o)\in\Omega_T$, assume that $u(x_o,t_o)>0$ and construct the truncated *'paraboloid'* of two sheets

$$\mathcal{P}_{\{s,\mu\}}(x_o,t_o) \equiv \left\{s \ge |t-t_o| > c\left[u(x_o,t_o)\right]^{2-p}|x-x_o|^p\mu^p\right\},$$

where c is the number claimed by Theorem 1.1 and μ and s are positive parameters. A consequence of (1.3) is the following:

COROLLARY 1.1. *Let u be a non-negative local weak solution of (1.1) and let p be in the range (1.2). There exist constants $c \in (0,1)$ and $\gamma > 1$ that can be determined a priori only in terms of N and p, such that*

$$\forall (x_o, t_o) \in \Omega_T, \ \forall s > 0 \ \text{such that} \ \mathcal{P}_{\{s,4\}}(x_o, t_o) \subset \Omega_T,$$

$$u(x_o, t_o) \le \gamma u(x,t), \quad \forall (x,t) \in \mathcal{P}_{\{s,1\}}(x_o, t_o). \tag{1.3'}$$

In particular for solutions of the Cauchy problem, we have

COROLLARY 1.2. *Let u be a non-negative local weak solution of (1.1) in $\mathbf{R}^N \times \mathbf{R}^+$ and let p satisfy (1.2). There exist constants $c \in (0,1)$ and $\gamma > 1$ that can be determined a priori only in terms of N and p, such that $(x_o, t_o) \in \mathbf{R}^N \times \mathbf{R}^+$,*

$$u(x_o, t_o) \le \gamma u(x_o, t), \quad \forall t \in [t_o, 2t_o]. \tag{1.3''}$$

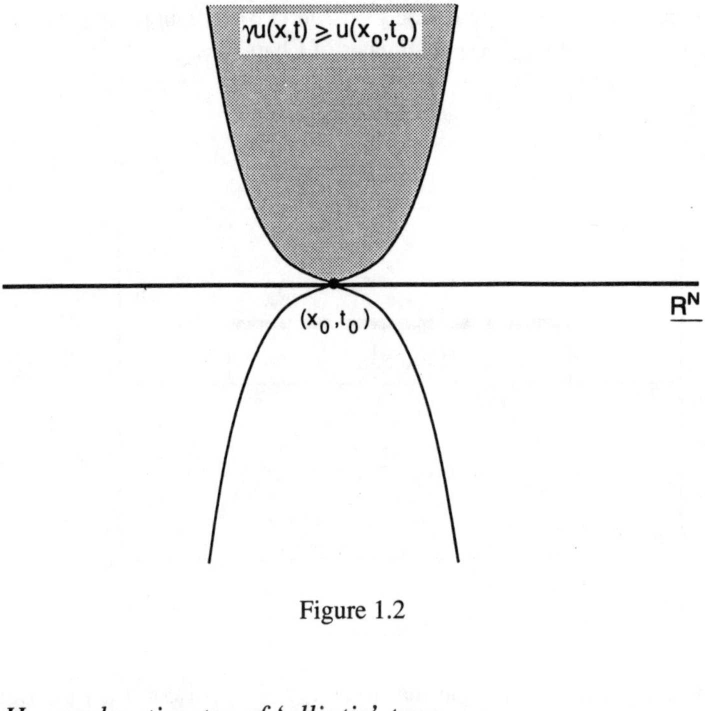

Figure 1.2

1-(i). Harnack estimates of 'elliptic' type

The p.d.e in (1.1) is singular in the sense that the modulus of ellipticity of its principal part becomes infinite at points where $|Du| = 0$. At these points the *'elliptic'* nature of the diffusion dominates the *'time-evolution'* of the process; that is, the positivity of u at some point (x_o, t_o) 'spreads' at the same time level over the full domain of definition of $x \to u(x,t_o)$. This is the content of Theorem 14.1 of Chap. IV and holds for non-negative solutions of (1.1) for the whole range $1<p<2$. When p is in the range (1.2), such a property can be made *quantitative* and takes the form of an *elliptic* Harnack inequality.

THEOREM 1.2. *Let u be a non-negative weak solution of (1.1) and let (1.2) hold. Fix any $(x_o,t_o)\in\Omega_T$ and construct the cylinder*

$$(1.6)\qquad \begin{cases} Q_{4\rho}(\theta) \equiv \{|x-x_o|<4\rho\}\times\{t_o-4\theta, t_o+4\theta\}, \\ \theta = c\,[u(x_o,t_o)]^{2-p}\,\rho^p, \qquad c>0, \end{cases}$$

where c is the constant of (1.4). There exist a constants $\gamma>1$, depending only upon N and p, such that

$$(1.7)\qquad \gamma^{-1}\sup_{B_\rho(x_o)} u(\cdot,t_o) \le u(x_o,t_o) \le \gamma \inf_{B_\rho(x_o)} u(\cdot,t_o),$$

provided $Q_{4\rho}(\theta)\subset\Omega_T$. The constant $\gamma\nearrow\infty$ as either $p\searrow\frac{2N}{N+1}$ or $p\nearrow 2$.

Remark 1.3. The strict positivity of $u(x_o,t_o)$ is not required and the Harnack estimate (1.7) holds at the *same time level.*

Remark 1.4. While Theorem 1.1 is '*stable*' as $p\nearrow 2$, this is not the case of Theorem 1.2. Indeed (1.7) fails for solutions of the heat equation. To verify this consider the heat kernel in 1−space dimension

$$\Gamma(x,t) = \frac{1}{\sqrt{4\pi\,t}}\,e^{-\frac{x^2}{4t}}, \qquad x\in\mathbf{R},$$

and apply (1.7) for the sequence of points $(x_o,t_o)\equiv(n,1)$, $n\in\mathbf{N}$. If Theorem 1.2 were to hold for $p=2$, we would have for some $\rho>0$

$$\Gamma(n,1)\le\gamma\,\Gamma(n+\rho,1).$$

Letting $n\to\infty$ we get a contradiction.

1-(ii). Generalisations

The theorems generalise to the case when the right hand side of (1.1) contains a forcing term $f(x,t,u)$ provided

$$(1.8)\qquad 0\le f(x,t,u,Du)\le f_o(x,t)+F\,u,$$

for a constant F and a function f_o satisfying

$$(1.9)\qquad f_o\in L^q_{loc}(\Omega_T), \qquad q>\frac{N+p}{p}.$$

2. Extinction in finite time (bounded domains)

PROPOSITION 2.1. *Let Ω be a bounded domain in $\mathbf{R}^N$ and let u be the unique non-negative weak solution of*

$$(2.1)\qquad \begin{cases} u \in C\left(\mathbf{R}^+; L^2(\Omega)\right) \cap L^p\left(\mathbf{R}^+; W_o^{1,p}(\Omega)\right), & 1<p<2, \\ u_t - \operatorname{div} |Du|^{p-2} Du = 0, & \text{in } \Omega_T, \\ u(\cdot,0) = u_o \in L^\infty(\Omega), & u_o \geq 0. \end{cases}$$

There exists a finite time T^* depending only upon N, p and u_o, such that

$$(2.2)\qquad u(\cdot,t) \equiv 0, \qquad \text{for all } t \geq T^*.$$

Moreover

$$(2.3)\qquad 0<T^* \leq \begin{cases} \gamma^* \|u_o\|_{2,\Omega}^{2-p} |\Omega|^{\frac{N(p-2)+2p}{2N}} & \text{if } \max\left\{1; \frac{2N}{N+2}\right\}<p<2 \\ \gamma^{**} \|u_o\|_{s,\Omega}^{2-p}, \quad s = \frac{N(2-p)}{p} & \text{if } 1<p\leq \frac{2N}{N+1}, \quad N\geq 2 \end{cases}$$

where γ^* and γ^{**} are two constants depending only upon N and p.

Remark 2.1. There is an overlap in the range of p in the two estimates of (2.3). For $1 < p < \frac{2N}{N+1}$, $N \geq 2$, the upper estimate of T^* does not depend upon the measure of Ω.

PROOF OF LEMMA 2.1: The solution of (2.1) is bounded in $\Omega \times [0,\infty)$ and Hölder continuous in $\overline{\Omega} \times [\varepsilon, \infty)$ for all $\varepsilon > 0$. Assume first that $p \geq \frac{2N}{N+1}$. In keeping with the notation of Chap. V we let

$$\lambda_r \equiv N(p-2) + rp, \ \forall r > 1 \ \text{ and } \lambda \equiv N(p-2) + p \ \text{ for } r = 1.$$

In the weak formulation of (2.1) take u as a testing function, modulo a Steklov average. This gives

$$(2.4)\qquad \frac{d}{dt}\|u(\cdot,t)\|_{2,\Omega}^2 + 2\|Du(\cdot,t)\|_{p,\Omega}^p = 0 \quad \text{in } \mathcal{D}'(\mathbf{R}^+).$$

By the Hölder inequality and the embedding of Corollary 2.1 of Chap. I, we have

$$\|u(\cdot,t)\|_{2,\Omega} \leq |\Omega|^{\frac{\lambda_2}{2Np}} \|u(\cdot,t)\|_{\frac{Np}{N-p},\Omega} \leq \gamma |\Omega|^{\frac{\lambda_2}{2Np}} \|Du(\cdot,t)\|_{p,\Omega}.$$

These remarks in (2.4) yield the differential inequality

$$\frac{d}{dt}\|u(\cdot,t)\|_{2,\Omega} + \gamma_1 \|u(\cdot,t)\|_{2,\Omega}^{p-1} \leq 0 \quad \text{in } \mathcal{D}'(\mathbf{R}^+),$$

where

$$\gamma_1 \equiv \gamma^{-p} |\Omega|^{-\frac{\lambda_2}{2N}}.$$

By integration

$$\|u(\cdot,t)\|_{2,\Omega}^{2-p} \leq \|u_o\|_{2,\Omega}^{2-p} - (2-p)\gamma_1 t, \tag{2.5}$$

as long as the right hand side is non-negative. From this,

$$\|u(\cdot,t)\|_{2,\Omega}^{\leq}\|u_o\|_{2,\Omega}\left\{1-\frac{(2-p)\gamma_1 t}{\|u_o\|_{2,\Omega}^{2-p}}\right\}_+^{\frac{1}{2-p}} \tag{2.6}$$

and

$$0 < T^* \leq \gamma^p |\Omega|^{\frac{\lambda_2}{2N}} \|u_o\|_{2,\Omega}^{2-p}.$$

Remark 2.2. The estimate (2.6) is 'stable' as $p \searrow \frac{2N}{N+1}$, i.e., as $\lambda \searrow 0$. As $p \nearrow 2$, the boundary value problem (2.1) tends to the corresponding boundary value problem for the heat equation$^{(1)}$ for which there is no extinction in finite time. Accordingly, letting $p \nearrow 2$ in (2.6) gives

$$\|u(\cdot,t)\|_{2,\Omega} \leq \|u_o\|_{2,\Omega}\, e^{-t/\gamma^2|\Omega|^{2/N}},$$

where γ is the constant of the embedding of Corollary 2.1 of Chap. I. Next we take p in the range

$$1 < p < \frac{2N}{N+1}, \qquad N \geq 2.$$

In the weak formulation of (2.1) we select, modulo a Steklov average, the testing function u^{s-1}, where

$$s = N\frac{2-p}{p} > 1. \tag{2.9}$$

This gives

$$\frac{1}{s}\frac{d}{dt}\|u(\cdot,t)\|_{s,\Omega}^{s} + \gamma_2 \|Du^{\frac{s+(p-2)}{p}}(\cdot,t)\|_{p,\Omega}^{p} = 0 \quad \text{in } \mathcal{D}'(\mathbf{R}^+),$$

where

$$\gamma_2 \equiv (s-1)\left(\frac{p}{s+(p-2)}\right)^p.$$

By the embedding of Corollary 2.1 of Chap. I and the specific choice of s, we have

---

$^{(1)}$ If $u^{(p)}$ are the solutions of (2.1) and u is the solution of (2.1) with $p=2$, the convergence takes place in the sense

$$u^{(p)},\, u_{x_i}^{(p)} \longrightarrow u,\, u_{x_i} \quad \text{in } C^\alpha\left[\overline{\Omega}\times(\varepsilon,\infty)\right],\ \forall\varepsilon>0, \quad i=1,2,\ldots,N,$$

and

$$\|u^{(p)}(\cdot,t) - u(\cdot,t)\|_{2,\Omega} \longrightarrow 0 \qquad \text{uniformly in } [0,T],\ \forall T>0.$$

Estimates of $u_{x_i}^{(p)}$ in $C^\alpha\left(\overline{\Omega}\times(\varepsilon,\infty)\right)$ uniform in $p>2N/(N+2)$ will be given in Chap. X.

$$\int_\Omega u^s(x,t)dx = \int_\Omega u^{\frac{s+(p-2)}{p}\frac{Np}{N-p}}(x,t)dx$$

$$\leq \gamma^{\frac{Np}{N-p}}\left(\int_\Omega \left|Du^{\frac{s+(p-2)}{p}}\right|^p(x,t)dx\right)^{\frac{N}{N-p}}.$$

We conclude that

$$\frac{d}{dt}\|u(\cdot,t)\|_{s,\Omega} + \gamma_3\|u(\cdot,t)\|_{s,\Omega}^{p-1} \leq 0 \ \text{ in } \mathcal{D}'(\mathbf{R}^+), \ \ \gamma_3 \equiv \gamma^{-p}\gamma_2.$$

From this, by integration

$$\|u(\cdot,t)\|_{s,\Omega} \leq \|u_o\|_{s,\Omega}\left\{1 - \frac{(2-p)\gamma_3\, t}{\|u_o\|_{s,\Omega}^{2-p}}\right\}_+^{\frac{1}{2-p}}.$$

This in turn implies (2.3).

Remark 2.3. These estimates deteriorate as $p \nearrow \frac{2N}{N+1}$ and are *'stable'* as $p \searrow 1$. However we cannot infer the convergence of (2.1) to a boundary value problem, in some reasonable topology, since the Hölder estimates of Chap. IV deteriorate as $p \searrow 1$ and (2.1) only gives $|Du| \in L^1(\Omega_T)$ uniformly in p.

Remark 2.4. Proposition 2.1 holds for solutions of variable sign. The only modification in the proof occurs in the case $1<p<\frac{2N}{N+1}$, $N \geq 2$. For this it suffices to take the testing functions $|u|^{s-2}u$.

2-(i). The Harnack inequality and the rate of extinction

The extinction profile is defined as the set $\partial\,[u>0] \cap \Omega_\infty$. By Theorem 16.1 of Chap. IV the extinction profile of the solution of (2.1) is the portion of hyperplane $\Omega \times \{t = T^*\}$. The Harnack estimate cannot hold in a *'parabolic geometry'* independent of u, say, for example, within a cylinder of the type

$$Q_\rho(x_o,t_o) \equiv Q_\rho^+(x_o,t_o) \cup Q_\rho^-(x_o,t_o), \quad Q_\rho^\pm(x_o,t_o) \equiv B_\rho(x_o) \times \{t_o \pm \rho^p\}.$$

Indeed if (x_o,t_o) belongs to the extinction profile and ρ is so small that $Q_\rho(x_o,t_o) \subset \Omega_\infty$, the solution u of (2.1) is positive in $Q_\rho^-(x_o,t_o)$ and it vanishes identically in $Q_\rho^+(x_o,t_o)$.
The intrinsic geometry of the Harnack inequality (1.3) implies an estimate of the rate of extinction of $u(\cdot,t)$ as $t \nearrow T^*$. We let

$$M = \|u\|_{\infty,\Omega_\infty}.$$

LEMMA 2.1. *Let u be the unique non-negative weak solution of (2.1) and let p be in the range (1.2). There exists a constant γ depending only upon N and p, such that for all $(x,t) \in \Omega \times \left(\frac{T^*}{2}, T^*\right)$*

(2.7) $$u(x,t)\leq \gamma \max\left\{M^{2-p};\frac{T^*}{[\mathrm{dist}\{x,\partial\Omega\}]^p}\right\}^{\frac{1}{2-p}}\left(\frac{T^*-t}{T^*}\right)^{\frac{1}{2-p}}.$$

PROOF: Fix $x\in\Omega$ and $\frac{T^*}{2}\leq t\leq T^*$, assume that $u(x,t)>0$ and set

(2.8) $$4\rho\equiv \min\left\{\mathrm{dist}\{x,\partial\Omega\};\left(\frac{T^*}{2M^{2-p}}\right)^{1/p}\right\}.$$

We apply (1.3) over the ball $B_\rho(x)$ and the cylinder

$$Q_{4\rho}(x,t)\equiv B_{4\rho}(x)\times\{t-[u(x,t)]^{2-p}(4\rho)^p, t+[u(x,t)]^{2-p}(4\rho)^p\}.$$

By virtue of the choice (2.8) such a cylinder is contained in Ω_∞ and (1.3) holds for it. We must have

$$T^*-t\geq c[u(x,t)]^{2-p}\rho^p,$$

otherwise, by the Harnack estimate, $u(x,t)=0$ against the assumption. This in turn implies the lemma.

3. Extinction in finite time (in $\mathbf{R}^N$)

PROPOSITION 3.1. *Let u be the unique non-negative weak solution of the Cauchy problem*

(3.1) $$\begin{cases} u\in C\left(\mathbf{R}^+;L^2(\mathbf{R}^N)\right)\cap L^p\left(\mathbf{R}^+;W^{1,p}(\mathbf{R}^N)\right),\\ u_t-\mathrm{div}\,|Du|^{p-2}Du=0 \quad \textit{in } \Sigma_\infty\equiv\mathbf{R}^N\times\mathbf{R}^+,\\ u(\cdot,0)=u_o\geq 0 \textit{ and } \begin{cases}\in C(\overline{B_r}) & \textit{for some } r>0\\ \equiv 0 & \textit{in } \mathbf{R}^N\backslash\overline{B_r}.\end{cases}\end{cases}$$

Then, if

(3.2) $$1<p<\frac{2N}{N+1},\qquad N\geq 2,$$

there exists a positive number T^* depending only upon N, p and u_o such that

$$u(\cdot,t)\equiv 0,\qquad \forall t\geq T^*.$$

Moreover

(3.3) $$0<T^*\leq\gamma^{**}\|u_o\|_{s,B_r}^{2-p},\qquad s=N\frac{2-p}{p},$$

for a constant γ^{**} depending only upon N and p.

PROOF: The solution of (3.1) can be constructed as the uniform limit in Σ_∞ of the sequence $\{u_n\}_{n\in\mathbf{N}}$ of the solutions of the problems in *bounded* domains[(1)]

---

(1) See §12 of Chap. VI.

$$(3.1)_n \qquad \begin{cases} u_n \in C\left(\mathbf{R}^+; L^2(B_n)\right) \cap L^p\left(\mathbf{R}^+; W_o^{1,p}(B_n)\right), \\ \dfrac{\partial}{\partial t} u_n - \operatorname{div} |Du_n|^{p-2} Du_n = 0 \quad \text{in } B_n \times (0,n) \\ u_n(\cdot,0) = u_o \in L^2(B_n). \end{cases}$$

If p is within the range (3.2), by Proposition 2.1, the extinction time T_n^* of u_n is estimated independent of meas B_n. Moreover since $u_n \le u_{n+1}$, we also have $T_n^* \le T_{n+1}^*$. This proves (3.3).

Remark 3.1. The proposition holds for data of variable sign. Also u_o need not be of compact support and it would suffice to assume

$$u_o \in L^s(\mathbf{R}^N), \qquad s = N\tfrac{2-p}{p}.$$

The proof is the same except for making precise in what sense the solutions of $(3.1)_n$ converge to the solution of (3.1).(2)

3-(i). The range (1.2) is optimal for a Harnack estimate to hold

Fix $(x_o, t_o) \in \mathbf{R}^N \times (0, T^*)$, where t_o is so close to T^* to satisfy

$$(3.4) \qquad T^* - t_o < \frac{c}{4^p}\, t_o,$$

where c is the constant appearing in (1.4). Now choose $\rho > 0$ so large that

$$(3.5) \qquad c\,[u(x_o,t_o)]^{2-p}\, \rho^p = T^* - t_o.$$

By the choice (3.4), the box

$$Q_{4\rho}(x_o,t_o) \equiv B_{4\rho}(x_o) \times \left\{ t_o - [u(x_o,t_o)]^{2-p}\,(4\rho)^p, t_o + [u(x_o,t_o)]^{2-p}\,(4\rho)^p \right\}$$

is contained in Σ_∞. If (1.3) were to hold for $1 < p < \frac{2N}{N+1}$, $N \ge 2$, for some constants c and γ independent of ρ, it would give

$$0 < u(x_o,t_o) \le \gamma \inf_{x \in B_\rho(x_o)} u(x, T^*) = 0.$$

Remark 3.2. The choice (3.5) is possible in the whole Σ_∞.

The same arguments imply that if p is in the range (1.2), no extinction in finite time can occur, for solutions of (3.1) . Within such a range, the Harnack estimate holds. Therefore if a finite extinction time T^* were to exist, the choices (3.4)-(3.5) would give $u(x, T^*) > 0$.

---

(2) We have $u_n \in L^s(\Sigma_\infty)$ uniformly in n. However such an order of integrability is not sufficient to guarantee that the solutions u_n are bounded. The number $\lambda_s = N(p-2) + sp$ is zero and the condition (5.1)-(5.3) of Chap. V are violated. Questions of convergence for general initial data in $L^1_{loc}\left(\mathbf{R}^N\right)$ will be discussed in Chap. XII.

4. An integral Harnack inequality for all $1<p<2$

A weak integral form of (1.3) holds for any non-negative weak solution of (1.1) for p in the whole range $1<p<2$, and it is crucial in the proof of the *pointwise* estimate (1.3). In the estimates to follow we denote with $\gamma=\gamma(N,p)$ a generic positive constant, which can be determined a priori only in terms of N and p and which can be different in different contexts.

PROPOSITION 4.1. *Let u be a non-negative weak solution of (1.1) and let $1<p<2$. There exists a constant $\gamma=\gamma(N,p)$ such that*

$$\forall(x_o,t_o)\in\Omega_\infty,\ \forall\rho>0 \ \text{ such that } \ B_{4\rho}(x_o)\subset\Omega,\ \forall t>t_o$$

$$\sup_{t_o\le\tau\le t}\int_{B_\rho(x_o)}u(x,\tau)dx\le\gamma\inf_{t_o\le\tau\le t}\int_{B_{2\rho}(x_o)}u(x,\tau)dx+\gamma\left(\frac{t-t_o}{\rho^\lambda}\right)^{\frac{1}{2-p}}. \tag{4.1}$$

Since $1<p<2$, the number $\lambda=N(p-2)+p$ might be of either sign. The proposition can be regarded as a weak form of a Harnack estimate, in that the L^1–norm of $u(\cdot,t)$ over a ball controls the L^1–norm of $u(\cdot,\tau)$ over a smaller ball, for any previous or later time. It could be stated over any pair of balls $B_\rho(x_o)$ and $B_{\sigma\rho}(x_o)$ for $\sigma\in(0,1)$. The constant $\gamma=\gamma(N,p,\sigma)$ would depend also on σ and $\gamma(N,p,\sigma)\nearrow\infty$ as $\sigma\nearrow 1$.

Remark 4.1. The proof shows that the constant $\gamma(N,p)$ deteriorates as $p\nearrow 2$. The proof depends on some local integral estimates of the gradient $|Du|$ which we derive next.

4-(i). Estimating the gradient of u

PROPOSITION 4.2. *Let u be a non-negative weak solution of (1.1) and let $1<p<2$. There exists a constant $\gamma=\gamma(N,p)$ such that*

$$\forall(x_o,t_o)\in\Omega_\infty,\ \forall\rho>0 \ \text{ such that } \ B_{4\rho}(x_o)\subset\Omega,\ \forall t>t_o,\ \forall\nu>0,\ \forall\sigma\in(0,1),$$

there holds

$$\int_{t_o}^{t}\int_{B_{\sigma\rho}(x_o)}(\tau-t_o)^{\frac{1}{p}}(u+\nu)^{-\frac{2}{p}}|Du|^p dxd\tau \tag{4.2}$$

$$\le\frac{\gamma\rho}{(1-\sigma)^p}\left[1+\left(\frac{t-t_o}{\rho^p}\right)\nu^{p-2}\right]\left(\frac{t-t_o}{\rho^\lambda}\right)^{\frac{1}{p}}$$

$$\times\left\{\sup_{t_o\le\tau\le t}\int_{B_\rho(x_o)}u(x,\tau)dx+\nu\rho^N\right\}^{\frac{2(p-1)}{p}},$$

$$\text{(4.3)}\qquad \frac{1}{\rho}\int_{t_o}^{t}\int_{B_{\sigma\rho}(x_o)}|Du|^{p-1}dxd\tau$$
$$\le \frac{\gamma}{(1-\sigma)^{p-1}}\left(\frac{t-t_o}{\rho^\lambda}\right)^{\frac{1}{p}}\left\{\sup_{t_o\le\tau\le t}\int_{B_\rho(x_o)}u(x,\tau)dx\right\}^{\frac{2(p-1)}{p}}$$
$$+\frac{\gamma}{(1-\sigma)^{p-1}}\left(\frac{t-t_o}{\rho^\lambda}\right)^{\frac{1}{2-p}},$$

$$\text{(4.4)}\qquad \frac{1}{\rho}\int_{t_o}^{t}\int_{B_{\sigma\rho}(x_o)}|Du|^{p-1}dxd\tau$$
$$\le\gamma\left\{\sup_{t_o\le\tau\le t}\int_{B_\rho(x_o)}u(x,\tau)dx\right\}+\gamma(\sigma)\left(\frac{t-t_o}{\rho^\lambda}\right)^{\frac{1}{2-p}}$$

Remark 4.2. The estimates (4.1)-(4.4) have been stated *'locally'*. However they continue to hold for $t_o=0$, i.e. for cylinders $B_\rho(x_o)\times(0,t)$ carrying the *'initial data'*.

Remark 4.3. The constant $\gamma(N,p)$ in (4.2)-(4.4) tends to infinity as $p\nearrow 2$.

PROOF OF PROPOSITION 4.2: We translate the coordinates so that (x_o,t_o) coincides with the origin and will work with non-negative weak solutions of

$$\text{(4.5)}\qquad u_t-\operatorname{div}|Du|^{p-2}Du=0,\ \text{ in } B_{4\rho}\times(0,\infty),\quad p\in(1,2).$$

Fix $\sigma\in(0,1)$ and let $x\to\zeta(x)$ be a non-negative piecewise smooth cutoff function in B_ρ that equals one on $B_{\sigma\rho}$ and such that $|D\zeta|\le 1/(1-\sigma)\rho$. In the weak formulation of (4.5) take the testing function

$$\varphi=-t^{\frac{1}{p}}\,(u+\nu)^{1-\frac{2}{p}}\,\zeta^p,\qquad \nu>0,$$

modulo a Steklov averaging process. We obtain for all $t>0$

$$(4.6)\qquad \frac{2-p}{p}\int_0^t\int_{B_\rho}\tau^{\frac{1}{p}}(u+\nu)^{-\frac{2}{p}}|Du|^p\zeta^p dxd\tau$$
$$=\frac{p}{2(p-1)}t^{\frac{1}{p}}\int_{B_\rho}(u+\nu)^{\frac{2(p-1)}{p}}(x,t)\zeta^p dx$$
$$+\frac{1}{2(p-1)}\int_0^t\int_{B_\rho}\tau^{\frac{1}{p}-1}(u+\nu)^{\frac{2(p-1)}{p}}\zeta^p dxd\tau$$
$$+p\int_0^t\int_{B_\rho}\tau^{\frac{1}{p}}|Du|^{p-2}Du\cdot D\zeta\zeta^{p-1}(u+\nu)^{1-\frac{2}{p}}dxd\tau.$$

We estimate the various terms on the right hand side in terms of the quantity

$$\mathcal{S}\equiv\sup_{0\le\tau\le t}\int_{B_\rho}u(x,\tau)\,dx.$$

Since $p\in(1,2)$ we have $\frac{2(p-1)}{p}<1$. Therefore by the Hölder inequality

$$\text{(i)}\qquad \frac{p}{2(p-1)}t^{\frac{1}{p}}\int_{B_\rho}(u+\nu)^{\frac{2(p-1)}{p}}\zeta^p dx$$
$$\le\gamma t^{\frac{1}{p}}\rho^{\frac{N(2-p)}{p}}\left(\sup_{0\le\tau\le t}\int_{B_\rho}u(x,\tau)dx+\nu\rho^N\right)^{\frac{2(p-1)}{p}}$$
$$=\gamma\rho\left(\frac{t}{\rho^\lambda}\right)^{\frac{1}{p}}\left(\mathcal{S}+\nu\rho^N\right)^{\frac{2(p-1)}{p}}.$$

$$\text{(ii)}\qquad \frac{1}{2(p-1)}\int_0^t\int_{B_\rho}\tau^{\frac{1}{p}-1}(u+\nu)^{\frac{2(p-1)}{p}}dxd\tau$$
$$\le\gamma\int_0^t\tau^{\frac{1}{p}-1}d\tau\sup_{0\le\tau\le t}\int_{B_\rho}(u+\nu)^{\frac{2(p-1)}{p}}(x,\tau)dx$$
$$\le\gamma\rho\left(\frac{t}{\rho^\lambda}\right)^{\frac{1}{p}}\left(\mathcal{S}+\nu\rho^N\right)^{\frac{2(p-1)}{p}}.$$

Next, by Young's inequality

(iii)
$$p\left|\int_0^t\!\!\int_{B_\rho} \tau^{\frac{1}{p}}|Du|^{p-2}Du\cdot D\zeta\zeta^{p-1}\,(u+\nu)^{1-\frac{2}{p}}\,dxd\tau\right|$$
$$\le \frac{1}{2^{\frac{2-p}{p}}}\int_0^t\!\!\int_{B_\rho} \tau^{\frac{1}{p}}\,(u+\nu)^{-\frac{2}{p}}\,|Du|^p\zeta^p dxd\tau$$
$$+\frac{\gamma(p)}{(1-\sigma)^p\rho^p}\int_0^t\!\!\int_{B_\rho} \tau^{\frac{1}{p}}\,(u+\nu)^{p-2}\,(u+\nu)^{\frac{2(p-1)}{p}}\,dxd\tau.$$

This last integral is estimated above by

$$\rho\,\frac{\gamma(N,p)}{(1-\sigma)^p}\left(\frac{t}{\rho^p}\right)\nu^{p-2}\left(\frac{t}{\rho^\lambda}\right)^{\frac{1}{p}}\left(\mathcal{S}+\nu\rho^N\right)^{\frac{2(p-1)}{p}}.$$

Combining these estimates in (4.6) proves (4.2). To prove (4.3), write (4.2) with $(x_o,t_o)\equiv(0,0)$ and select $\nu^{2-p}=(t/\rho^p)$. Then by the Hölder inequality

(4.7)
$$\int_0^t\!\!\int_{B_{\sigma\rho}}|Du|^{p-1}dxd\tau$$
$$=\int_0^t\!\!\int_{B_{\sigma\rho}}\tau^{\frac{1}{p}\frac{p-1}{p}}\,(u+\nu)^{-\frac{2}{p}\frac{p-1}{p}}\,|Du|^{p-1}\tau^{-\frac{1}{p}\frac{p-1}{p}}\,(u+\nu)^{\frac{2}{p}\frac{p-1}{p}}\,dxd\tau$$
$$\le\left(\int_0^t\!\!\int_{B_{\sigma\rho}}\tau^{\frac{1}{p}}(u+\nu)^{-\frac{2}{p}}|Du|^pdxd\tau\right)^{\frac{p-1}{p}}\left(\int_0^t\!\!\int_{B_{\sigma\rho}}\tau^{\frac{1}{p}-1}(u+\nu)^{\frac{2(p-1)}{p}}\,dxd\tau\right)^{\frac{1}{p}}.$$

The last integral is estimated above by

(a)
$$\gamma\rho\left(\frac{t}{\rho^\lambda}\right)^{\frac{1}{p}}\left\{\mathcal{S}+\left(\frac{t}{\rho^p}\right)^{\frac{1}{2-p}}\rho^N\right\}^{\frac{2(p-1)}{p}}$$
$$\le\gamma\rho\left(\frac{t}{\rho^\lambda}\right)^{\frac{1}{p}}\mathcal{S}^{\frac{2(p-1)}{p}}+\gamma\rho\left(\frac{t}{\rho^\lambda}\right)^{\frac{1}{2-p}}.$$

The first integral on the right hand side of (4.7) is estimated by (4.2) with the indicated choice of ν and it is majorised by the same quantity on the right hand side of (a), apart for a factor $(1-\sigma)^{1-p}$. Combining these remarks in (4.7) proves (4.3). Finally (4.4) follows from (4.3) by a further application of Young's inequality.

4-(ii). Proof of Proposition 4.1

Assume that (x_o, t_o) coincides with the origin and consider the family of expanding concentric balls

$$B_n \equiv \{|x| < \rho_n\}, \qquad \rho_n = \rho \sum_{i=0}^{n} 2^{-i}, \qquad n = 0, 1, 2, \ldots.$$

We have $B_o \equiv B_\rho$ and $B_\infty \equiv B_{2\rho}$. Introduce also the *'intermediate'* spheres

$$\widetilde{B_n} \equiv \left\{ |x| < \frac{1}{2}(\rho_n + \rho_{n+1}) \right\},$$

and let $x \to \zeta_n(x)$ be a piecewise smooth non-negative cutoff function in $\widetilde{B_n}$ that equals one on B_n and such that $|D\zeta_n| \le 2^{n+2}/\rho$. In the weak formulation of (4.5) take ζ_n as a testing function to obtain

$$\int_{\widetilde{B_n}} u(x,\tau_1)\zeta_n dx \le \int_{\widetilde{B_n}} u(x,\tau_2)\zeta_n dx + \frac{2^{n+2}}{\rho} \int_{\tau_1}^{\tau_2}\int_{\widetilde{B_n}} |Du|^{p-1} dx d\tau, \tag{4.8}$$

for any two time levels τ_1 and τ_2 in $[0,t]$. We take as τ_2, a time level in $[0,t]$ such that

$$\inf_{0\le\tau\le t} \int_{B_{2\rho}} u(x,\tau)dx = \int_{B_{2\rho}} u(x,\tau_2)dx \equiv \mathcal{I}.$$

We also set

$$\mathcal{S}_n \equiv \sup_{0\le\tau\le t} \int_{B_n} u(x,\tau)dx.$$

Since $\tau_1 \in [0,t]$ is arbitrary, (4.8) implies

$$\mathcal{S}_n \le \mathcal{I} + \frac{2^{n+2}}{\rho} \int_0^t\int_{\widetilde{B_n}} |Du|^{p-1} dx d\tau.$$

Next we apply (4.3) over the pair of balls $\widetilde{B_n}$ and B_{n+1} for which $(1-\sigma) \ge 2^{-(n+2)}$. This and Young's inequality give

$$\begin{aligned}\frac{2^{n+2}}{\rho} \int_0^t\int_{\widetilde{B_n}} |Du|^{p-1} dx d\tau &\le \gamma b^n \left(\frac{t}{\rho^\lambda}\right)^{\frac{1}{p}} (\mathcal{S}_{n+1})^{\frac{2(p-1)}{p}} + \gamma b^n \left(\frac{t}{\rho^\lambda}\right)^{\frac{1}{2-p}} \\ &\le \varepsilon \mathcal{S}_{n+1} + \gamma(N,p,\varepsilon)\, b^n \left(\frac{t}{\rho^\lambda}\right)^{\frac{1}{2-p}}, \qquad b = 2^{p^2/(2-p)}.\end{aligned}$$

valid for every $\varepsilon \in (0,1)$, for some constant $\gamma(N,p,\varepsilon)$ depending only upon N, p

and ε. Combining these estimates, we conclude that for every $\varepsilon\in(0,1)$ there exists a constant $\gamma(N,p,\varepsilon)$ such that

$$\mathcal{S}_n \le \varepsilon\mathcal{S}_{n+1} + \gamma(N,p,\varepsilon)\left\{\mathcal{I} + \left(\frac{t}{\rho^\lambda}\right)^{\frac{1}{2-p}}\right\} b^n.$$

The Proposition now follows from the interpolation Lemma 4.3 of Chap. I.

5. Sup-estimates for $\frac{2N}{N+2}<p<2$

We now combine the L^∞_{loc}–estimates of §5 of Chap. V with the integral inequality (4.1). If p is in the range (1.2), we may take $r=1$ in (5.1)-(5.4) of Chap. V and rewrite the latter as

$$\forall (x_o,t_o)\in\Omega_\infty,\ \ \forall\rho>0 \ \text{ such that } \ B_{4\rho}(x_o)\subset\Omega,\ \ \forall t>t_o$$

$$\|u(\cdot,t)\|_{\infty,B_\rho(x_o)} \le \gamma\,(t-t_o)^{-\frac{N}{\lambda}}\left(\sup_{t_o\le\tau\le t}\int_{B_{3\rho/2}(x_o)} u(x,\tau)dx\right)^{p/\lambda} + \gamma\left(\frac{t-t_o}{\rho^p}\right)^{\frac{1}{2-p}}. \tag{5.1}$$

This and Proposition 4.1 imply

LEMMA 5.1. *Let u be a non-negative local weak solution of (1.1) in Ω_∞ and let (1.2) hold. There exists a constant $\gamma(N,p)$ such that*

$$\forall (x_o,t_o)\in\Omega_\infty,\ \ \forall\rho>0 \ \textit{ such that } \ B_{4\rho}(x_o)\subset\Omega,\ \ \forall t>t_o$$

$$\sup_{x\in B_\rho(x_o)} u(x,t) \le \gamma\,(t-t_o)^{-\frac{N}{\lambda}}\left(\inf_{t_o\le\tau\le t}\int_{B_{2\rho}(x_o)} u(x,\tau)dx\right)^{p/\lambda} + \gamma\left(\frac{t-t_o}{\rho^p}\right)^{\frac{1}{2-p}}. \tag{5.2}$$

The constant $\gamma(N,p)$ tends to infinity as either $p\searrow\frac{2N}{N+1}$ or as $p\nearrow 2$.

Remark 5.1. The lemma continues to hold also for $t_o=0$, i.e., for cylinders $B_\rho(x_o)\times\mathbf{R}^+$ carrying the *'initial'* data.

The peculiar feature of this estimate is that the supremum of the solution over a ball at some time level is bounded above by the L^1–norm of u over a larger ball at either the *same* time level or some *'future'* time. This is in contrast with the behaviour of non-negative solutions of the heat equation. Accordingly, the constant $\gamma(N,p)$ deteriorates as $p\nearrow 2$.

5-(i). A special form of (5.2)

We will use this fact in the following form. Let u be a non-negative weak solution of the p.d.e. in (1.1) in some space-time domain and let p be in the range (1.2). Let $R>0$ and assume that the cylinder

$$\widetilde{Q}_{4R} \equiv B_{4R} \times \{-4,\, 0\}$$

is all contained in the domain of definition of u. Then

$$\sup_{x\in B_R} u(x,\tau) \le \gamma \left(\int_{B_{2R}} u(x,0)\,dx \right)^{p/\lambda} + \gamma R^{-\frac{p}{2-p}},\ \forall \tau \in [-2,0]. \tag{5.3}$$

6. Local subsolutions

As in the degenerate case, the proof of Theorem 1.1 is based on expanding the positivity set of the solution u by means of suitable comparison functions. Let b, k, μ be positive parameters satisfying

$$\frac{\mu}{b^{p-1}}\, k^{p-2} < 1. \tag{6.1}$$

Consider the cylindrical domain with annular cross section

$$Q(\theta) \equiv \left\{ \frac{\mu}{b^{p-1}}\, k^{p-2} < |x|^p < 1 \right\} \times \{0, \theta\}, \tag{6.2}$$

and the function

$$\Psi(x;t) \equiv \frac{k(1-|x|^2)_+^{\frac{p}{p-1}}}{\left\{ 1 + k^{\frac{2-p}{p-1}} b \left(\frac{|x|^p}{t} \right)^{\frac{1}{p-1}} \right\}^{\frac{p-1}{2-p}}}. \tag{6.3}$$

LEMMA 6.1. *Assume that p is in the range (1.2), i.e.,*

$$\lambda \equiv N(p-2) + p > 0.$$

Then the constant $b = b(N,p)$ can be chosen a priori only dependent upon N and p, so that

$$\begin{gathered} \forall k > 0, \quad \forall \mu > 0 > \textit{ satisfying (6.1)}, \\ \Psi_t - \operatorname{div}(|D\Psi|^{p-2} D\Psi) \le 0 \qquad \textit{a.e. in } Q(\theta), \\ \theta = \min \left\{ \left(\frac{\lambda}{2p} \right)^{p-1} \mu\, ;\, k^{2-p} \right\}. \end{gathered} \tag{6.4}$$

Remark 6.1. The proof below shows that the constant $b>1$ is *'stable'* as $p \nearrow 2$.

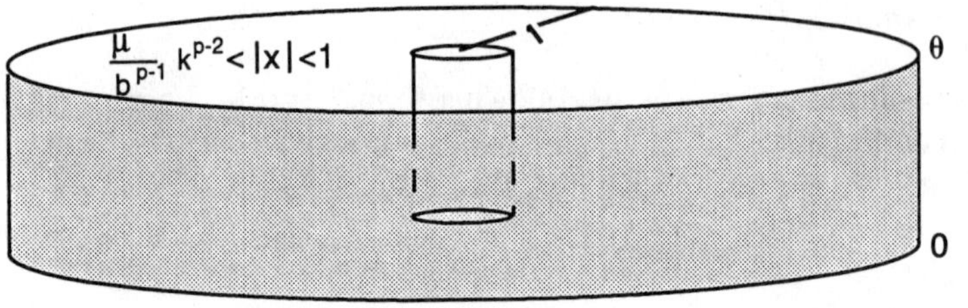

Figure 6.1

PROOF OF LEMMA 6.1: The function $x \to \Psi(x,t)$ is radial and decreasing with respect to $|x|$, so that writing (6.4) in polar coordinates we have

$$\begin{aligned}\mathcal{L}(\Psi) &\equiv \Psi_t - \operatorname{div}(|D\Psi|^{p-2}D\Psi)\\ &= \Psi_t + \left(\frac{N-1}{\rho}\right)(-\Psi')^{p-1} - (p-1)(-\Psi')^{p-2}\Psi'',\end{aligned}$$

where

$$\rho = |x|, \qquad \Psi' = \frac{d}{d\rho}\Psi\,; \quad \Psi'' = \frac{d^2}{d\rho^2}\Psi.$$

We write

$$\begin{aligned}(-\Psi')^{2-p}\mathcal{L}(\Psi) &= (-\Psi')^{2-p}\Psi_t + \mathcal{R}(\Psi),\\ \mathcal{R}(\Psi) &\equiv \frac{N-1}{\rho}(-\Psi') - (p-1)\Psi'',\end{aligned} \tag{6.5}$$

and calculate $\mathcal{R}(\Psi)$ as follows. First we set

$$\|z\| = k^{\frac{2-p}{p-1}}\, b\left(\frac{|x|^p}{t}\right)^{\frac{1}{p-1}}; \qquad \mathcal{F} = 1 + \|z\|$$

$$w = \frac{k}{\mathcal{F}^{\frac{p-1}{2-p}}} \qquad v = (1-|x|^2)^{\frac{p}{p-1}}.$$

Then by direct calculation

$$
(6.6)\qquad \begin{cases}
w' &= -\dfrac{p}{2-p}\left(\dfrac{w}{\rho}\right)\dfrac{\|z\|}{\mathcal{F}},\\
w'' &= \left(\dfrac{p}{2-p}\right)^2\left(\dfrac{w}{\rho^2}\right)\dfrac{\|z\|^2}{\mathcal{F}^2}+\dfrac{p}{2-p}\left(\dfrac{w}{\rho^2}\right)\dfrac{\|z\|}{\mathcal{F}}\\
&\qquad -\dfrac{p^2}{(2-p)(p-1)}\left(\dfrac{w}{\rho^2}\right)\dfrac{\|z\|}{\mathcal{F}^2},\\
v' &= -\dfrac{2p}{p-1}\,\rho\,(1-\rho^2)^{\frac{1}{p-1}},\\
v'' &= \dfrac{4p}{(p-1)^2}\rho^2(1-\rho^2)^{\frac{2-p}{p-1}}-\dfrac{2p}{p-1}\,(1-\rho^2)^{\frac{1}{p-1}}\\
&\geq -\dfrac{2p}{p-1}(1-\rho^2)^{\frac{1}{p-1}}.
\end{cases}
$$

We calculate the expressions $\Psi'=w'v+wv'$ and $\Psi''=w''v+2w'v'+wv''$ from (6.6) and combine them into $\mathcal{R}(\Psi)$ to obtain

$$
(6.7)\qquad \begin{aligned}
\mathcal{R}(\Psi) \leq \frac{p}{2-p}(1-\rho^2)^{\frac{p}{p-1}}\left(\frac{w}{\rho^2}\right)\frac{\|z\|}{\mathcal{F}}\Big\{N-1\\
-(p-1)\left[\frac{p}{2-p}\frac{\|z\|}{\mathcal{F}}+1\right]+p\frac{1}{\mathcal{F}}\Big\}\\
+2pw(1-\rho^2)^{\frac{1}{p-1}}\left\{\frac{(N-1)}{p-1}-\frac{2p}{2-p}\frac{\|z\|}{\mathcal{F}}+1\right\}.
\end{aligned}
$$

Rewrite the first factor in braces on the right-hand side of (6.7) as

$$
\{\cdots\}=\left(N-\frac{p}{2-p}\frac{\|z\|}{\mathcal{F}}\right).
$$

We will impose on $\|z\|$ to be so large that

$$
(6.8)\qquad N-\frac{p}{2-p}\frac{\|z\|}{\mathcal{F}}<0.
$$

This is possible since $N(p-2)+p\equiv\lambda>0$. The second term in braces on the right hand side of (6.7) is negative if we choose $\|z\|$ to satisfy (6.8). If $N=1,2$, this is a direct consequence of (6.8). If $N\geq 3$,

$$
\begin{aligned}
(p-1)\left\{\frac{N-1}{p-1}-\frac{2p}{2-p}\frac{\|z\|}{\mathcal{F}}+1\right\}=\left(N-\frac{p}{2-p}\frac{\|z\|}{\mathcal{F}}\right)\\
+\frac{p}{2-p}(3-2p)\frac{\|z\|}{\mathcal{F}}+(p-2).
\end{aligned}
$$

The first term is negative in view of (6.8) and the second is negative since $p>\frac{2N}{N+1}>\frac{3}{2}$ if $N\geq 3$. We drop the last negative term on the right hand side of (6.7) and estimate

$$\mathcal{R}(\Psi) \le \frac{p}{2-p} v \left(\frac{w}{\rho^2}\right) \frac{\|z\|}{\mathcal{F}} \left[N - \frac{p}{2-p}\frac{\|z\|}{\mathcal{F}}\right]. \tag{6.9}$$

We return to (6.5) and estimate above the term $(-\Psi')^{2-p}\Psi_t$. First using (6.6)

$$-\Psi' \le \frac{w}{\rho}\left(\frac{2p}{p-1} + \frac{p}{2-p}\right) \equiv \gamma\left(\frac{w}{\rho}\right).$$

Also

$$\Psi_t = \frac{1}{2-p} vw \frac{\|z\|}{\mathcal{F}}\frac{1}{t}.$$

Therefore

$$(-\Psi')^{2-p}\Psi_t \le \frac{\bar{\gamma}}{2-p}\left(\frac{w}{\rho}\right)^{2-p} vw\frac{\|z\|}{\mathcal{F}}\frac{1}{t}.$$

We combine this with (6.9) into (6.5) and set

$$\mathcal{L}^*(\Psi) = (-\Psi')^{2-p}[\Psi_t - \operatorname{div}(|D\Psi|^{p-2}D\Psi)]\frac{(2-p)\mathcal{F}\rho^2}{vw\|z\|}$$

to obtain

$$\mathcal{L}^*(\Psi) \le \gamma w^{2-p}\frac{\rho^p}{t} + p\left[N - \frac{p}{2-p}\frac{\|z\|}{\mathcal{F}}\right].$$

From the definition of w and $\|z\|$

$$w^{2-p}\frac{\rho^p}{t} = \left(\frac{\|z\|}{1+\|z\|}\right)^{p-1} b^{1-p} \le \frac{1}{b^{p-1}}$$

and

$$\mathcal{L}^*(\Psi) \le \frac{\gamma}{b^{p-1}} + \frac{p}{2-p}\left[-\lambda + \frac{p}{\mathcal{F}}\right].$$

We will choose $\|z\|$ so large that $\frac{p}{\mathcal{F}} \le \frac{\lambda}{2}$, and then select b from

$$\frac{\gamma}{b^{p-1}} - \frac{p}{2-p}\frac{\lambda}{2} = 0.$$

We will have $\frac{p}{\mathcal{F}} \equiv \frac{p}{1+\|z\|} \le \frac{\lambda}{2}$ if for example $\|z\| > \frac{2p}{\lambda}$, i.e. if

$$k^{2-p}\, b^{p-1}\, |x|^p \left(\frac{\lambda}{2p}\right)^{p-1} > t > 0.$$

From the construction of the cylinder $Q(\theta)$ in (6.2), we have

$$k^{2-p}\, b^{p-1}\, |x|^p \ge \mu.$$

Therefore to prove the lemma it suffices to take

$$0 < t \le \theta = \left(\frac{\lambda}{2p}\right)^{p-1} \mu.$$

7. Time expansion of positivity

The next subsolution of (1.1) will be employed to expand the set of positivity of u in the direction of increasing t. Set

$$\Phi(x;t) \equiv \frac{k\,\rho^{p\xi}}{R^{\xi}(t)}\left\{1-\left(\frac{|x|^p}{R(t)}\right)^{\frac{1}{p-1}}\right\}_+^2, \tag{7.1}$$

$$R(t) \equiv k^{p-2}\,t+\rho^p, \quad \mathcal{F} \equiv 1-\|z\|^{\frac{1}{p-1}}, \quad \|z\| \equiv \frac{|x|^p}{R(t)}.$$

Here k and ρ are positive parameters and $\xi > 1$ is a number to be chosen independent of k and ρ. For $t=0$ the function $\Phi(\cdot,0)$ is supported in the ball B_ρ and for $t>0$ the support of $x \to \Phi(x,t)$ is the *'expanding'* ball

$$|x| < \left(k^{p-2}t+\rho^p\right)^{\frac{1}{p}}.$$

We will consider Φ only within the domain

$$\mathcal{D}_{\{k,\xi\}} \equiv \left\{|x|^p < k^{p-2}t+\rho^p\right\} \times \left\{0 < t < \frac{k^{2-p}\rho^p}{\xi}\right\}. \tag{7.2}$$

The function Φ is continuous in $\overline{\mathcal{D}}_{\{k,\xi\}}$, vanishes on the *'lateral'* boundary of $\mathcal{D}_{\{k,\xi\}}$ and it is of class C^∞ in the interior of $\mathcal{D}_{\{k,\xi\}}$.

LEMMA 7.1. *The number $\xi = \xi(N,p)$ can be determined a priori only in terms of N and p so that*

$$\Phi_t - \operatorname{div}|D\Phi|^{p-2}D\Phi \le 0 \qquad \textit{in } \mathcal{D}_{\{k,\xi\}}.$$

Remark 7.1. The constant ξ is *'stable'* as $p \nearrow 2$.

PROOF OF LEMMA 7.1: By direct calculation within $\mathcal{D}_{\{k,\xi\}}$ we have

$$\Phi_t = -\frac{\xi k^{p-1}\rho^{p\xi}}{R^{\xi+1}(t)}\mathcal{F}^2 + \frac{2}{p-1}\frac{k^{p-1}\rho^{p\xi}}{R^{\xi+1}(t)}\mathcal{F}\|z\|^{\frac{1}{p-1}},$$

$$D\Phi = -\frac{2p}{p-1}\frac{k\rho^{p\xi}}{R^{\xi}(t)}\mathcal{F}\left(\frac{|x|}{R(t)}\right)^{\frac{1}{p-1}}\frac{x}{|x|},$$

$$|D\Phi|^{p-2}D\Phi = -\left(\frac{2p}{p-1}\right)^{p-1}\left[\frac{k\rho^{p\xi}}{R^{\xi}(t)}\mathcal{F}\right]^{p-1}\frac{x}{R(t)},$$

$$-\operatorname{div}|D\Phi|^{p-2}D\Phi = \left(\frac{2p}{p-1}\right)^{p-1}\frac{1}{R(t)}\left[\frac{k\rho^{p\xi}}{R^{\xi}(t)}\mathcal{F}\right]^{p-1}\left(N-p\frac{\|z\|^{\frac{1}{p-1}}}{\mathcal{F}}\right).$$

Setting

$$\mathcal{L}^*(\Phi) \equiv \frac{R^{\xi+1}}{k^{p-1}\rho^{p\xi}\mathcal{F}}\left\{\Phi_t - \operatorname{div}|D\Phi|^{p-2}D\Phi\right\},$$

the previous calculations give

$$
\begin{aligned}
(7.3) \qquad \mathcal{L}^*(\Phi) = \Big\{ & -\xi\mathcal{F} + \frac{2}{p-1}\|z\|^{\frac{1}{p-1}} \\
& + \left(\frac{2p}{p-1}\right)^{p-1}\left[\frac{\rho^{p\xi}}{R^{\xi}(t)}\mathcal{F}\right]^{p-2}\left(N - p\frac{\|z\|^{\frac{1}{p-1}}}{\mathcal{F}}\right)\Big\}.
\end{aligned}
$$

Introduce the two sets

$$
\mathcal{E}_1 \equiv \big\{(x,t) \in \mathcal{D}_{\{k,\xi\}} \mid \mathcal{F} < \delta\big\}, \quad \mathcal{E}_2 \equiv \big\{(x,t) \in \mathcal{D}_{\{k,\xi\}} \mid \mathcal{F} \ge \delta\big\}.
$$

Here δ is a small positive number to be determined so that, within $\mathcal{E}_1$, the last term on the right hand side of (7.3) is negative, i.e.,

$$
\left(N - p\frac{\|z\|^{\frac{1}{p-1}}}{\mathcal{F}}\right) = N + p - \frac{p}{\mathcal{F}} \le (N+p) - \frac{p}{\delta} < 0.
$$

With such a choice we have in $\mathcal{E}_1$

$$
(7.4) \qquad \mathcal{L}^*(\Phi) \le \frac{2}{p-1} + \left(\frac{2p}{p-1}\right)^{p-1}\left(N - \frac{1}{\delta}\right),
$$

where in estimating the term containing $R(t)$ on the right hand side of (7.3) we have used the fact that $1<p<2$. We determine δ so that the right hand side of (7.4) is non$-$positive and observe that such a choice can be made independent of ξ. Next, having determined δ, within $\mathcal{E}_2$ we have

$$
(7.5) \qquad \mathcal{L}^*(\Phi) \le -\xi\delta + \frac{2}{p-1} + N\left(\frac{2p}{p-1}\right)^{p-1}\left[\frac{R^{\xi}(t)}{\rho^{p\xi}}\frac{1}{\delta}\right]^{2-p}.
$$

Within the range (7.2) of t we estimate

$$
\left[\frac{R^{\xi}(t)}{\rho^{p\xi}}\frac{1}{\delta}\right]^{2-p} \le \left(1 + \frac{1}{\xi}\right)^{\xi(2-p)} \delta^{p-2} \le \left(\frac{e}{\delta}\right)^{2-p}.
$$

We substitute this estimate in (7.5) and choose ξ so large that the right hand side is non-positive.

8. Space-time configurations

Locally bounded weak solutions of (1.1) are locally Hölder continuous in the interior of their domain of definition, $\forall p \in (0,1)$. This is the content of Theorem 1.1 of Chap. IV. The proof consists of controlling the essential oscillation of a local solution over a family of nested and shrinking cylinders. Such a control is established in Proposition 2.1 of Chap. IV, by working with cylinders whose *'space dimensions'* are rescaled in terms of the solution itself. As observed in Remark 2.2

of Chap. IV, such a geometry is not the only possible. A version of Proposition 2.1 holds for an intrinsic parabolic geometry where the scaling occurs in the *'time dimension'*. We restate the proposition for such a geometry in the context of (1.1) and in a form convenient for the proof of the Harnack inequality. Let u be a local weak solution of (1.1). Fix $(\bar{x},\bar{t})\in\Omega_T$ and suppose that we can find a cylinder of the type

$$(8.1)\quad [(\bar{x},\bar{t})+Q\left(a_oR^p,R\right)]\equiv\{|x-\bar{x}|<R\}\times\{\bar{t}-a_oR^p\},\ a_o\equiv\left(\frac{\omega}{A}\right)^{2-p},$$

where A is an absolute constant, R is so small that $[(\bar{x},\bar{t})+Q\left(a_oR^p,R\right)]\subset\Omega_T$, and ω is any positive number satisfying

$$(8.2)\qquad \sup_{[(\bar{x},\bar{t})+Q(a_oR^p,R)]}|u|\le\omega.$$

PROPOSITION 8.1. *There exist constants $\varepsilon_o,\eta\in(0,1)$ and $C,A>1$ that can be determined a priori depending only upon N and p, satisfying the following. Construct the sequences $R_o=R$, $\omega_o=\omega$*

$$R_n=C^{-n}R,\qquad \omega_{n+1}=\eta\omega_n,\qquad n=1,2,\dots,$$

and the boxes

$$Q^{(n)}\equiv\{|x-\bar{x}|<R_n\}\times\{\bar{t}-a_nR_n^p,\bar{t}\},\quad a_n=\left(\frac{\omega_n}{A}\right)^{2-p}.$$

Then for all $n=0,1,2,\dots$

$$Q^{(n+1)}\subset Q^{(n)}\qquad \textit{and}\qquad \operatorname*{ess\,osc}_{Q^{(n)}}u\le\omega_n.$$

A consequence is the Hölder continuity of u at $(\bar{x},\bar{t})$. A particular case is

LEMMA 8.1. *There exist constants $\gamma>1$ and $\alpha\in(0,1)$ that can be determined a priori only in terms of the data, such that for all $0<\rho\le R$*

$$\operatorname*{ess\,osc}_{|x-\bar{x}|<\rho}u(x,\bar{t})\le\gamma\omega_o\left(\frac{\rho}{R}\right)^{\alpha}.$$

Remark 8.1. This is a version of Lemma 2.1 of Chap. IV, stated for a *'fixed time'* $\bar{t}$.

Remark 8.2. The constants A and C depend only upon N and p and are independent of u. Moreover they are *'stable'* as $p\nearrow 2$. This follows from the remarks of §3-(I) of Chap. IV.

9. Proof of the Harnack inequality

We let u be a non-negative local weak solution of (1.1) in Ω_T and let p be in the range (1.2). Let $(x_o, t_o) \in \Omega_T$, assume that $u(x_o, t_o) > 0$ and construct the cylinder

$$Q_{4\rho}(x_o, t_o) \equiv \{|x - x_o| < 4\rho\} \times \left\{t_o - [u(x_o, t_o)]^{2-p} (4\rho)^p, t_o + [u(x_o, t_o)]^{2-p} (4\rho)^p\right\},$$

where we assume that ρ is so small that $Q_{4\rho}(x_o, t_o) \subset \Omega_T$. The change of variables

$$x \longrightarrow \frac{x - x_o}{\rho}, \qquad t \longrightarrow \frac{t - t_o}{[u(x_o, t_o)]^{2-p} \rho^p},$$

maps $Q_{4\rho}(x_o, t_o)$ into the box $Q \equiv Q^+ \cup Q^-$, where

$$Q^+ \equiv B_4 \times [0, 4^p), \qquad Q^- \equiv B_4 \times (-4^p, 0].$$

Denoting again with x and t the new variables, the rescaled function

$$v(x, t) = \frac{1}{u(x_o, t_o)} u\big(x_o + \rho x, t_o + [u(x_o, t_o)]^{2-p} t \rho^p\big)$$

is a bounded non-negative weak solution of

$$\begin{cases} v_t - \operatorname{div} |Dv|^{p-2} Dv = 0 & \text{in } Q, \\ v(0,0) = 1. \end{cases}$$

To prove the theorem it suffices to determine constants c and γ_o in $(0, 1)$, depending only upon N and p such that

$$\inf_{x \in B_1} v(x, c) \geq \gamma_o. \tag{9.1}$$

9–(i). *Locating the sup of u in Q*

For $\tau \in (0, 1)$ construct the family of nested expanding cylinders

$$Q_\tau \equiv \{|x| < \tau\} \times \{-\delta\tau, 0\}$$

and the numbers

$$M_\tau \equiv \sup_{Q_\tau} v, \qquad N_\tau \equiv (1 - \tau)^{-\frac{p}{2-p}}.$$

Here $\delta \in (0, 1)$ is a small number to be chosen later and has the effect of rendering '*flat*' the boxes Q_τ.

Remark 9.1. This construction is similar to that in the proof of Theorem 2.1 of Chap. VI. The cylinders Q_τ however are *'thin'* in the $t-$dimension. Also the exponent of $(1-\tau)$ in the definition of N_τ is fixed and depends on the singularity of the p.d.e.

For $\tau=0$, we have $M_o=N_o$. Moreover as $\tau\nearrow 1$

$$N_\tau \to \infty \qquad \text{and} \qquad M_\tau < \infty,$$

since $v \in L^\infty_{loc}(Q)$. Therefore the equation $M_\tau = N_\tau$ has a largest root, say τ_o, which satisfies

$$M_{\tau_o} = (1-\tau_o)^{-\frac{p}{2-p}}; \qquad M_{\frac{1+\tau_o}{2}} \le 2^{\frac{p}{2-p}}(1-\tau_o)^{-\frac{p}{2-p}}.$$

Since v is Hölder continuous in Q, it achieves the value M_{τ_o} at some point $(\bar{x},\bar{t}) \in \overline{Q}_{\tau_o}$ and

$$\sup_{|x-\bar{x}|<\frac{1-\tau_o}{2}} v(x,\bar{t}) \le 2^{\frac{p}{2-p}}(1-\tau_o)^{-\frac{p}{2-p}}. \tag{9.2}$$

LEMMA 9.1. *There exist a positive number ε that can be determined a priori only in terms of N and p, such that*

$$v(x,\bar{t}) \ge \frac{1}{2}(1-\tau_o)^{-\frac{p}{2-p}}, \qquad \forall |x-\bar{x}| < \varepsilon(1-\tau_o).$$

Remark 9.2. The proof employs the estimates of Lemma 5.1 in the form (5.3). Therefore $\varepsilon \searrow 0$ as $p \nearrow 2$.

PROOF OF LEMMA 9.1: Construct the box

$$(\bar{x},\bar{t}) + \widetilde{Q}_{4R} \equiv \{|x-\bar{x}| < 4R\} \times \{\bar{t}-4,\bar{t}\}, \quad \text{where } 4R = \frac{1-\tau_o}{2}.$$

Apply to such a box the estimate (5.3) with the appropriate change of variables to obtain

$$\sup_{|x-\bar{x}|<R} v(x,t) \le \gamma\Big(\int_{B_{2R}} v(x,\bar{t})dx\Big)^{p/\lambda} + \gamma R^{-\frac{p}{2-p}}, \quad \forall \bar{t}-1 \le t \le \bar{t}.$$

In view of (9.2) and the definition of R

$$\sup_{B_{\frac{(1-\tau_o)}{8}}} v(x,t) \le \gamma_1(1-\tau_o)^{-\frac{p}{2-p}}, \qquad \forall \bar{t}-1 \le t \le \bar{t},$$

where $\gamma_1=\gamma_1(N,p)$ is a constant that can be determined a priori only in terms of N and p. Next consider the cylinder

$$\begin{cases} Q_{R_o} \equiv \{|x-\bar{x}| < R_o\} \times \{\bar{t} - \gamma_1^{2-p}(1-\tau_0)^{-p} R_o^p, \bar{t}\}, \\ R_o = \left(8\gamma_1^{\frac{2-p}{p}}\right)^{-1} (1-\tau_o). \end{cases}$$

By virtue of such a construction we have

$$\sup_{Q_{R_o}} v \leq \gamma_1(1-\tau_o)^{-\frac{p}{2-p}}.$$

The '*vertical size*' of Q_{R_o} is larger than

$$\left[\sup_{Q_{R_o}} v\right]^{2-p} R_o^p.$$

Therefore Q_{R_o} satisfies the space-time configuration of (8.1)-(8.2). We conclude that

$$\begin{aligned} &\forall 0<\rho<R_o, \quad \forall |x-\bar{x}|<\rho, \quad \text{at the level } \bar{t} \\ &v(x,\bar{t}) \geq v(\bar{x},\bar{t}) - \gamma\gamma_1(1-\tau_o)^{-\frac{p}{2-p}}\left(\frac{\rho}{R_o}\right)^{\alpha}. \end{aligned}$$

Since $v(\bar{x},\bar{t})=(1-\tau_o)^{-\frac{p}{2-p}}$, by taking $\rho=\eta R_o$, $\eta\in(0,1)$ we find

$$\begin{aligned} &v(x,\bar{t}) \geq (1-\tau_o)^{-\frac{p}{2-p}}(1-\gamma\gamma_1\eta^{\alpha}), \\ &\forall |x-\bar{x}| \leq \eta R_o \equiv \eta(8\gamma_1^{\frac{2-p}{p}})^{-1}(1-\tau_o) \end{aligned}$$

and the lemma follows by taking η so small that enough $(1-\gamma\gamma_1\eta^{\alpha})=\frac{1}{2}$ and then choosing

$$\varepsilon = \eta(8\gamma_1^{\frac{2-p}{p}})^{-1}.$$

9–(ii). *Time-expansion of positivity*

The previous arguments are independent of the number δ. We will now determine δ.

LEMMA 9.2. *There exist small positive numbers c_o, δ that can be determined a priori only in terms of N and p, such that*

$$v(x,t) \geq c_o(1-\tau_o)^{-\frac{p}{2-p}}, \quad \forall |x-\bar{x}|<\varepsilon(1-\tau_o), \ \forall \delta \leq t \leq 2\delta. \tag{9.3}$$

PROOF: Consider the comparison function $\Phi(x-\bar{x};t-\bar{t})$ in the domain $\mathcal{D}_{\{k,\xi\}}(\bar{x},\bar{t})$ defined in (7.1)–(7.2) with the choices,

$$k=\frac{1}{2}(1-\tau_o)^{-\frac{p}{2-p}}, \qquad \rho=\varepsilon(1-\tau_o).$$

The function Φ is a subsolution of (1.1) for a time interval

$$\frac{k^{2-p}\rho^p}{\xi} = \frac{\varepsilon^p 2^{-(2-p)}}{\xi} \equiv 3\delta.$$

For $t=\bar{t}$ by virtue of Lemma 9.1, $v \geq \Phi(x-\bar{x};0)$. Therefore by the comparison principle

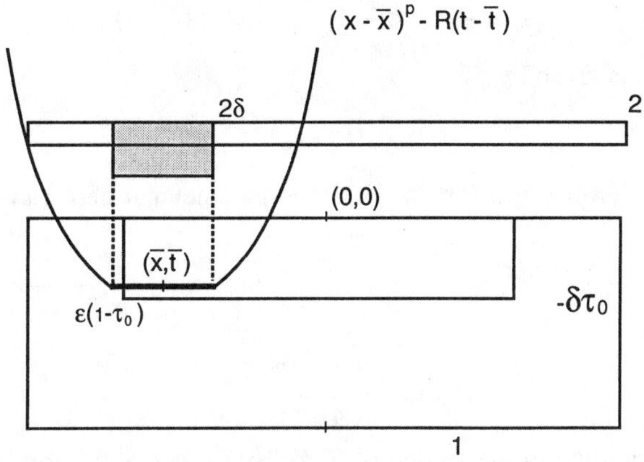

Figure 9.1

$$v \geq \Phi \quad \text{in } \{|x-\bar{x}|^p < R(t-\bar{t})\} \times \{0 < (t-\bar{t}) < 3\delta\}.$$

In particular for $\delta < t - \bar{t} < 3\delta$ and $|x| \leq \varepsilon(1-\tau_o)$,

$$\begin{aligned} v(x,t) &\geq \frac{\frac{1}{2}(1-\tau_o)^{-\frac{p}{2-p}}}{[1/\xi+1]^{\xi}} \left\{1 - \left(\frac{3\xi}{3\xi+1}\right)^{\frac{1}{p-1}}\right\}_+^2 \\ &\equiv c_o(1-\tau_o)^{-\frac{p}{2-p}}. \end{aligned}$$

The location of $\bar{t}$ in the box Q_{τ_o} is only known *qualitatively*. However, as $(t-\bar{t})$ ranges over $[\delta, 3\delta]$, the intervals $[\bar{t}+\delta < t < \bar{t}+3\delta]$ have the common intersection $[\delta \leq t \leq 2\delta]$ and the lemma is proved.

Remark 9.3. The number $\delta \searrow 0$ as $p \nearrow 2$. This follows from Remark 9.2 and the choice of δ above.

9–(iii). Sidewise expansion of positivity

We will expand the positivity set of v over the ball $\{|x| < 1\}$ at the time level $t = 2\delta$. For this we will prove that there exist a constant $\gamma_o = \gamma_o(N,p)$ such that

$$v(x, 2\delta) \geq \gamma_o, \qquad \forall\, |x-\bar{x}| < 2.$$

Consider the comparison function

$$\Psi\left(\frac{x-\bar{x}}{3}; \frac{t-\delta}{3^p}\right), \tag{9.4}$$

introduced in (6.3), in the annular cylindrical domain

$$\{\varepsilon(1-\tau_o)<|x-\bar{x}|<3\}\times\{\delta,2\delta\}.$$

The number k is given by

$$k=c_o(1-\tau_o)^{-\frac{p}{2-p}},$$

where c_o is determined in Lemma 9.2. The parameter μ here can be chosen by imposing

$$\frac{\mu}{b^{p-1}}\,k^{p-2}\leq\frac{\varepsilon^p(1-\tau_o)^p}{3^p},\qquad\text{i.e.,}\quad\mu\leq\frac{b^{p-1}}{c_o^{2-p}}\frac{\varepsilon^p}{3^p}.$$

We choose

$$\mu=\min\left\{\frac{1}{4};\,\frac{b^{p-1}}{c_o^{2-p}}\frac{\varepsilon^p}{3^p}\right\},$$

and pick θ according to the second of (6.4). By further restricting either μ or the number δ of Lemma 9.2 we may assume that $\theta=\delta$. The function Ψ in (9.4) vanishes for $|x-\bar{x}|=3$ and for $t=\delta$. Moreover for $|x-\bar{x}|=\varepsilon(1-\tau_o)$ and $\delta<t\leq 2\delta$,

$$\Psi\left(\frac{x-\bar{x}}{3},\frac{t-\delta}{3^p}\right)\leq c_o(1-\tau_o)^{-\frac{p}{2-p}}\leq v(x,t),$$

by Lemma 9.2. Therefore by the comparison principle, we have for $t=2\delta$ and $\forall|x-\bar{x}|<2$

$$\begin{aligned}v(x,2\delta)&\geq\frac{c_o(1-\tau_o)^{-\frac{p}{2-p}}(3^2-2^2)^{\frac{p}{p-1}}3^{-\frac{2(p-1)}{p}}}{\left\{1+\left[c_o(1-\tau_o)^{-\frac{p}{2-p}}\right]^{\frac{2-p}{p-1}}2b\delta^{-\frac{1}{p-1}}\right\}^{\frac{p-1}{2-p}}}\\&\geq\inf_{0\leq\tau\leq 1}\frac{c_o(1-\tau)^{-\frac{p}{2-p}}(3^2-2^2)^{\frac{p}{p-1}}3^{-\frac{2(p-1)}{p}}}{\left\{1+\left[c_o(1-\tau)^{-\frac{p}{2-p}}\right]^{\frac{2-p}{p-1}}2b\delta^{-\frac{1}{p-1}}\right\}^{\frac{p-1}{2-p}}}\\&\equiv\gamma_o(N,p).\end{aligned}$$

Remark 9.4. Because of the choice of δ and Remarks 9.2 and 9.3, the number $\gamma_o(N,p)$ tends to zero as $p\nearrow 2$.

9-(iv). Proof of Theorem 1.1 for p near 2

The proof is very similar to that of Theorem 2.1 of Chap. VI for p close to 2. We only indicate the main differences.

As before, construct the family of expanding cylinders $Q_\tau\equiv\{|x|<\tau\}\times\{-\tau,0\}$ and the numbers

$$M_\tau\equiv\|v\|_{\infty,Q_\tau},\qquad N_\tau\equiv(1-\tau)^{-\beta},$$

where β is a positive number to be chosen. The definition of the numbers N_τ differs from that in §9 since β is arbitrary. Let $\tau_o\in[0,1)$ be the largest root of the equation $M_\tau=N_\tau$, so that

$$M_{\tau_o} = (1-\tau_o)^{-\beta}, \qquad M_{\frac{1+\tau_o}{2}} \le 2^\beta (1-\tau_o)^{-\beta}.$$

If $(\bar{x}, \bar{t})$ is a point in Q_{τ_o} where v achieves the value M_{τ_o}, we have

$$v(x,t) \le 2^\beta (1-\tau_o)^{-\beta}; \ |x-\bar{x}| < \frac{1-\tau_o}{2}; \ \bar{t} - \frac{1-\tau_o}{2} < t < \bar{t}. \tag{9.5}$$

Let

$$R_o = \frac{1}{2} 2^{-\beta \frac{2-p}{p}} (1-\tau_o)^{\beta \frac{2-p}{p}} (1-\tau_o)$$

and consider the box

$$Q_o(\bar{x}, \bar{t}) \equiv \{|x-\bar{x}| < R_o\} \times \left\{ \bar{t} - \left[2^\beta (1-\tau_o)^{-\beta}\right]^{\frac{2-p}{p}} R_o^p, \bar{t} \right\}.$$

From the definitions of Q_τ and R_o we have $Q_o(\bar{x}, \bar{t}) \subset Q_{\frac{1+\tau_o}{2}}$, so that by (9.5),

$$\|v\|_{\infty, Q_o(\bar{x}, \bar{t})} \le 2^\beta (1-\tau_o)^{-\beta}$$

Therefore $Q_o(\bar{x}, \bar{t})$ satisfies the space-time configuration (8.1)-(8.2). It follows that

$$\forall |x-\bar{x}| < R_o, \quad |v(x,\bar{t}) - v(\bar{x}, \bar{t})| < \gamma 2^\beta (1-\tau_o)^{-\beta} \left(\frac{\rho}{R_o}\right)^\alpha .$$

By taking $\rho = \varepsilon R_o$ and then ε sufficiently small we have

LEMMA 9.1$'$. *There exists a small positive number $\varepsilon \in (0,1)$ that can be determined a priori only in terms of N, p such that*

$$v(x,\bar{t}) \ge \frac{1}{2}(1-\tau_o)^{-\beta}, \qquad \forall |x-\bar{x}| < \varepsilon (1-\tau_o)^{\beta \frac{2-p}{p} + 1}.$$

Remark 9.5. The constant ε depends upon β but it is *'stable'* as $p \nearrow 2$ since no use has been made of (5.3).

The proof can now be completed by expanding the positivity set of v with the aid of the comparison function $\mathcal{G}_{k,\rho}$ introduced in §3-(i) of Chap. VI.

10. Proof of Theorem 1.2

Fix a point $(x_*, t_*) \in \Omega_T$ assume that $u(x_*, t_*) > 0$ and for $R > 0$, construct the cylinder

$$Q_{8R}(x_*, t_*) \equiv \{|x - x_*| < 8R\} \\ \times \left\{ t_* - c\left[u(x_*, t_*)\right]^{2-p} 8R^p, \ t_* + c\left[u(x_*, t_*)\right]^{2-p} 8R^p \right\},$$

where $c = c(N,p)$ is the number appearing in Theorem 1.1, and R is any positive number such that $Q_{8R}(x_*, t_*)$ is contained in the domain of definition of u.

We first establish an auxiliary proposition, then we will prove that it implies the theorem.

PROPOSITION 10.1. *There exists constants $C=C(N,p)$ and $\eta=\eta(N,p)$ that can be determined a priori only in terms of N and p, such that*

$$C^{-1}\sup_{B_{\eta R}(x_*)} u(\cdot,t_o)\le u(x_o,t_o)\le C\inf_{B_{\eta R}(x_*)} u(\cdot,t_o), \tag{10.1}$$

where $x_o\equiv x_*$ and

$$t_o\equiv t_*+c\left[u\left(x_*,t_*\right)\right]^{2-p}R^p. \tag{10.2}$$

PROOF: The change of variables

$$x\longrightarrow\frac{x-x_*}{R},\qquad t\longrightarrow\frac{t-t_*}{\left[u\left(x_*,t_*\right)\right]^{2-p}R^p}$$

maps $Q_{8R}\left(x_*,t_*\right)$ into $Q\equiv B_8\times(-8,8)$. Denoting again with x and t the new variables, the rescaled function

$$v(x,t)\equiv\frac{1}{u\left(x_*,t_*\right)}u\left(x_*+Rx,t_*\left[u\left(x_*,t_*\right)\right]^{2-p}tR^p\right)$$

is a bounded non-negative solution of

$$\begin{cases} v_t-\operatorname{div}|Dv|^{p-2}Dv=0 \ \text{ in } Q.\\ v(0,0)=1,\end{cases}$$

We first prove that there exist a quantitative constant $\overline{C}=\overline{C}(N,p)$, such that

$$\frac{1}{\overline{C}}\le v(x,c)\le\overline{C},\qquad\forall|x|<1. \tag{10.3}$$

By the Harnack inequality (1.3)

$$v(x,c)\ge\gamma_o,\qquad\forall|x|<2, \tag{10.4}$$

for a quantitative constant $\gamma_o=\gamma_o(N,p)$. This proves the estimate below in (10.3). For the estimate above we require the following lemma.

LEMMA 10.1. *There exists a quantitative constant $\eta\in(0,1)$ depending only upon N and p, such that*

$$v(x,-c)\le 2/\gamma_o,\qquad\forall|x|<2\eta. \tag{10.5}$$

PROOF: For $\bar{x}$ ranging over the ball B_4 consider the closed truncated *'paraboloid'*

$$t+c\ge c\left[v(\bar{x},-c)\right]^{2-p}|x-\bar{x}|^p,\qquad -c\le t\le 0.$$

By the Harnack estimate

$$v(\bar{x},-c)\le\frac{1}{\gamma_o}v(x,0),\quad |x-\bar{x}|^p<\left[v(\bar{x},-c)\right]^{p-2}, \tag{10.6}$$

and in particular $v(0,-c)=1/\gamma_o$. Since v is Hölder continuous, the set

$$\{x \mid u(x,-c) < 2/\gamma_o\}$$

is non-empty and contains a ball about the origin. We claim that in particular it contains the ball $B_{2\eta}$, where

$$(2\,\eta)^p = \left(\frac{\gamma_o}{2}\right)^{2-p}.$$

If not, there would exist some $\bar{x}\in B_{2\eta}$ such that $v(\bar{x},-c)=2/\gamma_o$. It follows that the ball

$$|x-\bar{x}|^p < [v(\bar{x},-c)]^{p-2} = (2\,\eta)^p$$

covers the origin, and (10.6) for $x=0$ gives

$$\frac{2}{\gamma_o} = v(\bar{x},-c) \le \frac{1}{\gamma_o}.$$

The contradiction proves the lemma.

To prove the estimate above in (10.3), we combine the quantitative bound (10.5) with Proposition 4.1. This gives

$$\begin{aligned}\sup_{B_\eta} v(\cdot,c) &\le \gamma\, c^{-N/\lambda}\left(\int_{B_{2\eta}} v(x,-c)dx\right)^{p/\lambda} + \gamma\left(\frac{c}{\eta}\right)^{\frac{1}{2-p}}\\ &\le \bar{\gamma}(N,p).\end{aligned}$$

We return to the original coordinates and write the estimate above in (10.3) as

$$u(x,t_o) \le \overline{C}\,u\,(x_*,t_*)\,, \qquad \forall\,|x-x_o|<\eta R,\;\; x_o\equiv x_*.$$

Since, by Corollary 1.1, $u\,(x_*,t_*)\le\gamma u(x_o,t_o)$ the left estimate in (10.1) is proved. The estimate below in (10.3) reads

$$u(x,t_o) \ge \gamma_o u\,(x_*,t_*)\,, \qquad \forall|x-x_*|<\eta R.$$

On the other hand the estimate above in (10.3) for $x = x_*$ gives $u(x_o,t_o) \le \overline{C}\,u\,(x_*,t_*)$. Combining these last two estimates proves the bound above in (10.1) and the proposition follows.

10-(i). Proof of Theorem 1.2

Fix $(x_o,t_o)\in\Omega_T$ and $\rho>0$. Let η be the constant claimed by Lemma 10.1 and let γ be the constant of the Harnack estimate (1.3). Set

$$r = \gamma^{\frac{2-p}{p}}\,\rho/\eta,$$

and construct the cylinder

$$Q_{8r}(x_o,t_o) \equiv \{|x-x_o|<8r\}$$
$$\times\left\{t_o - c\,[u(x_o,t_o)]^{2-p}\,8r^p,\ t_o + c\,[u(x_o,t_o)]^{2-p}\,8r^p\right\}.$$

Wthout loss of generality, we assume that $Q_{8r}(x_o,t_o) \subset \Omega_T$. First fix the time level

$$t_* = t_o - c\,[u(x_o,t_o)]^{2-p}\,r^p,$$

and choose $R>0$ from

$$t_o - t_* = c\,[u\,(x_*,t_*)]^{2-p}\,R^p, \qquad x_* \equiv x_o.$$

The definitions of t_* and R give

$$c\,[u(x_o,t_o)]^{2-p}\,r^p = t_o - t_* = c\,[u\,(x_*,t_*)]^{2-p}\,R^p.$$

By Corollary 1.1, $u\,(x_*,t_*) \le \gamma u(x_o,t_o)$. Therefore

$$R \ge \gamma^{\frac{p-2}{p}}\,r^p \equiv \rho/\eta.$$

Applying Proposition 10.1 with such a choice of the point (x_*,t_*) and radius R proves the theorem.

11. Bibliographical notes

Theorem 1.1 and its proof is taken from [44]. The form of Theorem 1.2 was conceived by Nash [84], who believed it to be true for solutions of the heat equation. Moser [83] pointed out that (1.7) is not dilation invariant for solutions of the heat equation. It becomes scalar invariant in a specific *intrinsic* geometry. The results adapt to equations of porous medium–type and its generalisations (see [44]). In the context of the plasma equations estimates of the rate of extinctions were derived by Berryman–Holland [13,14]. Proposition 3.1 is due to Bénilan and Crandall [9]. The estimates of §§ 4 and 5 are taken from [42]. The subsolution Ψ of §6 appears in [44]. The subsolution Φ of §7 is a modification of a subsolution introduced in [4]. It is natural to ask whether an intrinsic Harnack estimate continues to hold for non-negative solutions of p.d.e.'s with full quasilinear structure. This is the case if $p=2$ and it remains an open issue for degenerate $(p>2)$ and singular $(1<p<2)$ equations. A step in this direction is in [29]. It is shown that Theorem 1.1 holds true for non-negative weak solutions of

$$v_t - \left(|Dv|^{p-2}a_{ij}(x,t)\,u_{x_i}\right)_{x_j} = 0,$$

where $(x,t)\to a_{ij}(x,t)$ are only bounded and measurable and the matrix (a_{ij}) is positive definite.

VIII

Degenerate and singular parabolic systems

1. Introduction

We turn now to quasilinear systems whose principal part becomes either degenerate or singular at points where $|Du|=0$. To present a streamlined cross section of the theory, we refer to the model system

$$(1.1)\qquad \begin{cases} \mathbf{u} \equiv (u_1, u_2, \ldots, u_m)\,, \;\; m \in \mathbf{N}, \\ u_i \in C_{loc}\left(0,T; L^2_{loc}(\Omega)\right) \cap L^p\left(0,T; W^{1,p}_{loc}(\Omega)\right), \; i=1,2,\ldots,m, \\ \mathbf{u}_t - \operatorname{div} |D\mathbf{u}|^{p-2} D\mathbf{u} = 0 \;\; \text{in } \Omega_T. \end{cases}$$

The solutions are meant in the weak sense

$$(1.2)\qquad \int_\Omega u_i \varphi_i(x,\tau)dx \Big|_{t_1}^{t_2} + \int_{t_1}^{t_2}\!\!\int_\Omega \left\{-u_i\varphi_{i,t} + |D\mathbf{u}|^{p-2} Du_i \cdot D\varphi_i\right\} dx d\tau = 0,$$

for all intervals $[t_1, t_2] \subset (0,T]$ and all testing functions $\varphi \equiv (\varphi_1, \varphi_2, \ldots, \varphi_m)$ satisfying

$$(1.3)\qquad \varphi_i \in W^{1,2}_{loc}\left(0,T; L^2(\Omega)\right) \cap L^p_{loc}\left(0,T; W^{1,p}_o(\Omega)\right), \quad i = 1, 2, \ldots, m.$$

For these we derive local sup-bounds on the modulus of the solution $|\mathbf{u}|$ and its space gradient $|D\mathbf{u}|$ and establish the estimate

$$(1.4)\qquad u_{i,x_j} \in C^\alpha_{loc}(\Omega_T), \quad i=1,2,\ldots,m, \;\; j=1,2,\ldots,N,$$

for some $\alpha\in(0,1)$. This is the focal point of the theory. Weak solutions of *elliptic* systems in general are not continuous *everywhere* within their domain of definition. We refer to [48] for counterexamples and an account of the theory. Solutions of (1.1) are regular everywhere in Ω_T because of the special nature of the system. If $\mathbf{u}$ solves (1.1), then the function $|D\mathbf{u}|^2$ is a non-negative subsolution of a parabolic p.d.e.[(1)] It is precisely such a property, which for elliptic systems is called *'quasi–subharmonicity'*,[(2)] that permits one to prove (1.4) everywhere in Ω_T.

These estimates can be extended up to $t=0$ if the system in (1.1) is associated with a *smooth* initial datum $\mathbf{u}_o$. They also carry over to the lateral boundary of Ω_T if (1.1) is associated with *homogeneous* either Dirichlet or Neumann data on $S_T\equiv\partial\Omega\times(0,T)$. If the data are not homogeneous, the theory is fragmented and incomplete. In the case of non-homogeneous Dirichlet data, we will show that

$$u_i\in C^\delta\left(\overline{\Omega}\times(\varepsilon,T)\right)\quad\text{for arbitrary }\ \delta\in(0,1),\ \forall\varepsilon\in(0,T),$$

provided $p>\max\left\{1\,;\,\frac{2N}{N+2}\right\}$. However the key estimate (1.4) is not known to hold in such a case, and it is a major open problem in the theory.

The $C^{1,\alpha}$ regularity (1.4) requires a preliminary estimation of the type

$$(1.5)\qquad \|D\mathbf{u}\|_{\infty,\mathcal{K}}\le\text{const},\qquad \mathcal{K}\ \text{a compact subset of}\ \Omega_T.$$

The degenerate case $p>2$ and the singular case $p\in(1,2)$ are rather different with respect to such an estimate. The function class in (1.1) implies that[(3)]

$$(1.6)\qquad |\mathbf{u}|\in L^q_{loc}\left(\Omega_T\right),\quad q=p\,\frac{N+2}{N}.$$

If $p>2$, such *integrability* suffices to establish (1.5). If $1<p<2$, the sup-bound (1.5) can be derived only if further *'integrability'* is assumed on $|\mathbf{u}|$. Precisely,

$$(1.7)\qquad |\mathbf{u}|\in L^r_{loc}(\Omega_T),\ \text{where}\ r\ge2\ \text{satisfies}\ \lambda_r\equiv N(p-2)+rp>0.$$

This is analogous to the condition imposed in Theorem 5.1 of Chap. V. It implies (1.4) and in addition

$$(1.8)\qquad u_{i,x_\ell,x_h}\in L^p_{loc}(\Omega_T).$$

1-(i). About the singular case $1<p<2$

In the degenerate case $p>2$, the behaviour of the solutions of (1.1) is entirely a *local* fact. In particular the sup-bound (1.5) and the estimate (1.4) are a sole consequence of $\mathbf{u}$ being a weak solution of (1.1). If $1<p<2$ due to the singular

---

(1) See (3.3) in the Preface or (1.8) of Chap. IX.

(2) We refer to Meier [77] for some sufficient conditions for an elliptic system to be quasi-subharmonic.

(3) See Proposition 3.1 of Chap. I.

nature of the p.d.e. some global information is needed. This is not related to systems. Indeed it occurs also in Theorem 5.1 of Chap. V to establish a sup-bound for solutions of a *single* equation. Since our estimates involve $\mathbf{u}$ and $D\mathbf{u}$, the global information needed regards both the solution and its space gradient. Let $r \geq 2$ satisfy (1.7) and let $\mathbf{u}$ be a local weak solution of (1.1) for $p \in (1,2)$. We assume that

$$\text{(1.9)} \qquad \begin{cases} \mathbf{u} \text{ can be constructed as the weak limit in } L^r_{loc}(\Omega_T) \text{ of a} \\ \text{sequence of } \textit{bounded} \text{ subsolutions } \{\mathbf{u}_n\}_{n\in\mathbf{N}} \text{ of (1.1)} \\ \text{satisfying in addition } |D\mathbf{u}_n| \in L^2_{loc}(\Omega_T). \end{cases}$$

We stress however that all our estimates will depend only upon the quantities

$$\|\mathbf{u}\|_{r,\mathcal{K}}, \quad \|D\mathbf{u}\|_{p,\mathcal{K}}, \quad \mathcal{K} \text{ a compact subset of } \Omega_T.$$

Such an assumption is not restrictive in view of the available existence theory[(1)] and the special form of (1.1).

1-(ii). General structures

We will develop the theory for the homogeneous system (1.1). The same results however continue to hold for the following general class of quasilinear systems

$$\text{(1.10)} \qquad \frac{\partial}{\partial t}u_i - \operatorname{div} \mathbf{A}^{(i)}(x,t,D\mathbf{u}) = B^{(i)}(x,t,\mathbf{u},D\mathbf{u}) \text{ in } \Omega_T, \quad i = 1,2,\dots,m,$$

where the functions

$$\mathbf{A}^{(i)} \equiv \left(A_1^{(i)}, A_2^{(i)}, \dots, A_N^{(i)}\right) : \Omega_T \times \mathbf{R}^{Nm} \longrightarrow \mathbf{R}^N,$$
$$B^{(i)} : \Omega_T \times \mathbf{R} \times \mathbf{R}^{Nm} \longrightarrow \mathbf{R}, \quad i = 1,2,\dots,m,$$

satisfy the structure condition

$$(S_1) \qquad C_o |D\mathbf{u}|^p - \varphi_o \leq \mathbf{A}^{(i)} \cdot Du_i \leq C_1 |D\mathbf{u}|^p + \varphi_1,$$

$$(S_2) \qquad \frac{\partial \mathbf{A}^{(i)}}{\partial u_{\ell,x_k}} u_{\ell,x_k x_j} \cdot Du_{i,x_j} \geq C_o |D\mathbf{u}|^{p-2} \sum_{j=1}^{N} \sum_{i=1}^{m} |Du_{i,x_j}|^2 - \varphi_o,$$

$$(S_3) \qquad \sum_{k=1}^{N} \sum_{i,\ell=1}^{m} \left| \frac{\partial \mathbf{A}^{(i)}}{\partial u_{\ell,x_k}} \right| \leq C_1 |D\mathbf{u}|^{p-2} + \varphi_1,$$

---

(1) See Lions [73].

$$(\mathcal{S}_4) \qquad \sum_{k=1}^{N}\sum_{i=1}^{m}\left|\frac{\partial \mathbf{A}^{(i)}}{\partial x_k}\right| \le C_1|D\mathbf{u}|^{p-1} + \varphi_2,$$

$$(\mathcal{S}_5) \qquad \sum_{i=1}^{m}\left|B^{(i)}\right| \le C_1|D\mathbf{u}|^{p-1} + \varphi_2,$$

where C_i, $i=0,1$, are given positive constants and φ_i, $i=0,1,2$, are given non-negative functions satisfying

$$(\mathcal{S}_6) \qquad \varphi_o + \varphi_1^{\frac{p}{p-1}} + \varphi_2^2 \in L^q_{loc}(\Omega_T), \quad q > \frac{N+2}{2}.$$

Remark 1.1. The structure condition $(\mathcal{S}_2)$ is somewhat formal since there is no stipulation that u_{ℓ,x_kx_j} have meaning at all. More correctly it should be written with u_{ℓ,x_kx_j} and Du_{i,x_j} replaced by tensors $\xi_{\ell,k,j}$. Neverthless we prefer the formal but suggestive form of $(\mathcal{S}_2)$.

We will develop the main points of the theory for the model system (1.1) and indicate later how to modify the arguments to include (1.10).

2. Boundedness of weak solutions

We will use the notation of §3 of Chap. II. Thus $Q(\theta,\rho)$ is the cylinder with *'vertex'* at the origin. Its cross sections are the cubes K_ρ and its height is θ. The cylinder $[(x_o,t_o)+Q(\theta,\rho)]$ has the *'vertex'* at (x_o,t_o) and is congruent to $Q(\theta,\rho)$. With ζ we denote a piecewise smooth non-negative cutoff function in $Q(\theta,\rho)$ vanishing on the parabolic boundary of $Q(\theta,\rho)$.

THEOREM 2.1 (THE CASE $p>2$). *Let* $\mathbf{u}$ *be a local weak solution of (1.1), and let* $p>2$. *Then for all* $\varepsilon\in(0,2]$ *there exists a constant* γ *depending only upon* N,p,m *and* ε, *such that for every cylinder* $[(x_o,t_o)+Q(\theta,\rho)]\subset\Omega_T$ *and for ever* $\sigma\in(0,1)$,

$$(2.1) \qquad \sup_{[(x_o,t_o)+Q(\sigma\theta,\sigma\rho)]}|\mathbf{u}| \le \frac{\gamma\,(\theta/\rho^p)^{1/\varepsilon}}{(1-\sigma)^{(N+p)/\varepsilon}}\left(\iint_{[(x_o,t_o)+Q(\theta,\rho)]}|\mathbf{u}|^{p-2+\varepsilon}dxd\tau\right)^{1/\varepsilon} \wedge \left(\frac{\rho^p}{\theta}\right)^{\frac{1}{p-2}}.$$

THEOREM 2.1 (THE CASE $1<p<2$). *Let* $\mathbf{u}$ *be a local weak solution of (1.1) for* $1<p<2$. *Assume moreover that*

$$(2.2) \qquad |\mathbf{u}|\in L^r_{loc}(\Omega_T), \;\; r\ge 1 \quad \lambda_r\equiv N(p-2)+rp>0,$$

and that (1.9) holds. There exists a constant γ *depending only upon* N, p, m *and* r *such that for every cylinder* $[(x_o, t_o) + Q(\theta, \rho)] \subset \Omega_T$ *and for every* $\sigma \in (0,1)$,

$$\sup_{[(x_o,t_o)+Q(\sigma\theta,\sigma\rho)]} |\mathbf{u}| \le \frac{\gamma\,(\rho^p/\theta)^{N/\lambda_r}}{(1-\sigma)^{p(N+p)/\lambda_r}} \left(\iint_{[(x_o,t_o)+Q(\theta,\rho)]} |\mathbf{u}|^r dx d\tau \right)^{p/\lambda_r} \wedge \left(\frac{\theta}{\rho^p}\right)^{\frac{1}{2-p}}, \tag{2.3}$$

2-(i). An auxiliary proposition

The arguments are similar to the proof of local boundedness of solutions of a single equation and are based on local energy inequalities which we derive next. We set

$$|\mathbf{u}| \equiv w. \tag{2.4}$$

PROPOSITION 2.1. *Let* $\mathbf{u}$ *be a local weak solution of the system (1.1) in* Ω_T, *and let* $f(\cdot)$ *be a non-negative, bounded, Lipschitz function in* R^+. *There exists a constant* $\gamma \equiv \gamma(N, p, m)$, *such that*

$$\forall (x_o, t_o) \in \Omega_T\ \forall \rho, \theta > 0 \text{ such that } [(x_o, t_o) + Q(\theta, \rho)] \subset \Omega_T \tag{2.5}$$

$$\sup_{t_o-\theta \le t \le 0} \int_{[x_o+K_\rho]} \left(\int_0^w s f(s) ds \right) \zeta^p(x,t) dx$$

$$+ \iint_{[(x_o,t_o)+Q(\theta,\rho)]} |Dw|^p f(w) \zeta^p dx d\tau + \iint_{[(x_o,t_o)+Q(\theta,\rho)]} |D\mathbf{u}|^{p-2} |Dw|^2 w f'(w) \zeta^p dx d\tau$$

$$\le \gamma \iint_{[(x_o,t_o)+Q(\theta,\rho)]} w^p f(w) |D\zeta|^p dx d\tau + \gamma \iint_{[(x_o,t_o)+Q(\theta,\rho)]} \left(\int_0^w s f(s) ds \right) \zeta^{p-1} \zeta_t dx d\tau.$$

PROOF: The weak formulation (1.2) can be rewritten in terms of Steklov averages, as

$$\int_\Omega \left\{ \frac{\partial}{\partial t} u_{i,h} \varphi_i + \left[|D\mathbf{u}|^{p-2} D u_i \right]_h \cdot D\varphi_i \right\} dx d\tau = 0,\ \forall h \in (0,T), \tag{2.6}$$

$$\forall\, 0 < t \le T - h, \quad \forall \varphi_i \in W_o^{1,p}(\Omega) \cap L^2(\Omega)\ i = 1, 2, \ldots, m$$

Since $\frac{\partial}{\partial t} u_{i,h} \in L^2_{loc}(\Omega_T)$, this implies

$$\frac{\partial}{\partial t} u_{i,h} - \operatorname{div} \left[|D\mathbf{u}|^{p-2} D u_i \right]_h = 0 \quad \text{a.e. in } \Omega_T. \tag{2.6)$'$}$$

Without loss of generality we may assume that (x_o, t_o) coincides with the origin. In (2.6), take the testing function

$$\varphi_i = u_{i,h}\, f\left(|u_h|\right) \zeta^p .$$

We add over $i = 1, 2, \ldots, m$ and integrate in dt, over the interval $-\theta \le t \le 0$, to obtain

$$\begin{aligned}
&\int_{-\theta}^{t} \frac{\partial}{\partial \tau} \int_{K_\rho} \left(\int_0^{|u_h|} s f(s) ds \right) \zeta^p dx d\tau \\
&+ \int_{-\theta}^{t}\int_{K_\rho} \left[|D\mathbf{u}|^{p-2} Du_i \right]_h \cdot Du_{i,h} f\left(|u_h|\right) \zeta^p dx d\tau \\
&+ \int_{-\theta}^{t}\int_{K_\rho} \left[|D\mathbf{u}|^{p-2} \frac{\partial}{\partial x_\ell} u_{i,h} \right]_h \frac{u_{i,h} f'(|u_h|)}{|u_h|} u_{j,h} \frac{\partial}{\partial x_\ell} u_{j,h} \zeta^p dx d\tau \\
&= -p \int_{-\theta}^{t}\int_{K_\rho} \left[|D\mathbf{u}|^{p-2} Du_i \right]_h u_{i,h} f\left(|u_h|\right) \zeta^{p-1} D\zeta \, dx d\tau .
\end{aligned}$$

We perform an integration by parts in the first integral and then let $h \to 0$. The various limits are justified since $|D\mathbf{u}| \in L^p_{loc}(\Omega_T)$ and $|\mathbf{u}| \in C_{loc}\left(0, T; L^2_{loc}(\Omega_T)\right)$. This gives

$$\begin{aligned}
(2.7)\quad &\sup_{-\theta \le t \le 0} \int_{K_\rho} \left(\int_0^w s f(s) ds \right) \zeta^p(x,t) dx \\
&+ \iint_{Q(\theta,\rho)} |D\mathbf{u}|^p f(w) \zeta^p dx d\tau + \iint_{Q(\theta,\rho)} |D\mathbf{u}|^{p-2} |Dw|^2 w f'(w) \zeta^p dx d\tau \\
&\le p \iint_{Q(\theta,\rho)} |D\mathbf{u}|^{p-1} w f(w) \zeta^{p-1} |D\zeta| dx d\tau + p \iint_{Q(\theta,\rho)} \left(\int_0^w s f(s) ds \right) \zeta^{p-1} \zeta_t dx d\tau .
\end{aligned}$$

By Young's inequality for every $\eta > 0$

$$\begin{aligned}
\iint_{Q(\theta,\rho)} |D\mathbf{u}|^{p-1} w f(w) \zeta^{p-1} |D\zeta| \, dx d\tau \le \eta & \iint_{Q(\theta,\rho)} |D\mathbf{u}|^p f(w) \zeta^p dx d\tau \\
&+ \gamma(\eta) \iint_{Q(\theta,\rho)} w^p f(w) |D\zeta|^p dx d\tau .
\end{aligned}$$

Next by Schwartz inequality

$$|Dw|^2 = w^{-2}\sum_{j=1}^{N}\left(u_\ell\, u_{\ell,x_j}\right)^2 \leq w^{-2}\sum_{\ell=1}^{m} u_\ell^2 \sum_{j=1}^{N}\sum_{\ell=1}^{m} u_{\ell,x_j}^2 \equiv |D\mathbf{u}|^2.$$

Therefore $|D\mathbf{u}|^p \geq |Dw|^p$. Combining these estimates in (2.7) proves the proposition.

COROLLARY 2.1. *The integral inequality (2.5) continues to hold for non-negative, non-decreasing functions f in $\mathbf{R}^+$, satisfying*

$$\sup_{0\leq s\leq k} f'(s) < \infty, \quad \textit{for all } k > 0,$$

provided

(2.8) $$w^p f(w) \quad \textit{and} \quad \int_0^w s f(s) ds \in L^1_{loc}(\Omega_T).$$

PROOF: Fix $k > 0$ and write (2.5) for the truncated functions

$$f_k(s) \equiv \begin{cases} f(s) & \text{for } 0 \leq s \leq k \\ f(k) & \text{for } \qquad s \geq k. \end{cases}$$

Letting $k \to \infty$ gives (2.5) for such an f. The limit of the various terms on the left hand side follows from Fatou's Lemma and the limit of the terms on the right hand side is justified by virtue of (2.8).

2-(ii). Proof of Theorems 2.1

The starting point is the energy estimate (2.5) where we assume, up to a translation, that (x_o, t_o) coincides with the origin. Fix $\sigma \in (0,1)$ and consider the family of nested cylinders $Q_n \equiv Q(\theta_n, \rho_n)$, where

(2.9) $$\begin{cases} \rho_n = \sigma\rho + \dfrac{(1-\sigma)}{2^n}\rho, & n = 0, 1, 2\ldots, \\ \theta_n = \sigma\theta + \dfrac{(1-\sigma)}{2^n}\theta. \end{cases}$$

It follows from the definition that

(2.10) $$Q_o = Q(\theta, \rho) \quad \text{and} \quad Q_\infty = Q(\sigma\theta\ \sigma\rho).$$

Consider also the family of boxes

(2.11) $$\tilde{Q}_n = Q\left(\tilde{\theta}_n, \tilde{\rho}_n\right)$$

where for $n = 0, 1, 2, \ldots$

(2.12) $$\begin{cases} \tilde{\rho}_n = \dfrac{\rho_n + \rho_{n+1}}{2} = \sigma\rho + \dfrac{3(1-\sigma)}{2^{n+2}}\rho, \\ \tilde{\theta}_n = \dfrac{\theta_n + \theta_{n+1}}{2} = \sigma\theta + \dfrac{3(1-\sigma)}{2^{n+2}}\theta. \end{cases}$$

For these boxes we have the inclusion

$$Q_{n+1} \subset \tilde{Q}_n \subset Q_n \qquad n = 0, 1, 2, \ldots.$$

Introduce the sequence of increasing levels

$$k_n = k - \frac{k}{2^n} \tag{2.13}$$

where k is a positive number to be chosen. We will work with the inequalities (2.5) written for the functions $(u - k_{n+1})_+$, over the boxes Q_n. The cutoff function ζ_n is taken to satisfy

$$\begin{cases} \zeta_n \text{ vanishes on the parabolic boundary of } Q_n \\ \zeta_n \equiv 1 \quad \text{in } \tilde{Q}_n \\ |D\zeta_n| \le \dfrac{2^{n+2}}{(1-\sigma)\rho}, \quad 0 \le \zeta_{n,t} \le \dfrac{2^{n+2}}{(1-\sigma)\theta}. \end{cases} \tag{2.14}$$

Set

$$f_\varepsilon(s) \equiv \begin{cases} 1 & \text{if} \quad (s - k_{n+1}) \ge \varepsilon \\ \varepsilon^{-1}(s - k_{n+1}) & \text{if } 0 < (s - k_{n+1}) < \varepsilon \\ 0 & \text{if} \quad (s - k_{n+1}) \le 0, \end{cases} \tag{2.15}$$

and as a function $f(w)$ take $f_\varepsilon\left[(w - k_{n+1})_+\right]$. We put these choices in (2.5) and neglect the non-negative term involving $|D\mathbf{u}|^{p-2}$ since $f'_\varepsilon(s) \ge 0$. Letting $\varepsilon \to 0$ we obtain

$$\begin{aligned} \sup_{-\theta_n < t < 0} \int_{K_{\rho_n}} (w - k_{n+1})_+^2 \zeta_n^p(x,t)dx + \iint_{Q_n} |D(w - k_{n+1})_+ \zeta_n|^p dxd\tau \\ \le \frac{\gamma 2^{np}}{(1-\sigma)^p \rho^p} \iint_{Q_n} w^p \chi\left[(w - k_{n+1})_+ > 0\right] dxd\tau \\ + \frac{\gamma 2^n}{(1-\sigma)\theta} \iint_{Q_n} w(w - k_{n+1})_+ dxd\tau. \end{aligned}$$

We estimate the two integrals on the right hand side as in (7.2)-(7.5) of Chap. V. This gives the inequalities

$$\begin{aligned} \sup_{-\theta_n < t < 0} \int_{K_{\tilde{\rho}_n}} (w - k_{n+1})_+^2 (x,t)dx + \iint_{\tilde{Q}_n} |D(w - k_{n+1})_+|^p dxd\tau \\ \le \frac{\gamma 2^{n\delta}}{(1-\sigma)^p} \left(\frac{1}{\rho^p k^{\delta - p}} + \frac{1}{\theta k^{\delta - 2}} \right) \iint_{Q_n} (w - k_n)_+^\delta \, dxd\tau, \end{aligned} \tag{2.16}$$

valid for all $\delta \ge \max\{p; 2\}$. If $p > 2$, the proof is now concluded as in the proof of Theorem 4.1 in §12 of Chap. V. If $1 < p < 2$, we may take $\delta = r$ in (2.16) and obtain the analog of the recursive integral inequalities (10.3) of Chap. V. The proof of Theorem 2.1 for the singular case $1<p<2$ is now concluded as in the proof of Theorem 5.1 in §16 of Chap. V.

3. Weak differentiability of $|D\mathbf{u}|^{\frac{p-2}{2}} D\mathbf{u}$ and energy estimates for $|D\mathbf{u}|$

The main tool in investigating the local behaviour of the of the space-gradient of the solutions of (1.1) are certain *local* energy estimates for u_{i,x_j}. These are derived by first '*differentiating*' (1.1) and then by taking testing functions roughly speaking of the type

$$\varphi_i = u_{i,x_j}\, f(|D\mathbf{u}|),$$

up to some localising cutoff function. Here $f(\cdot)$ is a non-negative Lipschitz function in $\mathbf{R}^+$. In this section we discuss a rigorous way of carrying the indicated calculations.

PROPOSITION 3.1 (THE DEGENERATE CASE $p>2$). *Let* $\mathbf{u}$ *be a local weak solution in* Ω_T *of the degenerate system (1.1). Then*

$$|D\mathbf{u}|^{\frac{p-2}{2}} u_{i,x_j} \in L^2_{loc}\left(0,T;W^{1,2}_{loc}(\Omega)\right),\ i=1,2,\dots,m,\ j=1,2,\dots,N,$$

and there exists a constant $\gamma=\gamma(N,p)$*, such that*

$$\forall (x_o,t_o)\in\Omega_T,\ \forall\,[(x_o,t_o)+Q(\theta,\rho)]\subset\Omega_T,\ \forall\sigma\in(0,1),$$

$$\iint\limits_{[(x_o,t_o)+Q(\sigma\theta,\sigma\rho)]} |D\mathbf{u}|^{p-2}|D^2\mathbf{u}|^2 dxd\tau \leq \frac{\gamma}{(1-\sigma)^2}\left[\rho^{-2}+\theta^{-1}\right] \iint\limits_{[(x_o,t_o)+Q(\theta,\rho)]} (1+|Du|^p)\, dxd\tau \tag{3.1}$$

where

$$|D^2\mathbf{u}|^2 \equiv \sum_{i=1}^{m}\sum_{j,k=1}^{N} u^2_{i,x_jx_k}.$$

Moreover

$$u_{i,x_j}\in C_{loc}\left(0,T;L^2_{loc}(\Omega)\right),\ i=1,2,\dots,m,\ j=1,2,\dots,N. \tag{3.2}$$

PROPOSITION 3.1 (THE SINGULAR CASE $1<p<2$). *Let* $\mathbf{u}$ *be a local weak solution of the singular system (1.1) in* Ω_T *and let the approximation assumption (1.9) hold. Then*

$$|D\mathbf{u}|^{\frac{p-2}{2}} u_{i,x_j} \in L^2_{loc}\Big(0,T; W^{1,2}_{loc}(\Omega)\Big)\,,\ i=1,2,\dots,m\,,\ j=1,2,\dots,N,$$

and there exists a constant $\gamma=\gamma(N,p)$, *such that*

$$\forall (x_o,t_o)\in \Omega_T,\ \forall\,[(x_o,t_o)+Q\,(\theta,\rho)]\subset \Omega_T,\ \forall\sigma\in(0,1),$$

(3.3)
$$\iint\limits_{[(x_o,t_o)+Q(\sigma\theta,\sigma\rho)]} |D\mathbf{u}|^{p-2}|D^2\mathbf{u}|^2 dx d\tau$$
$$\le \frac{\gamma}{(1-\sigma)^2}\left[\rho^{-2}+\theta^{-2}\right](1+M_o^2)\iint\limits_{[(x_o,t_o)+Q(\theta,\rho)]}(1+|Du|^p)\,dx d\tau,$$

where

$$M_o\equiv \|\mathbf{u}\|_{\infty,[(x_o,t_o)+Q(\theta,\rho)]}.$$

Moreover

$$u_{i,x_j}\in L^p_{loc}\left(0,T;W^{1,p}_{loc}(\Omega)\right),$$

and there exists a constant $\gamma=\gamma(N,p)$ *such that*

$$\forall (x_o,t_o)\in \Omega_T,\ \forall\,[(x_o,t_o)+Q\,(\theta,\rho)]\subset \Omega_T,\ \forall\sigma\in(0,1),$$

(3.4)
$$\iint\limits_{[(x_o,t_o)+Q(\sigma\theta,\sigma\rho)]} |D^2\mathbf{u}|^p dx d\tau$$
$$\le \frac{\gamma}{(1-\sigma)^p}\left[\rho^{-p}+\theta^{-p}\right](1+M_o^p)\iint\limits_{[(x_o,t_o)+Q(\theta,\rho)]}(1+|Du|^p)\,dx d\tau.$$

Finally

(3.5)
$$u_{i,x_j}\in C_{loc}\big(0,T;L^2_{loc}(\Omega)\big)\,,\ i=1,2,\dots,m,\ j=1,2,\dots,N.$$

This local regularity permits to derive local energy estimates for $D\mathbf{u}$. To simplify the symbolism we set

(3.6)
$$v\equiv|D\mathbf{u}|.$$

Given a cylinder $[(x_o,t_o)+Q\,(\theta,\rho)]\subset\Omega_T$ we let ζ denote a non-negative piecewise smooth cutoff function in $[(x_o,t_o)+Q\,(\theta,\rho)]$ that vanishes on the boundary of the cube $[x_o+K_\rho]$. In particular we are not requiring in general that ζ vanishes for $t=t_o-\theta$.

PROPOSITION 3.2 (LOCAL ENERGY ESTIMATES). *Let* $\mathbf{u}$ *be a local weak solution of (1.1) for* $p>1$. *In the singular case* $1<p<2$ *assume in addition that the approximation assumption (1.9) be in force. Let also* $f(\cdot)$ *denote a non-negative, non-decreasing Lipschitz function in* $\mathbf{R}^+$. *There exists a constant* $\gamma=\gamma(N,p)$ *such that*

(3.7) $\forall\,(x_o,t_o)\in\Omega_T,\ \forall\,[(x_o,t_o)+Q(\theta,\rho)]\subset\Omega_T$

$$\sup_{t_o-\theta\le t\le 0}\int_{[x_o+K_\rho]}\left(\int_0^v sf(s)ds\right)\zeta^2(x,t)dx\Bigg|_{t_o-\theta}^{t}$$

$$+\iint_{[(x_o,t_o)+Q(\theta,\rho)]} v^{p-2}|D^2\mathbf{u}|^2 f(v)\zeta^2 dxd\tau+\iint_{[(x_o,t_o)+Q(\theta,\rho)]} v^{p-1}\left|Dv\right|^2 f'(v)\zeta^2 dxd\tau$$

$$+(p-2)\sum_{i=1}^{m}\iint_{[(x_o,t_o)+Q(\theta,\rho)]} v^{p-3}\left|Dv\cdot Du_i\right|^2 f'(v)\zeta^2 dxd\tau$$

$$\le\gamma\iint_{[(x_o,t_o)+Q(\theta,\rho)]} v^p f(v)|D\zeta|^2 dxd\tau+\gamma\iint_{[(x_o,t_o)+Q(\theta,\rho)]}\left(\int_0^v sf(s)ds\right)\zeta\zeta_t dxd\tau.$$

COROLLARY 3.1. *The integral inequalities (3.7) continue to hold for non-negative, non-decreasing functions* f *in* $\mathbf{R}^+$, *satisfying*

$$\sup_{0\le s\le k} f'(s)<\infty,\quad \textit{for all } k>0,$$

provided

$$v^p f(v)\quad \textit{and}\quad \int_0^v sf(s)ds\in L^1_{loc}(\Omega_T). \tag{3.8}$$

PROOF: Analogous to that of Corollary 2.1.

3-(i). Taking discrete derivatives of (1.1)

For a function $F\in L^p_{loc}(\Omega_T)$ and $\eta\in\mathbf{R}\backslash\{0\}$, we introduce the discrete derivative with respect to the x_j variable

$$\delta_j F(x,t)\equiv\eta^{-1}\{F(x_1,\dots,x_j+\eta,\dots,x_N)-F(x_1,\dots,x_j,\dots,x_N)\}.$$

This is defined for

$$x\in\Omega^{|\eta|}\equiv\left\{x\in\Omega\,\big|\,\operatorname{dist}(x,\partial\Omega)>|\eta|\right\},$$

where we let $|\eta|$ be so small that $\Omega^{|\eta|}$ is not empty. We also let δF denote the discrete gradient of F,i.e.,

$$\delta F \equiv (\delta_1 F, \delta_2 F, \dots, \delta_N F)\,.$$

The discrete derivative of (2.6)′, with respect to x_j, takes the form

$$\begin{aligned}(3.9)\qquad &\frac{\partial}{\partial t}\delta_j u_{i,h} - \operatorname{div}\left[\delta_j|D\mathbf{u}|^{p-2}Du_i\right]_h = 0,\\ &i=1,2,\dots,m, \qquad \text{a.e. } \Omega^{|\eta|}\times(0,T-h).\end{aligned}$$

In transforming the term $\left[\delta_j|D\mathbf{u}|^{p-2}Du_i\right]$, we only specify the x_j variable for simplicity of symbolism. We have

$$\begin{aligned}(3.10)\qquad &\delta_j|D\mathbf{u}|^{p-2}Du_i\\ &\equiv \frac{1}{\eta}\int_0^1 \frac{d}{d\sigma}\Big\{\Big|\sigma D\mathbf{u}(x_j+\eta)+(1-\sigma)D\mathbf{u}(x_j)\Big|^{p-2}\\ &\qquad\times\Big(\sigma Du_i(x_j+\eta)+(1-\sigma)Du_i(x_j)\Big)\Big\}d\sigma\\ &= D\delta_j u_i\int_0^1\Big|\sigma D\mathbf{u}(x_j+\eta)+(1-\sigma)D\mathbf{u}(x_j)\Big|^{p-2}d\sigma\\ &+(p-2)\delta_j u_{\ell,x_k}\int_0^1\Big|\sigma Du(x_j+\eta)+(1-\sigma)Du(x_j)\Big|^{p-4}\\ &\qquad\times\Big(\sigma u_{\ell,x_k}(x_j+\eta)+(1-\sigma)u_{\ell,x_k}(x_j)\Big)\\ &\qquad\times\Big(\sigma Du_i(x_j+\eta)+(1-\sigma)Du_i(x_j)\Big)d\sigma.\end{aligned}$$

To simplify the symbolism we let $\Delta_i^{(j)}(\sigma)$ denote the $N-$dimensional vector

$$\Delta_i^{(j)}(\sigma) \equiv \sigma Du_i(x_j+\eta)+(1-\sigma)Du_i(x_j)$$

and let $\Delta^{(j)}(\sigma)$ be the $N\times m$ matrix

$$\Delta^{(j)}(\sigma) \equiv \sigma D\mathbf{u}(x_j+\eta)+(1-\sigma)D\mathbf{u}(x_j).$$

Having fixed the point $(x_o,t_o)\in\Omega_T$, if $[(x_o,t_o)+Q(\theta,\rho)]\subset\Omega_T$ we may assume, up to a translation, that (x_o,t_o) coincides with the origin, and then by choosing $|\eta|$ and h sufficiently small we may assume that $Q(\theta,\rho)\subset\Omega^{|\eta|}\times(0,T-h)$. We multiply (3.9) by the testing function

$$\delta_j u_{i,h}\, f\left(|\delta\mathbf{u}_h|\right)\zeta^2,$$

where ζ is a standard non-negative cutoff function that vanishes on the boundary of K_ρ. We integrate over $(-\theta,t)$ for arbitrary $-\theta<t\le 0$, and add over $i=1,2,\dots,m$ and $j=1,2,\dots,N$. This gives

$$\int_{K_\rho}\left(\int_0^{|\delta \mathbf{u}_h|} sf(s)\,ds\right)\zeta^2(x,t)\,dx\Bigg|_{-\theta}^{t}$$
$$+\int_{-\theta}^{t}\int_{K_\rho}\left[\delta_j|D\mathbf{u}|^{p-2}Du_i\right]_h\cdot D\delta_j u_{i,h} f\left(|\delta\mathbf{u}_h|\right)\zeta^2 dxd\tau$$
$$+\int_{-\theta}^{t}\int_{K_\rho}\left[\delta_j|D\mathbf{u}|^{p-2}Du_i\right]_h\cdot \delta_j u_{i,h} Df\left(|\delta\mathbf{u}_h|\right)\zeta^2 dxd\tau$$
$$=-2\int_{-\theta}^{t}\int_{K_\rho}\left[\delta_j|D\mathbf{u}|^{p-2}Du_i\right]_h\cdot \delta_j u_{i,h} f\left(|\delta\mathbf{u}_h|\right)\zeta D\zeta\, dxd\tau$$
$$+2\int_{-\theta}^{t}\int_{K_\rho}\left(\int_0^{|\delta\mathbf{u}_h|} sf(s)ds\right)\zeta\zeta_t\, dxd\tau.$$

In this equality we first let $h\to 0$, while $|\eta|>0$ remains fixed. The various limits are justified since $|D\mathbf{u}|\in L^p_{loc}(\Omega_T)$ and $u\in C_{loc}\left(0,T;L^2_{loc}(\Omega)\right)$. Making use also of (3.10) we obtain

$$\sup_{-\theta<t<0}\int_{K_\rho}\left(\int_0^{|\delta\mathbf{u}|} sf(s)\,ds\right)\zeta^2(x,t)\,dx\Bigg|_{-\theta}^{t} \tag{3.11}$$
$$+\iint_{Q(\theta,\rho)}\left(\int_0^1|\Delta^{(j)}(\sigma)|^{p-2}d\sigma\right)\left|D\delta_j\mathbf{u}\right|^2 f(|\delta\mathbf{u}|)\zeta^2 dxd\tau$$
$$+(p-2)\iint_{Q(\theta,\rho)}\left(\int_0^1|\Delta^{(j)}(\sigma)|^{p-4}|\Delta^{(j)}(\sigma)\cdot D\delta_j\mathbf{u}|^2 d\sigma\right) f\left(|\delta\mathbf{u}|\right)\zeta^2 dxd\tau$$
$$+\iint_{Q(\theta,\rho)}\left(\int_0^1|\Delta^{(j)}(\sigma)|^{p-2}d\sigma\right)\left|D|\delta\mathbf{u}|\right|^2|\delta\mathbf{u}|f'\left(|\delta\mathbf{u}|\right)\zeta^2 dxd\tau$$
$$+(p-2)\iint_{Q(\theta,\rho)}\left(\int_0^1|\Delta^{(j)}(\sigma)|^{p-4}\Delta^{(j)}(\sigma)\cdot D\delta_j\mathbf{u}\Delta_i^{(j)}(\sigma)\delta_j u_i d\sigma\right)$$
$$\times D|\delta\mathbf{u}|f'\left(|\delta\mathbf{u}|\right)\zeta^2 dxd\tau$$
$$\le 2(p-1)\iint_{Q(\theta,\rho)}\left(\int_0^1|\Delta^{(j)}(\sigma)|^{p-2}d\sigma\right)\left|D\delta_j\mathbf{u}\right||\delta\mathbf{u}|\, f\left(|\delta\mathbf{u}|\right)\zeta|D\zeta|dxd\tau$$
$$+2\iint_{Q(\theta,\rho)}\left(\int_0^{|\delta\mathbf{u}|} sf(s)ds\right)\zeta\zeta_t\, dxd\tau.$$

First we observe that the sum of the first two integrals over $Q(\theta,\rho)$ on the left hand side, is bounded below by

$$\min\{1;(p-1)\}\iint_{Q(\theta,\rho)}\left(\int_0^1|\Delta^{(j)}(\sigma)|^{p-2}d\sigma\right)\left|D\delta_j\mathbf{u}\right|^2 f(|\delta\mathbf{u}|)\,\zeta^2 dx d\tau.$$

If $p>2$, this is obtained by discarding the coefficient $(p-2)$. If $1<p<2$, we estimate below

$$\begin{aligned}(p-2)&\iint_{Q(\theta,\rho)}\left(\int_0^1|\Delta^{(j)}(\sigma)|^{p-4}\left|\Delta^{(j)}(\sigma)\cdot D\delta_j\mathbf{u}\right|^2 d\sigma\right) f(|\delta\mathbf{u}|)\,\zeta^2 dx d\tau\\ &\geq (p-2)\iint_{Q(\theta,\rho)}\left(\int_0^1|\Delta^{(j)}(\sigma)|^{p-2}d\sigma\right)\left|D\delta_j\mathbf{u}\right|^2 f(|\delta\mathbf{u}|)\,\zeta^2 dx d\tau.\end{aligned}$$

Next by Young's inequality, for all $\varepsilon>0$,

$$\begin{aligned}&\iint_{Q(\theta,\rho)}\left(\int_0^1|\Delta^{(j)}(\sigma)|^{p-2}d\sigma\right)\left|D\delta_j\mathbf{u}\right|\,|\delta\mathbf{u}|\,f(|\delta\mathbf{u}|)\,\zeta|D\zeta|\,dx d\tau\\ &\leq \varepsilon\iint_{Q(\theta,\rho)}\left(\int_0^1|\Delta^{(j)}(\sigma)|^{p-2}d\sigma\right)\left|D\delta_j\mathbf{u}\right|^2 f(|\delta\mathbf{u}|)\,\zeta^2 dx d\tau\\ &+\gamma_\varepsilon\iint_{Q(\theta,\rho)}\left(\int_0^1|\Delta^{(j)}(\sigma)|^{p-2}d\sigma\right)|\delta\mathbf{u}|^2 f(|\delta\mathbf{u}|)\,|D\zeta|^2 dx d\tau.\end{aligned}$$

These remarks in (3.11) give the integral inequality involving discrete derivatives

$$
\begin{aligned}
(3.12)\qquad & \sup_{-\theta<t<0} \int_{K_\rho} \left(\int_0^{|\delta \mathbf{u}|} s f(s)\, ds \right) \zeta^2(x,t)\, dx \Bigg|_{-\theta}^{t} \\
&+ [\min\{1;(p-1)\} - \varepsilon] \iint_{Q(\theta,\rho)} \left(\int_0^1 |\Delta^{(j)}(\sigma)|^{p-2} d\sigma \right) \\
&\qquad\qquad \times \left| D\delta_j \mathbf{u} \right|^2 f(|\delta \mathbf{u}|)\, \zeta^2 dx d\tau \\
&+ \iint_{Q(\theta,\rho)} \left(\int_0^1 |\Delta^{(j)}(\sigma)|^{p-2} d\sigma \right) \left| D|\delta \mathbf{u}| \right|^2 |\delta \mathbf{u}| f'(|\delta \mathbf{u}|)\, \zeta^2 dx d\tau \\
&+(p-2) \iint_{Q(\theta,\rho)} \left(\int_0^1 |\Delta^{(j)}(\sigma)|^{p-4} \left(\Delta^{(j)}(\sigma) \cdot D\delta_j \mathbf{u} \right) \Delta_i^{(j)}(\sigma) \delta_j u_i d\sigma \right) \\
&\qquad\qquad \times D|\delta \mathbf{u}| f'(|\delta \mathbf{u}|)\, \zeta^2 dx d\tau \\
&\le \gamma \iint_{Q(\theta,\rho)} \left(\int_0^1 |\Delta^{(j)}(\sigma)|^{p-2} d\sigma \right) |\delta \mathbf{u}|^2 f(|\delta \mathbf{u}|)\, |D\zeta|^2 dx d\tau \\
&+\gamma \iint_{Q(\theta,\rho)} \left(\int_0^{|\delta \mathbf{u}|} s f(s) ds \right) \zeta \zeta_t\, dx d\tau,
\end{aligned}
$$

for a constant $\gamma = \gamma(p, \varepsilon)$.

3-(ii). Weak differentiability of $|D\mathbf{u}|^{\frac{p-2}{2}} u_{i,x_j}$

In (3.12) take $f \equiv 1$ and select a cutoff function that vanishes on the parabolic boundary of $Q(\theta, \rho)$. In particular, $\zeta(\cdot, -\theta) = 0$. We discard the first term and observe that the integrand in the remaining integral on the right hand side is non-negative. Therefore letting $\eta \to 0$ with the aid of Fatou's Lemma gives

$$
(3.13)\qquad \iint_{Q(\theta,\rho)} v^{p-2} \left| D^2 \mathbf{u} \right|^2 \zeta^2 dx d\tau \le \gamma \iint_{Q(\theta,\rho)} \left(v^p |D\zeta|^2 + v^2 \zeta \zeta_t \right) dx d\tau
$$

for a constant $\gamma = \gamma(p)$. If $p > 2$, the inequality (3.1) follows from (3.13) by choosing ζ, a cutoff function that equals one on $Q(\sigma\theta, \sigma\rho)$ and such that

$$
|D\zeta| \le \frac{1}{(1-\sigma)\rho}, \qquad 0 \le \zeta_t \le \frac{1}{(1-\sigma)\theta}.
$$

To prove (3.3) for the singular case, we transform the last integral in (3.13) by means of an integration by parts as follows.

$$\begin{aligned}
\iint\limits_{Q(\theta,\rho)} v^2\zeta dxd\tau =& \iint\limits_{Q(\theta,\rho)} \mathbf{Du}\cdot D\mathbf{u}\, v^{\frac{p-2}{2}} v^{\frac{2-p}{2}}\zeta\, dxd\tau \\
=& \iint\limits_{Q(\theta,\rho)} u_i D\left[v^{\frac{p-2}{2}} Du_i\right] v^{\frac{2-p}{2}}\zeta\, dxd\tau \\
+& \iint\limits_{Q(\theta,\rho)} u_i v^{\frac{p-2}{2}} Du_i Dv^{\frac{2-p}{2}}\zeta\, dxd\tau \\
+& \iint\limits_{Q(\theta,\rho)} u_i v^{\frac{p-2}{2}} Du_i v^{\frac{2-p}{2}} D\zeta\, dxd\tau \\
\le& \gamma\|\mathbf{u}\|_{\infty,Q(\theta,\rho)} \iint\limits_{Q(\theta,\rho)} \left(v^{p-2}|D^2\mathbf{u}|^2\zeta^2\right)^{\frac{1}{2}} v^{\frac{2-p}{2}}\, dxd\tau \\
&+\gamma\|\mathbf{u}\|_{\infty,Q(\theta,\rho)} \iint\limits_{Q(\theta,\rho)} (1+|D\mathbf{u}|^p)\,|D\zeta|\, dxd\tau.
\end{aligned}$$

Finally (3.4) follows from (3.3) and Hölder inequality, since

$$\iint\limits_{Q(\theta,\rho)} |D^2\mathbf{u}|^p dxd\tau = \iint\limits_{Q(\theta,\rho)} \left(v^{p-2}|D^2\mathbf{u}|^2\right)^{p/2} v^{\frac{p(2-p)}{2}}\, dxd\tau.$$

Since $v^{\frac{p-2}{2}}|D^2\mathbf{u}|\in L^2_{loc}(\Omega_T)$, the energy inequality (3.7) follows from (3.12) by letting $\eta\to 0$.

3-(iv). Continuity of $u_{i,x_j}(t)$ in $L^2_{loc}(\Omega)$ and energy estimates

By virtue of (3.1) and (3.3) the system in (1.1) can be written in the differentiated form

$$\text{(3.14)}\qquad \begin{aligned} &\frac{\partial}{\partial t}u_{i,x_j} - \operatorname{div}\left(|D\mathbf{u}|^{p-2}Du_i\right)_{x_j} = 0 \quad \text{in } \mathcal{D}'(\Omega_T). \\ &i=1,2,\dots,N, \quad j=1,2,\dots,m. \end{aligned}$$

Moreover (3.12) implies that

$$\text{(3.15)}\qquad u_{i,x_j} \in L^\infty_{loc}\left(0,T;L^2_{loc}(\Omega)\right).$$

These two facts imply that $t\to u_{i,x_j}(t)$ is weakly continuous in $L^2_{loc}(\Omega)$. Indeed let $\varphi\in L^2(K_\rho)$ and let $\{\varphi_n\}$ be a sequence of functions in $C_o^\infty(K_\rho)$ such that

$$\|\varphi-\varphi_n\|_{2,K_\rho} \longrightarrow 0 \quad \text{as } n\to\infty.$$

Taking φ_n as a testing function in (3.14) and integrating over $K_\rho\times(t_1,t_2)$ gives

$$\int_{K_\rho} [u_{i,x_j}(t_2) - u_{i,x_j}(t_1)]\, \varphi_n dx = \int_{t_1}^{t_2}\int_{K_\rho} \left(|D\mathbf{u}|^{p-2} Du_i\right) D\varphi_{n,x_j}\, dx d\tau$$

for almost all $-\theta < t_1 < t_2 \le 0$. Therefore

$$\limsup_{|t_2-t_1|\to 0} \int_{K_\rho} [u_{i,x_j}(t_2) - u_{i,x_j}(t_1)]\, \varphi_n\, dx = 0.$$

From this and (3.15)

$$\begin{aligned} \limsup_{|t_2-t_1|\to 0} &\left| \int_{K_\rho} [u_{i,x_j}(t_2) - u_{i,x_j}(t_1)]\, \varphi\, dx \right| \\ &\le \limsup_{|t_2-t_1|\to 0} \left| \int_{K_\rho} [u_{i,x_j}(t_2) - u_{i,x_j}(t_1)]\, \varphi_n dx \right| \\ &\quad + 2 \sup_{-\theta\le t\le 0} \|u_{i,x_j}\|_{2,K_\rho}(t) \|\varphi - \varphi_n\|_{2,K_\rho}. \end{aligned}$$

To prove that u_{i,x_j} is strongly continuous in $L^2_{loc}(\Omega)$ it suffices to prove that

$$\limsup_{|t_2-t_1|\to 0} \left(\|u_{i,x_j}\zeta\|_{2,K_\rho}(t_2) - \|u_{i,x_j}\zeta\|_{2,K_\rho}(t_1)\right) \longrightarrow 0, \tag{3.16}$$

where ζ is a piecewise smooth cutoff functions in K_ρ vanishing on ∂K_ρ. In (3.14) take the testing function

$$u_{i,x_j}\, \zeta^2$$

and integrate over $K_\rho \times (t_1, t_2)$. By calculations similar to those leading to (3.12) we obtain

$$\begin{aligned} &\left| \int_{K_\rho} [v^2(t_2) - v^2(t_1)]\, \zeta^2 dx \right| \\ &\quad \le \int_{t_1}^{t_2}\int_{K_\rho} v^{p-2} \left|D^2\mathbf{u}\right|^2 dx d\tau + \gamma \int_{t_1}^{t_2}\int_{K_\rho} v^p |D\zeta|^2 dx d\tau. \end{aligned}$$

4. Boundedness of $|D\mathbf{u}|$. Qualitative estimates

Using the weak differentiability of $|D\mathbf{u}|^{\frac{p-2}{2}} u_{i,x_j}$ we first prove that $|D\mathbf{u}|$ is in $L^q_{loc}(\Omega_T)$ for all $q \ge 1$. If $\mathcal{K}_o \subset \mathcal{K}_1$ are compact subsets of Ω_T, we will show that the norm $\|D\mathbf{u}\|_{q,\mathcal{K}_o}$ is bounded only in terms of q, dist $\{\mathcal{K}_o; \mathcal{K}_1\}$ and the norm

$\|D\mathbf{u}\|_{p,\mathcal{K}_1}$. We will do this in a *qualitative* way and with no precise specification of the functional dependence. We will use such qualitative information to prove still *qualitatively* that $|D\mathbf{u}| \in L^\infty_{loc}(\Omega_T)$, with bounds only dependent on local L^p–norms of $|D\mathbf{u}|$. Finally, in the next section, we will turn such qualitative information into precise *quantitative* estimates of $\|D\mathbf{u}\|_{\infty,\mathcal{K}_o}$ over compact subsets $\mathcal{K}_o \subset \Omega_T$.

LEMMA 4.1. *Let* $\mathbf{u}$ *be a local weak solution of (1.1). Moreover in the singular case* $1<p<2$ *let the approximation assumption (1.9) be in force. Then*

$$|D\mathbf{u}| \in L^q_{loc}(\Omega_T), \qquad \textit{for every } q \in [1,\infty).$$

PROOF: Consider first the degenerate case $p>2$. Let $Q(\theta,\rho) \subset \Omega_T$ and let ζ be a standard non-negative cutoff function vanishing on the parabolic boundary of $Q(\theta,\rho)$. Thus, in particular, $\zeta(\cdot,-\theta) \equiv 0$. In (3.7), take $f(v)=v^\beta$, where $\beta \geq 0$ is to be chosen. Proceeding formally we obtain

$$\sup_{-\theta\leq t\leq 0} \int_{K_\rho} v^{2+\beta}\zeta^2(x,t)dx \leq \gamma \iint_{Q(\theta,\rho)} \left(1+v^{p+\beta}\right) dx d\tau \tag{4.1}$$

$$\int_{-\theta}^{0}\int_{K_\rho} \left|D\, v^{\frac{p+\beta}{2}}\right|^2 dx d\tau \leq \gamma \iint_{Q(\theta,\rho)} \left(1+v^{p+\beta}\right) dx d\tau, \tag{4.2}$$

where $\gamma = \gamma(N,p,\beta,\zeta_t,D\zeta)$. These are rigorous if the right hand side is finite. We apply the embedding Theorem 2.1 of Chap. I to the functions

$$x \longrightarrow \left(v^{\frac{p+\beta}{2}}\zeta\right)(x,t), \qquad \text{a.e. } t \in (-\theta,0),$$

over the cubes K_ρ. It suffices to consider the case $N>2$. Indeed if $N=1,2$, we may consider $\mathbf{u}$ as a vector field defined in $\mathbf{R}^N$ $N \geq 3$, up to a localisation, and deduce inequalities (4.1)-(4.2) for it. Let δ be a positive number to be chosen. Then by Corollary 2.1 of Chap. I and Hölder's inequality

$$\begin{aligned}\int_{K_\rho} \left[v^{p+\beta+\delta}\zeta^2\right](x,t)dx &\leq \left(\int_{K_\rho}\left[v^{\frac{p+\beta}{2}}\zeta\right]^{\frac{2N}{N-2}} dx\right)^{\frac{N-2}{N}} \left(\int_{K_\rho} v^{\delta\frac{N}{2}}\zeta^2(x,t)dx\right)^{2/N} \\ &\leq \int_{K_\rho}\left|Dv^{\frac{p+\beta}{2}}\zeta\right|^2 dx \left(\sup_{-\theta\leq t\leq 0}\int_{K_\rho} v^{\delta\frac{N}{2}}\zeta^2(x,t)dx\right)^{2/N}.\end{aligned}$$

We integrate over $(-\theta,0)$, to obtain

$$\iint_{Q(\theta,\rho)} v^{p+\beta+\delta}\zeta^2 dxd\tau \leq \iint_{Q(\theta,\rho)} \left|Dv^{\frac{p+\beta}{2}}\zeta\right|^2 dxd\tau \left(\sup_{-\theta\leq t\leq 0}\int_{K_\rho} v^{\delta\frac{N}{2}}\zeta^2(x,t)dx\right)^{2/N}.$$

Choosing $\delta = 2\frac{\beta+2}{N}$ and combining this with (4.1)-(4.2) gives the recursive inequalities

$$\iint_{Q(\theta,\rho)} v^{p+\beta\frac{N+2}{N}+\frac{4}{N}}\zeta^2 dxd\tau \leq \gamma \iint_{Q(\theta,\rho)} \left(1+v^{p+\beta}\right) dxd\tau, \tag{4.3}$$

for a constant $\gamma = \gamma(N,p,\beta,\zeta_t,D\zeta)$. The right hand side is finite for $\beta = 0$. Therefore $|D\mathbf{u}| \in L^{p+4/N}_{loc}(\Omega_T)$. We may now again apply (4.3) with $\beta = 4/N$ and proceed in this fashion to prove the lemma.

We now turn to the singular case $1 < p < 2$. In (3.7) assume that $(x_o,t_o) \equiv (0,0)$ and choose a cutoff function ζ that vanishes on the parabolic boundary of $Q(\theta,\rho)$. Take also $f(v) = v^\beta$, where $\beta \geq 0$ is to be chosen. By working with the approximations claimed by (1.9) we will use the *qualitative* information that $|D\mathbf{u}| \in L^2_{loc}(\Omega_T)$. Our estimates however will be only in terms of $\|D\mathbf{u}\|_{p,Q(\theta,\rho)}$. Proceeding formally we obtain from (3.7)

$$\iint_{Q(\theta,\rho)} \left|v^{\frac{\beta+p-2}{2}}D^2\mathbf{u}\zeta\right|^2 dxd\tau \leq \gamma\left\{\iint_{Q(\theta,\rho)} v^{p+\beta}dxd\tau + \iint_{Q(\theta,\rho)} v^{2+\beta}\zeta dxd\tau\right\}, \tag{4.4}$$

where $\gamma = \gamma(N,p,\beta,\zeta_t,D\zeta)$. Also by a formal integration by parts

$$\begin{aligned}
\iint_{Q(\theta,\rho)} v^{\beta+2}\zeta dxd\tau &= \iint_{Q(\theta,\rho)} v^{\frac{\beta+p-2}{2}} D\mathbf{u}\cdot D\mathbf{u}\, v^{\frac{\beta+2-p}{2}}\zeta\, dxd\tau \\
&= \iint_{Q(\theta,\rho)} \mathbf{u}\, D\left(v^{\frac{\beta+p-2}{2}}D\mathbf{u}\right) v^{\frac{\beta+2-p}{2}}\zeta\, dxd\tau \\
&+ \iint_{Q(\theta,\rho)} \mathbf{u}\, v^{\frac{\beta+p-2}{2}} D\mathbf{u}\, Dv^{\frac{\beta+2-p}{2}}\zeta\, dxd\tau \\
&+ \iint_{Q(\theta,\rho)} \mathbf{u}\, v^{\frac{\beta+p-2}{2}} D\mathbf{u}\, v^{\frac{\beta+2-p}{2}} D\zeta\, dxd\tau \\
&\leq \gamma\|\mathbf{u}\|_{\infty,Q(\theta,\rho)} \iint_{Q(\theta,\rho)} \left|v^{\frac{\beta+p-2}{2}}D^2\mathbf{u}\right|\zeta\, v^{\frac{\beta+2-p}{2}} dxd\tau \\
&+ \gamma\|\mathbf{u}\|_{\infty,Q(\theta,\rho)} \iint_{Q(\theta,\rho)} v^{\beta+1}\, dxd\tau.
\end{aligned}$$

We combine this with (4.4) and make use of the Schwartz inequality to arrive at

(4.5) $$\iint_{Q(\theta,\rho)} v^{\beta+2}\zeta dxd\tau \le \gamma \iint_{Q(\theta,\rho)} \left(1 + v^{\beta+(2-p)} + v^{\beta+1}\right) dxd\tau,$$

for a constant
$$\gamma=\gamma\left(N,p,\beta,\zeta_t,D\zeta,\|\mathbf{u}\|_{\infty,Q(\theta,\rho)}\right).$$

This inequality is indeed rigorous as long as the right hand side is finite. We apply it first with $\beta=p-1$ to deduce that $|D\mathbf{u}|\in L^{p+1}_{loc}(\Omega_T)$, with bounds only dependent on $\|D\mathbf{u}\|_{p,Q(\theta,\rho)}$. Then we apply it again with $\beta=p$ to deduce that $|D\mathbf{u}|\in L^{p+2}_{loc}(\Omega_T)$. Proceeding this way proves the lemma.

LEMMA 4.2. *Let* $\mathbf{u}$ *be a local weak solution of (1.1). Moreover in the singular case* $1<p<2$ *let the approximation assumption (1.9) be in force. Then*

$$|D\mathbf{u}| \in L^\infty_{loc}(\Omega_T).$$

PROOF: Consider first the degenerate case $p>2$. Let $Q(\theta,\rho)\subset\Omega_T$ and let Q_n and $\tilde{Q}_n$ be the family of cylinders introduced in (2.9)-(2.12). Let also k_n and ζ_n be respectively the increasing levels defined in (2.13) and the cutoff functions in Q_n introduced in (2.14). We put these choices in the energy estimates (3.7) and as a function $f(v)$ take
$$f(v)\equiv(v-k_{n+1})_+^{p-2}.$$

By virtue of Lemma 4.1 and Corollary 3.1, such a choice is admissible. The term involving $D^2\mathbf{u}$ is estimated below by

$$\iint_{Q_n} v^{p-2}|D^2\mathbf{u}|^2 f(v)\zeta^2 dxd\tau \ge \left(\frac{k}{2}\right)^{p-2}\iint_{Q_n}\left|D\,(v-k_{n+1})_+^{\frac{p}{2}}\right|^2\zeta_n^2 dxd\tau.$$

These choices yield the inequalities

(4.6) $$\begin{aligned}&\sup_{-\theta_n\le t\le 0}\int_{K_{\rho_n}}\left[(v-k_{n+1})_+^{\frac{p}{2}}\zeta_n\right]^2(x,t)dx\\&\quad+k^{p-2}\iint_{Q_n}\left|D\left[(v-k_{n+1})_+^{\frac{p}{2}}\zeta_n\right]\right|^2 dxd\tau\\&\le\frac{\gamma\,4^n}{(1-\sigma)^2\rho^2}\iint_{Q_n}v^{2(p-1)}\chi[(v-k_{n+1})_+>0]\,dxd\tau\\&\quad+\frac{\gamma\,2^n}{(1-\sigma)\theta}\iint_{Q_n}v^p\chi[(v-k_{n+1})_+>0]\,dxd\tau,\end{aligned}$$

for a constant $\gamma=\gamma(N,p)$. To simplify the symbolism let us set

(4.7) $$Y_n\equiv\iint_{Q_n}(v-k_n)_+^p\,dxd\tau.$$

Then we have[1]

$$\tag{4.8} \iint_{Q_n} \chi[(v-k_{n+1})_+ > 0]\, dx d\tau \le \gamma \frac{2^{np}}{k^p} \iint_{Q_n} (v-k_n)_+^p \, dx d\tau \equiv \gamma 2^{np} k^{-p} Y_n.$$

By Proposition 3.1 of Chap. I with $m=p-2$ and $q=2(N+2)/N$,

$$\tag{4.9} \begin{aligned} Y_{n+1} &\le \iint_{Q_n} \left[(v-k_{n+1})_+^{p/2} \zeta_n\right]^2 dx d\tau \\ &\le \left(\iint_{Q_n} \left[(v-k_{n+1})_+^{p/2} \zeta_n\right]^{2\frac{N+2}{N}} dx d\tau \right)^{\frac{N}{N+2}} \\ &\quad \times \left(\iint_{Q_n} \chi[(v-k_{n+1})_+ > 0]\, dx d\tau \right)^{\frac{2}{N+2}} \\ &\le \frac{\gamma\, 2^{pn}}{(1-\sigma)^2} k^{-\frac{\lambda_2}{N+2}} Y_n^{\frac{2}{N+2}} \left\{ \rho^{-2} \iint_{Q_n} v^{2(p-1)} \chi[(v-k_{n+1})_+ > 0]\, dx d\tau \right. \\ &\quad \left. + \theta^{-1} \iint_{Q_n} v^p \chi[(v-k_{n+1})_+ > 0]\, dx d\tau \right\}, \end{aligned}$$

where $\lambda_2 \equiv N(p-2)+2p$. These are the key recursive inequalities needed to derive a quantitative sup-bound for $|D\mathbf{u}|$. We will use them first in a *qualitative* way as follows. First let A denote a lump constant depending upon θ, σ, ρ and the quantities

$$\|D\mathbf{u}\|_{2,Q(\theta,\rho)} \quad \|D\mathbf{u}\|_{q,Q(\theta,\rho)} \quad q=(N+2)(p-2)+p.$$

Then we estimate

$$\begin{aligned} &\iint_{Q_n} v^{2(p-1)} \chi[(v-k_{n+1})_+ > 0]\, dx d\tau \\ &= \iint_{Q_n} v^{\frac{q}{N+2}} v^{p\frac{N+1}{N+2}} \chi[(v-k_{n+1})_+ > 0]\, dx d\tau \\ &\le \left(\iint_{Q_n} v^q dx d\tau \right)^{\frac{1}{N+2}} \left(\iint_{Q_n} v^p \chi[(v-k_{n+1})_+ > 0]\, dx d\tau \right)^{\frac{N+1}{N+2}}. \end{aligned}$$

---

[1] See §7-(i) of Chap. V and, in particular, estimate (7.2).

We have also[1]

$$\iint_{Q_n} v^p \chi[(v-k_{n+1})_+ > 0]\, dx d\tau \le \gamma 2^{np}\, Y_n. \tag{4.10}$$

Therefore

$$\iint_{Q_n} v^{2(p-1)} \chi[(v-k_{n+1})_+ > 0]\, dx d\tau \le A 2^{np} Y_n^{\frac{N+1}{N+2}}.$$

These remarks in (4.9) give the recursive inequalities

$$Y_{n+1} \le A\, k^{-\frac{p}{N+2}} b^n Y_n^{1+\frac{1}{N+2}}, \quad n=1,2,\dots, \quad b = 4^{p+1}$$

where we have also used the choice $k \ge 1$ and the inequality

$$Y_n^{1+\frac{2}{N+2}} \le A\, Y^{1+\frac{1}{N+2}}.$$

It follows from Lemma 4.1 of Chap. I that $Y_n \to 0$ as $n \to \infty$ if

$$Y_o = \left(A\, k^{-\frac{p}{N+2}}\right)^{-(N+2)} b^{-(N+2)^2}.$$

Therefore

$$\begin{aligned} \|D\mathbf{u}\|_{\infty, Q(\sigma\theta, \sigma\rho)} &\le \max\{1; k\} \\ &\le 1 + A^{\frac{N+2}{p}} b^{(N+2)^2/p} \|D\mathbf{u}\|_{p, Q(\theta,\rho)}. \end{aligned}$$

We now turn to the singular case $1<p<2$. The starting point is still the energy estimate (3.7) where we choose $Q_n, \tilde{Q}_n, \zeta_n$ and the levels k_n as before. As a function $f(\cdot)$ we take

$$f(v) \equiv \begin{cases} v^{2-p}\,(v-k_{n+1})_+^{r-2} & \text{if } r>2 \\ v^{2-p} f_\varepsilon\left[(v-k_{n+1})_+\right] & \text{if } r=2, \end{cases}$$

where $f_\varepsilon(\cdot)$ is the Lipschitz approximation to the Heaviside graph, introduced in (2.15). After we let $\varepsilon \to 0$, the first term on the left hand side is bounded below for all $t \in (-\theta, 0)$ by the quantity

$$\begin{aligned} &\int_{K_{\rho_n}} \left(\int_{k_{n+1}}^{v} s^{2-p} s\,(s-k_{n+1})^{r-2}\, ds\right)_+ \zeta_n^2(x,t) dx \\ &\ge \frac{1}{r-1}\left(\frac{k}{2}\right)^{2-p} \int_{K_{\rho_n}} \left[(v-k_{n+1})_+^{r/2}\,\zeta_n\right]^2 dx. \end{aligned}$$

---

[1] See for example estimate (7.5) of Chap. V.

The term involving $D^2\mathbf{u}$ is estimated below by

$$\iint_{Q_n} |D^2\mathbf{u}|^2 \, (v - k_{n+1})_+^{r-2} \, \chi[(v-k_{n+1})_+ > 0] \, \zeta_n^2 \, dx d\tau$$

$$\geq \frac{4}{r^2} \iint_{Q_n} \left| D \, (v - k_{n+1})_+^{r/2} \right|^2 \zeta_n^2 \, dx d\tau.$$

Combining these estimates in (3.7) we arrive at

$$k^{2-p} \sup_{-\theta \leq t \leq 0} \int_{K_{\rho n}} \left[(v - k_{n+1})_+^{r/2} \, \zeta_n(x,t) \right]^2 dx \tag{4.11}$$

$$+ \iint_{Q_n} \left| D \, (v - k_{n+1})_+^{r/2} \, \zeta_n \right|^2 dx d\tau$$

$$\leq \frac{\gamma \, 4^n}{(1-\sigma)^2 \rho^2} \iint_{Q_n} v^r \chi[(v-k_{n+1})_+ > 0] \, dx d\tau$$

$$+ \frac{\gamma 4^n}{(1-\sigma)\theta} \iint_{Q_n} v^{r+2-p} \chi[(v-k_{n+1})_+ > 0] \, dx d\tau,$$

where $\gamma = \gamma(N, p, r)$. To simplify the symbolism we set

$$S_n \equiv \iint_{Q_n} (v - k_n)_+^r \, dx d\tau, \tag{4.12}$$

and combine (4.11) with the embedding of Proposition 3.1 of Chap. I, with $m = p = 2$ and $q = 2(N+2)/N$. This gives

$$S_{n+1} \leq \frac{\gamma \, 2^{(r+2)n}}{(1-\sigma)^2} k^{-2\frac{(2-p)+r}{N+2}} S_n^{\frac{2}{N+2}} \tag{4.13}$$

$$\times \left\{ \rho^{-2} \iint_{Q_n} v^r \chi[(v-k_{n+1})_+ > 0] \, dx d\tau \right.$$

$$\left. + \, \theta^{-1} \iint_{Q_n} v^{r+(2-p)} \chi[(v-k_{n+1})_+ > 0] \, dx d\tau \right\},$$

where we have used a version of (4.8). These are the key inequalities needed to derive a quantitative bound of $|D\mathbf{u}|$ in the singular case $1 < p < 2$. As before, we will use them first in a qualitative way. We choose $k \geq 1$ and let A denote a lump constant depending upon ρ, θ, σ and the norms

$$\|v\|_{2,Q(\theta,\rho)}, \quad \|v\|_{q,Q(\theta,\rho)}, \quad q = (N+2)(2-p) + r.$$

Then we estimate

$$
\begin{aligned}
&\iint_{Q_n} v^{r+(2-p)}\chi\big[(v-k_{n+1})_+>0\big]\,dxd\tau \\
&= \iint_{Q_n} v^{(2-p)+\frac{r}{N+2}}\, v^{r\frac{N+1}{N+2}}\chi\big[(v-k_{n+1})_+>0\big]\,dxd\tau \\
&\le \left(\iint_{Q_n} v^q dxd\tau\right)^{\frac{1}{N+2}} \left(\iint_{Q_n} v^r\chi\big[(v-k_{n+1})_+>0\big]\,dxd\tau\right)^{\frac{N+1}{N+2}} \\
&\le A\,2^{nr}\, S_n^{\frac{N+1}{N+2}}.
\end{aligned}
$$

In deriving the last inequality we have used a version of (4.10). These estimates in (4.13) yield the recursive inequalities

$$
S_{n+1} \le A\, b^n\, k^{-\frac{r}{N+2}}\, S_n^{1+\frac{1}{N+1}}, \qquad b=4^{r+1}.
$$

Here we have used the choice $k\ge 1$ and the inequality

$$
S_n^{1+\frac{1}{N+2}} \le A\, S_n^{1+\frac{1}{N+1}}.
$$

The proof is now concluded as in the degenerate case.

5. Quantitative sup-bounds of $|D\mathbf{u}|$

THEOREM 5.1 (THE CASE $p>2$). *Let* $\mathbf{u}$ *be a local weak solution of the degenerate system (1.1). There exists a constant* $\gamma=\gamma(N,p)$ *such that*

$$
\forall (x_o,t_o)\in\Omega_T,\ \forall\,[(x_o,t_o)+Q\,(\theta,\rho)]\subset\Omega_T,\ \forall\sigma\in(0,1),
$$

$$
(5.1)\qquad \sup_{[(x_o,t_o)+Q(\sigma\theta,\sigma\rho)]} |D\mathbf{u}| \le \frac{\gamma\sqrt{(\theta/\rho^2)}}{(1-\sigma)^{(N+2)/2}} \left(\fint\!\!\!\fint_{[(x_o,t_o)+Q(\theta,\rho)]} |D\mathbf{u}|^p dxd\tau\right)^{1/2} \wedge \left(\frac{\rho^2}{\theta}\right)^{\frac{1}{p-2}}.
$$

THEOREM 5.2 (THE SINGULAR CASE $1<p<2$). *Let* $\mathbf{u}$ *be a local weak solution of the singular system (1.1) and let the approximation assumption (1.9) be in force. Moreover let* $r\ge 2$ *satisfy*

$$
(5.2)\qquad \nu_r \equiv N(p-2)+2r>0.
$$

Then there exists a constant $\gamma=\gamma(N,p,r)$ *such that*

$$\forall(x_o,t_o)\in\Omega_T,\ \forall\,[(x_o,t_o)+Q(\theta,\rho)]\subset\Omega_T,\ \forall\sigma\in(0,1),$$

$$\sup_{[(x_o,t_o)+Q(\sigma\theta,\sigma\rho)]}|D\mathbf{u}|\le\frac{\gamma\,(\rho^2/\theta)^{N/\nu_r}}{(1-\sigma)^{2(N+2)/\nu_r}}\left(\iint_{[(x_o,t_o)+Q(\theta,\rho)]}|D\mathbf{u}|^r\,dxd\tau\right)^{2/\nu_r}\wedge\left(\frac{\theta}{\rho^2}\right)^{\frac{1}{2-p}}. \tag{5.3}$$

Remark 5.1. The constant $\gamma(N,p,r)$ in (5.3) tends to infinity as $\nu_r\to 0$.

5-(i). *Proof of Theorem 5.1*

We start from the recursive inequalities (4.9) and estimate the first integral on the right hand side as follows:

$$\begin{aligned}\iint_{Q_n}&v^{2(p-1)}\chi[(v-k_{n+1})_+>0]\,dxd\tau\\&\le\left(\sup_{Q_n}v\right)^{2-p}\iint_{Q_n}v^p\chi[(v-k_{n+1})_+>0]\,dxd\tau\\&\le 2^{np}\left(\sup_{Q_n}v\right)^{2-p}\iint_{Q_n}(v-k_n)_+^p\,dxd\tau.\end{aligned}$$

Therefore (4.9) yields

$$Y_{n+1}\le\frac{\gamma\,b^n}{(1-\sigma)^2}k^{-\frac{\lambda_2}{N+2}}\left\{\left(\sup_{Q_n}v\right)^{p-2}\rho^{-2}+\theta^{-1}\right\}Y_n^{1+\frac{2}{N+2}},\quad b=4^p. \tag{5.4}$$

If for some $n=0,1,2,\dots$ we have

$$\sup_{Q_n}v^{p-2}\le\frac{\rho^2}{\theta},$$

there is nothing to prove. Otherwise we rewrite (5.4) as

$$Y_{n+1}\le\frac{\gamma\,b^n}{[(1-\sigma)\rho]^2}\left(\sup_{Q(\theta,\rho)}v\right)^{p-2}k^{-\frac{\lambda_2}{N+2}}\,Y_n^{1+\frac{2}{N+2}}.$$

It follows from Lemma 4.1 of Chap. I that $\{Y_n\}_{n\in\mathbf{N}}\to 0$ as $n\to\infty$, if k is chosen from

$$k^{\lambda_2/2}\equiv\left(\frac{\gamma\,b^{(N+2)/2}}{[(1-\sigma)\rho]^2}\right)^{(N+2)/2}\left(\sup_{Q(\theta,\rho)}v\right)^{\frac{(N+2)(p-2)}{2}}\iint_{Q(\theta,\rho)}v^p\,dxd\tau.$$

We conclude that there exists a constant $\gamma=\gamma(N,p)$ such that

$$\sup_{Q(\sigma\theta,\sigma\rho)} v \le \frac{\gamma}{[(1-\sigma)\rho]^{2(N+2)/\lambda_2}} \left(\sup_{Q(\theta,\rho)} v\right)^{1-4/\lambda_2} \times \left(\iint_{Q(\theta,\rho)} v^p dxd\tau\right)^{2/\lambda_2}. \tag{5.5}$$

If $\sigma\in(0,1)$ is fixed, consider the family of boxes $\mathcal{Q}^{(n)} \equiv \mathcal{Q}(\theta_n,\rho_n)$, where

$$\rho_o \equiv \sigma\rho \qquad \theta_o = \sigma\theta$$

and for $n=1,2,\dots$

$$\rho_n = \sigma\rho + (1-\sigma)\rho\sum_{i=1}^n 2^{-i} \qquad \theta_n = \sigma\theta + (1-\sigma)\theta\sum_{i=1}^n 2^{-i}.$$

By construction,

$$\mathcal{Q}^{(o)} \equiv Q(\sigma\theta,\sigma\rho) \quad \text{and} \quad \mathcal{Q}^{(\infty)} \equiv Q(\theta,\rho).$$

Set

$$M_n = \operatorname*{ess\,sup}_{\mathcal{Q}^{(n)}} v$$

and write (5.5) for the pair of boxes $\mathcal{Q}^{(n)}$ and $\mathcal{Q}^{(n+1)}$. This gives

$$M_n \le M_{n+1}^{1-4/\lambda_2} d^{n\,4/\lambda_2} (B\,d)^{4/\lambda_2}, \tag{5.6}$$

where

$$B \equiv \frac{\gamma^{\lambda_2/4}}{[(1-\sigma)\rho]^{(N+2)/2}} \left(\iint_{Q(\theta,\rho)} v^p dxd\tau\right)^{1/2}, \quad d = 2^{(N+2)/2}.$$

The proof is now concluded by the interpolation Lemma 4.3 of Chap. I.

5-(ii). *Proof of Theorem 5.2*

We start from the recursive inequalities (4.13) and estimate

$$\iint_{Q_n} v^{r+(2-p)}\chi[(v-k_{n+1})_+>0]\,dxd\tau$$

$$\le \left(\sup_{Q_n} v\right)^{2-p} \iint_{Q_n} v^r\chi[(v-k_{n+1})_+>0]\,dxd\tau$$

$$\le 2^{nr}\left(\sup_{Q_n} v\right)^{2-p} S_n.$$

We may assume that

$$\sup_{Q_n} v^{2-p} \geq \frac{\theta}{\rho^2}, \qquad \text{for all } n = 0, 1, 2, \ldots .$$

Otherwise there is nothing to prove. Taking this into account, we rewrite (4.13) as

$$S_{n+1} \leq \frac{\gamma\, b^n}{(1-\sigma)^2} k^{\frac{2}{N+2}(r+2-p)} \frac{\left(\sup_{Q(\theta,\rho)} v\right)^{2-p}}{\theta} S_n^{1+\frac{2}{N+2}}$$

where $b = 4^{r+1}$. By an argument analogous to that in the degenerate case, this implies

$$\sup_{Q(\sigma\theta,\sigma\rho)} v \leq \frac{\gamma}{(1-\sigma)^{\frac{N+2}{r+2-p}}} \left(\frac{\sup_{Q(\theta,\rho)} v^{2-p}}{\theta} \right)^{\frac{N+2}{2(r+2-p)}} \times \left(\iint_{Q(\theta,\rho)} v^r \, dx d\tau \right)^{1/(r+2-p)} . \tag{5.7}$$

The proof is now concluded with an interpolation process as in the degenerate case. This is possible if the power of the term $\sup_{Q(\theta,\rho)} v$ on the right hand side of (5.7) is less than one. Since

$$\frac{(2-p)(N+2)}{2(r+2-p)} = 1 - \frac{\nu_r}{2(r+2-p)},$$

this occurs if (5.2) holds. We also remark that the interpolation process applied to (5.7) generates a dependence of the type of $1/\nu_r$ in the constant $\gamma(N, p, r)$ appearing in (5.3).

5-(iii). Interpolation inequalities

The inequality of Theorem 5.1 can be interpolated. For example consider (5.1) for $(x_o, t_o) \equiv (0, 0)$ and rewrite it as

$$\sup_{Q(\sigma\theta,\sigma\rho)} v \leq \frac{\gamma \sqrt{(\theta/\rho^2)}}{(1-\sigma)^{(N+2)/2}} \left(\sup_{Q(\theta,\rho)} v \right)^{\frac{2-\varepsilon}{2}} \left(\iint_{Q(\theta,\rho)} v^{p-2+\varepsilon} \, dx d\tau \right)^{1/2} \wedge \left(\frac{\rho^2}{\theta} \right)^{\frac{1}{p-2}} .$$

Such an inequality can be interpolated as long as $\varepsilon \in (0, 2]$ and proves the following:

THEOREM 5.1′ (THE CASE $p > 2$). *Let* $\mathbf{u}$ *be a local weak solution of the degenerate system (1.1). Then for every* $\varepsilon \in (0,2]$, *there exists a constant* $\gamma = \gamma(N,p,\varepsilon)$ *such that*

$$\forall (x_o,t_o)\in\Omega_T,\ \forall\,[(x_o,t_o)+Q(\theta,\rho)]\subset\Omega_T,\ \forall\sigma\in(0,1),$$

$$\sup_{[(x_o,t_o)+Q(\sigma\theta,\sigma\rho)]} |D\mathbf{u}| \le \frac{\gamma\,(\theta/\rho^2)^{1/\varepsilon}}{(1-\sigma)^{(N+2)/\varepsilon}} \left(\iint_{[(x_o,t_o)+Q(\theta,\rho)]} |D\mathbf{u}|^{p-2+\varepsilon}\,dx d\tau \right)^{1/\varepsilon} \wedge \left(\frac{\rho^2}{\theta}\right)^{\frac{1}{p-2}}. \tag{5.8}$$

Remark 5.2. The constant $\gamma=\gamma(N,p,\varepsilon)\nearrow\infty$ as $\varepsilon\searrow 0$.

Also, (5.3) can be interpolated. We rewrite it for $(x_o,t_o)\equiv(0,0)$ and in the form

$$\sup_{Q(\sigma\theta,\sigma\rho)} v \le \frac{\gamma\,(\rho^2/\theta)^{N/\nu_r}}{(1-\sigma)^{2(N+2)/\nu_r}} \left(\sup_{Q(\theta,\rho)} v \right)^{\frac{2(r-q)}{\nu_r}} \left(\iint_{Q(\theta,\rho)} v^q\,dx d\tau \right)^{2/\nu_r} \wedge \left(\frac{\theta}{\rho^2}\right)^{\frac{1}{2-p}}.$$

This can be interpolated as long as $q\in(0,r]$ satisfies $\frac{2(r-q)}{\nu_r}<1$. This occurs if

$$\nu_q \equiv N(p-2)+2q > 0. \tag{5.9}$$

The interpolation process gives

THEOREM 5.2′ (THE SINGULAR CASE $1<p<2$). *Let* $\mathbf{u}$ *be a local weak solution of the singular system (1.1) and let the approximation assumption (1.9) be in force. Moreover let* $r\ge 2$ *satisfy (5.2). Then for every* $q\in(0,r]$ *satisfying (5.9) there exists a constant* $\gamma=\gamma(N,p,r,q)$ *such that*

$$\forall (x_o,t_o)\in\Omega_T,\ \forall\,[(x_o,t_o)+Q(\theta,\rho)]\subset\Omega_T,\ \forall\sigma\in(0,1),$$

$$\sup_{[(x_o,t_o)+Q(\sigma\theta,\sigma\rho)]} |D\mathbf{u}| \le \frac{\gamma\,(\rho^2/\theta)^{N/\nu_q}}{(1-\sigma)^{2(N+2)/\nu_q}} \left(\iint_{[(x_o,t_o)+Q(\theta,\rho)]} |D\mathbf{u}|^{q}\,dx d\tau \right)^{2/\nu_q} \wedge \left(\frac{\theta}{\rho^2}\right)^{\frac{1}{2-p}}. \tag{5.10}$$

Remark 5.3. The constant $\gamma(N,p,r,q)$ in (5.10) tends to infinity as $\nu_q\to 0$.

Remark 5.4. Estimate (5.10) is formally equivalent to (5.3), the only difference being that q is not required to be larger or equal to 2. The only condition is that (5.9) be verified. In particular, (5.10) holds for $q=p$ provided

$$p > \frac{2N}{N+2}. \tag{5.11}$$

COROLLARY 5.1. *Let* $1<p<2$. *Then* $|D^2\mathbf{u}| \in L^2_{loc}(\Omega_T)$.

PROOF: From (3.3), for every $[(x_o,t_o)+Q(\theta,\rho)] \subset \Omega_T$,

$$\begin{aligned}\iint_{[(x_o,t_o)+Q(\theta,\rho)]} |D^2\mathbf{u}|^2 dx d\tau &= \iint_{[(x_o,t_o)+Q(\theta,\rho)]} |D\mathbf{u}|^{2-p} |D\mathbf{u}|^{p-2} |D^2\mathbf{u}|^2 dx d\tau \\ &\le \sup_{[(x_o,t_o)+Q(\theta,\rho)]} |D\mathbf{u}|^{2-p} \iint_{[(x_o,t_o)+Q(\theta,\rho)]} |D\mathbf{u}|^{p-2} |D^2\mathbf{u}|^2 dx d\tau < \infty.\end{aligned}$$

6. General structures

Let $\mathbf{u}$ be a local weak solution of the non-linear system (1.10) subject to the structure conditions $(\mathcal{S}_1)$-$(\mathcal{S}_6)$. The local boundedness of $\mathbf{u}$ can be established as in the proof of Theorem 2.1. The main modification occurs in the handling of the *'perturbation terms'* φ_i, $i=0,1,2$. These contribute to the energy inequalities (2.5) with an extra term of the type

$$\iint_{[(x_o,t_o)+Q(\theta,\rho)]} \{\varphi_o (f(w) + wf'(w)) + (\varphi_1 |D\zeta| + \varphi_2)\, wf(w)\}\, dx d\tau.$$

Given the choice (2.15) of $f(\cdot)$, these terms are estimated as in the sup-bounds established in Chap.V for general equations.[1] The weak differentiability of the term $|D\mathbf{u}|^{p-2} D\mathbf{u}$ follows from the structure conditions $(\mathcal{S}_1)$-$(\mathcal{S}_2)$. We proceed as before by working first with the discrete derivatives. All the terms involving the *'derivatives'* $\delta_j u_{i,x_\ell}$, are dominated by the terms arising from the right hand side of $(\mathcal{S}_2)$.[2] Following the same process of §3 yields local energy estimates similar to (3.7) with constants $\gamma=\gamma(N,p,C_o,C_1)$ and with the right hand side augmented by the extra integral

$$\iint_{[(x_o,t_o)+Q(\theta,\rho)]} \{\varphi_o (f(v) + vf'(v)) + (\varphi_1 |D\zeta| + \varphi_2)\, vf(v)\}\, dx d\tau.$$

(1) See for example Theorem 3.1 of Chap. V and its proof.

(2) See also Remark 1.1.

These energy estimates imply that $|D\mathbf{u}| \in L^{\infty}_{loc}(\Omega_T)$ be the same iterative techniques of §4. The *'perturbation terms'* are dealt with as in Chap. V.

7. Bibliographical notes

In the case of a single equation the estimate (1.5) up to S_T has been established by Lieberman [68]. Estimates in the norm $C^{1,\alpha}$ up to S_T for Dirichlet data, are not known even for *elliptic* systems. Results for a single elliptic equations are due to Lieberman [69] and Lin [72]. The general structures of §1-(II) have been introduced first by Tolksdorff [95]. The arguments of finite differences to prove that $|D\mathbf{u}|^{p-2}u_{i,x_j}$ is weakly differentiable were introduced by Uhlenbeck [99] in the context of elliptic systems. The sup-bound of Theorem 2.1 for the degenerate case $p > 2$ is new. The same theorem for the singular case $1 < p < 2$ is due to Choe [31]. The qualitative Lemmas 4.1 and 4.2 appear in [36] for all $p > \frac{2N}{N+2}$, and in Choe [31] for all $p > 1$ provided (1.7) holds. Even though some *quantitative* estimates of the gradient appear in a variety of forms in [27,36,37], Chen [25] and Choe [30], the precise form of Theorems 5.1 and 5.2 as well as their *interpolated* version in §5-(III), seems to be new.

IX

Parabolic p-systems: Hölder continuity of $D\mathbf{u}$

1. The main theorem

The space gradient $D\mathbf{u}$ of local weak solutions of the quasilinear system (1.10) of Chap. VIII are locally Hölder continuous in Ω_T provided the structure conditions $(\mathcal{S}_1)$-$(\mathcal{S}_6)$ are in force. We will show this first for the homogeneous system (1.1) and then will indicate how to extend it to the general systems (1.10). The estimates of this chapter hold in the interior of Ω_T and deteriorate near its parabolic boundary Γ. If $\mathcal{K}$ is a compact subset of Ω_T we let dist$(\mathcal{K};\Gamma)$ denote the parabolic distance from $\mathcal{K}$ to the parabolic boundary Γ of Ω_T, i.e.,

$$\text{dist}\,(\mathcal{K};\Gamma) \equiv \inf_{\substack{(x,t)\in\mathcal{K} \\ (y,s)\in\Gamma}} \left(|x-y| + \sqrt{|t-s|}\right).$$

THEOREM 1.1. *Let* $\mathbf{u}$ *be a local weak solution of (1.1) of Chap. VIII. Moreover if* $1<p<2$ *let the approximation condition (1.9) be in force. Then*

$$(x,t)\to u_{i,x_j}(x,t)\in C^{\alpha}_{loc}(\Omega_T),\quad \textit{for some}\ \ \alpha\in(0,1),$$

for all $i=1,2,\dots,m$ *and all* $j=1,2,\dots,N$. *Moreover for every compact subset* $\mathcal{K}$ *of* Ω_T, *there exist constants* $\alpha=\alpha(N,p)\in(0,1)$ *and* $\gamma=\gamma(N,p,\|D\mathbf{u}\|_{\infty,\mathcal{K}})>1$, *such that*

$$(1.1)\qquad \left|u_{i,x_j}(x_1,t_1)-u_{i,x_j}(x_2,t_2)\right| \le \gamma\left(\frac{|x_1-x_2|+|t_1-t_2|^{1/2}}{\text{dist}\,(\mathcal{K}\,;\,\Gamma)}\right)^{\alpha},$$

for every pair of points $(x_1,t_1),\ (x_2,t_2)\in\mathcal{K}$.

Remark 1.1. The constants γ and a are independent of dist $(\mathcal{K}\,;\,\Gamma)$. They however deteriorate as $p\searrow 1$, i.e.,

$$\liminf_{p\searrow 1} \gamma\left(N,p,\|D\mathbf{u}\|_{\infty,\mathcal{K}}\right),\ \alpha^{-1}\left(N,p\right) \longrightarrow \infty.$$

Remark 1.2. The functional dependence of γ upon $\|D\mathbf{u}\|_{\infty,\mathcal{K}}$ will be given in §§3 and 4.

1-(i). Some notation and the two basic propositions

The proof of Theorem 1.1 is based on estimating the essential oscillation of u_{i,x_j} in cylindrical domains of the type $[(x_o,t_o)+Q(\theta,\rho)]\subset\Omega_T$. After a translation we may assume that (x_o,t_o) coincides with origin. Let μ and R be positive numbers and consider the cylinders

$$\text{(1.2)}\qquad \begin{cases} Q_R(\mu)\equiv K_R\times\left\{-\mu^{2-p}R^2,0\right\}, & \text{satisfying}\\ \sup\limits_{Q_R(\mu)} |D\mathbf{u}|\le\mu. \end{cases}$$

The geometry of $Q_R(\mu)$ is *intrinsic* in that the $t-$dimension is '*stretched*' by a factor, loosely speaking, of the order of $|D\mathbf{u}|^{2-p}$. Let us assume for the moment that such boxes can be constructed. Then Theorem 1.1 is a consequence of the following two propositions.

PROPOSITION 1.1. *There exist numbers ν,κ,δ in $(0,1)$ that can be determined a priori only in terms of N and p, such that if*

$$\text{(1.3)}\qquad \operatorname{meas}\left\{(x,t)\in Q_R(\mu)\ \middle|\ |D\mathbf{u}|(x,t)<(1-\nu)\mu\right\}<\nu\left|Q_R(\mu)\right|,$$

there holds

$$\text{(1.4)}\qquad \iint\limits_{Q_{\delta^{n+1}R}(\mu)}\left|D\mathbf{u}-(D\mathbf{u})_{n+1}\right|^2dxd\tau\le\kappa\,\delta^{N+2}\iint\limits_{Q_{\delta^nR}(\mu)}\left|D\mathbf{u}-(D\mathbf{u})_n\right|^2dxd\tau,$$

for all $n=1,2,\ldots$, where

$$(D\mathbf{u})_n=\fint\!\!\fint_{Q_{\delta^nR}(\mu)} D\mathbf{u}\,dxd\tau.$$

PROPOSITION 1.2. *There exists numbers $\frac{1}{2}<\sigma<\eta<1$ and such that if*

$$\text{(1.5)}\qquad \operatorname{meas}\left\{(x,t)\in Q_R(\mu)\ \middle|\ |D\mathbf{u}|(x,t)<(1-\nu)\mu\right\}\ge\nu\left|Q_R(\mu)\right|,$$

there holds

$$|\mathbf{Du}|(x,t) \le \eta\mu, \qquad \forall (x,t) \in Q_{\sigma R}(\mu). \tag{1.6}$$

These two facts will be used to establish the following:

THEOREM 1.2. *Assume that the cylinder $Q_R(\mu)$ satisfies (1.2) for some $\mu > 0$. There exist constants $\gamma > 1$ and $\alpha \in (0,1)$ that can be determined a priori only in terms of N and p, such that*

$$\operatorname*{ess\,osc}_{Q_\rho(\mu)} u_{i,x_j} \le \gamma\,\mu \left(\frac{\rho}{R}\right)^\alpha, \quad \forall 0<\rho\le R, \tag{1.7}$$

for all $i=1,2,\dots,m$ and all $j=1,2,\dots,N$.

1-(ii). Constructing $Q_R(\mu)$

Assume first that $p>2$. The number $R>0$ being fixed, let μ_o be the smallest value of the parameter μ such that $Q_R(\mu_o) \subset \Omega_T$. If $\|\mathbf{Du}\|_{\infty,Q_R(\mu_o)} \le \mu_o$, then $Q_R(\mu_o)$ satisfies (1.2). Otherwise we take as μ the largest root of the equation

$$\|\mathbf{Du}\|_{\infty,Q_R(\mu)} = \mu.$$

Such an equation has finite roots since $\|\mathbf{Du}\|_{\infty,Q_R(\mu_o)} > \mu_o$, and

$$\mu \to \|\mathbf{Du}\|_{\infty,Q_R(\mu)} \text{ remains bounded as } \mu\to\infty.$$

These arguments are based on the fact that, since $p>2$, the *'vertical size'* of $Q_R(\mu)$ decreases as μ increases. In the singular case we consider instead boxes of the type

$$(1.2)' \qquad \begin{cases} \mathcal{Q}_R(\mu) \equiv \left\{|x| \le \mu^{\frac{p-2}{2}} R\right\} \times \{-R^2, 0\}, \text{ satisfying} \\ \sup\limits_{\mathcal{Q}_R(\mu)} |\mathbf{Du}| \le \mu. \end{cases}$$

As μ increases, the cubes $\left\{|x| < \mu^{\frac{p-2}{2}} R\right\}$ shrink. Therefore there exist some μ for which (1.2)′ holds.

Remark 1.3. The previous propositions could be stated and proved in the geometry of the boxes (1.2)′. Indeed setting $\mu^{\frac{p-2}{2}} R = r$ permits one to recast the *'space scaling'* of (1.2)′ in terms of the *'time scaling'* of (1.2).

1-(iii). More about the intrinsic geometry

Take formally the x_j–derivative of (1.1) of Chap. VIII and multiply the ith equation of the system so obtained, by u_{i,x_j}. Adding over $i=1,2,\dots,m$ and $j=1,2,\dots,N$, and setting $w=|D\mathbf{u}|^2$ we arrive at the formal differential inequality

$$\frac{\partial}{\partial t}w - \left(a_{\ell,k}w^{\frac{p-2}{2}}w_{x_k}\right)_{x_\ell} \le 0 \quad \text{in } \Omega_T, \tag{1.8}$$

where

$$a_{\ell,k} \equiv \left\{\delta_{\ell,k} + (p-2)\frac{u_{i,x_\ell}u_{i,x_k}}{|D\mathbf{u}|^2}\right\}. \tag{1.9}$$

The matrix $(a_{\ell,k})$ is positive definite and w is a non-negative weak solution of an equation of the porous medium type. This is a parabolic version of the *quasi-subharmonicity*. The degeneracy of (1.8) is of the order of $w^{\frac{p-2}{2}}$ and it is overcome by the choice of the parabolic geometry of $Q_R(\mu)$.

2. Estimating the oscillation of $D\mathbf{u}$

We assume Propositions 1.1 and 1.2 for the moment and proceed to prove Theorem 1.1. Let $Q_R(\mu)$ be a cylinder satisfying (1.2) and define the two sequences

$$\begin{cases} \mu_o = \mu,\ R_o = R \text{ and for } n = 1,2,\dots, \\ \mu_{n+1} = \eta\mu_n, \quad R_{n+1} = c_o R_n, \end{cases} \tag{2.1}$$

where η and σ are the numbers claimed by Proposition 1.2 and

$$c_o = \frac{1}{2}\sigma\eta^{\frac{p-2}{2}}. \tag{2.2}$$

Since $\eta \in (\frac{1}{2},1)$, we have $c_o \in (0,1)$ for all $p>1$. Suppose the assumption (1.5) holds with $R=R_o$ and $\mu=\mu_o$. Then

$$\sup_{Q_{\sigma R_o}(\mu_o)} |D\mathbf{u}| \le \eta\,\mu_o \equiv \mu_1.$$

From the definitions (2.1) and (2.2) it follows that $R_1<R_o$ and

$$\frac{R_1^2}{\mu_1^{p-2}} = \frac{\sigma^2\eta^{p-2}}{4}\,\frac{R_o^2}{\eta^{p-2}\mu_o^{p-2}} \equiv \left(\frac{\sigma}{2}\right)^2\frac{R_o^2}{\mu^{p-2}}.$$

This implies that the cylinder $Q_{R_1}(\mu_1)$ is contained in $Q_{\sigma R_o}(\mu_o)$ and

$$\sup_{Q_{R_1}(\mu_1)} |D\mathbf{u}| \le \mu_1.$$

Therefore $Q_{R_1}(\mu_1)$ satisfies (1.2) and if the assumption (1.5) of Proposition 1.2 is verified again for such a box we have

$$\sup_{Q_{R_2}(\mu_1)} |D\mathbf{u}| \le \mu_2.$$

Proceeding in this fashion, suppose the assumption of Proposition 1.2 is verified for the cylinders

$$Q_{R_n}(\mu_n),\quad n = 0, 1, 2, \ldots, n_o - 1 \text{ for some positive integer } n_o.$$

Then

$$\sup_{Q_{R_n}(\mu_n)} |D\mathbf{u}| \le \mu_n \equiv \eta^n \mu_o, \quad n = 0, 1, 2, \ldots, n_o. \tag{2.3}$$

From the definitions (2.1) and (2.2) it follows that

$$R_n = c_o^n R_o, \quad \eta^n = \left(\frac{R_n}{R_o}\right)^{\alpha_1}, \quad \alpha_1^{-1} = -\log_{1/\eta} c_o.$$

One verifies that $c_o < \eta$ for all $p > 1$, and consequently $\alpha_1 \in (0, 1)$. We rewrite (2.3) as

$$\sup_{Q_{R_n}(\mu_n)} |D\mathbf{u}| \le \mu_o \left(\frac{R_n}{R_o}\right)^{\alpha_1}, \quad \text{for } n = 0, 1, 2, \ldots, n_o. \tag{2.4}$$

Suppose now that the assumption (1.5) of Proposition 1.2 fails for n_o. We call R_{n_o} the *switching radius*. Then for the box $Q_{R_{n_o}}(\mu_{n_o})$ the assumption (1.3) of Proposition 1.1 holds and we conclude that

$$\begin{aligned}\fint\!\!\fint_{Q_{\delta^i R_{n_o}}(\mu_{n_o})} \left|D\mathbf{u} - (D\mathbf{u})_i\right|^2 dx d\tau &\le \kappa^i \fint\!\!\fint_{Q_{R_{n_o}}(\mu_{n_o})} \left|D\mathbf{u} - (D\mathbf{u})_o\right|^2 dx d\tau \\ &\le \kappa^i \mu_{n_o}^2, \quad i = 1, 2, \ldots\end{aligned} \tag{2.5}$$

Writing

$$\left|(D\mathbf{u})_{i+1} - (D\mathbf{u})_i\right|^2 \le 2\left|D\mathbf{u} - (D\mathbf{u})_{i+1}\right|^2 + 2\left|D\mathbf{u} - (D\mathbf{u})_i\right|^2 \tag{2.6}$$

and taking the integral average over $Q_{\delta^{i+1} R_{n_o}}(\mu_{n_o})$ gives

$$\left|(D\mathbf{u})_{i+1} - (D\mathbf{u})_i\right|^2 \le \gamma \kappa^i \mu_{n_o}^2, \quad \gamma = 2\left(\kappa + \delta^{-(N+2)}\right).$$

Therefore $\{(D\mathbf{u})_i\}_{i \in \mathbf{N}}$ is a Cauchy sequence whose limit we denote with $D\mathbf{u}(x_o, t_o)$. To motivate this terminology we recall that our arguments are carried over an arbitrary cylinder $[(x_o, t_o) + Q_R(\mu)]$ with *vertex* at (x_o, t_o). Therefore if n_o is the *switching radius* of the box $[(x_o, t_o) + Q_R(\mu)]$, the limit of the averages,

$$\fint\!\!\fint_{[(x_o,t_o)+Q_{\delta^i R_{n_o}}(\mu_{n_o})]} D\mathbf{u}\, dx d\tau,$$

is $D\mathbf{u}(x_o,t_o)$ for almost all $(x_o,t_o)\in\Omega_T$. It follows from (2.6) that

$$\left|D\mathbf{u}(x_o,t_o)-(D\mathbf{u})_i\right|^2\le\gamma\,\kappa^i\mu_{n_o}^2,\quad i=1,2,\ldots.$$

Fix $0<\rho<R_{n_o}$ and denote with $(D\mathbf{u})_\rho$ the integral average of $D\mathbf{u}$ over $Q_\rho(\mu_{n_o})$. Let i be a positive integer such that

$$\delta^i R_{n_o}\le\rho\le\delta^{i-1}R_{n_o},\tag{2.7}$$

and estimate

$$\begin{aligned}\left|(D\mathbf{u})_\rho-(D\mathbf{u})_i\right|^2 &\le \iint\limits_{Q_\rho(\mu_{n_o})}\!\!\!\!\!\!-\;\left|D\mathbf{u}-(D\mathbf{u})_i\right|^2dxd\tau\\ &\le\gamma\delta^{-(N+2)}\kappa^i\mu_{n_o}^2.\end{aligned}\tag{2.8}$$

Therefore

$$\begin{aligned}\left|D\mathbf{u}(x_o,t_o)-(D\mathbf{u})_\rho\right|^2 &\le 2\left|D\mathbf{u}(x_o,t_o)-(D\mathbf{u})_i\right|^2\\ &\quad+2\left|(D\mathbf{u})_\rho-(D\mathbf{u})_i\right|^2\\ &\le\gamma(\delta)\,\kappa^i\mu_{n_o}^2.\end{aligned}\tag{2.9}$$

It follows from (2.7) and (2.9) that

$$\delta^i\le\left(\frac{\rho}{R_{n_o}}\right),\quad \kappa^i\le\left(\frac{\rho}{R_{n_o}}\right)^{\alpha_2},\quad \alpha_2^{-1}\equiv\left|\log_{1/\kappa}\delta\right|.$$

Let $2\alpha_o=\min\{\alpha_1;\alpha_2\}$. Then combining (2.9) and (2.4) we conclude

LEMMA 2.1. *There exist constants $\gamma>1$ and $\alpha_o\in(0,1)$ that can be determined a priori only in terms of N and p, such that for almost all $(x_o,t_o)\in\Omega_T$ such that $[(x_o,t_o)+Q_R(\mu)]\subset\Omega_T$, and for all $0<\rho\le R$, there holds*

$$\left|D\mathbf{u}(x_o,t_o)-(D\mathbf{u})_\rho\right|\le\gamma\,\mu_o\left(\frac{\rho}{R}\right)^{\alpha_o}.\tag{2.10}$$

Moreover

$$\iint\limits_{Q_\rho(\mu_{n_o})}\!\!\!\!\!\!-\;\left|D\mathbf{u}-(D\mathbf{u})_\rho\right|^2dxd\tau\le\gamma\,\mu_o^2\left(\frac{\rho}{R}\right)^{2\alpha_o}.\tag{2.11}$$

Remark 2.1. The lemma holds also in the geometry of the boxes $[(x_o,t_o)+\mathcal{Q}_R(\mu)]$ introduced in (1.2)$'$. Indeed we may set $\mu^{\frac{p-2}{2}}R=r$ and work within the cylinder $[(x_o,t_o)+Q_r(\mu)]$. We arrive at a version of (2.10) that reads

$$\left|D\mathbf{u}(x_o,t_o)-(D\mathbf{u})_s\right|\le\gamma\mu_o\left(\frac{s}{r}\right)^{\alpha_o},\quad\forall\,0<s\le r.\tag{2.10$'$}$$

Returning to the geometry of $[(x_o, t_o) + \mathcal{Q}_R(\mu)]$ proves the assertion. Analogous considerations hold for (2.11).

3. Hölder continuity of $\mathbf{D}u$ (the case $p>2$)

We assume that $|\mathbf{D}u| \in L^\infty(\Omega_T)$ and set

$$\mu \equiv \|\mathbf{D}u\|_{\infty,\Omega_T}.$$

This is no loss of generality, by possibly working with another compact set $\mathcal{K}'$ satisfying $\mathcal{K} \subset \mathcal{K}' \subset \Omega_T$, and

$$\text{dist}\,(\mathcal{K}; \mathcal{K}') \geq \tfrac{1}{2} d.$$

We will prove the Hölder continuity of $\mathbf{D}u$ in the time and space variables separately.

3-(i). Hölder continuity in t

Fix two points $(x_o, t_i) \in \mathcal{K}$, $i = 0, 1$, with the same *'abscissa'* x_o. We let $t_1 > t_o$ and construct the cylinders

$$[(x_o, t_i) + \mathcal{Q}_R(\mu)] \equiv \left\{ |x - x_o| < \mu^{\frac{p-2}{2}} R \right\} \times \left\{ t_i - R^2, t_i \right\}.$$

The box $[(x_o, t_1) + \mathcal{Q}_R(\mu)]$ intersects $[(x_o, t_o) + \mathcal{Q}_R(\mu)]$ at the point (x_o, t_o), if $(t_1 - t_o) < R^2$. Moreover they are contained in Ω_T if

$$\max \left\{ \mu^{\frac{p-2}{2}} R;\, R \right\} \leq \text{dist}\,(\mathcal{K}; \Gamma). \tag{3.1}$$

LEMMA 3.1. *Let (3.1) hold. There exist constants $\gamma > 1$ and $\alpha \in (0,1)$ that can be determined a priori only in terms of N and p such that for all pairs $(x_o, t_i) \in \mathcal{K}$, $i=0,1$,*

$$\left| \mathbf{D}u(x_o, t_1) - \mathbf{D}u(x_o, t_o) \right| \leq \gamma \mu \left(\frac{\sqrt{t_1 - t_o}}{R} \right)^\alpha. \tag{3.2}$$

PROOF: Let R_{n_i} be the switching radii of the cylinders $[(x_o, t_i) + \mathcal{Q}_R(\mu)]$ and introduce the two boxes

$$\begin{aligned} \mathcal{Q}_i &\equiv \left[(x_o, t_i) + \mathcal{Q}_{R_{n_i}}(\mu_{n_i}) \right] \\ &= \left\{ |x - x_i| < \mu_{n_i}^{\frac{p-2}{2}} R_{n_i} \right\} \times \left\{ t_i - R_{n_i}^2, t_i \right\}. \end{aligned}$$

Assume first that

$$2(t_1 - t_o) \leq \min \left\{ R_{n_o}^2;\, R_{n_1}^2 \right\}, \tag{3.3}$$

and for $i=0,1$ construct the two cylinders

$$\begin{aligned}\mathcal{C}_i &\equiv \left[(x_o,t_i)+\mathcal{Q}_{\sqrt{2(t_1-t_o)}}(\mu_{n_i})\right] \\ &= \left\{|x-x_o|<\mu_{n_i}^{\frac{p-2}{2}}\sqrt{2(t_1-t_o)}\right\}\times\{t_i-2(t_1-t_o),t_i\}.\end{aligned}$$

By virtue of (3.3) we have the inclusions $\mathcal{C}_i\subset\mathcal{Q}_i$, $i=0,1$. Moreover $\mathcal{C}_o$ and $\mathcal{C}_1$ intersect in a box satisfying

$$\text{(3.4)} \qquad \operatorname{meas}\left[\mathcal{C}_o\cap\mathcal{C}_1\right]\ge\min\{\mu_{n_o};\mu_{n_1}\}\,(t_1-t_o)^{(N+2)/2}.$$

Set

$$(D\mathbf{u})_{\mathcal{C}_i}\equiv \fint\!\!\fint_{\mathcal{C}_i} D\mathbf{u}\,dxd\tau,\qquad i=0,1,$$

and estimate

$$\begin{aligned}\left|D\mathbf{u}(x_o,t_1)-D\mathbf{u}(x_o,t_o)\right| &\le \left|D\mathbf{u}(x_o,t_1)-(D\mathbf{u})_{\mathcal{C}_1}\right| \\ &\quad+\left|D\mathbf{u}(x_o,t_o)-(D\mathbf{u})_{\mathcal{C}_o}\right| \\ &\quad+\left|(D\mathbf{u})_{\mathcal{C}_1}-(D\mathbf{u})_{\mathcal{C}_o}\right|.\end{aligned}$$

By (2.10) and Remark 2.1 we have, for $i=0,1$,

$$\left|D\mathbf{u}(x_o,t_i)-(D\mathbf{u})_{\mathcal{C}_i}\right|\le\gamma\,\mu\left(\frac{\sqrt{t_1-t_o}}{R}\right)^{\alpha_o}.$$

To estimate the last term we add and subtract $D\mathbf{u}(x,t)$ where $(x,t)\in\mathcal{C}_o\cap\mathcal{C}_1$, and then take the integral average over such intersection, i.e.,

$$\begin{aligned}\left|(D\mathbf{u})_{\mathcal{C}_o}-(D\mathbf{u})_{\mathcal{C}_o}\right| &\le \fint\!\!\fint_{\mathcal{C}_o\cap\mathcal{C}_1}\left|(D\mathbf{u})_{\mathcal{C}_1}-D\mathbf{u}(x,t)\right|dxd\tau \\ &\quad+\fint\!\!\fint_{\mathcal{C}_o\cap\mathcal{C}_1}\left|D\mathbf{u}(x,t)-(D\mathbf{u})_{\mathcal{C}_o}\right|dt.\end{aligned}$$

Without loss of generality we may assume that $\min\{\mu_{n_o};\mu_{n_1}\}=\mu_{n_1}$. Then we estimate the first integral by extending the integration over the larger set $\mathcal{C}_1$. Taking into account the definition of $\mathcal{C}_1$, (3.4) and (2.11), we obtain

$$\begin{aligned}\fint\!\!\fint_{\mathcal{C}_o\cap\mathcal{C}_1}\left|(D\mathbf{u})_{\mathcal{C}_1}-D\mathbf{u}(x,t)\right|dxd\tau &\le\gamma\fint\!\!\fint_{\mathcal{C}_1}\left|(D\mathbf{u})_{\mathcal{C}_1}-D\mathbf{u}(x,t)\right|dxd\tau \\ &\le\gamma\,\mu\left(\frac{\sqrt{t_1-t_o}}{R}\right)^{\alpha_o}.\end{aligned}$$

To estimate the second integral, let β be a small positive number to be chosen and assume that

(3.5)
$$\mu_{n_1} \geq \mu_{n_o} \left(\frac{\sqrt{t_1 - t_o}}{R} \right)^{\beta} .$$

Then using again (2.11) and (3.4),

$$\iint_{\mathcal{C}_o \cap \mathcal{C}_1} |D\mathbf{u}(x,t) - (D\mathbf{u})_{\mathcal{C}_o}| dt$$
$$\leq \gamma \left(\frac{\mu_{n_o}}{\mu_{n_1}} \right)^{N(p-2)/2} \iint_{\mathcal{C}_o} |D\mathbf{u}(x,t) - (D\mathbf{u})_{\mathcal{C}_o}| dt$$
$$\leq \gamma \left(\frac{\mu_{n_o}}{\mu_{n_1}} \right)^{N(p-2)/2} \mu \left(\frac{\sqrt{t_1 - t_o}}{R} \right)^{\alpha_o}$$
$$\leq \gamma \mu \left(\frac{\sqrt{(t_1 - t_o)}}{R} \right)^{\alpha_o - \beta N(p-2)/2} .$$

Therefore if (3.3) and (3.5) hold, the assertion (3.2) follows by taking $\beta = \alpha_o / N(p-2)$ and then choosing $\alpha = \alpha_o/2$. If (3.5) is violated,

$$\left| D\mathbf{u}(x_o, t_o) - D\mathbf{u}(x_1, t_1) \right| \leq 2\mu \left(\frac{\sqrt{t_1 - t_o}}{R} \right)^{\alpha_o / N(p-2)} ,$$

and the assertion follows by suitably modifying the definition of α.

We consider next the case when (3.3) is violated, i.e.,

$$2(t_1 - t_o) > \min \left\{ R_{n_o}^2 \, ; \, R_{n_1}^2 \right\} .$$

If $R_{n_i}^2 \leq 2(t_1 - t_o)$ for $i=0,1$, then by (2.4)

$$\left| D\mathbf{u}(x_o, t_1) - D\mathbf{u}(x_o, t_o) \right| \leq \mu_{n_o} + \mu_{n_1} \leq \gamma \mu \left(\frac{\sqrt{t_1 - t_o}}{R} \right)^{\alpha} .$$

Therefore we may assume that, say,

$$R_{n_1}^2 \leq 2(t_1 - t_o) < R_{n_o}^2 .$$

We conclude the proof by reducing this case to the situation (3.3). Let $n_* \leq n_1$ be a positive integer satisfying

$$R_{n_*}^2 \geq 2(t_1 - t_o) \geq R_{n_*+1}^2 ,$$

and introduce the cylinders Q_o and $\mathcal{Q}_*$, where

$$\mathcal{Q}_* \equiv (x_o, t_1) + \mathcal{Q}_{R_{n_*}}(\mu_{n_*}) \equiv \left\{ |x - x_o| < \mu_{n_*}^{\frac{p-2}{2}} R_{n_*} \right\} \times \left\{ t_1 - R_{n_*}^2, t_1 \right\} .$$

Since we have

$$2(t_1 - t_o) \leq \min \left\{ R_{n_o}^2 \, ; \, R_{n_*}^2 \right\} ,$$

the box $\mathcal{Q}_*$ will now play the same role as the cylinder $\mathcal{Q}_1$ in the case (3.3). The proof is now concluded as before, observing that for $\mathcal{Q}_*$ the two inequalities (2.10) and (2.11) hold true.

LEMMA 3.1$'$. *There exist constants $\gamma>1$ and $\alpha\in(0,1)$ that can be determined a priori only in terms of N and p such that for every pair of points $(x_i,t_o)\in\mathcal{K}$, $i=0,1$,*

$$\left|D\mathbf{u}(x_o,t_1)-D\mathbf{u}(x_o,t_o)\right|\le\gamma\mu\left(\frac{\max\left\{1;\mu^{\frac{p-2}{2}}\right\}\sqrt{t_1-t_o}}{\operatorname{dist}(\mathcal{K};\Gamma)}\right)^{\alpha}. \tag{3.2$'$}$$

PROOF: If $\mu\le1$ we take $R=\operatorname{dist}(\mathcal{K};\Gamma)$ in (3.1). Otherwise we take $\mu^{\frac{p-2}{2}}R=\operatorname{dist}(\mathcal{K};\Gamma)$.

3-(ii). Hölder continuity in x

Fix two points (x_o,t_o) and (x_1,t_o) in $\mathcal{K}$, at the same time level t_o, and let

$$[(x_i,t_o)+Q_R(\mu)]\equiv\{|x-x_i|<R\}\times\{t_o-\mu^{2-p}R^2,\,t_o\},\quad i=0,1,$$

be two boxes satisfying (1.2). The box $[(x_o,t_o)+Q_R(\mu)]$ will intersect x_1 if $|x_o-x_1|<R$. Moreover they are contained in Ω_T if

$$\max\left\{\mu^{\frac{2-p}{2}}R;\,R\right\}\le\operatorname{dist}(\mathcal{K};\Gamma). \tag{3.6}$$

LEMMA 3.2. *Let (3.6) hold. There exist constants $\gamma>1$ and $\alpha\in(0,1)$ that can be determined a priori only in terms of N and p, such that for all $(x_i,t_o)\in\mathcal{K}$, $i=0,1$,*

$$\left|D\mathbf{u}(x_1,t_o)-D\mathbf{u}(x_o,t_o)\right|\le\gamma\mu\left(\frac{|x_o-x_1|}{R}\right)^{\alpha}. \tag{3.7}$$

PROOF: Let R_{n_i} be the switching radii corresponding to the two boxes $[(x_i,t_o)+Q_R(\mu)]$, and construct the two cylinders

$$\begin{aligned}Q_i&\equiv\left[(x_i,t_o)+Q_{R_{n_i}}(\mu_{n_i})\right]\\&\equiv\{|x-x_i|<R_{n_i}\}\times\left\{t_o-\mu_{n_i}^{2-p}R_{n_i}^2,t_o\right\}.\end{aligned}$$

Consider separately the following two cases:

$$2\,|x_o-x_1|\le\min\left\{R_{n_o};\,R_{n_1}\right\}, \tag{3.8}$$

$$2\,|x_o-x_1|>\min\left\{R_{n_o};\,R_{n_1}\right\}. \tag{3.8$'$}$$

If (3.8) holds, construct the two boxes

$$\begin{aligned}C_i&\equiv\left[(x_i,t_o)+Q_{2|x_o-x_1|}(\mu_{n_i})\right]\\&\equiv\{|x-x_i|<2|x_o-x_1|\}\times\left\{t_o-\mu_{n_i}^{2-p}4|x_o-x_1|^2,t_o\right\}.\end{aligned}$$

By construction these are contained in Q_i and they overlap in a box satisfying

$$\text{meas}\,[C_o \cap C_1] \geq |x_o - x_1|^{N+2} \min\{\mu_{n_o}; \mu_{n_1}\}^{2-p}. \tag{3.9}$$

Set

$$(D\mathbf{u})_{C_i} \equiv \fint\!\!\!\!\fint_{C_i} D\mathbf{u}\,dt,$$

and estimate

$$\begin{aligned}\Big|D\mathbf{u}(x_1,t_o) - D\mathbf{u}(x_o,t_o)\Big| &\leq \Big|D\mathbf{u}(x_1,t_o) - (D\mathbf{u})_{C_1}\Big| \\ &+ \Big|D\mathbf{u}(x_o,t_o) - (D\mathbf{u})_{C_o}\Big| \\ &+ \Big|(D\mathbf{u})_{C_1} - (D\mathbf{u})_{C_o}\Big|.\end{aligned}$$

The proof now proceeds as for the Hölder continuity in t with minor changes.

LEMMA 3.2$'$. *There exist constants $\gamma>1$ and $\alpha\in(0,1)$ that can be determined a priori only in terms of N and p such that for every pair of points $(x_i,t_o)\in\mathcal{K}$, $i=0,1$,*

$$\Big|D\mathbf{u}(x_1,t_o) - D\mathbf{u}(x_o,t_o)\Big| \leq \gamma\mu\left(\frac{|x_o-x_1|}{\text{dist}\,(\mathcal{K};\Gamma)}\right)^{\alpha}. \tag{3.10}$$

PROOF: If $\mu \geq 1$, in (3.6) we take $R = \text{dist}\,(\mathcal{K};\Gamma)$. If $\mu < 1$, we take $R = \mu^{\frac{p-2}{2}}\,\text{dist}\,(\mathcal{K};\Gamma)$. Then (3.7) reads

$$\Big|D\mathbf{u}(x_1,t_o) - D\mathbf{u}(x_o,t_o)\Big| \leq \gamma\mu\left(\frac{|x_o-x_1|}{\mu^{\frac{p-2}{2}}\,\text{dist}\,(\mathcal{K};\Gamma)}\right)^{\alpha}. \tag{3.7$'$}$$

If

$$\mu^{\frac{p-2}{2}} \leq \left(\frac{|x_1-x_o|}{\text{dist}\,(\mathcal{K};\Gamma)}\right)^{\alpha/2},$$

there is nothing to prove. Otherwise (3,7)$'$ gives

$$\Big|D\mathbf{u}(x_1,t_o) - D\mathbf{u}(x_o,t_o)\Big| \leq \gamma\mu\left(\frac{|x_o-x_1|}{\text{dist}\,(\mathcal{K};\Gamma)}\right)^{\alpha/2}.$$

Thus (3.10) follows by suitably redefining the number α.

3-(iii). A version of Theorem 1.1

Combining Lemmas 3.1$'$ and 3.2$'$ gives the following form of Theorem 1.1:

THEOREM 1.1′ (THE DEGENERATE CASE $p>2$). *Let $\mathbf{u}$ be a weak solution in Ω_T of the degenerate system (1.1) of Chap. VIII, and assume that $\mu = \|D\mathbf{u}\|_{\infty,\Omega_T} < \infty$. There exist constants $\gamma > 1$ and $\alpha \in (0,1)$ that can be determined a priori only in terms of N and p such that, for every compact subset $\mathcal{K}$ of Ω_T,*

$$(1.1') \quad \left| D\mathbf{u}(x_o,t_o) - D\mathbf{u}(x_1,t_1) \right| \le \gamma\mu \left(\frac{|x_o - x_1| + \max\{1; \mu^{\frac{p-2}{2}}\}\sqrt{|t_o - t_1|}}{\operatorname{dist}(\mathcal{K};\Gamma)} \right)^{\alpha},$$

for every pair of points $(x_i,t_i) \in \mathcal{K}$, $i=0,1$.

Remark 3.1. The constants γ and α are independent of dist $(\mathcal{K};\Gamma)$ and μ.

The form of (1.1)′ suggest we reduce the system (1.1) of Chap. VIII to another for which $\mu=1$. Introduce the change of variables

$$v_i \equiv \frac{u_i}{\mu}, \qquad i = 1,2,\ldots,m, \qquad \text{and } \tau = t\,\mu^{p-2}.$$

Then $\mathbf{v} \equiv (v_1, v_2, \ldots, v_m)$ satisfies

$$\mathbf{v}_t - \operatorname{div} |D\mathbf{v}|^{p-2} D\mathbf{v} = 0, \qquad \text{in } \widetilde{\Omega_T} \equiv \Omega \times (0, T\,\mu^{p-2}),$$

and $\|Dv\|_{\infty,\widetilde{\Omega_T}} = 1$. We write (1.1)′ for $\mathbf{v}$ in the variables (x,τ) and return to the original coordinates. This gives

$$\left| D\mathbf{u}(x_o,t_o) - D\mathbf{u}(x_1,t_1) \right| \le \gamma \|D\mathbf{u}\|_{\infty,\Omega_T} \left(\frac{|x_1 - x_o| + \mu^{\frac{p-2}{2}} \sqrt{|t_1 - t_o|}}{\mu - \operatorname{dist}(\mathcal{K};\Gamma)} \right)^{\alpha}$$

for all pairs $(x_i,t_i) \in \mathcal{K}$, $i=0,1$, where

$$\mu - \operatorname{dist}(\mathcal{K};\Gamma) \equiv \inf_{\substack{(x,t)\in\mathcal{K} \\ (y,s)\in\Gamma}} \left(|x-y| + \mu^{\frac{p-2}{2}} \sqrt{|t-s|} \right)$$

is the *intrinsic* parabolic distance from $\mathcal{K}$ to Γ.

4. Hölder continuity of $D\mathbf{u}$ (the case $1<p<2$)

In the singular case $1<p<2$, the role of the parabolic geometry is reversed. As before, we investigate separately the Hölder continuity in the space and in the t, variables. The arguments are similar to those in the degenerate case and we only indicate the main differences.

4-(i). Hölder continuity in t

Fix two points $(x_o, t_i) \in \mathcal{K}$, $i=0,1$, with the same *'abscissa'* x_o. We let $t_1 > t_o$ and construct the cylinders

$$[(x_o, t_i) + \mathcal{Q}_R(\mu)] \equiv \left\{|x - x_o| < \mu^{\frac{p-2}{2}} R\right\} \times \left\{t_i - R^2, t_i\right\}, \quad i = 0, 1.$$

The box $[(x_o, t_1) + \mathcal{Q}_R(\mu)]$ intersects $[(x_o, t_o) + \mathcal{Q}_R(\mu)]$ if $(t_1 - t_o) < R^2$. Moreover they are contained in Ω_T if

$$\max\left\{\mu^{\frac{p-2}{2}} R;\ R\right\} \le \operatorname{dist}(\mathcal{K}; \Gamma). \tag{4.1}$$

Proceeding as in the case $p>2$ we have

LEMMA 4.1. *Let (4.1) hold. There exist constants* $\gamma > 1$ *and* $\alpha \in (0,1)$ *that can be determined a priori only in terms of N and p such that*

$$\left|D\mathbf{u}(x_o, t_1) - D\mathbf{u}(x_o, t_o)\right| \le \gamma \mu \left(\frac{\sqrt{t_1 - t_o}}{R}\right)^{\alpha}. \tag{4.2}$$

Next if $\mu \ge 1$, we take $R = d$ in (4.1), and if $\mu < 1$, we rewrite (4.2) as

$$\left|D\mathbf{u}(x_o) - D\mathbf{u}(x_o, t_o)\right| \le \gamma \mu \left(\frac{\sqrt{t_1 - t_o}}{\mu^{\frac{2-p}{2}} \operatorname{dist}(\mathcal{K}; \Gamma)}\right)^{\alpha}. \tag{4.2$'$}$$

Arguing as in the proof of lemma 3.2$'$ and by possibly redefining the constants γ and α, we obtain

LEMMA 4.1$'$. *There exist constants* $\gamma > 1$ *and* $\alpha \in (0,1)$ *that can be determined a priori only in terms of N and p such that for every pair of points* $(x_o, t_i) \in \mathcal{K}$, $i = 0, 1$,

$$\left|D\mathbf{u}(x_o, t_1) - D\mathbf{u}(x_o, t_o)\right| \le \gamma \mu \left(\frac{\sqrt{t_1 - t_o}}{\operatorname{dist}(\mathcal{K}; \Gamma)}\right)^{\alpha}. \tag{4.3}$$

4-(ii). Hölder continuity in x

Fix two points (x_o, t_o) and (x_1, t_o) in $\mathcal{K}$, at the same time level t_o, and let

$$[(x_i, t_o) + Q_R(\mu)] \equiv \{|x - x_i| < R\} \times \left\{t_o - \mu^{2-p} R^2,\ t_o\right\}, \quad i = 0, 1,$$

be two boxes satisfying (1.2). The box $[(x_o, t_o) + Q_R(\mu)]$ intersects x_1 if $|x_o - x_1| < R$. Moreover they are contained in Ω_T if

$$\max\left\{\mu^{\frac{2-p}{2}} R;\ R\right\} \le \operatorname{dist}(\mathcal{K}; \Gamma). \tag{4.4}$$

We proceed as in the case $p>2$ and establish

LEMMA 4.2. *Let (4.4) hold. There exist constants $\gamma>1$ and $\alpha\in(0,1)$ that can be determined a priori only in terms of N and p, such that for all $(x_i,t_o)\in\mathcal{K}$, $i=0,1$,*

$$\left|D\mathbf{u}(x_1,t_o)-D\mathbf{u}(x_o,t_o)\right|\le\gamma\mu\left(\frac{\max\left\{1;\mu^{\frac{2-p}{2}}\right\}|x_o-x_1|}{\operatorname{dist}(\mathcal{K};\Gamma)}\right)^{\alpha}. \tag{4.5}$$

4-(iii). A version of Theorem 1.1

Combining Lemmas 4.1 and 4.2 gives the following form of Theorem 1.1

THEOREM 1.1″ (THE SINGULAR CASE $1<p<2$).. *Let $\mathbf{u}$ be a weak solution in Ω_T of the degenerate system (1.1) of Chap. VIII, and assume that $\mu=\|D\mathbf{u}\|_{\infty,\Omega_T}<\infty$. There exist constants $\gamma>1$ and $\alpha\in(0,1)$ that can be determined a priori only in terms of N and p such that, for every compact subset $\mathcal{K}$ of Ω_T,*

$$\begin{aligned}&\left|D\mathbf{u}(x_o,t_o)-D\mathbf{u}(x_1,t_1)\right|\\&\qquad\le\gamma\mu\left(\frac{\max\{1;\mu^{\frac{2-p}{2}}\}|x_o-x_1|+\sqrt{|t_o-t_1|}}{\operatorname{dist}(\mathcal{K};\Gamma)}\right)^{\alpha},\end{aligned} \tag{1.1''}$$

for every pair of points $(x_i,t_i)\in\mathcal{K}$, $i=0,1$.

Remark 4.1. The constants γ and α are independent of dist $(\mathcal{K};\Gamma)$ and μ.

Arguing as in §3–(III), the Hölder continuity of u_{i,x_j} can be expressed in terms of the *intrinsic* parabolic distance μ–dist $(\mathcal{K};\Gamma)$.

5. Some algebraic lemmas

We let $Q_R(\mu)\subset\Omega_T$ be a cylinder satisfying (1.2) and consider the system

$$\frac{\partial}{\partial t}u_i-\operatorname{div}|D\mathbf{u}|^{p-2}Du_i=0,\quad\text{in }\ Q_R(\mu),\ \ p>1, \tag{5.1}$$

and the one obtained by taking the derivative with respect to x_j, i.e.,

$$\begin{aligned}&\frac{\partial}{\partial t}u_{i,x_j}-\operatorname{div}\left(|D\mathbf{u}|^{p-2}Du_{i,x_j}+\frac{\partial}{\partial x_j}|D\mathbf{u}|^{p-2}Du_i\right)=0,\\&\text{in }Q_R(\mu),\ i=1,2,\dots,m,\ \ j=1,2,\dots,N.\end{aligned} \tag{5.2}$$

We let $\mathbf{V}$ denote a vector in $\mathbf{R}^{N\times m}$ satisfying

$$4^{-1}\mu\le|\mathbf{V}|\le4\mu. \tag{5.3}$$

We also let $\gamma=\gamma(N,p)$ denote a generic positive constant that can be determined a priori only in terms of the indicated quantities.

LEMMA 5.1. *There exists a constant* $\gamma = \gamma(N,p)$, *such that for every vector* $\mathbf{V} \in \mathbf{R}^{N \times m}$, *and for all* $p > 1$,

$$(5.4) \qquad (|D\mathbf{u}| + |\mathbf{V}|)^{\frac{p-2}{2}} |D\mathbf{u} - \mathbf{V}| \leq \gamma \Big| |D\mathbf{u}|^{\frac{p-2}{2}} D\mathbf{u} - |\mathbf{V}|^{\frac{p-2}{2}} \mathbf{V} \Big|.$$

LEMMA 5.2. *Let* $1 < p < 2$. *There exists a constant* $\gamma = \gamma(N,p)$ *such that for every vector* $\mathbf{V} \in \mathbf{R}^{N \times m}$,

$$(5.5) \qquad \Big| |D\mathbf{u}|^{p-2} D\mathbf{u} - |\mathbf{V}|^{p-2} \mathbf{V} \Big| \leq \gamma |D\mathbf{u} - \mathbf{V}|^{p-1}.$$

Moreover if the vector $\mathbf{V}$ *satisfies (5.3), then*

$$(5.6) \qquad \Big| |D\mathbf{u}|^{p-2} D\mathbf{u} - |\mathbf{V}|^{p-2} \mathbf{V} \Big| \leq \gamma \mu^{p-2} |D\mathbf{u} - \mathbf{V}|,$$

$$(5.7) \qquad \Big| |D\mathbf{u}|^{p-2} D\mathbf{u} - |\mathbf{V}|^{p-2} \mathbf{V} \Big|^2 |D\mathbf{u}|^{2-p} \leq \gamma \mu^{p-2} |D\mathbf{u} - \mathbf{V}|^2.$$

Remark 5.1. These lemmas are algebraic in nature and could be stated for any pair of vectors $\mathbf{U}$ and $\mathbf{V}$, provided (5.3) is replaced by

$$(5.3)' \qquad \tfrac{1}{4}|\mathbf{U}| \leq |\mathbf{V}| \leq 4|\mathbf{U}|.$$

Also in (5.4) the number $(p-2)/2$ could be replaced by any number and in (5.5)-(5.7) the number $(p-2)$ could be replaced by any negative number.

PROOF OF LEMMA 5.1: By calculation,

$$\begin{aligned}
&\Big| |D\mathbf{u}|^{\frac{p-2}{2}} D\mathbf{u} - |\mathbf{V}|^{\frac{p-2}{2}} \mathbf{V} \Big| \, |D\mathbf{u} - \mathbf{V}| \\
&\geq \left| \left\langle |D\mathbf{u}|^{\frac{p-2}{2}} D\mathbf{u} - |\mathbf{V}|^{\frac{p-2}{2}} \mathbf{V}, \, D\mathbf{u} - \mathbf{V} \right\rangle \right| \\
&= \left| \left\langle \int_0^1 \frac{d}{ds} \Big| sD\mathbf{u} + (1-s)\mathbf{V} \Big|^{\frac{p-2}{2}} (sD\mathbf{u} + (1-s)\mathbf{V}) \, ds, \, D\mathbf{u} - \mathbf{V} \right\rangle \right| \\
&= \int_0^1 \Big| sD\mathbf{u} + (1-s)\mathbf{V} \Big|^{\frac{p-2}{2}} |D\mathbf{u} - \mathbf{V}|^2 ds \\
&\quad + \frac{p-2}{2} \int_0^1 \Big| sD\mathbf{u} + (1-s)\mathbf{V} \Big|^{\frac{p-6}{2}} \Big| \langle sD\mathbf{u} + (1-s)\mathbf{V}, \, D\mathbf{u} - \mathbf{V} \rangle \Big|^2 ds \\
&\geq \min\{1; (p-1)\} |D\mathbf{u} - \mathbf{V}|^2 \int_0^1 \Big| sD\mathbf{u} + (1-s)\mathbf{V} \Big|^{\frac{p-2}{2}} ds.
\end{aligned}$$

If $1 < p < 2$,

$$\int_0^1 \Big| sD\mathbf{u} + (1-s)\mathbf{V} \Big|^{\frac{p-2}{2}} ds \geq (|D\mathbf{u}| + |\mathbf{V}|)^{\frac{p-2}{2}},$$

and the lemma follows in this case. If $p > 2$, assume for example that $|D\mathbf{u}| > |\mathbf{V}|$. Then

$$\int_0^1 \left| sD\mathbf{u} + (1-s)\mathbf{V} \right|^{\frac{p-2}{2}} ds \geq \int_{1/2}^1 (s|D\mathbf{u}| - (1-s)|\mathbf{V}|)^{\frac{p-2}{2}} ds$$
$$\geq \frac{1}{p} |D\mathbf{u}|^{\frac{p-2}{2}}.$$

PROOF OF LEMMA 5.2: For $1<p<2$, write

$$\begin{aligned} \left| |D\mathbf{u}|^{p-2}D\mathbf{u} - |\mathbf{V}|^{p-2}\mathbf{V} \right| &\leq |\mathbf{V}|^{p-2}|D\mathbf{u} - \mathbf{V}| \\ &\quad + \left| |D\mathbf{u}|^{p-2} - |\mathbf{V}|^{p-2} \right| |D\mathbf{u}| \\ &\leq |\mathbf{V}|^{p-2}|D\mathbf{u} - \mathbf{V}| \\ &\quad + (2-p)|\mathbf{V}|^{p-2}|D\mathbf{u} - \mathbf{V}| \left(\frac{|D\mathbf{u}|^{p-1}}{\xi|\mathbf{V}|^{p-1} + (1-\xi)|D\mathbf{u}|^{p-1}} \right), \end{aligned}$$

for some $\xi \in [0,1]$. Interchanging the role of $D\mathbf{u}$ and $\mathbf{V}$ gives

$$\begin{aligned} \text{(5.8)} \quad & \left| |D\mathbf{u}|^{p-2}D\mathbf{u} - |\mathbf{V}|^{p-2}\mathbf{V} \right| \\ & \leq |\mathbf{V}|^{p-2}|D\mathbf{u} - \mathbf{V}| \left\{ 1 + (2-p)\frac{|D\mathbf{u}|^{p-1}}{\xi|\mathbf{V}|^{p-1} + (1-\xi)|D\mathbf{u}|^{p-1}} \right\}, \end{aligned}$$

for some $\xi \in [0,1]$, and

$$\begin{aligned} \text{(5.8}'\text{)} \quad & \left| |D\mathbf{u}|^{p-2}D\mathbf{u} - |\mathbf{V}|^{p-2}\mathbf{V} \right| \\ & \leq |D\mathbf{u}|^{p-2}|D\mathbf{u} - \mathbf{V}| \left\{ 1 + (2-p)\frac{|\mathbf{V}|^{p-1}}{\eta|\mathbf{V}|^{p-1} + (1-\eta)|D\mathbf{u}|^{p-1}} \right\}, \end{aligned}$$

for some $\eta \in [0,1]$. To prove (5.5) assume first that

$$\text{(5.9)} \qquad |\mathbf{V}| > \frac{1}{2}|D\mathbf{u} - \mathbf{V}|.$$

This implies

$$|\mathbf{V}|^{p-2} \leq 2^{2-p}|D\mathbf{u} - \mathbf{V}|^{p-2} \quad \text{and} \quad |\mathbf{V}| > \tfrac{1}{3}|D\mathbf{u}|.$$

These inequalities in (5.8) prove (5.5). If (5.9) is false, its converse gives the two inequalities

$$|D\mathbf{u} - \mathbf{V}|^{2-p} \geq 2^{2-p}|\mathbf{V}|^{2-p} \quad \text{and} \quad |D\mathbf{u}| > \tfrac{1}{2}|\mathbf{V}|.$$

These in $(5.8)'$ imply that the term in braces on the right hand side is bounded above by an absolute constant. Moreover

$$|\mathbf{V}|^{p-2}\,|D\mathbf{u} - \mathbf{V}| \leq |D\mathbf{u}|^{p-2}\,|D\mathbf{u} - \mathbf{V}|^{p-1}\,|D\mathbf{u} - \mathbf{V}|^{2-p}.$$

The two inequalities (5.6) and (5.7) are an immediate consequence of (5.8) and the assumption (5.3).

Let $\mathbf{H}$ be the vector in $\mathbf{R}^{N\times m}$ defined by

$$H_i \equiv |D\mathbf{u}|^{p-2}Du_i - |\mathbf{V}|^{p-2}V_i - |\mathbf{V}|^{p-2}(Du_i - V_i) \\ - (p-2)|\mathbf{V}|^{p-4}V_{\ell,k}(u_{\ell,x_k} - V_{\ell,k})V_i, \quad i=1,2,\dots,m. \tag{5.12}$$

We will estimate $|\mathbf{H}|$ for all $p>1$. For this we first set

$$\mathbf{W}(t) \equiv tD\mathbf{u} + (1-t)\mathbf{V}, \quad \text{for } t\in[0,1], \tag{5.13}$$

and rewrite (5.12) in the form

$$\begin{aligned} H_i &= \int_0^1 \frac{d}{dt}\left\{|tD\mathbf{u}+(1-t)\mathbf{V}|^{p-2}(tDu_i+(1-t)V_i)\right\}dt \\ &\quad - |\mathbf{V}|^{p-2}(Du_i - V_i) - (p-2)|\mathbf{V}|^{p-4}V_{\ell,k}(u_{\ell,x_k} - V_{\ell,k})V_i \\ &= (Du_i - V_i)\int_0^1 \left\{|\mathbf{W}|^{p-2} - |\mathbf{V}|^{p-2}\right\}dt \\ &\quad + (p-2)(Du_\ell - V_\ell)\int_0^1 \left\{|\mathbf{W}|^{p-4}W_\ell W_i - |\mathbf{V}|^{p-4}V_\ell V_i\right\}dt \\ &\equiv H_i^{(1)} + H_i^{(2)}, \end{aligned}$$

where we have dropped the t-dependence from $\mathbf{W}$. From (5.13)

$$\mathbf{W} - \mathbf{V} = t(D\mathbf{u} - \mathbf{V}), \tag{5.14}$$

and for every $s\in[0,1]$,

$$s\mathbf{W} + (1-s)\mathbf{V} = \mathbf{V} + st(D\mathbf{u} - \mathbf{V}). \tag{5.15}$$

LEMMA 5.3. *There exists a constant $\gamma=\gamma(N,p)$ such that for every constant vector $\mathbf{V}\in\mathbf{R}^{N\times m}$ satisfying (5.3), and for all $p>1$,*

$$|\mathbf{H}| \le \frac{\gamma}{\mu}(|D\mathbf{u}| + |\mathbf{V}|)^{p-2}|D\mathbf{u} - \mathbf{V}|^2. \tag{5.16}$$

Remark 5.2. The lemma holds for every pair of vectors $\mathbf{U}$ and $\mathbf{V}$ satisfying (5.3)$'$.

PROOF OF LEMMA 5.3 ($p>2$): Assume first that $|\mathbf{V}|\le 2|D\mathbf{u}-\mathbf{V}|$. Then

$$\begin{aligned} |\mathbf{H}| &\le \gamma(p)|D\mathbf{u}-\mathbf{V}|\int_0^1\left(|\mathbf{W}(t)|^{p-2} + |\mathbf{V}|^{p-2}\right)dt \\ &\le \gamma|D\mathbf{u}-\mathbf{V}|^{p-1} \le \frac{\gamma}{|\mathbf{V}|}(|D\mathbf{u}|+|\mathbf{V}|)^{p-2}|D\mathbf{u}-\mathbf{V}|^2. \end{aligned}$$

Therefore (5.16) follows in this case since $\mathbf{V}$ satisfies (5.3). If

(5.17)
$$|\mathbf{V}| > 2|D\mathbf{u} - \mathbf{V}|,$$

then by the mean value theorem and (5.14)-(5.15),

$$|\mathbf{H}| \le \gamma(p)|D\mathbf{u} - \mathbf{V}|^2 \int_0^1 |s\mathbf{W} + (1-s)\mathbf{V}|^{p-3} dt,$$

for some $s \in [0,1]$. By (5.15) and (5.17) we have

$$2^{-1}|\mathbf{V}| \le |s\mathbf{W} + (1-s)\mathbf{V}| \le \frac{3}{2}|\mathbf{V}|,$$

and this implies the lemma.

PROOF OF LEMMA 5.3 $(1<p<2)$: It suffices to prove that

(5.18)
$$|\mathbf{H}| \le \gamma\mu^{p-3}|D\mathbf{u} - \mathbf{V}|^2.$$

Assume first that

(5.19)
$$|\mathbf{V}| \le |D\mathbf{u} - \mathbf{V}|,$$

and let $t^* \in [0,1]$ be defined by

$$t^* = \frac{|\mathbf{V}|}{|D\mathbf{u} - \mathbf{V}|}.$$

Then

$$\begin{aligned}
|\mathbf{H}| &\le \gamma(p) \int_0^1 \Big| t|D\mathbf{u} - \mathbf{V}| - |\mathbf{V}| \Big|^{p-2} |D\mathbf{u} - \mathbf{V}|\, dt + \gamma|D\mathbf{u} - \mathbf{V}|\,|\mathbf{V}|^{p-2} \\
&\le \gamma \left\{ \int_{t^*}^0 \frac{d}{dt}\Big| t|D\mathbf{u} - \mathbf{V}| - |\mathbf{V}| \Big|^{p-1} dt + \int_{t^*}^1 \frac{d}{dt}\Big| t|D\mathbf{u} - \mathbf{V}| - |\mathbf{V}| \Big|^{p-1} dt \right\} \\
&\quad + \gamma|D\mathbf{u} - \mathbf{V}|\,|\mathbf{V}|^{p-2} \\
&\le \gamma\left(|\mathbf{V}|^{p-1} + |D\mathbf{u} - \mathbf{V}|^{p-1}\right).
\end{aligned}$$

Therefore taking into account (5.19) and (5.3), the lemma follows in this case. Consider now the case when (5.19) is violated, i.e.,

(5.19)′
$$|\mathbf{V}| > |D\mathbf{u} - \mathbf{V}|.$$

Then for $i = 1, 2, \ldots, m$,

$$H_i = (Du_i - V_i) \int_0^1\int_0^1 \frac{d}{ds} |s\mathbf{W} + (1-s)\mathbf{V}|^{p-2} ds\, dt$$

$$+ (Du_\ell - V_\ell) \int_0^1\int_0^1 \frac{d}{ds} \Big\{ |s\mathbf{W} + (1-s)\mathbf{V}|^{p-4} (sW_\ell + (1-s)V_\ell)$$

$$\times (sW_i + (1-s)V_i) \Big\} ds\, dt.$$

Therefore

$$(5.20) \qquad |\mathbf{H}| \le \gamma(p) |D\mathbf{u} - \mathbf{V}|^2 \int_0^1\int_0^1 t\, |s\mathbf{W} + (1-s)\mathbf{V}|^{p-3} ds\, dt.$$

Next, by (5.15) and (5.19)′

$$\begin{aligned} |s\mathbf{W} + (1-s)\mathbf{V}| &= \Big| \mathbf{V} + st(D\mathbf{u} - \mathbf{V}) \Big| \\ &\ge \Big| |\mathbf{V}| - st|D\mathbf{u} - \mathbf{V}| \Big| \\ &\ge |\mathbf{V}|(1 - st). \end{aligned}$$

This in (5.20) gives

$$|\mathbf{H}| \le \gamma |D\mathbf{u} - \mathbf{V}|^2 |\mathbf{V}|^{p-3} \int_0^1\int_0^1 t(1-st)^{p-3} ds\, dt.$$

Since $1 < p < 2$, the last integral is finite and the Lemma follows.

6. Linear parabolic systems with constant coefficients

Let $\mathbf{V}$ be any vector in $\mathbf{R}^{N\times m}$ satisfying (5.3). To the system (5.1) we associate its linearised version

$$(6.1) \qquad \frac{\partial}{\partial t} v_i - \left(|\mathbf{V}|^{p-2} v_{i,x_\ell} + (p-2)|\mathbf{V}|^{p-4} V_{j,k} v_{j,x_k} V_{i,\ell} \right)_{x_\ell}, \quad \text{in } Q_R(\mu), \quad i = 1, 2, \ldots, m.$$

Let

$$\mathbf{v} \equiv (v_1, v_2, \ldots, v_m)$$

and for $0 < \rho \le R$ we let $(D\mathbf{v})_\rho$ denote the integral average of $D\mathbf{v}$ over $Q_\rho(\mu)$.

THEOREM 6.1. *There exists a constant* $\gamma=\gamma(N,p)$, *such that for all* $0<\rho\le R$, *and for every constant vector* $\mathbf{W}\in\mathbf{R}^{N\times m}$,

$$\iint_{Q_\rho(\mu)} |D\mathbf{v}-(D\mathbf{v})_\rho|^2 dxd\tau \le \gamma\left(\frac{\rho}{R}\right)^2 \iint_{Q_R(\mu)} |D\mathbf{v}-\mathbf{W}|^2 dxd\tau. \tag{6.2}$$

To prove the theorem we introduce the change of variables

$$t\longrightarrow t\,\mu^{p-2},\qquad \mathbf{v}(x,t)\longrightarrow \mathbf{v}\left(x,t\mu^{2-p}\right).$$

This transforms $Q_\rho(\mu)$ into $Q_\rho(1)\equiv Q_\rho$ for all $0<\rho\le R$, and tranforms (6.1) into a system for which

$$4^{-1}\le|\mathbf{V}|\le 4.$$

In a precise way, the transformed vector $\mathbf{v}$ is a solution of

$$\frac{\partial}{\partial t}v_i-\left(a^{i,j}_{\ell,k}\,v_{i,x_k}\right)_{x_\ell}=0,\quad \text{in } Q_R, \tag{6.3}$$

where the coefficients

$$a^{i,j}_{\ell,k}\equiv|\mathbf{V}|^{p-2}\left\{\delta^{ij}_{\ell,k}+(p-2)\frac{V_{j,k}V_{i,\ell}}{|\mathbf{V}|^2}\right\}$$

satisfy the ellipticity condition

$$C_o(N,p)|\xi|^2\le a^{ij}_{\ell,k}\xi_{i,\ell}\xi_{j,k}\le C_1(N,p)|\xi|^2,\quad \forall\xi\in\mathbf{R}^{N\times m}, \tag{6.4}$$

for two given constants $C_o<C_1$ depending only upon N and p. Therefore it will suffice to prove Theorem 6.1 for $\mu=1$. In the remainder of the section we let $\mathbf{v}$ be a solution of (6.3) in Q_R and let (6.4) hold. Let α denote a multiindex of size $|\alpha|$, i.e.,

$$\alpha\equiv(\alpha_1,\alpha_2,\dots,\alpha_N),\ \ \alpha_j\in\mathbf{N}\cup\{0\},\ j=1,2,\dots,N;\ \ |\alpha|=\sum_{j=1}^N\alpha_j,$$

and for $f\in C^\infty(Q_R)$ let

$$D^\alpha_x f\equiv\frac{\partial^{\alpha_1}}{\partial x_1^{\alpha_1}}\frac{\partial^{\alpha_2}}{\partial x_2^{\alpha_2}}\cdots\frac{\partial^{\alpha_N}}{\partial x_N^{\alpha_N}}f.$$

For non-negative integers m and n we also set

$$|D^m_x f|\equiv\sum_{|\alpha|=m}|D^\alpha_x f|,\quad D^n_t f\equiv\frac{\partial^n}{\partial t^n}f,\quad |D^o_x f|=|D^o_t f|=|f|.$$

LEMMA 6.1. *There exists a constant* $\gamma=\gamma(N,p)$ *such that for all non-negative integers* m,n *and all* $0<\rho\le R$,

(6.5) $$\iint\limits_{Q_{\rho/2}} \left|D_x^{m+1} D_t^n \mathbf{v}\right|^2 dx d\tau \le \gamma \rho^{-2} \iint\limits_{Q_\rho} \left|D_x^m D_t^n \mathbf{v}\right|^2 dx d\tau,$$

(6.6) $$\iint\limits_{Q_{\rho/2}} \left|D_t^{n+1} D_x^m \mathbf{v}\right|^2 dx d\tau \le \gamma \rho^{-4} \iint\limits_{Q_\rho} \left|D_x^m D_t^n \mathbf{v}\right|^2 dx d\tau.$$

PROOF: The system (6.3) is also solved by the vectors $\mathbf{w} \equiv D_x^\alpha D_t^n \mathbf{v}$. Let ζ be a non-negative smooth cutoff function in Q_ρ vanishing on the parabolic boundary of Q_ρ and such that

$$\zeta \equiv 1 \text{ on } Q_{\rho/2}, \quad |D\zeta| \le 2/\rho, \quad 0 \le \zeta_t \le (2/\rho)^2.$$

Multiply the system (6.3), written for $\mathbf{w}$, by the testing function $\mathbf{w}\zeta^2$ and integrate over Q_ρ, to arrive at (6.5). To prove (6.6), multiply the same system by $\mathbf{w}_t\zeta^2$ and integrate over Q_ρ. This gives

(6.7) $$\begin{aligned}\iint\limits_{Q_\rho} |\mathbf{w}_t|^2 \zeta^2 dx d\tau + \iint\limits_{Q_\rho} \left(a_{\ell,k}^{i,j} w_{j,x_k} \frac{\partial}{\partial \tau} w_{i,x_\ell}\right) \zeta^2 dx d\tau \\ = -2 \iint\limits_{Q_\rho} a_{\ell,k}^{i,j} w_{j,x_k} w_{i,t} \zeta\, \zeta_{x_\ell}\, dx d\tau \\ + \frac{1}{2} \iint\limits_{Q_\rho} |\mathbf{w}_t|^2 \zeta^2 dx d\tau + \frac{\gamma}{\rho^2} \iint\limits_{Q_\rho} |D\mathbf{w}|^2 dx d\tau.\end{aligned}$$

The integral involving $a_{\ell,k}^{i,j}$ on the left hand side of (6.7) equals

$$\begin{aligned}\frac{1}{2} \int_{-\rho^2}^{0} \frac{d}{dt} \int_{K_\rho} a_{\ell,k}^{i,j} w_{j,k} w_{i,x_\ell}\, \zeta^2 dx d\tau - \iint\limits_{Q_\rho} a_{\ell,k}^{i,j} w_{j,k} w_{i,x_\ell} \zeta\, \zeta_t dx d\tau \\ \ge -\frac{\gamma}{\rho^2} \iint\limits_{Q_\rho} |D\mathbf{w}|^2 dx d\tau.\end{aligned}$$

These remarks in (6.7) give

$$\iint\limits_{Q_{\rho/2}} |\mathbf{w}_t|^2 dx d\tau \le \gamma \rho^{-2} \iint\limits_{Q_\rho} |D\mathbf{w}|^2 dx d\tau.$$

The lemma now follows by applying (6.5) and suitably modifying the scale of the radii ρ and $\rho/2$.

LEMMA 6.2. *There exists a constant $\gamma=\gamma(N,p)$ such that for all $0<\rho\le R$*

$$\iint_{Q_\rho} |\mathbf{v}|^2 dx d\tau \le \gamma \left(\frac{\rho}{R}\right)^{N+2} \iint_{Q_R} |\mathbf{v}|^2 dx d\tau. \tag{6.8}$$

PROOF: It suffices to prove the lemma for $0<\rho\le R/2^{N+2}$. Let ζ be the standard cutoff function in $Q_{R/2^{N+1}}$ that equals one on $Q_{R/2^{N+2}}$ and such that

$$|D\zeta| \le 2^{N+2}/R \quad \text{and} \quad 0<\zeta_t \le \left(2^{N+2}/R\right)^2.$$

If $0<\rho<R/2^{N+2}$, we have

$$\iint_{Q_\rho} |\mathbf{v}|^2 dx d\tau \le \gamma\rho^{N+2} \|\mathbf{v}\|^2_{\infty,Q_\rho} \le \gamma\rho^{N+2}\|\mathbf{v}\,\zeta\|^2_{\infty,Q_{R/2^{N+1}}}. \tag{6.9}$$

On the other hand for all $(x,t)\in Q_{R/2^{N+1}}$,

$$\begin{aligned}|\mathbf{v}\,\zeta|(x,t) &= \left| \int_{-R^2/2^{N+1}}^{t} D_t\left(\mathbf{v}\,\zeta\right)(x,\tau) d\tau \right| \le \gamma \int_{Q_{R/2^{N+1}}} |D_x^N D_t(\mathbf{v}\zeta)|\, dx d\tau \\ &\le R^{\frac{N+2}{2}} \left(\iint_{Q_{R/2^{N+1}}} |D_x^N D_t(\mathbf{v}\zeta)|^2\, dx d\tau \right)^{1/2}.\end{aligned}$$

Combining this with Lemma 6.1 we obtain the estimate

$$\|\mathbf{v}\,\zeta\|^2_{\infty,Q_{R/2^{N+1}}} \le \gamma R^{-(N+2)} \iint_{Q_R} |\mathbf{v}|^2 dx d\tau.$$

This in (6.9) proves the lemma.

PROOF OF THEOREM 6.1: Since the vectors $\mathbf{v}_{x_h,x_s}$ solve (6.3) for $h,s = 1,2,\dots,N$, we have from Lemma 6.2

$$\iint_{Q_\rho} |D^2\mathbf{v}|^2 dx d\tau \le \gamma \left(\frac{\rho}{R}\right)^{N+2} \iint_{Q_{R/2}} |D^2\mathbf{v}| dx d\tau, \tag{6.10}$$

for a constant $\gamma=\gamma(N,p)$ and for all $0<\rho\le R/2$. To estimate the right hand side of (6.10) observe that the vectors $\mathbf{v}_{x_s}$ solve the system

$$\frac{\partial}{\partial t} v_{i,x_s} - \left(a^{i,j}_{\ell,k} v_{j,x_s,x_k}\right)_{x_\ell} = 0, \quad \text{in } Q_R. \tag{6.11}$$

Let $\mathbf{W}$ be any constant vector in $\mathbf{R}^{N\times m}$ and multiply (6.11) by the testing function $(w_{i,x_s} - W_{i,s})\,\zeta^2$, where ζ is the standard cutoff function in Q_R that equals one in $Q_{R/2}$. This gives

$$\iint\limits_{Q_{R/2}} |D^2\mathbf{v}|^2 dxd\tau \leq \gamma R^{-2} \iint\limits_{Q_R} |D\mathbf{v} - \mathbf{W}|^2 dxd\tau.$$

To estimate the left hand side of (6.10) set

$$(D\mathbf{v})_\rho(t) \equiv \fint_{K_\rho} D\mathbf{v}(x,t)dx, \quad \forall 0 < \rho \leq R/2, \quad -\rho^2 \leq t \leq 0.$$

Then $x \to (Dv_i(x,t) - (Dv_i)_\rho(t))$ has zero average over K_ρ, and by the embedding Theorem 2.1 and Remark 2.1 of Chap. I,

$$\iint\limits_{Q_\rho} \left|D\mathbf{v} - (D\mathbf{v})_\rho(\tau)\right|^2 dxd\tau \leq \gamma\rho^2 \iint\limits_{Q_\rho} |D^2\mathbf{v}|^2 dxd\tau.$$

Write

$$\text{(6.12)} \qquad \iint\limits_{Q_\rho} \left|D\mathbf{v} - (D\mathbf{v})_\rho\right|^2 dxd\tau \leq \gamma \left(\frac{\rho}{R}\right)^{N+4} \iint\limits_{Q_R} |D\mathbf{v} - \mathbf{W}|^2 dxd\tau$$
$$+ \gamma\rho^N \int_{-\rho^2}^{0} \left|(D\mathbf{v})_\rho - (D\mathbf{v})_\rho(\tau)\right|^2 d\tau,$$

and estimate the last term by

$$\rho^N \int_{-\rho^2}^{0} \left|(D\mathbf{v})_\rho - (D\mathbf{v})_\rho(\tau)\right|^2 d\tau \leq \gamma\rho^{N+2} \sup_{-\rho^2 \leq t,\tau \leq 0} \left|(D\mathbf{v})_\rho(t) - (D\mathbf{v})_\rho(\tau)\right|^2.$$

Next integrate (6.11) over $K_\rho \times (\tau, t)$ and divide by meas$\{K_\rho\}$ to obtain

$$\left|(D\mathbf{v})_\rho(t) - (D\mathbf{v})_\rho(\tau)\right| \leq \gamma\rho^{-N} \iint\limits_{Q_\rho} |D^2\mathbf{v}| dxd\tau$$
$$\leq \rho^{-N}\rho^{\frac{N+2}{2}} \left(\iint\limits_{Q_\rho} \left|D^2\mathbf{v}\right|^2 dxd\tau\right)^{1/2}$$
$$\leq \gamma\rho^{-N/2} \left(\iint\limits_{Q_\rho} |D\mathbf{v} - \mathbf{W}|^2 dxd\tau\right)^{1/2}.$$

Therefore the last term on the right hand side of (6.12) is estimated by

$$\gamma\rho^2 \left(\frac{\rho}{R}\right)^{N+2} \iint\limits_{Q_{R/2}} |D^2\mathbf{v}|^2 dxd\tau \leq \gamma \left(\frac{\rho}{R}\right)^{N+4} \iint\limits_{Q_R} |D\mathbf{v} - \mathbf{W}|^2 dxd\tau.$$

7. The perturbation lemma

LEMMA 7.1. *There exists a constant $\gamma=\gamma(N,p)$ such that for every constant vector $\mathbf{V}$ in $\mathbf{R}^{N\times m}$ satisfying (5.3),*

$$\sup_{-\mu^{2-p}\frac{R^2}{4}\le t\le 0}\int_{K_{R/2}\times\{t\}}|D\mathbf{u}-\mathbf{V}|^2dx+\iint_{Q_{R/2}(\mu)}|D\mathbf{u}|^{p-2}|D^2\mathbf{u}|^2dxd\tau \le\gamma\frac{\mu^{p-2}}{R^2}\iint_{Q_R(\mu)}|D\mathbf{u}-\mathbf{V}|^2dxd\tau. \tag{7.1}$$

PROOF: Let ζ be a cutoff function in $Q_R(\mu)$ that equals one on $Q_{R/2}(\mu)$, and such that

$$|D\zeta|\le 2/R,\quad |D^2\zeta|\le(4/R)^2,\quad 0\le\zeta_t\le 2\mu^{p-2}/R^2.$$

In the weak formulation of (5.2) we take the testing functions

$$\left(u_{i,x_j}-V_{i,j}\right)\zeta^2,$$

modulo a Steklov time average. We obtain

$$\sup_{-\mu^{2-p}R^2\le t\le 0}\int_{Q_R(\mu)}|D\mathbf{u}-\mathbf{V}|^2\zeta^2(x,t)\,dx+\iint_{Q_R(\mu)}|D\mathbf{u}|^{p-2}|D^2u|^2\zeta^2dxd\tau \le\gamma\iint_{Q_R(\mu)}|D\mathbf{u}-\mathbf{V}|^2\,\zeta\zeta_t\,dxd\tau+J, \tag{7.2}$$

where

$$J\equiv\gamma\iint_{Q_R(\mu)}\left(|D\mathbf{u}|^{p-2}Du_i-|\mathbf{V}|^{p-2}V_{i,j}\right)_{x_j}\left(u_{i,x_j}-V_{i,j}\right)\zeta D\zeta dxd\tau.$$

If $p>2$, we have

$$J\le\frac14\iint_{Q_R(\mu)}|D\mathbf{u}|^{p-2}|D^2\mathbf{u}|^2\zeta^2dxd\tau+\gamma\frac{\mu^{p-2}}{R^2}\iint_{Q_R(\mu)}|D\mathbf{u}-\mathbf{V}|^2\,dxd\tau.$$

Putting this estimate in (7.2) proves the lemma in the degenerate case. To estimate J in the singular case $1<p<2$, we first integrate by parts in the variable x_i. This gives

(7.3)
$$J \le \gamma \iint_{Q_R(\mu)} \Big| |D\mathbf{u}|^{p-2} D\mathbf{u} - |\mathbf{V}|^{p-2}\mathbf{V} \Big| |D^2\mathbf{u}| \zeta |D\zeta| dx d\tau$$
$$+ \iint_{Q_R(\mu)} \Big| |D\mathbf{u}|^{p-2} D\mathbf{u} - |\mathbf{V}|^{p-2}\mathbf{V} \Big| |D\mathbf{u} - \mathbf{V}| \zeta |D^2\zeta| dx d\tau$$
$$\equiv I_1 + I_2.$$

By (5.7) of Lemma 5.2 and Schwartz inequality

$$I_1 \le \frac{1}{4} \iint_{Q_R(\mu)} |D\mathbf{u}|^{p-2} |D^2\mathbf{u}|^2 \zeta^2 dx d\tau + \gamma \frac{\mu^{p-2}}{R^2} \iint_{Q_R(\mu)} |D\mathbf{u} - \mathbf{V}|^2 dx d\tau,$$

and by (5.6)

$$I_2 \le \gamma \frac{\mu^{p-2}}{R^2} \iint_{Q_R(\mu)} |D\mathbf{u} - \mathbf{V}|^2 dx d\tau.$$

Combining these estimates in (7.2) proves the Lemma.

Let $\partial_p Q_{R/2}(\mu)$ denote the parabolic boundary of $Q_{R/2}(\mu)$. Consider the boundary value problem

(7.4)
$$\begin{cases} v_{i,t} - \left(|\mathbf{V}|^{p-2} v_{i,x_j} + (p-2)|\mathbf{V}|^{p-4} V_{\ell,k} v_{\ell,x_k} V_{i,j} \right)_{x_j}, & \text{in } Q_{R/2}(\mu) \\ v_i \big|_{\partial_p Q_{R/2}(\mu)} = u_i, \quad i = 1, 2, \dots, m. \end{cases}$$

The existence of a unique solution to (7.4) can be established for example by a Galerkin procedure.(1) The solution $\mathbf{v} \equiv (v_1, v_2, \dots, v_m)$ of (7.4) is *'regular'* in the interior of $Q_{R/2}(\mu)$, in the sense of Theorem 6.1. The next lemma compares $\mathbf{u}$ and $\mathbf{v}$.

LEMMA 7.2. *There exists a constant* $\gamma = \gamma(N, p)$*, such that for all* $0 < \rho < R/2$ *and for every vector* $\mathbf{V}$ *satisfying (5.3),*

(7.5)
$$\iint_{Q_\rho(\mu)} |D\mathbf{u} - D\mathbf{v}|^2 dx d\tau \le \gamma \left(\mu^{-2} \fint\!\!\fint_{Q_R(\mu)} |D\mathbf{u} - \mathbf{V}|^2 dx d\tau \right)^a \iint_{Q_R(\mu)} |D\mathbf{u} - \mathbf{V}|^2 dx d\tau,$$

where $a = \min\{\frac{1}{2}; \frac{2}{N}\}$.

PROOF: Write the system (5.1) in the form

$$\frac{\partial}{\partial t} u_i - \left(|\mathbf{V}|^{p-2} u_{i,x_j} + (p-2)|\mathbf{V}|^{p-4} V_{\ell,k} u_{\ell,x_k} V_{i,j} \right)_{x_j} = \operatorname{div} H_i,$$
$$i = 1, 2, \dots, m,$$

(1) See Lions [73].

where the vectors H_i are introduced in (5.12). From this, subtract (7.4), and in the weak formulation of the system so obtained, take the testing function $u_i - v_i$. This is admissible since it vanishes on $\partial_p Q_{R/2}(\mu)$. Adding over $i=1,2,\ldots,m$, gives

$$\mu^{p-2}\iint_{Q_{R/2}(\mu)} |D\mathbf{u} - D\mathbf{v}|^2 dx d\tau \le \gamma \iint_{Q_{R/2}(\mu)} |\mathbf{H}|\,|D\mathbf{u} - D\mathbf{v}| dx d\tau,$$

where we have taken into account the fact that $\mathbf{V}$ satisfies (5.3). Using Schwartz inequality on the right hand side and then Lemma 5.3 to estimate $|\mathbf{H}|^2$, we arrive at

$$\iint_{Q_{R/2}(\mu)} |D\mathbf{u} - D\mathbf{v}|^2 dx d\tau \le \gamma \mu^{-2(p-1)} \iint_{Q_{R/2}(\mu)} (|D\mathbf{u}| + |\mathbf{V}|)^{2(p-2)} |D\mathbf{u} - \mathbf{V}|^4 dx d\tau. \tag{7.6}$$

To estimate the right hand side of (7.6) assume first that $\mathbf{N} \ge 4$ so that

$$a \equiv \min\{\frac{1}{2}; \frac{2}{N}\} = \frac{2}{N}.$$

To simplify the symbolism we let $r = \mu^{2-p} R^2/4$. We have

$$\begin{aligned}
&\iint_{Q_{R/2}(\mu)} (|D\mathbf{u}| + |\mathbf{V}|)^{2(p-2)} |D\mathbf{u} - \mathbf{V}|^4 dx d\tau \\
&\le \int_{-r}^{0} \left(\int_{K_{R/2}} (|D\mathbf{u}| + |\mathbf{V}|)^{2(p-2)} |D\mathbf{u} - \mathbf{V}|^4 dx \right)^{\frac{2}{N}} \\
&\qquad \times \left(\int_{K_{R/2}} (|D\mathbf{u}| + |\mathbf{V}|)^{2(p-2)} |D\mathbf{u} - \mathbf{V}|^4 dx \right)^{\frac{N-2}{N}} d\tau \\
&\le \gamma \mu^{\frac{4}{N}(p-1)} \sup_{-r\le t\le 0} \left(\int_{K_{R/2}} |D\mathbf{u} - \mathbf{V}|^2 dx \right)^{\frac{2}{N}} \\
&\qquad \times \int_{-r}^{0} \left(\int_{K_{R/2}} (|D\mathbf{u}| + |\mathbf{V}|)^{2(p-2)} |D\mathbf{u} - \mathbf{V}|^4 dx \right)^{\frac{N-2}{N}} d\tau.
\end{aligned} \tag{7.7}$$

By Lemma 7.1

$$
(7.8)\quad \mu^{\frac{4}{N}(p-1)} \sup_{-r\le t\le 0}\left(\int_{K_{R/2}} |D\mathbf{u}-\mathbf{V}|^2 dt\right)^{\frac{2}{N}}
\le \gamma R^{-\frac{4}{N}} \mu^{\frac{4}{N}(p-1)+\frac{2}{N}(p-2)} \left(\iint_{Q_R(\mu)} |D\mathbf{u} - \mathbf{V}|^2 dx d\tau\right)^{\frac{2}{N}}.
$$

To estimate the last factor in (7.7) we majorise the integrand by means of Lemma 5.1. It gives

$$
\begin{aligned}
(|D\mathbf{u}|+|\mathbf{V}|)^{2(p-2)}\,|D\mathbf{u}-\mathbf{V}|^4 &= \left\{(|D\mathbf{u}|+|\mathbf{V}|)^{\frac{p-2}{2}}\,|D\mathbf{u}-\mathbf{V}|\right\}^4 \\
&\le \gamma\left||D\mathbf{u}|^{\frac{p-2}{2}} D\mathbf{u} - |\mathbf{V}|^{\frac{p-2}{2}}\mathbf{V}\right|^4 \\
&\le \mu^{p\frac{N-4}{N-2}}\left||D\mathbf{u}|^{\frac{p-2}{2}} D\mathbf{u} - |\mathbf{V}|^{\frac{p-2}{2}}\mathbf{V}\right|^{\frac{2N}{N-2}}.
\end{aligned}
$$

Let $x \to \zeta(x)$ be a non-negative piecewise smooth cutoff function in K_R that equals one on $K_{3R/4}$ and such that $|D\zeta| \le 4/R$. Then for a.e. $t\in\{-r,0\}$, by the embedding Corollary 2.1 of Chap. I, we have

$$
\begin{aligned}
&\left(\int_{K_{R/2}} (|D\mathbf{u}|+|\mathbf{V}|)^{2(p-2)}\,|D\mathbf{u}-\mathbf{V}|^4 dx\right)^{\frac{N-2}{N}} \\
&\le \gamma\,\mu^{p\frac{N-4}{N}}\left(\int_{K_{3R/4}}\left[\left||D\mathbf{u}|^{\frac{p-2}{2}} D\mathbf{u} - |\mathbf{V}|^{\frac{p-2}{2}}\mathbf{V}\right|\zeta\right]^{\frac{2N}{N-2}} dx\right)^{\frac{N-2}{N}} \\
&\le \gamma\,\mu^{p\frac{N-4}{N}} \int_{K_{3R/4}}\left|D\left[|D\mathbf{u}|^{\frac{p-2}{2}} D\mathbf{u} - |\mathbf{V}|^{\frac{p-2}{2}}\mathbf{V}\right]\zeta\right|^2 dx \\
&\le \gamma\,\mu^{p\frac{N-4}{N}}\left\{\int_{K_{3R/4}} |D\mathbf{u}|^{p-2}|D^2\mathbf{u}|^2 dx + \mu^{p-2} R^{-2}\int_{K_{3R/4}} |D\mathbf{u}-\mathbf{V}|^2 dx\right\}.
\end{aligned}
$$

Here in estimating the last term we have used the algebraic inequality

$$\left||D\mathbf{u}|^{\frac{p-2}{2}} D\mathbf{u} - |\mathbf{V}|^{\frac{p-2}{2}}\mathbf{V}\right|^2 \leq \gamma\mu^{p-2}|D\mathbf{u} - \mathbf{V}|^2,$$

which follows from (5.6) of Lemma 5.2 with p replaced by $(p+2)/2$. Therefore the last factor on the right hand side of (7.7) is estimated by

$$\int_{-r}^{0}\left(\int_{K_{R/2}}(|D\mathbf{u}| + |\mathbf{V}|)^{2(p-2)}\,|D\mathbf{u} - \mathbf{V}|^4 dx\right)^{\frac{N-2}{N}} d\tau$$

$$\leq \gamma\,\mu^{p\frac{N-4}{N}}\Bigg\{\iint_{Q_{3R/2}(\mu)}|D\mathbf{u}|^{p-2}|D^2\mathbf{u}|^2 dxd\tau$$

$$+ \mu^{p-2}\,R^{-2}\iint_{Q_R(\mu)}|D\mathbf{u} - \mathbf{V}|^2 dxd\tau\Bigg\}$$

$$\leq \mu^{2(p-1)-\frac{4p}{N}}\,R^{-2}\iint_{Q_R(\mu)}|D\mathbf{u} - \mathbf{V}|^2 dxd\tau,$$

where we have also used Lemma 7.1. We now combine these calculations in (7.7) and then in (7.6) to obtain

$$\iint_{Q_\rho(\mu)}|D\mathbf{u} - D\mathbf{v}|^2 dxd\tau$$

$$\leq \gamma\left(\mu^{-2}\fint\!\!\fint_{Q_R(\mu)}|D\mathbf{u} - \mathbf{V}|^2 dxd\tau\right)^{\frac{2}{N}}\iint_{Q_R(\mu)}|D\mathbf{u} - \mathbf{V}|^2 dxd\tau,$$

provided $N \geq 4$. If $N = 2, 3$, we transform the integral on the right hand side of (7.6) by Hölder's inequality as follows.

$$\begin{aligned}(7.9)\qquad &\iint\limits_{Q_{R/2}(\mu)} (|D\mathbf{u}|+|\mathbf{V}|)^{2(p-2)}\,|D\mathbf{u}-\mathbf{V}|^4 dx d\tau \\ &= \int_{-r}^{0}\int_{K_{R/2}} |D\mathbf{u}-\mathbf{V}|\,(|D\mathbf{u}|+|\mathbf{V}|)^{2(p-2)}\,|D\mathbf{u}-\mathbf{V}|^3 dx d\tau \\ &\le \int_{-r}^{0}\left(\int_{K_{R/2}} |D\mathbf{u}-\mathbf{V}|^2 dx\right)^{\frac12} \\ &\quad\times\left(\int_{K_{R/2}} (|D\mathbf{u}|+|\mathbf{V}|)^{4(p-2)}\,|D\mathbf{u}-\mathbf{V}|^6 dx\right)^{\frac12} d\tau \\ &\le \mu^{\frac{p-2}{2}} \sup_{-r\le t\le 0}\left(\int_{K_{R/2}} |D\mathbf{u}-\mathbf{V}|^2 dx\right)^{\frac12} \\ &\quad\times \int_{-r}^{0}\left(\int_{K_{R/2}} \left[(|D\mathbf{u}|+|\mathbf{V}|)^{\frac{p-2}{2}}\,|D\mathbf{u}-\mathbf{V}|\right]^6 dx\right)^{\frac12} d\tau.\end{aligned}$$

By Lemma 7.1

$$\begin{aligned}(7.10)\qquad &\mu^{\frac{p-2}{2}} \sup_{-r\le t\le 0}\left(\int_{K_{R/2}} |D\mathbf{u}-\mathbf{V}|^2 dx\right)^{\frac12} \\ &\qquad\le \gamma\mu^{p-2}R^{-1}\left(\iint\limits_{Q_R(\mu)} |D\mathbf{u}-\mathbf{V}|^2 dx d\tau\right)^{\frac12}.\end{aligned}$$

We estimate the last term on the right hand side of (7.9) separately for $N=3$ and $N=2$.

The case $N=3$

Let ζ be defined as before. Then for a.e. $t\in(-r,0)$,

$$\left(\int_{K_{R/2}}\left[(|D\mathbf{u}|+|\mathbf{V}|)^{\frac{p-2}{2}}|D\mathbf{u}-\mathbf{V}|\right]^6 dx\right)^{\frac{1}{2}}$$

$$\le \gamma\,\mu^{\frac{p}{2}}\,R^{\frac{1}{2}}\left(\int_{K_{R/2}}\left[(|D\mathbf{u}|+|\mathbf{V}|)^{\frac{p-2}{2}}|D\mathbf{u}-\mathbf{V}|\right]^6 dx\right)^{\frac{1}{3}}$$

$$\le \gamma\,\mu^{\frac{p}{2}}\,R^{\frac{1}{2}}\left(\int_{K_{3R/4}}\left[\left||D\mathbf{u}|^{\frac{p-2}{2}}D\mathbf{u}-|\mathbf{V}|^{\frac{p-2}{2}}\mathbf{V}\right|\zeta\right]^6 dx\right)^{\frac{1}{3}}$$

$$\le \gamma\,\mu^{\frac{p}{2}}\,R^{\frac{1}{2}}\int_{K_{3R/4}}\left|D\left[|D\mathbf{u}|^{\frac{p-2}{2}}D\mathbf{u}-|\mathbf{V}|^{\frac{p-2}{2}}\mathbf{V}\right]\zeta\right|^2 dx.$$

Therefore

$$\int_{-r}^{0}\left(\int_{K_{R/2}}\left[(|D\mathbf{u}|+|\mathbf{V}|)^{\frac{p-2}{2}}|D\mathbf{u}-\mathbf{V}|\right]^6 dx\right)^{\frac{1}{2}}d\tau$$

$$\le \gamma\,\mu^{2(p-1)-\frac{p}{2}}\,R^{-\frac{3}{2}}\iint_{Q_R(\mu)}|D\mathbf{u}-\mathbf{V}|^2 dx d\tau.$$

Combining these estimates in (7.9) and then in (7.6) proves the lemma for $N=3$.

The case $N=2$

We apply the embedding Theorem 2.1 of Chap. I with $q=6$, $\alpha=2/3$ and $s=1$. This gives for a.e. $t\in(-r,0)$

$$\left(\int_{K_{R/2}}\left[(|D\mathbf{u}|+|\mathbf{V}|)^{\frac{p-2}{2}}|D\mathbf{u}-\mathbf{V}|\right]^6 dx\right)^{\frac{1}{2}}$$

$$\le \gamma\left(\int_{K_{3R/4}}\left[\left||D\mathbf{u}|^{\frac{p-2}{2}}D\mathbf{u}-|\mathbf{V}|^{\frac{p-2}{2}}\mathbf{V}\right|\zeta\right]^6 dx\right)^{\frac{1}{2}}$$

$$\le \gamma\int_{K_{3R/4}}\left|D\left[|D\mathbf{u}|^{\frac{p-2}{2}}D\mathbf{u}-|\mathbf{V}|^{\frac{p-2}{2}}\mathbf{V}\right]\zeta\right|^2 dx$$

$$\times\left(\int_{K_{3R/4}}\left||D\mathbf{u}|^{\frac{p-2}{2}}D\mathbf{u}-|\mathbf{V}|^{\frac{p-2}{2}}\mathbf{V}\right|^2 dx\right)^{\frac{1}{2}}.$$

Therefore

$$\int_{-r}^{0}\left(\int_{K_{R/2}}\left[(|D\mathbf{u}|+|\mathbf{V}|)^{\frac{p-2}{2}}\,|D\mathbf{u}-\mathbf{V}|\right]^{6}dx\right)^{\frac{1}{2}}d\tau$$
$$\leq \gamma\mu^{\frac{p}{2}}R\iint_{Q_R(\mu)}\left|D\left[|D\mathbf{u}|^{\frac{p-2}{2}}D\mathbf{u}-|\mathbf{V}|^{\frac{p-2}{2}}\mathbf{V}\right]\zeta\right|^{2}dxd\tau.$$

We estimate these integrals by means of Lemma 7.1 and combine the calculations in (7.9) and in (7.6) to conclude that (7.5) holds with $a=\frac{1}{2}$.

8. Proof of Proposition 1.1-(i)

LEMMA 8.1. *There exist constants $\kappa, \delta, \varepsilon \in (0,1)$ that can be determined a priori only in terms of N and p, such that if $\mathbf{V}_o$ is a constant vector in $\mathbf{R}^{N\times m}$ satisfying*

(8.1) $$2^{-1}\mu \leq |\mathbf{V}_o| \leq \mu$$

(8.2) $$\fint\!\!\fint_{Q_R(\mu)}|D\mathbf{u}-\mathbf{V}_o|^2 dxd\tau \leq \varepsilon\mu^2,$$

then there exists a constant vector $\mathbf{V}_1 \in \mathbf{R}^{N\times m}$ such that

(8.3) $$\tfrac{1}{4}\mu \leq \left(\tfrac{1}{2}-\sqrt{2\kappa}\right)\mu \leq |\mathbf{V}_1| \leq \left(1+\sqrt{2\kappa}\right)\mu$$

(8.4) $$\iint_{Q_{\delta R}(\mu)}|D\mathbf{u}-\mathbf{V}_1|^2 dxd\tau \leq \kappa\,\delta^{N+2}\iint_{Q_R(\mu)}|D\mathbf{u}-\mathbf{V}_o|^2 dt,$$

(8.5) $$\fint\!\!\fint_{Q_{\delta R}(\mu)}|D\mathbf{u}-\mathbf{V}_1|^2 \leq \varepsilon\,\mu^2.$$

PROOF: Let $\mathbf{v}$ be the unique solution of (7.4) and set

$$\mathbf{V}_1 \equiv \fint\!\!\fint_{Q_{\delta R}(\mu)} D\mathbf{v}\,dt,$$

where $\delta \in (0,1)$ is to be chosen. The perturbation Lemma 7.2 with $\mathbf{V}=\mathbf{V}_o$, the triangle inequality and (8.2) give

$$\iint_{Q_{\delta R}(\mu)} |D\mathbf{u} - \mathbf{V}_1|^2 dx d\tau \le \gamma\, \varepsilon^a \iint_{Q_R(\mu)} |D\mathbf{u} - \mathbf{V}_o|^2 dx d\tau + \iint_{Q_{\delta R}(\mu)} |D\mathbf{v} - \mathbf{V}_1|^2 dx d\tau.$$

By Theorem 6.1

$$\iint_{Q_{\delta R}(\mu)} |D\mathbf{v} - \mathbf{V}_1|^2 dx d\tau \le \gamma \delta^{N+4} \iint_{Q_{R/2}(\mu)} |D\mathbf{v} - \mathbf{V}_o|^2 dx d\tau,$$

and again by Lemma 7.2 with $\mathbf{V} = \mathbf{V}_o$ and (8.2)

$$\iint_{Q_{R/2}(\mu)} |D\mathbf{v} - \mathbf{V}_o|^2 dx d\tau \le \gamma\, (1 + \varepsilon^a) \iint_{Q_R(\mu)} |D\mathbf{u} - \mathbf{V}_o|^2 dx d\tau,$$

for a constant $\gamma = \gamma(N, p)$. Combining these inequalities we obtain

$$\iint_{Q_{\delta R}(\mu)} |D\mathbf{u} - \mathbf{V}_1|^2 dx d\tau \le \gamma\, (\delta^{N+4} + \varepsilon^a) \iint_{Q_R(\mu)} |D\mathbf{u} - \mathbf{V}_o|^2 dx d\tau, \quad \delta \le 1/2.$$

To prove (8.4) choose $\varepsilon^a = \delta^{N+4}$, and then δ so small that

$$2\gamma\, \delta^2 \le \kappa.$$

Inequality (8.5) follows from (8.4) and the *smallness* assumption (8.2). To prove (8.3) write

$$\begin{aligned} \mathbf{V}_1 - \mathbf{V}_o &= \fint\!\!\fint_{Q_{\delta R}(\mu)} (D\mathbf{v} - \mathbf{V}_o)\, dx d\tau \\ &= \fint\!\!\fint_{Q_{\delta R}(\mu)} \{(D\mathbf{v} - D\mathbf{u}) + (D\mathbf{u} - \mathbf{V}_o)\}\, dx d\tau \end{aligned}$$

and

$$|\mathbf{V}_1 - \mathbf{V}_o|^2 \le 2 \fint\!\!\fint_{Q_{\delta R}(\mu)} |D\mathbf{u} - D\mathbf{v}|^2 dx d\tau + 2 \fint\!\!\fint_{Q_{\delta R}(\mu)} |D\mathbf{u} - \mathbf{V}_o|^2 dx d\tau.$$

By Lemma 7.2 and the indicated choices of ε and δ

$$\fint\!\!\fint_{Q_{\delta R}(\mu)} |D\mathbf{u} - D\mathbf{v}|^2 dx d\tau \le \kappa \fint\!\!\fint_{Q_{\delta R}(\mu)} |D\mathbf{u} - \mathbf{V}_o|^2 dx d\tau.$$

Therefore using again (8.2)

$$
\text{(8.6)} \qquad |\mathbf{V}_1 - \mathbf{V}_o|^2 \le 2\left(\kappa + \delta^{-(N+2)}\right) \iint\limits_{Q_R(\mu)} |D\mathbf{u} - \mathbf{V}_o|^2 dx d\tau
$$
$$
\le 2\left(\kappa + \delta^{-(N+2)}\right) dl^{2(N+4)} \mu^2 \le 2\kappa\, \mu^2.
$$

By choosing κ sufficiently small we may insure that

$$
|\mathbf{V}_1| \ge |\mathbf{V}_o| - \sqrt{2\kappa}\,\mu \ge \left(\tfrac{1}{2} - \sqrt{2\,\kappa}\right)\mu \ge \frac{1}{4}\mu
$$

and

$$
|\mathbf{V}_1| \le |\mathbf{V}_o| + \sqrt{2\kappa}\mu \le \left(1 + \sqrt{2\,\kappa}\right)\mu.
$$

LEMMA 8.2. *There exist constants $\kappa, \delta, \varepsilon \in (0,1)$ that can be determined a priori only in terms of N and p, such that if $\mathbf{V}_o$ is a constant vector in $\mathbf{R}^{N\times m}$ satisfying (8.1) and (8.2), then there exists a sequence of constant vectors $\{\mathbf{V}_i\}_{i\in\mathbf{N}}$ in $\mathbf{R}^{N\times m}$, satisfying*

$$
\text{(8.7)} \qquad 4^{-1}\mu \le |\mathbf{V}_i| \le 4\mu,
$$

$$
\text{(8.8)} \qquad \iint\limits_{Q_{\delta^i R}(\mu)} |D\mathbf{u} - \mathbf{V}_i|^2 dx d\tau \le \varepsilon\, \mu^2,
$$

$$
\text{(8.9)} \qquad \iint\limits_{Q_{\delta^{i+1} R}(\mu)} |D\mathbf{u} - \mathbf{V}_{i+1}|^2 dx d\tau \le \kappa\delta^{N+2} \iint\limits_{Q_{\delta^i R}(\mu)} |D\mathbf{u} - \mathbf{V}_i|^2 dx d\tau,
$$

for $i = 1, 2, \ldots$.

PROOF: The sequence is constructed inductively by using the procedure of the previous lemma. To prove that $|\mathbf{V}_i|$ are in the range (8.7), we refer back to (8.6), i.e.

$$
|\mathbf{V}_{i+1} - \mathbf{V}_i|^2 \le 2\left(\kappa + \delta^{-(N+2)}\right) \iint\limits_{Q_{\delta^i R}(\mu)} |D\mathbf{u} - \mathbf{V}_i|^2 dx d\tau.
$$

We iterate over i and use again the *smallness* assumption (8.2) to obtain

$$
|\mathbf{V}_{i+1} - \mathbf{V}_i|^2 \le 2\left(\kappa + \delta^{-(N+2)}\right) \kappa^i \iint\limits_{Q_R(\mu)} |D\mathbf{u} - \mathbf{V}_o|^2 dx d\tau
$$
$$
\le 2\mu^2 \delta^{2(N+4)} \left(\kappa + \delta^{-(N+2)}\right) \kappa^i.
$$

From this by taking roots and adding over i

$$
|\mathbf{V}_{i+1} - \mathbf{V}_o| \le \mu\delta \sum_{i=1}^{\infty} \sqrt{\kappa}^i \le \mu\, \frac{\sqrt{\kappa}}{1 - \sqrt{\kappa}},
$$

where we have used the specific choice of δ in terms of κ. Choosing now κ sufficiently small proves the Lemma.

9. Proof of Proposition 1.1-(ii)

The number ν in the assumption (1.3) can be chosen to insure the existence of a constant vector $\mathbf{V}_o \in \mathbf{R}^{N\times m}$ satisfying (8.1) and (8.2). This is the content of this section. Set $|D\mathbf{u}|=v$ and, for all $0<\rho\le R$,

$$A^\nu_\rho \equiv \big\{(x,t)\in Q_\rho(\mu) \mid v(x,t) > (1-\nu)\mu\big\}\,, \tag{9.1}$$

$$B^\nu_\rho \equiv \big\{(x,t)\in Q_\rho(\mu) \mid v(x,t) < (1-\nu)\mu\big\}\,. \tag{9.2}$$

We will choose $\nu\in(0,\frac14)$ and rewrite (1.3) as

$$\big|B^\nu_R\big| \le \nu\big|Q_R(\mu)\big|, \qquad \nu\in(0,\tfrac14). \tag{9.3}$$

LEMMA 9.1. *There exists a constant $\gamma=\gamma(N,p)$ such that for all $\sigma\in(0,1)$*

$$\iint_{A^\nu_{\sigma R}} |D\mathbf{u}|^{p-2}\big|D^2\mathbf{u}\big|^2 dx d\tau \le \frac{\gamma\,\mu^2\nu}{(1-\sigma)^2}\,R^N. \tag{9.4}$$

PROOF: Consider the differentiated equation (5.2) and in its weak formulation take the testing function

$$u_{i,x_j}\left(v^2-k^2\right)_+\zeta^2, \qquad k=(1-2\nu)\mu,$$

modulo a Steklov averaging process. Here ζ is a non-negative piecewise smooth cutoff function in $Q_R(\mu)$ that equals one on $Q_{\sigma R}(\mu)$ and such that

$$|D\zeta| \le \frac{1}{(1-\sigma)R}, \qquad 0\le\zeta_t\le\frac{\mu^{p-2}}{(1-\sigma)R^2}.$$

After we add over $i=1,2,\dots,m$ and $j=1,2,\dots,N$, we arrive at

$$\begin{aligned}
&\sup_{-\mu^{2-p}R^2\le t\le 0}\int_{K_R}\left(v^2-k^2\right)_+^2\zeta^2(x,t)\,dx \\
&\quad+\iint_{Q_R(\mu)} v^{p-2}\big|Dv^2\big|^2\zeta^2\chi\left[v>k\right]\,dxd\tau \\
&\quad+\sum_{i=1}^m\sum_{j=1}^N\iint_{Q_{\sigma R}(\mu)}|D\mathbf{u}|^{p-2}\big|Du_{i,x_j}\big|^2\left(v^2-k^2\right)_+\zeta^2 dxd\tau \\
&\le\gamma\iint_{Q_R(\mu)} v^{p-2}\big|Dv^2\big|\left(v^2-k^2\right)_+\zeta|D\zeta|\,dxd\tau \\
&\quad+\gamma\iint_{Q_R(\mu)}\left(v^2-k^2\right)_+^2\zeta\zeta_t\,dxd\tau
\end{aligned} \tag{9.5}$$

for a constant $\gamma=\gamma(N,p)$. By the Schwartz inequality

$$\gamma \iint_{Q_R(\mu)} v^{p-2}|Dv^2|\left(v^2-k^2\right)_+\zeta|D\zeta|dxd\tau$$

$$\leq \iint_{Q_R(\mu)} v^{p-2}|Dv^2|^2\zeta^2\chi\left[v>k\right]dxd\tau$$

$$+\gamma^2\iint_{Q_R(\mu)} v^{p-2}\left(v^2-k^2\right)_+^2|D\zeta|^2dxd\tau.$$

We put this in (9.5) and in the resulting inequality we discard all the non-negative terms on the left hand side except the integral containing Du_{i,x_j}. This gives

(9.6)
$$\iint_{Q_R(\mu)} |\mathbf{Du}|^{p-2}\left|D^2\mathbf{u}\right|^2\left(v^2-k^2\right)_+\zeta^2dxd\tau$$

$$\leq \gamma\iint_{Q_R(\mu)}\left(v^{p-2}|D\zeta|^2+\zeta_t\right)\left(v^2-k^2\right)_+^2dxd\tau.$$

Since $\left(v^2-k^2\right)_+\leq 4\nu\mu^2$,

$$\iint_{Q_R(\mu)}\left(v^2-k^2\right)_+^2\zeta_t\,dxd\tau \leq \frac{\gamma(\nu\mu^2)^2}{(1-\sigma)^2R^2}\,\mu^{p-2}|Q_R(\mu)|$$

$$\leq \frac{\gamma\,\nu^2\mu^4}{(1-\sigma)^2}\,R^N$$

where we have used the structure of ζ and the intrinsic geometry of $Q_R(\mu)$. Also

$$\iint_{Q_R(\mu)} v^{p-2}\left(v^2-k^2\right)_+^2|D\zeta|^2dxd\tau \leq \frac{\gamma\,\nu^2\mu^4}{(1-\sigma)^2}\,R^N.$$

This is obvious if $p>2$. If $1<p<2$, we observe that the integral is extended over the set $v>(1-2\nu)\mu$. We estimate below the integral on the left hand side of (9.6) by extending the integration over the smaller set $[v>(1-\nu)\mu]$. On such a set, $\left(v^2-k^2\right)_+\geq\nu\mu^2$. These remarks in (9.6) prove (9.4).

Set for all $0<\rho\leq R$ and all $t\in[-\mu^{2-p}\rho^2,0]$

$$(\mathbf{Du})_\rho(t)\equiv \fint_{K_\rho}\mathbf{Du}(x,t)\,dx.$$

LEMMA 9.2. *There exists a constant $\gamma=\gamma(N,p)$, such that for all $\sigma\in(\frac{1}{2},1)$*

(9.7) $$\iint_{Q_{\sigma R}(\mu)} \left| \mathbf{Du} - (\mathbf{Du})_{\sigma R}(\tau) \right|^2 dx d\tau \le \frac{\gamma\, \mu^2 \nu^{1/(N+1)}}{(1-\sigma)^{2N/(N+1)}}.$$

PROOF: Fix $\sigma \in (0,1)$, and for all $t \in \left[-\mu^{2-p}(\sigma R)^2, 0\right]$, set

$$\mathbf{V}(t) \equiv \fint_{K_{\sigma R}} |\mathbf{Du}|^{\frac{p-2}{2}} \mathbf{Du}(x,t)\, dx.$$

We apply the multiplicative embedding of Theorem 2.1 of Chap. I to the functions

$$x \longrightarrow |\mathbf{Du}|^{\frac{p-2}{2}} \mathbf{Du}(x,t) - \mathbf{V}(t), \qquad \forall t \in \left[-\mu^{2-p}(\sigma R)^2, 0\right],$$

which have zero average over $K_{\sigma R}$. For the choice of the parameters

$$\alpha = \frac{N}{N+1}, \qquad q = 2, \qquad s = 1,$$

we obtain

$$\int_{K_{\sigma R}} \left| |\mathbf{Du}|^{\frac{p-2}{2}} \mathbf{Du} - \mathbf{V}(t) \right|^2 dx \le \gamma \int_{K_{\sigma R}} \left(|\mathbf{Du}|^{p-2} |D^2\mathbf{u}|^2 \right)^{\frac{N}{N+1}} dx$$
$$\times \left(\int_{K_{\sigma R}} \left| |\mathbf{Du}|^{\frac{p-2}{2}} \mathbf{Du} - \mathbf{V}(t) \right| dx \right)^{\frac{2}{N+1}}.$$

The last integral is majorised by $\gamma \mu^{\frac{p}{N+1}} R^{\frac{2N}{N+1}}$. Therefore integrating this inequality over $\left[-\mu^{2-p}(\sigma R)^2, 0\right]$ gives

$$\left(\mu^{\frac{p}{N+1}} R^{\frac{2N}{N+1}} \right)^{-1} \iint_{Q_{\sigma R}(\mu)} \left| |\mathbf{Du}|^{\frac{p-2}{2}} \mathbf{Du} - \mathbf{V}(\tau) \right|^2 dx d\tau$$
$$\le \gamma \iint_{Q_{\sigma R}(\mu)} \left(|\mathbf{Du}|^{p-2} |D^2\mathbf{u}|^2 \right)^{\frac{N}{N+1}} dx d\tau$$
$$= \gamma \iint_{A^\nu_{\sigma R}} \left(|\mathbf{Du}|^{p-2} |D^2\mathbf{u}|^2 \right)^{\frac{N}{N+1}} dx d\tau$$
$$+ \iint_{B^\nu_{\sigma R}} \left(|\mathbf{Du}|^{p-2} |D^2\mathbf{u}|^2 \right)^{\frac{N}{N+1}} dx d\tau$$
$$\le \gamma \left| Q_{\sigma R}(\mu) \right|^{\frac{1}{N+1}} \left(\iint_{A^\nu_{\sigma R}} |\mathbf{Du}|^{p-2} |D^2\mathbf{u}|^2 dx d\tau \right)^{\frac{N}{N+1}}$$
$$+ \gamma \left| B^\nu_R \right|^{\frac{1}{N+1}} \left(\iint_{Q_{\sigma R}(\mu)} |\mathbf{Du}|^{p-2} |D^2\mathbf{u}|^2 dx d\tau \right)^{\frac{N}{N+1}},$$

where A^ν_ρ and B^μ_ρ are defined in (9.1)-(9.2). We estimate the first integral by Lemma 9.1 and the second by using the '*smallness condition*' (9.3) and Lemma 7.1. We conclude that there exists a constant $\gamma = \gamma(N,p)$ such that

$$\iint_{Q_{\sigma R}(\mu)} \left| |Du|^{\frac{p-2}{2}} Du - \mathbf{V}(\tau) \right|^2 dx d\tau \le \frac{\gamma \mu^2 \nu^{1/(N+1)}}{(1-\sigma)^{2N/(N+1)}} R^{N+2}. \tag{9.8}$$

Introduce the vectors $\mathbf{w}(t)$ by

$$\mathbf{V}(t) \equiv |\mathbf{w}(t)|^{\frac{p-2}{2}} \mathbf{w}(t),$$

and observe that

$$|\mathbf{w}(t)| \le \mu, \quad \forall t \in \left[-\mu^{2-p}(\sigma R)^2, 0 \right].$$

By the algebraic Lemma 5.1,

$$\begin{aligned} \iint_{Q_{\sigma R}(\mu)} \left| |Du|^{\frac{p-2}{2}} Du - \mathbf{V}(\tau) \right|^2 dx d\tau \qquad (9.9) \\ \ge \iint_{Q_{\sigma R}(\mu)} (|Du| + |\mathbf{w}(\tau)|)^{p-2} |Du - \mathbf{w}(\tau)|^2 dx d\tau. \end{aligned}$$

We treat separately the cases $p>2$ and $1<p<2$.

The degenerate case $p > 2$

We minorise the left hand side of (9.9) by extending the integration over the smaller set $A^\nu_{\sigma R}$. On such a set,

$$(|Du| + |\mathbf{w}(t)|)^{p-2} \ge |Du|^{p-2} \ge 2^{2-p} \mu^{p-2}.$$

This with (9.8) yields

$$\iint_{A^\nu_{\sigma R}} |Du - \mathbf{w}(\tau)|^2 dx d\tau \le \frac{\gamma \mu^2 \nu^{1/(N+1)}}{(1-\sigma)^{2N/(N+1)}} |Q_R(\mu)|. \tag{9.10}$$

Next write

$$\iint_{Q_{\sigma R}(\mu)} |Du - \mathbf{w}(t)|^2 dx d\tau = \iint_{A^\nu_{\sigma R}} |Du - \mathbf{w}(\tau)|^2 dx d\tau + \iint_{B^\nu_{\sigma R}} |Du - \mathbf{w}(\tau)|^2 dx d\tau.$$

The first integral is estimated in (9.10) and the second is majorised by $2\mu^2 \nu |Q_R(\mu)|$, in view of the '*smallness*' condition (9.3). We conclude that

$$\int_{-\mu^{2-p}(\sigma R)^2}^{0} \left(\min_{\mathbf{V} \in \mathbf{R}^{N \times m}} \int_{K_{\sigma R}} |Du - \mathbf{V}|^2 dx \right) d\tau \le \frac{\gamma \mu^2 \nu^{1/(N+1)}}{(1-\sigma)^{2N/(N+1)}} |Q_{\sigma R}(\mu)|,$$

for a constant $\gamma = \gamma(N, p)$. The minimum on the left hand side is achieved for $\mathbf{V} \equiv (D\mathbf{u})_{\sigma R}(t)$. This proves the lemma if $p > 2$.

The singular case $1 < p < 2$

Since $|\mathbf{w}(t)| \le \mu$, we have $(|D\mathbf{u}| + |\mathbf{w}(t)|)^{p-2} \ge 2^{p-2}\mu^{p-2}$. Putting this in (9.9) and combining it with (9.8) gives

$$\iint_{Q_{\sigma R}(\mu)} |D\mathbf{u} - \mathbf{w}(t)|^2 dx d\tau \le \frac{\gamma\, \mu^2 \nu^{1/(N+1)}}{(1-\sigma)^{2N/(N+1)}} |Q_{\sigma R}(\mu)|.$$

The proof is now concluded by a minimization procedure.

10. Proof of Proposition 1.1-(iii)

Let $(D\mathbf{u})_\rho$ denote the integral average of $D\mathbf{u}$ over $Q_\rho(\mu)$, i.e.,

$$(D\mathbf{u})_\rho \equiv \fint\!\!\fint_{Q_\rho(\mu)} D\mathbf{u}\, dx d\tau.$$

LEMMA 10.1. *There exists positive constants γ, a, b that can be determined a priori only in terms of N and p, such that for all $\sigma \in (\frac{1}{2}, 1)$,*

$$\fint\!\!\fint_{Q_{\sigma R}(\mu)} |D\mathbf{u} - (D\mathbf{u})_{\sigma R}|^2 dx d\tau \le \gamma\, \mu^2 \left\{ \frac{\nu^a}{(1-\sigma)^b} + (1-\sigma) \right\}. \tag{10.1}$$

PROOF: By Lemma 9.2

$$\fint\!\!\fint_{Q_{\sigma R}(\mu)} |D\mathbf{u} - (D\mathbf{u})_{\sigma R}|^2 dx d\tau \le \frac{\gamma\, \mu^2 \nu^{1/(N+1)}}{(1-\sigma)^{2N/(N+1)}} + \fint\!\!\fint_{Q_{\sigma R}(\mu)} |(D\mathbf{u})_{\sigma R} - (D\mathbf{u})_{\sigma R}(\tau)|^2 dx d\tau$$

and

$$\fint\!\!\fint_{Q_{\sigma R}(\mu)} |(D\mathbf{u})_{\sigma R} - (D\mathbf{u})_{\sigma R}(\tau)|^2 dx d\tau \le \sup_{-\mu^{2-p}(\sigma R)^2 \le t, s \le 0} \left| \fint_{K_{\sigma R}} \Big(D\mathbf{u}(x,t) - D\mathbf{u}(x,s) \Big) dx \right|^2. \tag{10.2}$$

Let $\tilde\sigma = (1+\sigma)/2$ and denote with $x \to \tilde\zeta(x)$ a non-negative smooth cutoff function in $K_{\tilde\sigma R}$ that equals one on $K_{\sigma R}$ and such that

$$|D\tilde{\zeta}| \le \frac{2}{(1-\tilde{\sigma})R} \equiv \frac{4}{(1-\sigma)R}, \quad |D^2\tilde{\zeta}| \le \frac{16}{(1-\sigma)R}.$$

Write

$$\int_{K_{\sigma R}} \Big(\mathbf{Du}(x,t) - \mathbf{Du}(x,s)\Big)\, dx = \int_{\tilde{\sigma} K_R} \Big(\mathbf{Du}(x,t) - \mathbf{Du}(x,\tau)\Big)\tilde{\zeta}^2\, dx - \int_{K_{\tilde{\sigma}R}\backslash K_{\sigma R}} \Big(\mathbf{Du}(x,t) - \mathbf{Du}(x,s)\Big)\tilde{\zeta}^2 dx.$$

The last integral is estimated above by $\gamma(1-\sigma)\mu R^N$. To estimate the first integral we integrate the differentiated system (5.2) over (τ, t), multiply by $\tilde{\zeta}$ and integrate over $K_{\tilde{\sigma}R}$. This gives

$$\left| \int_{K_{\tilde{\sigma}R}} \Big(u_{i,x_j}(t) - u_{i,x_j}(s)\Big)\tilde{\zeta}^2\, dx \right| = \left| \int_s^t \int_{K_{\tilde{\sigma}R}} \tilde{\zeta}^2 \operatorname{div}\left(v^{p-2} Du_{i,x_j} + v_{x_j}^{p-2} Du_i\right)\, dxds \right|. \tag{10.3}$$

The case $p > 2$

The right hand side of (10.3) is estimated by

$$\gamma \iint_{Q_{\tilde{\sigma}R}(\mu)} |\mathbf{Du}|^{p-2}|D^2\mathbf{u}||D\tilde{\zeta}|dxd\tau \le \frac{\gamma\,\mu^{\frac{p-2}{2}}}{(1-\sigma)R} \iint_{Q_{\tilde{\sigma}R}(\mu)} |\mathbf{Du}|^{\frac{p-2}{2}}|D^2\mathbf{u}|dxd\tau.$$

To estimate the last integral write

$$\begin{aligned}
\iint_{Q_{\tilde{\sigma}R}(\mu)} |\mathbf{Du}|^{\frac{p-2}{2}}|D^2\mathbf{u}|dxd\tau &= \iint_{A^{\nu}_{\tilde{\sigma}R}} |\mathbf{Du}|^{\frac{p-2}{2}}|D^2\mathbf{u}|dxd\tau \\
&\quad + \iint_{B^{\nu}_{\tilde{\sigma}R}} |\mathbf{Du}|^{\frac{p-2}{2}}|D^2\mathbf{u}|dxd\tau \\
&\le |Q_R(\mu)|^{\frac{1}{2}} \left(\iint_{A^{\nu}_{\tilde{\sigma}R}} |\mathbf{Du}|^{p-2}|D^2\mathbf{u}|^2 dxd\tau \right)^{\frac{1}{2}} \\
&\quad + |B^{\nu}_R|^{\frac{1}{2}} \left(\iint_{Q_{\tilde{\sigma}R}(\mu)} |\mathbf{Du}|^{p-2}|D^2\mathbf{u}|^2 dxd\tau \right)^{\frac{1}{2}}.
\end{aligned}$$

The first integral is estimated by Lemma 9.1 and the second term is estimated by the '*smallness*' condition (9.3) and Lemma 7.1. Combining these estimates in (10.2) proves the lemma.

The case $1<p<2$

Estimate the right hand side of (10.3) as follows.

$$\left|\int_{K_{\tilde\sigma R}}\Big(u_{i,x_j}(t)-u_{i,x_j}(s)\Big)\tilde\zeta^2\,dx\right|$$

$$=\left|\int_s^t\int_{K_{\tilde\sigma R}} D\tilde\zeta^2\left\{|D\mathbf{u}|^{p-2}D\mathbf{u}-|(D\mathbf{u})_{\tilde\sigma R}(s)|^{p-2}(D\mathbf{u})_{\tilde\sigma R}(s)\right\}_{x_j}dxd\tau\right|$$

$$\le\int_s^t\int_{K_{\tilde\sigma R}}|D^2\tilde\zeta|\,\left||D\mathbf{u}|^{p-2}D\mathbf{u}-|(D\mathbf{u})_{\tilde\sigma R}(s)|^{p-2}(D\mathbf{u})_{\tilde\sigma R}(s)\right|dxd\tau.$$

By the structure of the cutoff function $\tilde\zeta$ and (5.5) of Lemma 5.2, this is majorised by

$$\frac{4}{(1-\sigma)^2R^2}\int_s^t\int_{K_{\tilde\sigma R}}\left|Du-(Du)_{\tilde\sigma R}(s)\right|^{p-1}dxds$$

$$\le\gamma\frac{|Q_R(\mu)|}{(1-\sigma)^2R^2}\left(\fint\!\!\fint_{Q_{\tilde\sigma R}(\mu)}\left|D\mathbf{u}-(D\mathbf{u})_{\tilde\sigma R}(\tau)\right|^2dxd\tau\right)^{\frac{p-1}{2}}.$$

We estimate the last integral by Lemma 9.2 and combine it with (10.3) to prove the lemma.

11. Proof of Proposition 1.1 concluded

LEMMA 11.1. *Let $\varepsilon\in(0,1)$ be the number claimed by Lemma 8.1. There exists a number $\nu\in(0,\frac14)$ such that if (9.3) holds, then*

$$\fint\!\!\fint_{Q_R(\mu)}\left|D\mathbf{u}-(D\mathbf{u})_R\right|^2dxd\tau\le\varepsilon\mu^2,\tag{11.1}$$

$$2^{-1}\mu\le\left|(D\mathbf{u})_R\right|\le\mu.\tag{11.2}$$

PROOF: Write

$$\iint\limits_{Q_R(\mu)} \left|\mathbf{Du}-(\mathbf{Du})_R\right|^2 dxd\tau = \sigma^{N+2} \iint\limits_{Q_{\sigma R}(\mu)} \left|\mathbf{Du}-(\mathbf{Du})_{\sigma R}\right|^2 dxd\tau$$
$$+ \left|Q_R(\mu)\right|^{-1} \iint\limits_{Q_R(\mu)\setminus Q_{\sigma R}(\mu)} \left|\mathbf{Du}-(\mathbf{Du})_R\right|^2 dxd\tau$$
$$+ \sigma^{N+2} \iint\limits_{Q_{\sigma R}(\mu)} \left|(\mathbf{Du})_R-(\mathbf{Du})_{\sigma R}\right|^2 dxd\tau.$$

The first integral is estimated by Lemma 10.1 and the second is bounded above by $\gamma(1-\sigma)\mu^2$. To estimate the last integral write

$$(\mathbf{Du})_R-(\mathbf{Du})_{\sigma R} = \left|Q_{\sigma R}(\mu)\right|^{-1} \left\{ \sigma^{N+2} \iint\limits_{Q_R(\mu)} \mathbf{Du} dxd\tau - \iint\limits_{Q_{\sigma R}(\mu)} \mathbf{Du} dxd\tau \right\}$$
$$= \left|Q_{\sigma R}(\mu)\right|^{-1} \left\{ (\sigma^{N+2}-1) \iint\limits_{Q_R(\mu)} \mathbf{Du} dxd\tau + \iint\limits_{Q_R(\mu)\setminus Q_{\sigma R}(\mu)} \mathbf{Du} dxd\tau \right\}.$$

This implies that

$$\left|(\mathbf{Du})_R-(\mathbf{Du})_{\sigma R}\right|^2 \le \gamma(1-\sigma)\mu^2$$

and

$$\iint\limits_{Q_R(\mu)} \left|\mathbf{Du}-(\mathbf{Du})_R\right|^2 dxd\tau \le \gamma\,\mu^2 \left\{ \frac{\nu^a}{(1-\sigma)^b} + (1-\sigma) \right\}.$$

To prove (11.1) choose σ so close to one that $\gamma(1-\sigma) \le \frac{1}{2}\varepsilon$, and then ν so small that $\gamma\nu^a(1-\sigma)^{-b} \le \frac{1}{2}\varepsilon$. To prove (11.2) we first observe that the definitions (9.1)–(9.2) imply

$$\left|A_R^\nu\right| = \left|Q_R(\mu)\setminus B_R^\nu\right| \ge (1-\nu)\left|Q_R(\mu)\right|.$$

Then by the *'smallness'* assumption (9.3)

$$\iint\limits_{Q_R(\mu)} |\mathbf{Du}|^2 dxd\tau \ge \iint\limits_{A_R^\nu} |\mathbf{Du}|^2 dxd\tau \ge \mu^2(1-\nu)^3\left|Q_R(\mu)\right|.$$

Using now (11.1)

$$\iint\limits_{Q_R(\mu)} |\mathbf{Du}|^2 dxd\tau - \iint\limits_{Q_R(\mu)} (\mathbf{Du})_R^2 dxd\tau = \iint\limits_{Q_R(\mu)} \left|\mathbf{Du}-(\mathbf{Du})_R\right|^2 dxd\tau$$
$$\le \varepsilon\mu^2.$$

From this

$$\left|(D\mathbf{u})_R\right|^2 \geq \iint\limits_{Q_R(\mu)} |D\mathbf{u}|^2 dx d\tau - \varepsilon\mu^2 \geq \left\{(1-\nu)^3 - \varepsilon\right\}\mu^2.$$

Proof of Proposition 1.1: Let $\varepsilon, \delta\, \kappa \in (0,1)$ be fixed as in Lemma 8.1. We start the iteration process of Lemma 8.2 with $\mathbf{V}_o \equiv (D\mathbf{u})_R$, and let $\{\mathbf{V}_i\}_{i\in\mathbf{N}}$ be the corresponding sequence of constant vectors in $\mathbf{R}^{N\times m}$ satisfying (8.7). It is apparent that, by an application of the triangle inequality, the vectors $\mathbf{V}_i$ can be replaced by $(D\mathbf{u})_i$, by possibly modifying the number κ.

12. Proof of Proposition 1.2-(i)

We assume that the *smallness* condition (9.3) does not hold, i.e.,

$$|A_R^\nu| < (1-\nu)\left|Q_R(\mu)\right|. \tag{12.1}$$

Lemma 12.1. *Let (12.1) hold. There exists some t_*,*

$$-\mu^{2-p}R^2 \leq t_* \leq -\frac{\nu}{2}\mu^{2-p}R^2, \tag{12.2}$$

such that

$$\operatorname{meas}\left\{x \in K_R \;\middle|\; v(x,t_*) > (1-\nu)\mu\right\} < \frac{1-\nu}{1-\nu/2}\left|K_R\right|, \quad v = |D\mathbf{u}|. \tag{12.3}$$

Proof: Indeed if not,

$$\begin{aligned}|A_R^\nu| &\geq \int_{-\mu^{2-p}R^2}^{-\mu^{2-p}(\nu/2)R^2} \operatorname{meas}\left\{x \in K_R \;\middle|\; v(x,\tau) > (1-\nu)\mu\right\} d\tau \\ &\geq (1-\nu)\left|Q_R(\mu)\right|,\end{aligned}$$

contradicting (12.1).

We will work with the function $w \equiv |D\mathbf{u}|^2$, which satisfies (1.8) within the cylinder $K_R \times (t_*, 0)$. Introduce the change of variables

$$\tau = -t/t_*, \quad \xi = x/R, \quad \tilde{w}(\xi,\tau) = w(R\xi, -t_*\tau)$$

and the convex function of $\tilde{w}$

$$z \equiv \max\left\{\frac{\tilde{w}}{\mu^2}\,;\,\frac{1}{2}\right\}.$$

Then $K_R \times (t_*, 0)$ is mapped into $Q_1 \equiv K_1 \times (-1,0)$ and, denoting again with (x,t) the transformed variables, z satisfies

$$z_t - \left(A_{\ell,k} z_{x_\ell}\right)_{x_k} \leq 0 \;\text{ in } Q_1 \quad \text{and} \quad 0 < z \leq 1, \tag{12.4}$$

where the matrix $(A_{\ell,k})$ is uniformly elliptic with eigenvalues bounded above and below independent of μ. Indeed it follows from (1.9) and the range (12.2) of t_* that

$$c_o(p)|\xi|^2 \le A_{\ell,k}\xi_\ell\xi_k \le C_o(p)|\xi|^2, \quad \forall \xi \in \mathbf{R}^N, \text{ a.e. } Q_1 \tag{12.5}$$

for two constants $c_o(N,p,\nu) \le C_o(N,p,\nu)$. The information of Lemma 12.1 in terms of z implies

$$\text{meas}\left\{x \in K_1 \mid z(x,-1) > (1-\nu)\right\} \le \frac{1-\nu}{1-\nu/2}|K_1|. \tag{12.6}$$

Without loss of generality we may assume that z satisfies (12.4) in a slightly larger box, say Q_2. This can be achieved by starting for example with $Q_{2R}(\mu)$. Proposition 1.2 is a consequence of the following:

THEOREM 12.1. *Let* $z \in C\left(-2,0;L^2(K_2)\right) \cap L^2\left(-2,0;W^{1,2}(K_2)\right)$ *be a subsolution of (12.4)-(12.5), and let (12.6) hold. There exists* $\eta = \eta(N,p,\nu) \in (0,1)$, *such that*

$$\text{meas}\left\{(x,t) \in Q_{\frac{1}{2}} \mid z(x,t) > (1-\eta)\right\} = 0.$$

In view of (12.4), the proof of the theorem uses techniques typical of a *single* equation. Even though these methods have been presented in various forms in Chapters II and III, we reproduce here the main points, to render the theory self-contained.

12-(i). Some energy estimates for z

LEMMA 12.2. *Let* $0 < \eta_o < \nu$ *and consider the function*

$$\Psi(z) \equiv \ln^+\left\{\frac{\nu}{\nu - (z-(1-\nu))_+ + \eta_o}\right\}.$$

There exists a constant $\gamma = \gamma(N,p,\nu)$ *such that for all* $t \in (-1,0)$ *and for all* $0 < \sigma < 1$,

$$\int_{K_\sigma \times \{t\}} \Psi^2(z)\,dx \le \int_{K_1 \times \{-1\}} \Psi^2(z)\,dx + \frac{\gamma}{(1-\sigma)^2} \iint_{Q_1} \Psi(z)\,dx d\tau. \tag{12.7}$$

PROOF: Let $x \to \zeta(x)$ be a cutoff function in K_1 that equals one on K_σ, and in the weak formulation of (12.4) take the testing function $\Psi\Psi'\zeta^2$, modulo a Steklov averaging process. Then (12.7) follows by estimates analogous to those in Proposition 3.2 of Chap. II.

LEMMA 12.3. *For* $0<\rho\le 1$ *and* $0<\sigma<1$ *let* $(x,t)\to\zeta(x,t)$ *be a cutoff function in* Q_ρ *that equals one on* $Q_{\sigma\rho}$ *and vanishes on the parabolic boundary of* ρ. *There exists a constant* $\gamma=\gamma(N,p,\nu)$ *such that for all* $k\ge 1/2$

$$\| (z-k)_+ \zeta\|^2_{V^{2,2}(Q_\rho)} \le \frac{\gamma}{(1-\sigma)^2\rho^2}\| (z-k)_+ \|^2_{2,Q_\rho}. \tag{12. 8}$$

The proof of (12.8) is analogous to the proof of the energy estimates of Proposition 3.1 of Chap. II. The spaces $V^{m,p}(Q_\rho)$ for $m,p\ge 1$ are introduced in §3 of Chap. I.

13. Proof of Proposition 1.2 concluded

LEMMA 13.1. *There exists a constant* $\eta_o\in(0,\nu)$ *depending only upon* N,p,ν *such that for all* $t\in(-1,0)$

$$\text{meas}\left\{x\in K_1 \mid z(x,t)>(1-\eta_o)\right\} < \left(1-\nu^2/4\right)|K_1|. \tag{13.1}$$

PROOF: We will use the logarithmic inequality of Lemma 12.2. Since $\Psi(z)$ vanishes on the set $[z<(1-\nu)]$, by virtue of (12.6), the first term on the right hand side of (12.7) is majorised by

$$\frac{1-\nu}{1-\nu/2}\ln^2\left(\frac{\nu}{\eta_o}\right)|K_1|.$$

The second term is majorised by

$$\frac{\gamma}{(1-\sigma)^2}|K_1|\ln\left(\frac{\nu}{\eta_o}\right).$$

We estimate below the right hand side by extending the integration to the smaller set $[z(\cdot,t)>(1-\eta_o)]$. On such a set

$$\Psi(z)\ge \ln(\nu/2\eta_o).$$

Combining these estimates in (12.7) gives

$$\begin{aligned}\text{meas}\left\{x\in K_\sigma \mid z(x,t)>(1-\eta_o)\right\} &< \frac{1-\nu}{1-\nu/2}|K_1|\left\{\frac{\ln(\nu/\eta_o)}{\ln(\nu/2\eta_o)}\right\}^2\\ &\quad+\frac{\gamma}{(1-\sigma)^2}|K_1|\frac{\ln(\nu/\eta_o)}{\ln^2(\nu/2\eta_o)}.\end{aligned}$$

Also

$$\begin{aligned}
\text{meas}\left\{x\in K_1 \mid z(x,t)>(1-\eta_o)\right\}&\\
&\le \text{meas}\left\{x\in K_\sigma \mid z(x,t)>(1-\eta_o)\right\}+(1-\sigma)|K_1|\\
&\le \frac{1-\nu}{1-\nu/2}|K_1|\frac{\ln^2(\nu/\eta_o)}{\ln^2(\nu/2\eta_o)}\\
&+\frac{\gamma}{(1-\sigma)^2}|K_1|\frac{\ln(\nu/\eta_o)}{\ln^2(\nu/2\eta_o)}+(1-\sigma)|K_1|.
\end{aligned}$$

Choose σ so that $(1-\sigma)\le \nu^2/8$ and then η_o so that

$$\frac{\gamma}{(1-\sigma)^2}\frac{\ln(\nu/\eta_o)}{\ln^2(\nu/2\eta_o)}\le\frac{\nu^2}{8}.$$

By choosing η_o even smaller if necessary, we may insure that

$$\frac{1-\nu}{1-\nu/2}\frac{\ln^2(\nu/\eta_o)}{\ln^2(\nu/2\eta_o)}\le 1-\frac{\nu^2}{2}.$$

Having determined η_o, let s_o be the largest positive integer such that $2^{-s_o}\ge\eta_o$. For $s\ge s_o$, set

$$A_s(t)\equiv\left\{x\in K_1 \mid z(x,t)>(1-2^{-s})\right\},\qquad A_s\equiv\int_{-1}^{0}\left|A_s(\tau)\right|d\tau.$$

Then Lemma 13.1 implies that

$$\left|K_1\backslash A_s(t)\right|\ge(\nu/2)^2|K_1|,\qquad \forall t\in(-1,0). \tag{13.2}$$

LEMMA 13.2. *For every $\nu_*\in(0,1)$ there exists a positive integer $s_*>s_o$ such that*

$$A_{s_*}\le\nu_*|Q_1|. \tag{13.3}$$

PROOF: Apply Lemma 2.2 of Chap. I to the functions $x\to z(x,t)$ for $t\in(-1,0)$, and for the levels

$$\ell=1-2^{-(s+1)},\qquad k=1-2^{-s},\quad s\ge s_o.$$

Taking into account (13.2), we obtain

$$\begin{aligned}
2^{-s}\left|A_{s+1}\right|&\le\frac{\gamma}{\left|K_1\backslash A_s(t)\right|}\int_{A_s(t)\backslash A_{s+1}(t)}|Dz|\,dx\\
&\le\gamma(N,p,\nu)\left(\int_{K_1}\left|D\left(z-(1-2^{-s})\right)_+\right|^2dx\right)^{\frac12}\\
&\quad\times\left(\left|A_s(t)\right|-\left|A_{s+1}(t)\right|\right)^{\frac12}.
\end{aligned}$$

We square both sides of this inequality, integrate in dt over $(-1,0)$ and estimate the resulting integral on the right hand side by the energy inequalities (12.8) written over the pair of cylinders Q_1 and Q_2. This gives

$$4^{-s}A_{s+1}^2 \leq \gamma 4^{-s}\left(A_s - A_{s+1}\right).$$

Divide through by 4^{-s} and add these inequalities for $s=s_o, s_o+1, \dots, s_*-1$ to obtain

$$(s_* - s_o - 1)A_{s_*} \leq \gamma \sum_{s=s_o}^{s_*} (A_s - A_{s+1}) \leq \gamma|Q_1|.$$

Therefore

$$A_{s_*} \leq \frac{\gamma}{(s_* - s_o - 1)}|Q_1|.$$

PROOF OF THEOREM 12.1: Consider the family of nested boxes

$$Q_n \equiv K_{\rho_n} \times \{-\rho_n^2, 0\}, \quad \rho_n \equiv \frac{1}{2} + \frac{1}{2^{n+1}}, \quad n = 0,1,\dots,$$

and the increasing levels

$$k_n \equiv 1 - 2^{-s_*-1} - 2^{-s_*-1-n}, \qquad n = 0,1,\dots,$$

and set

$$Y_n \equiv \text{meas}\left\{(x,t) \in Q_n \mid z(x,t) > k_n\right\}.$$

Write the energy inequality (12.8) over the boxes Q_n for the functions $(z-k_n)_+$, where ζ is the standard cutoff function in Q_n that equals one on Q_{n+1}. By the embedding Proposition 3.1 of Chap. I, with $m=p=2$,

$$\begin{aligned}
\iint_{Q_{n+1}} (z-k_n)_+^2\, dx d\tau &\leq \iint_{Q_n} \left[(z-k_n)_+\zeta\right]^2 dx d\tau \\
&\quad\times \left(\iint_{Q_n} \left[(z-k_n)_+\zeta\right]^{\frac{2N}{N+2}} dx d\tau\right)^{\frac{N}{N+2}} Y_n^{\frac{2}{N+2}} \\
&\leq \|(z-k_n)_+\zeta\|_{V^{2,2}(Q_n)}^2 Y_n^{\frac{2}{N+2}} \\
&\leq 4^{s_*} Y_n^{1+\frac{2}{N+2}}.
\end{aligned}$$

On the other hand

$$Y_{n+1} \leq \gamma 4^{n+s_*} \iint_{Q_{n+1}} (z-k_n)_+^2\, dx d\tau.$$

Therefore

$$Y_{n+1} \leq \gamma 4^n Y_n^{1+\frac{2}{N+2}}, \qquad n = 0,1,2,\dots.$$

It follows from Lemma 4.1 of Chap. I that $\{Y_n\} \to 0$ as $n \to \infty$ provided

$$(13.4) \qquad Y_o \equiv A_{s_*} \leq \gamma^{-\frac{N+2}{2}} 4^{-\left(\frac{N+2}{2}\right)^2} \equiv \nu_* .$$

To prove the theorem we have only to pick s_* by the procedure of Lemma 13.2 so that (13.4) is satisfied and then set $\eta = 2^{-(s_*+1)}$.

14. General structures

Consider the general non-linear system (1.10) of Chap. VIII subject to the structure conditions $(\mathcal{S}_1)$-$(\mathcal{S}_6)$. The proof of Propositions 1.1 and 1.2 for these systems is analogous to that in §§6-11. The corresponding *'linear'* system about a point $(x_o, t_o) \in \Omega_T$ is

$$\frac{\partial}{\partial t} u_i - \operatorname{div}\left(\frac{\partial}{\partial u_{k,x_\ell}} \mathbf{A}^{(i)}(x_o, t_o, \mathbf{V})\, u_{k,x_\ell}\right) = 0.$$

For this, the *linear* analysis of §6 can be carried with minor changes. The analog of the 'algebraic' lemmas of §5 are a direct consequence of the structure conditions $(\mathcal{S}_1)$-$(\mathcal{S}_6)$. In the proof of Propositions 1.1 and 1.2, when working within cylinders $[(x_o, t_o) + Q(\theta, \rho)]$, the *'perturbation terms'* φ_i, $i = 0, 1, 2$, contribute terms that are infinitesimal with ρ of higher order with respect to those generated by the principal part. This is due to the integrability condition $(\mathcal{S}_6)$. Further details on the estimation of the lower order terms can be carried out with an analysis similar to the estimation of the lower order terms in Chaps. II–V.

15. Bibliographical notes

The content of this Chapter is essentially taken from [36,37]. The estimation of the oscillation of $D\mathbf{u}$ in §§2 and 3 builds on [37] but it is essentially new. The algebraic Lemmas of §5 are scattered in the literature mainly without proofs. We have attempted to rephrase them in the context of $p-$systems. The theory of *linear* parabolic systems of §6 is taken from Campanato [23]. The rest of the Chapter follows [36,37].

X

Parabolic p-systems: boundary regularity

1. Introduction

We will establish everywhere regularity up the boundary for weak solutions of the parabolic system

$$
(1.1) \qquad \begin{cases} \mathbf{u} \equiv (u_1, u_2, \ldots, u_m)\,, \; m \in \mathbf{N}, \\ u_i \in C\left(\varepsilon, T; L^2(\Omega)\right) \cap L^p\left(\varepsilon, T; W^{1,p}(\Omega)\right), \\ u_{i,t} - \operatorname{div} |D\mathbf{u}|^{p-2} Du_i = B_i(x, t, \mathbf{u}, D\mathbf{u}), \text{ in } \Omega \times (\varepsilon, T), \\ \varepsilon \in (0, T), \; i = 1, 2, \ldots, m, \; p > \max\{1; \frac{2N}{N+2}\}, \end{cases}
$$

associated with Dirichlet boundary data

$$
(1.2) \qquad u_i(\cdot, t) = g_i(\cdot, t) \quad \text{on } \partial\Omega \times (\varepsilon, T),
$$

in the sense of the traces on $\partial\Omega$, of functions in $W^{1,p}(\Omega)$. The basic assumptions on $\partial\Omega$, the boundary data $\mathbf{g}$ and the forcing term $\mathbf{B}$

$$
\mathbf{g} \equiv (g_1, g_2, \ldots, g_m)\,, \qquad \mathbf{B} \equiv (B_1, B_2, \ldots, B_m)\,,
$$

are the following:

($\mathbf{A}_1$) $\partial\Omega$ is of class $C^{1,\lambda}$ for some $\lambda \in (0,1)$, in the sense of (1.2) of Chap. I. Thus the norm $\|\partial\Omega\|_{1+\lambda}$ is finite.

($\mathbf{A}_2$) The functions g_i, $i = 1, 2, \ldots, m$, are restrictions to $\partial\Omega$ of functions $\tilde{g}_i$, defined in the whole Ω_T, and satisfying

$$\tilde{g}_{i,x_j} \in C^\lambda(\overline{\Omega}_T), \quad g_{i,t} \in L^\infty(\Omega_T), \qquad i = 1,2,\ldots,m\,, \quad j = 1,2,\ldots,N. \tag{1.3}$$

We set[1]

$$\|\mathbf{g}\| \equiv \sum_{i=1}^{m}\sum_{j=1}^{N}\left\{\|\tilde{g}_i\|_{\infty,\Omega_T} + \|\tilde{g}_{i,t}\|_{\infty,\Omega_T} + \left[\tilde{g}_{i,x_j}\right]_{\lambda,\Omega_T}\right\}. \tag{1.4}$$

$$\left|\mathbf{B}(x,t,\mathbf{u},D\mathbf{u})\right| \le B_o\left(1+|D\mathbf{u}|^{p-1}\right), \quad \text{a.e. in } \Omega_T, \tag{$\mathbf{A}_3$}$$

for some given constant B_o. We say that a constant $\gamma = \gamma\,(\text{data})$ depends only upon the data if it can be determined a priori only in terms of

$$(\text{data}) \equiv (N, p, B_o, \|\partial\Omega\|_{1+\lambda}, \|\mathbf{g}\|)\,.$$

THEOREM 1.1. *Let* $\mathbf{u}$ *be a weak solution of (1.1)-(1.2) in* $\Omega\times(\varepsilon,T)$*, and let* $(\mathbf{A}_1)$-$(\mathbf{A}_3)$ *hold. Then*

$$u_i \in C^{1-\alpha}\left(\overline{\Omega}\times(\varepsilon,T]\right), \ \textit{for every } \alpha\in(0,1), \ i = 1,2,\ldots,m.$$

Moreover for every $\alpha\in(0,1)$ *and every* $\varepsilon\in(0,T)$*, there exists a constant*

$$\gamma = \gamma\left(\alpha, \varepsilon, \|D\mathbf{u}\|_{p,\Omega\times(\varepsilon,T)}, \textit{data}\right),$$

such that

$$[u_i]_{(1-\alpha),\,\overline{\Omega}\times[\varepsilon,T]} \le \gamma. \tag{1.5}$$

The constant γ *tends to infinity as either* $\varepsilon\searrow 0$ *or as* $\alpha\searrow 0$.

Remark 1.1. The constant γ is '*stable*' as $p\to 2$.

THEOREM 1.2 (HOMOGENEOUS BOUNDARY DATA). *Let* $\mathbf{u}$ *be a weak solution of (1.1)–(1.2) with* $\mathbf{g}\equiv 0$ *and let* $(\mathbf{A}_1)-(\mathbf{A}_2)$ *hold. For every* $\varepsilon\in(0,T)$ *there exist constants*

$$\gamma = \gamma\left(\varepsilon, \|D\mathbf{u}\|_{p,\Omega\times(\varepsilon,T)}, \textit{data}\right) > 1 \quad \textit{and} \quad \alpha = \alpha(\textit{data}) \in (0,1)$$

such that

$$\left[u_{i,x_j}\right]_{\alpha,\overline{\Omega}\times[\varepsilon,T]} \le \gamma, \quad i = 1,2,\ldots,m, \ j = 1,2,\ldots,N.$$

The constant $\gamma\nearrow\infty$ *as* $\varepsilon\searrow 0$.

We will only carry the proof of Theorem 1.1. The proof of Theorem 1.2 follows exactly the same arguments, where in the various estimates the contributions coming from $\|\mathbf{g}\|$ are discarded.

---

(1) For a smooth function ϕ, the norm $[\phi]_{\lambda,\mathcal{K}}$ is defined in (1.3) of Chap. I.

2. Flattening the boundary

Let $\varepsilon \in (0,T)$ be fixed. We will estimate the oscillation of u_i about each point $(x_o,t_o) \in \partial\Omega \times (\varepsilon, T)$. For this we first introduce a change of coordinates that maps a small portion of $\partial\Omega$ about (x_o,t_o) into a portion of an hyperplane. After a translation we may assume that (x_o,t_o) coincides with the origin. We will work within the cylinder

$$\mathcal{Q}_{\mathcal{R}} \equiv K_{\mathcal{R}} \times \{-\mathcal{R}, 0\}, \quad 2\mathcal{R} = \min\{\rho_o\,;\,\varepsilon\},$$

where ρ_o is the number that determines the structure of $\partial\Omega$ as in (1.2) of Chap. I. The portion of the boundary $\partial\Omega \cap K_{\mathcal{R}}$ is represented by

$$x_N = \Phi(\bar{x}), \quad \bar{x} \equiv (x_1, x_2, \ldots, x_{N-1}),$$

where Φ is a function of class $C^{1,\lambda}$ in the $(N-1)$–dimensional ball $\mathcal{B}_{\mathcal{R}}$, satisfying

$$|D\Phi(0)| = 0 \quad \text{and} \quad \|D\Phi\|_{\infty,\mathcal{B}_{\mathcal{R}}} \le 1/2. \tag{2.1}$$

The last condition can be realised by taking a smaller ρ_o if necessary. With respect to the new variables

$$\tilde{x}_i = x_i, \quad i = 1, 2, \ldots, N-1; \qquad \tilde{x}_N = x_N - \Phi(\bar{x}),$$

the portion $\partial\Omega \cap K_{\mathcal{R}}$ coincides with the portion of the hyperplane $\tilde{x}_N = 0$ within $K_{\mathcal{R}}$. We orient $\tilde{x}_N$ so that, say, $\Omega \cap K_{\mathcal{R}} \subset \{\tilde{x}_N > 0\}$ and set

$$\mathcal{Q}^+_{\mathcal{R}} \equiv \mathcal{Q}_{\mathcal{R}} \cap \{\tilde{x}_N > 0\}.$$

Denoting again by x the transformed variables $\tilde{x}$ and with u_i, B_i, Φ, etc., the transformed functions, the system (1.1) takes the form

$$\frac{\partial}{\partial t} u_i - \left(a_{\ell,k}(x, D\mathbf{u})\, u_{i,x_k}\right)_{x_\ell} = B_i(x,t,\mathbf{u},D\mathbf{u}), \ \text{ in } \mathcal{Q}^+_{\mathcal{R}}, \tag{2.2}$$

$$(a_{\ell,k}(x,D\mathbf{u})) \equiv \mathbf{A}(x) \Big| |D\mathbf{u}|^2 + u^2_{x_N} |D\Phi|^2 - 2u_{x_N} \langle D\mathbf{u}, D\Phi\rangle \Big|^{\frac{p-2}{2}}, \tag{2.3}$$

$$\mathbf{A}(x) \equiv \begin{pmatrix} \mathbf{I}_{N-1} & -D\Phi(x) \\ -D\Phi(x) & (1 + |D\Phi|^2(x)) \end{pmatrix}, \tag{2.4}$$

where $\mathbf{I}_{N-1}$ is the $(N-1)\times(N-1)$ identity matrix. To reduce (2.2) to a system with homogeneous boundary data on $\mathcal{Q}_{\mathcal{R}} \cap \{x_N = 0\}$ set

$$w_i \equiv u_i - \tilde{g}_i, \quad i = 1, 2, \ldots, m,$$

and rewrite (2.2) in the form

$$\frac{\partial}{\partial t} w_i - \operatorname{div} \mathbf{A}_i(x,t,D\mathbf{w}) = B_i + \frac{\partial}{\partial x_\ell} f_{i,\ell}, \quad \text{in } \mathcal{Q}^+_{\mathcal{R}}, \tag{2.5}$$

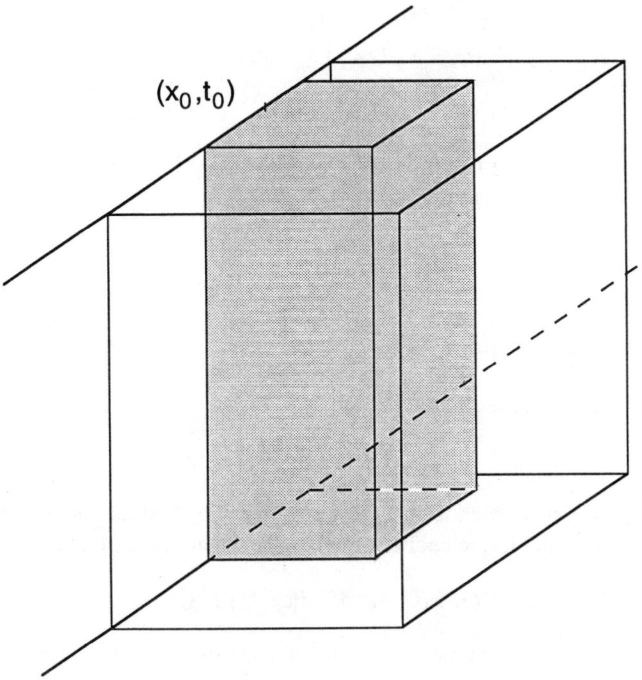

Figure 2.1

(2.6) $$A_{i,\ell}\left(x,t,D\mathbf{w}\right)=a_{\ell,k}\left(x,D\mathbf{w}+D\tilde{\mathbf{g}}\right)w_{i,x_k},$$

(2.7) $$B_i=B_i\left(x,t,\mathbf{w}+\mathbf{g},D\mathbf{w}+D\tilde{\mathbf{g}}\right)-\frac{\partial}{\partial t}\tilde{g}_i,$$

(2.8) $$f_{i,\ell}=a_{\ell,k}\left(x,D\mathbf{w}+D\tilde{\mathbf{g}}\right)\tilde{g}_{i,x_k}.$$

Using the assumptions $(\mathbf{A}_1)$-$(\mathbf{A}_3)$ we find the following structure conditions and regularity properties on the various terms of (2.5):

(2.9) $$\begin{cases} A_{i,\ell}(x,t,D\mathbf{w})w_{i,x_\ell}\geq\gamma_o\left|D\mathbf{w}+D\tilde{\mathbf{g}}\right|^{p-2}\left|D\mathbf{w}\right|^2\\ A_{i,\ell}(x,t,D\mathbf{w})w_{i,x_\ell}\leq\gamma_1\left|D\mathbf{w}+D\tilde{\mathbf{g}}\right|^{p-2}\left|D\mathbf{w}\right|^2,\end{cases}$$

for two positive constants $\gamma_o\leq\gamma_1$ depending only upon the data. Moreover for all $i=1,2,\ldots,m$ and $k=1,2,\ldots,N$,

(2.10) $$\left|A_{i,k}(x,t,\xi)-A_{i,k}(y,\tau,\xi)\right|\leq\gamma\left(1+|\xi|^{p-1}\right)\left(|x-y|+|t-\tau|\right)^{\lambda}$$
$$\forall\xi\in\mathbf{R}^{N\times m},\ \text{ and for a.e. } (x,t),(y,\tau)\in Q_{\mathcal{R}}^{+},$$

$$|B_i(x,t,D\mathbf{w})| \le \gamma\left(1+|D\mathbf{w}|^{p-1}\right), \tag{2.11}$$

$$|f_{i,\ell}(x,t,\mathbf{w},D\mathbf{w})| \le \gamma|D\mathbf{u}|^{p-2}. \tag{2.12}$$

From (2.1) and the definitions (2.3)-(2.4) and (2.6), it follows that

$$A_{i,\ell}(0,0,\xi) = |\xi+\mathbf{b}|^{p-2}\xi_{i,k}\,\delta_{k,\ell}, \quad \forall\xi\in\mathbf{R}^{N\times m}, \tag{2.13}$$

$$\mathbf{b}\equiv(D\tilde{\mathbf{g}})(0,0). \tag{2.14}$$

2-(i). Comparison functions

Consider cylindrical domains of the type

$$\left[(x_o,t_o)+Q\left(R^{2+\eta},R\right)\right],$$

where $\eta\in(-1,1)$ is to be chosen, $x_o\in K_{\mathcal{R}}\cap\{x_N=0\}$ and the faces of the cubes $[x_o+K_R]$ are parallel to the coordinate axes. These boxes are contained in $\mathcal{Q}_{\mathcal{R}}$ if

$$0<R<\tfrac{1}{2}\mathcal{R} \quad\text{and}\quad (x_o,t_o)\in\mathcal{Q}_{\frac{1}{2}\mathcal{R}}, \tag{2.15}$$

which from now on we assume. The proof of Theorem 1.1 is based on comparing $\mathbf{w}$ in a neighborhood of each point $(x_o,t_o)\in\mathcal{Q}_{\mathcal{R}/2}\cap\{x_N\ge0\}$, with the solution of

$$\begin{cases} \mathbf{v}\equiv(v_1,v_2,\dots,v_m),\ m\in\mathbf{N}, \\ v_{i,t}-\operatorname{div}|D\mathbf{v}|^{p-2}Dv_i=0, \text{ in } \left[(x_o,t_o)+Q\left(R^{2+\eta},R\right)\right]\cap\mathcal{Q}^+_{\mathcal{R}}, \\ v_i=w_i \text{ on } \partial_p\left[(x_o,t_o)+Q\left(R^{2+\eta},R\right)\right]\cap\mathcal{Q}^+_{\mathcal{R}}, \end{cases} \tag{2.16}$$

where ∂_pQ denotes the parabolic boundary of a cylindrical domain Q. The existence of a unique weak solution of (2.16) can be established by a Galerkin procedure.(1) Denote by

$$(\bar{x},x_N),\quad \bar{x}\equiv(x_1,x_2,\dots,x_{N-1}),$$

the coordinates in $K_{\mathcal{R}}$. Then since $\mathbf{w}$ vanishes for $x_N=0$, we also have $\mathbf{v}(\bar{x},0,t)=0$. We let $\tilde{\mathbf{v}}$ and $\tilde{\mathbf{w}}$ denote the odd extensions of $\mathbf{v}$ and $\mathbf{w}$ in the cylinder $[(x_o,t_o)+Q(R^{2+\eta},R)]\cap\{x_N\le0\}$, i.e.,

$$\tilde{\mathbf{v}}\equiv\begin{cases} \mathbf{v}(\bar{x},x_N,t), & \text{in } \left[(x_o,t_o)+Q\left(R^{2+\eta},R\right)\right]\cap\{x_N\ge0\} \\ -\mathbf{v}(\bar{x},-x_N,t), & \text{in } \left[(x_o,t_o)+Q\left(R^{2+\eta},R\right)\right]\cap\{x_N\le0\}, \end{cases}$$

$$\tilde{\mathbf{w}}\equiv\begin{cases} \mathbf{w}(\bar{x},x_N,t), & \text{in } \left[(x_o,t_o)+Q\left(R^{2+\eta},R\right)\right]\cap\{x_N\ge0\} \\ -\mathbf{w}(\bar{x},-x_N,t), & \text{in } \left[(x_o,t_o)+Q\left(R^{2+\eta},R\right)\right]\cap\{x_N\le0\}. \end{cases}$$

Then, by the reflexion principle $\tilde{\mathbf{v}}$ is the unique solution of

(1) See, for example, [73].

$$(2.17)\qquad \begin{cases} \tilde{\mathbf{v}} \equiv (\tilde{v}_1, \tilde{v}_2, \dots, \tilde{v}_m)\,, \ m \in \mathbf{N}, \\ \tilde{v}_{i,t} - \operatorname{div} |D\tilde{\mathbf{v}}|^{p-2} D\tilde{v}_i = 0, \ \text{ in } \left[(x_o, t_o) + Q\left(R^{2+\eta}, R\right)\right], \\ \tilde{v}_i = \tilde{w}_i \ \text{ on } \partial_p \left[(x_o, t_o) + Q\left(R^{2+\eta}, R\right)\right]. \end{cases}$$

It follows from the *interior* estimates of Theorems 5.1 and 5.2′ of Chap. VIII that $|D\tilde{\mathbf{v}}|$ is bounded in the interior of $\left[(x_o,t_o)+Q\left(R^{2+\eta},R\right)\right]$ and it satisfies the sup-bounds (5.1) and (5.3). We restate these bounds for the special geometry of $\left[(x_o,t_o)+Q\left(R^{2+\eta},R\right)\right]$.

THEOREM 2.1 (THE DEGENERATE CASE $p>2$). *Let $\tilde{\mathbf{v}}$ be the weak solution of (2.17). There exists a constant $\gamma=\gamma(N,p)$ such that for all $0<\rho\le\frac{1}{2}R$*

$$(2.18)\qquad \sup_{[(x_o,t_o)+Q(\rho^{2+\eta},\rho)]} |D\tilde{\mathbf{v}}| \le \gamma \left(R^{\eta} \iint_{[(x_o,t_o)+Q(R^{2+\eta},R)]} |D\mathbf{v}|^p \, dx d\tau \right)^{1/2} + R^{-\frac{\eta}{p-2}}.$$

THEOREM 2.2 (THE SINGULAR CASE $\max\left\{1; \frac{2N}{N+2}\right\} < p < 2$). *Let $\tilde{\mathbf{v}}$ be a weak solution of (2.17). There exists a constant $\gamma=\gamma(N,p)$ such that for all $0<\rho\le\frac{1}{2}R$*

$$(2.19)\qquad \sup_{[(x_o,t_o)+Q(\rho^{2+\eta},\rho)]} |D\tilde{\mathbf{v}}| \le \gamma \left(R^{-N\eta} \iint_{[(x_o,t_o)+Q(R^{2+\eta},R)]} |D\mathbf{v}|^p \, dx d\tau \right)^{1/\nu_p} + R^{\frac{\eta}{2-p}},$$

where $\nu_p = N(p-2)+2p$.

Remark 2.1. Theorem 2.2 is a restatement of Theorem 5.2′ of Chap. VIII with $q=p$. By Remark 5.4 of the same Chapter, such a choice is admissible.

3. An iteration lemma

LEMMA 3.1. *Let $s\to\varphi(s)$ be a non-negative non-decreasing function defined in $[0,1]$ and satisfying*

$$(3.1)\qquad \varphi(\rho) \le A\left(\frac{\rho}{R}\right)^{\beta} \varphi(R) + \frac{1}{2} A\left(R^{\beta-\nu\kappa} + \rho^{\beta} R^{-\kappa}\right), \quad \forall 0<\rho\le R\le 1,$$

for given positive constants A, β, ν, κ satisfying in addition $\beta>\kappa$ and $\nu\in(0,1)$. Then for every

$$(3.2)\qquad 0 \le \delta < \kappa \left(\frac{(1-\nu)\kappa}{(1-\nu)\kappa+\beta} \right),$$

there exists a constant γ depending only upon A, β, ν and δ, such that

$$\varphi(\rho) \le \gamma\left(\frac{\rho}{R^q}\right)^{\beta-\kappa+\delta}(\varphi(R)+1) \tag{3.3}$$

$$\forall 0<\rho\le R\le 1, \quad \textit{where } q = 1+\frac{(1-\nu)\kappa}{\beta}.$$

PROOF: Choose $R_o < 1$ and define the sequence $R_{n+1} = R_n^q$, $n = 0, 1, 2, \dots$. Then

$$\begin{aligned}\varphi(R_{n+1}) &\le AR_n^{(1-\nu)\kappa}\varphi(R_n) + \frac{1}{2}A\left(R_{n+1}^{\frac{\beta-\nu\kappa}{q}} + R_{n+1}^{\beta-\kappa/q}\right)\\ &= AR_n^{(1-\nu)\kappa}\varphi(R_n) + AR_{n+1}^{\beta-\kappa/q}.\end{aligned}$$

Iteration of these inequalities gives

$$\begin{aligned}\varphi(R_{n+1}) \le{}& A^{n+1}\left(\prod_{j=0}^{n} R_{n-j}^{(1-\nu)\kappa}\right)\varphi(R_o)\\ &+\sum_{j=1}^{n} A^{j+1}R_{n+1-j}^{\beta-\kappa/q}\left(\prod_{i=0}^{j-1} R_{n-i}^{(1-\nu)\kappa}\right) + AR_{n+1}^{\beta-\kappa/q}.\end{aligned}$$

Now

$$\prod_{j=0}^{n} R_{n-j}^{(1-\nu)\kappa} = \left(\frac{R_{n+1}}{R_o}\right)^{\beta}$$

and

$$\begin{aligned}R_{n+1-j}^{\beta-\kappa/q}\left(\prod_{i=0}^{j-1} R_{n-i}^{(1-\nu)\kappa}\right) &= R_o^{(\beta-\kappa/q)q^{n+1-j}} R_o^{\beta\left(q^{n+1}-q^{n+1-j}\right)}\\ &\le R_o^{(\beta-\kappa/q)q^{n+1}} \equiv R_{n+1}^{\beta-\kappa/q}.\end{aligned}$$

Therefore if $A\ge 2$

$$\varphi(R_{n+1}) \le A^{n+1}\left(\frac{R_{n+1}}{R_o}\right)^{\beta}\varphi(R_o) + 2A^{n+1}R_{n+1}^{\beta-\kappa/q}.$$

Fix δ in the range (3.2) and set

$$\varepsilon = \kappa\left(\frac{(1-\nu)\kappa}{(1-\nu)\kappa+\beta}\right) - \delta; \quad \beta - \frac{\kappa}{q} = \beta-\kappa+\delta+\varepsilon.$$

Then

$$\varphi(R_{n+1}) \le \left(AR_o^{\varepsilon\frac{q^{n+1}}{n+1}}\right)^{n+1}\left(\frac{R_{n+1}}{R_o}\right)^{\beta-\kappa+\delta}(\varphi(R_o)+1). \tag{3.4}$$

The first coefficient in (3.3) is independent of n if n is so large that $AR_o^{\varepsilon\frac{q^{n+1}}{n+1}} \le 1$. Let n_o be the smallest integer satisfying

$$\frac{q^{n_o+1}}{n_o+1} \geq \frac{\ln A}{\left|\ln R_o^{\varepsilon}\right|}.$$

It follows from (3.3) that if $n \geq n_o$,

$$\varphi(R_{n+1}) \leq \left(\frac{R_{n+1}}{R_o}\right)^{\beta-\kappa+\delta} (\varphi(R_o)+1).$$

If $R_o < 1$ is fixed, for every $\rho \in (0, R_o]$ there exist some $n \in \mathbf{N}$ such that $R_o^{q^{n+1}} \leq \rho \leq R_o^{q^n}$. Therefore the equation $\rho = R_o^{\theta q^n}$ has a root $\theta \in [1, q]$. Starting the process with R_o replaced by R_o^{θ} gives

$$\begin{aligned}
\varphi(\rho) &\leq \left(\frac{\rho}{R_o^{\theta}}\right)^{\beta-\kappa-\delta} (\varphi(R_o^{\theta})+1) \\
&\leq \left(\frac{\rho}{R_o^{q}}\right)^{\beta-\kappa-\delta} (\varphi(R_o)+1).
\end{aligned}$$

Remark 3.1. The lemma continues to hold for $\nu = 0$. The constant γ on the right hand side of (3.3) is *'stable'* as $\nu \searrow 0$.

4. Comparing w and v (the case $p > 2$)

We start by comparing $\mathbf{w}$ solution of (2.5) with the solution $\mathbf{v}$ of (2.16). Having fixed $(x_o, t_o) \in \mathcal{Q}_{\frac{1}{2}R} \cap \{x_N = 0\}$, we may assume, after a translation, that it coincides with the origin. Setting

$$Q_R^{\eta} \equiv Q\left(R^{2+\eta}, R\right), \quad {}^{+}Q_R^{\eta} \equiv Q\left(R^{2+\eta}, R\right) \cap \{x_N > 0\},$$

the vectors $\mathbf{w}$ and $\mathbf{v}$ satisfy

$$\begin{aligned}
(4.1) \qquad & \frac{\partial}{\partial t}(v_i - w_i) - \operatorname{div}\left(|D\mathbf{v}|^{p-2} Dv_i - |D\mathbf{w}|^{p-2} Dw_i\right) \\
&= -\operatorname{div}\left(\mathbf{A}_i(x,t,D\mathbf{w}) - \mathbf{A}_i(0,0,D\mathbf{w})\right) \\
&\quad - \operatorname{div}\left(\mathbf{A}_i(0,0,D\mathbf{w}) - |D\mathbf{w}|^{p-2} Dw_i\right) \\
&\quad - B_i(x,t,D\mathbf{w}) - \frac{\partial}{\partial x_\ell} f_{i,\ell}(x,t,\mathbf{w},D\mathbf{w}), \text{ in } {}^{+}Q_R^{\eta},
\end{aligned}$$

$$(4.2) \qquad v_i - w_i = 0 \text{ on the parabolic boundary of } {}^{+}Q_R^{\eta}.$$

From (2.10) it follows that

$$\left|A_{i,\ell}(x,t,D\mathbf{w}) - A_{i,\ell}(0,0,D\mathbf{w})\right| \leq \gamma R^{\lambda}\left(1 + |D\mathbf{w}|^{p-1}\right),$$

for $i = 1, 2, \ldots, m$ and $\ell = 1, 2, \ldots, N$. Moreover from (2.13) and (2.14)

$$\left|\mathbf{A}_i(0,0,D\mathbf{w}) - \left|D\mathbf{w}\right|^{p-2} Dw_i\right| \le \gamma \left| |D\mathbf{w}+\mathbf{b}|^{p-2} - |D\mathbf{w}|^{p-2}\right| |D\mathbf{w}|$$
$$\le \gamma \left(1+|D\mathbf{w}|^{p-2}\right).$$

In the weak formulation of (4.1) take the testing function $v_i - w_i$ modulo a Steklov average, and add over $i=1,2,\dots,m$. We estimate the terms on the right hand side by the remarks above and the left hand side by making use of the algebraic Lemma 4.4 of Chap. I to obtain

$$\iint_{{}^+Q_R^\eta} |D\mathbf{w} - D\mathbf{v}|^p dx d\tau \le \gamma R^\lambda \iint_{{}^+Q_R^\eta} \left(1+|D\mathbf{w}|^{p-1}\right) \left|D\mathbf{w} - D\mathbf{v}\right| dx d\tau$$
$$+\gamma \iint_{{}^+Q_R^\eta} \left(1+|D\mathbf{w}|^{p-2}\right) \left|D\mathbf{w} - D\mathbf{v}\right| dx d\tau$$
$$+\gamma \iint_{{}^+Q_R^\eta} \left(1+|D\mathbf{w}|^{p-1}\right) |\mathbf{w} - \mathbf{v}| \, dx d\tau$$
$$= I^{(1)} + I^{(2)} + I^{(3)}.$$

In the estimates below we integrate over the boxes Q_R^η, rather than ${}^+Q_R^\eta$. In doing so we think of $\mathbf{v}$ and $\mathbf{w}$ as defined in the whole Q_R^η through an odd extension as indicated in §2. By the Schwartz inequality

$$I^{(1)} \le \frac{1}{4} \iint_{Q_R^\eta} |D\mathbf{w} - D\mathbf{v}|^p \, dx d\tau + \gamma R^{\lambda \frac{p}{p-1}} \iint_{Q_R^\eta} (1+|D\mathbf{w}|^p) \, dx d\tau,$$
$$I^{(2)} \le \frac{1}{4} \iint_{Q_R^\eta} |D\mathbf{w} - D\mathbf{v}|^p \, dx d\tau + \gamma \iint_{Q_R^\eta} (1+|D\mathbf{w}|^p)^{\frac{p-2}{p-1}} \, dx d\tau.$$

Since $\mathbf{v}-\mathbf{w}$ vanishes on the lateral boundary of Q_R^η, by the Sobolev embedding,[(1)]

$$I^{(3)} \le \gamma R \left(\iint_{Q_R^\eta} (1+|D\mathbf{w}|^p) \, dx d\tau \right)^{\frac{p-1}{p}} \left(\iint_{Q_R^\eta} |D\mathbf{w} - D\mathbf{v}|^p \, dx d\tau \right)^{\frac{1}{p}}$$
$$\le \frac{1}{4} \iint_{Q_R^\eta} |D\mathbf{w} - D\mathbf{v}|^p \, dx d\tau + \gamma R^{\frac{p}{p-1}} \iint_{Q_R^\eta} (1+|D\mathbf{w}|^p) \, dt.$$

Combining these estimates gives

(1) Corollary 2.1 of Chap. I.

$$\iint\limits_{Q_R^\eta} |D\mathbf{w}-D\mathbf{v}|^p\,dxd\tau \le \gamma R^{\lambda\frac{p}{p-1}} \iint\limits_{Q_R^\eta} (1+|D\mathbf{w}|^p)\,dxd\tau + \gamma \iint\limits_{Q_R^\eta} (1+|D\mathbf{w}|^p)^{\frac{p-2}{p-1}}\,dxd\tau. \tag{4.3}$$

From this we deduce two inequalities. First, since

$$\frac{p-2}{p-1}<1 \quad \text{and} \quad (1+|D\mathbf{w}|^p)^{\frac{p-2}{p-1}} \le (1+|D\mathbf{w}|^p),$$

we have

$$\fint\!\!\fint\limits_{Q_R^\eta} |D\mathbf{v}|^p dxd\tau \le \gamma \fint\!\!\fint\limits_{Q_R^\eta} (1+|D\mathbf{w}|^p)\,dxd\tau. \tag{4.4}$$

Second, for $\alpha>0$ set

$$\mathcal{F}(\alpha,\eta,R) \equiv R^{\alpha p} \fint\!\!\fint\limits_{Q_R^\eta} (1+|D\mathbf{w}|^p)\,dxd\tau, \tag{4.5}$$

and observe that

$$\iint\limits_{Q_R^\eta} (1+|D\mathbf{w}|^p)^{\frac{p-2}{p-1}}\,dxd\tau \le \gamma R^{N+2+\eta-\alpha p\frac{p-2}{p-1}} \left[\mathcal{F}(\alpha,\eta,R)\right]^{\frac{p-2}{p-1}}.$$

This implies that $\forall\, 0<\rho\le R\le \frac{1}{2}\mathcal{R}$

$$\iint\limits_{Q_\rho^\eta} (1+|D\mathbf{w}|^p)\,dxd\tau \le \gamma R^{\lambda\frac{p}{p-1}} \iint\limits_{Q_R^\eta} (1+|D\mathbf{w}|^p)\,dxd\tau + \iint\limits_{Q_\rho^\eta} |D\mathbf{v}|^p dxd\tau + \gamma R^{N+2+\eta-\alpha p\frac{p-2}{p-1}} \left[\mathcal{F}(\alpha,\eta,R)\right]^{\frac{p-2}{p-1}}. \tag{4.6}$$

By taking R sufficiently small and by interpolation, the first term on the right hand side of (4.6) can be eliminated. This is the content of the following lemma:

LEMMA 4.1. *There exists a constant $\gamma=\gamma(data)$ such that for all $\forall\, 0<\rho\le\frac{1}{4}R\le \frac{1}{8}\mathcal{R}$*

$$\iint\limits_{Q_\rho^\eta} (1+|D\mathbf{w}|^p)\,dxd\tau \le \gamma \iint\limits_{Q_{2\rho}^\eta} |D\mathbf{v}|^p dxd\tau + \gamma R^{N+2+\eta-\alpha p\frac{p-2}{p-1}} \left[\mathcal{F}(\alpha,\eta,R)\right]^{\frac{p-2}{p-1}}. \tag{4.6'}$$

PROOF: It suffices to prove the lemma for $R\le R_o$ where R_o is so small that

$$\gamma\,(2R_o)^{\lambda\frac{p}{p-1}} \le \frac{1}{2}.$$

Let $0<\rho\le\frac{1}{2}R_o$ and consider the sequence of radii

$$\rho_n \equiv \rho + 2^{-n}\rho, \quad n=0,1,2,\ldots.$$

Write (4.6) for $R=\rho_{n-1}$ and $\rho=\rho_{n,n\ge1}$, and set

$$Y_n \equiv \iint\limits_{Q^\eta_{\rho_n}} (1+|D\mathbf{w}|^p)\,dt$$

$$Z \equiv \gamma \iint\limits_{Q^\eta_{2\rho}} |D\mathbf{v}|^p dx d\tau + \gamma R^{N+2+\eta-\alpha p\frac{p-2}{p-1}} \left[\mathcal{F}(\alpha,\eta,R)\right]^{\frac{p-2}{p-1}}.$$

Then by iteration from (4.6),

$$Y_\infty \equiv \iint\limits_{Q^\eta_\rho} (1+|D\mathbf{w}|^p)\,dx d\tau$$

$$\le \lim_{n\to\infty} \left\{ 2^{-n} \iint\limits_{Q^\eta_{2\rho}} (1+|D\mathbf{w}|^p)\,dx d\tau + \gamma\, Z \sum_{i=1}^n 2^{-i} \right\}.$$

We return to (4.6)$'$ and estimate the integral involving $D\mathbf{v}$ in terms of $D\mathbf{u}$. Let $0<\rho\le\frac{1}{4}R$. Then by Theorem 2.1 and (4.4)

$$\iint\limits_{Q^\eta_\rho} |D\mathbf{v}|^p dx d\tau \le \gamma \rho^{N+2+\eta} \|D\mathbf{v}\|^p_{\infty,Q^\eta_\rho}$$

$$\le \gamma \rho^{N+2+\eta} \left\{ R^{\eta p/2} \left(\fint\!\!\fint_{Q^\eta_R} |D\mathbf{v}|^p dx d\tau \right)^{p/2} + R^{\eta\frac{p}{p-2}} \right\}$$

$$\le \gamma \rho^{N+2+\eta} \Bigg\{ R^{\eta p/2} \left(\fint\!\!\fint_{Q^\eta_R} |D\mathbf{v}|^p dx d\tau \right)^{\frac{p-2}{2}}$$

$$\times \left(\fint\!\!\fint_{Q^\eta_R} |D\mathbf{v}|^p dx d\tau \right) + R^{-\eta\frac{p}{p-2}} \Bigg\}.$$

Next choose

$$\eta = \alpha(p-2), \qquad \text{for some } \alpha > 0.$$

Then

$$R^{\eta p/2}\left(\fint\!\!\fint_{Q_R^\eta} |D\mathbf{v}|^p dx d\tau\right)^{\frac{p-2}{2}} \le \gamma\left[\mathcal{F}(\alpha,\eta,R)\right]^{\frac{p-2}{2}},$$

and by suitably modifying the constant γ we deduce from (4.6)′ that for all $0 < \rho \le R \le \frac{1}{2}\mathcal{R}$,

$$\iint_{Q_\rho^\eta} (1+|D\mathbf{w}|^p)\,dxd\tau \tag{4.7}$$
$$\le \gamma\left[\mathcal{F}(\alpha,\eta,R)\right]^{\frac{p-2}{2}}\left(\frac{\rho}{R}\right)^{N+2+\eta}\iint_{Q_R^\eta}(1+|D\mathbf{w}|^p)\,dxd\tau$$
$$+\gamma\left[\mathcal{F}(\alpha,\eta,R)\right]^{\frac{p-2}{p-1}} R^{N+2+\eta-\alpha\frac{p-2}{p-1}} + \gamma\rho^{N+2+\eta}R^{-\alpha p},$$

for a constant $\gamma = \gamma\,(\text{data})$. Set

$$\mathcal{F}(\alpha) \equiv \sup_{\substack{(x_o,t_o)\in\mathcal{Q}_{\mathcal{R}/2}\\ 0<\rho\le R\le\mathcal{R}/2}}\left\{\rho^{\alpha p}\fint\!\!\fint_{[(x_o,t_o)+Q_\rho^\eta]}(1+|D\mathbf{w}|^p)\,dt\right\} \tag{4.8}$$

and

$$\mathcal{G}(\alpha) \equiv \max\left\{\mathcal{F}^{\frac{p-2}{2}}(\alpha)\,;\ \mathcal{F}^{\frac{p-2}{p-1}}(\alpha)\right\}. \tag{4.9}$$

We summarise:

PROPOSITION 4.1. *Let $\alpha > 0$ and $\eta = \alpha(p-2)$. There exists a constant $\gamma = \gamma\,(data)$, independent of α, η, ρ, R, such that*

for all $(x_o,t_o)\in\mathcal{Q}^+_{\mathcal{R}/2}$ *and for all* $0<\rho\le R\le\mathcal{R}/2$,

$$\iint_{[(x_o,t_o)+Q_\rho^\eta]}(1+|D\mathbf{w}|^p)dxd\tau \tag{4.10}$$
$$\le \gamma\,\mathcal{G}(\alpha)\left(\frac{\rho}{R}\right)^{N+2+\eta}\iint_{[(x_o,t_o)+Q_R^\eta]}(1+|D\mathbf{w}|^p)dxd\tau$$
$$+\gamma\,\mathcal{G}(\alpha)R^{N+2+\eta-\alpha p\frac{p-2}{p-1}}+\gamma\rho^{N+2+\eta}R^{-\alpha p}.$$

PROOF: The previous arguments prove the proposition for those points $(x_o,t_o)\in\mathcal{Q}_{\frac{1}{2}\mathcal{R}}\cap\{x_N=0\}$.

The estimate is obvious for boxes $[(x_o,t_o)+Q_R^\eta]\subset\mathcal{Q}^+_{\mathcal{R}}$, by interior estimates. If $[(x_o,t_o)+Q_R^\eta]$ intersects $\{x_N=0\}$, then either

$$\left[(x_o,t_o)+Q^\eta_{\frac{1}{4}R}\right]\subset\mathcal{Q}_{\mathcal{R}} \quad\text{or}\quad \left[(x_o,t_o)+Q^\eta_{\frac{1}{4}R}\right]\cap\{x_N=0\}\ne\emptyset. \tag{4.11}$$

In the first case we may establish (4.10) with R replaced by $\frac{1}{4}R$. The general case follows by suitably modifying the constant γ. If the second of (4.11) holds, we let

$$x_* \equiv \left(x_{o,1}, x_{o,2}, \ldots, x_{o,(N-1)}, 0\right) \tag{4.12}$$

and observe that

$$\left[(x_*, t_o) + Q^{\eta}_{\frac{1}{2}R}\right] \subset [(x_o, t_o) + Q^{\eta}_{R}] .$$

We carry on the process leading to (4.10) for such a new box, for all $2x_{o,N} \le \rho < \frac{1}{2}R$. This implies that (4.10) holds for all $x_{o,N} < \rho \le \frac{1}{2}R$. If $\rho \le x_{o,N}$, we consider the cylinder $\left[(x_o, t_o) + Q^{\eta}_{x_{o,N}}\right]$, which satisfies the inclusion

$$\left[(x_o, t_o) + Q^{\eta}_{\frac{1}{4}x_{o,N}}\right] \subset \mathcal{Q}^{+}_{\mathcal{R}}.$$

Then by interior estimates, (4.10) holds with R replaced by $x_{o,N}$. Combining the two cases and suitably modifying the constant γ we conclude that (4.10) holds for all $(x_o, t_o) \in \overline{\mathcal{Q}}^{+}_{\frac{1}{2}\mathcal{R}}$ and all $0 < \rho \le R \le \frac{1}{2}\mathcal{R}$.

5. Estimating the local average of $|D\mathbf{w}|$ (the case $p > 2$)

LEMMA 5.1. *For every $\alpha \in (0,1)$ there exists a constant $\gamma = \gamma(\alpha, data)$, such that*

$$\text{for all } (x_o, t_o) \in \mathcal{Q}^{+}_{\mathcal{R}/2} \quad \text{and for all} \quad 0 < \rho \le R \le \mathcal{R}/2,$$

$$\fint\!\!\fint_{[(x_o,t_o)+Q^{\eta}_{\rho}]} (1 + |D\mathbf{w}|^p)\, dx d\tau \le \gamma(\alpha, data)\, \rho^{-\alpha p}, \quad \eta = \alpha(p-2). \tag{5.1}$$

PROOF: Define the sequences $\alpha_o = (N+2)/2$ and for $n = 1, 2, \ldots$,

$$\alpha_{n+1} = \alpha_n(1 - \bar{\delta}_n); \quad \bar{\delta}_n = \frac{\alpha_n}{\alpha_n + N + 2 + \eta_n}; \quad \eta_n = \alpha_n(p-2).$$

We will prove inductively that

$$\mathcal{F}(\alpha_n) \le \gamma(\alpha_n, \text{data}), \quad n = 0, 1, \ldots. \tag{5.2}$$

Since $|D\mathbf{w}| \in L^p(\Omega_T)$,

$$\fint\!\!\fint_{[(x_o,t_o)+Q^{\eta_o}_{\rho}]} (1 + |D\mathbf{w}|^p)\, dx d\tau \le \gamma \rho^{-\frac{N+2}{2}p} \left(1 + \|D\mathbf{w}\|^p_{p,\Omega_T}\right).$$

Therefore

$$\mathcal{F}(\alpha_o) \le \gamma_o \equiv \left(1 + \|D\mathbf{w}\|^p_{p,\Omega_T}\right).$$

Suppose the lemma holds for α_n and let us show that it continues to hold for α_{n+1}. If $\mathcal{F}(\alpha_n)\le\gamma(\alpha_n)$, the quantity $\mathcal{G}(\alpha_n)$ introduced in (4.9) is bounded and we may use (4.10) with $\alpha=\alpha_n$ and $\eta=\eta_n$. We apply the iterative Lemma 3.1 to the function

$$\varphi(\rho)=\iint\limits_{[(x_o,t_o)+Q_\rho^\eta]}(1+|D\mathbf{w}|^p)\,dxd\tau,$$

with the choice of parameters

$$(5.3)\qquad \beta=N+2+\eta_n,\quad \kappa=\alpha_n p,\quad \nu=\frac{p-2}{p-1},\quad \delta=\bar{\delta}_n\,\alpha_n p.$$

We obtain

$$\fint\!\!\!\fint\limits_{[(x_o,t_o)+Q_\rho^{\eta_n}]}(1+|D\mathbf{w}|^p)\,dxd\tau\le\gamma(\alpha_{n+1})\,\rho^{-\alpha_{n+1}p}\,(\varphi(R)+1)\,.$$

Let $0<\rho\le\mathcal{R}/2$ be fixed and consider the point $(x_o,t_o)\equiv(0,0)$. Without loss of generality assume that

$$\rho^{\eta_{n+1}}/\rho^{\eta_n}\ \text{ is an integer},$$

and partition the cylinder $\left[(x_o,t_o)+Q_\rho^{\eta_{n+1}}\right]$ into $s=\rho^{\eta_{n+1}-\eta_n}$ adjacent boxes with '*vertices*', say $(0,t_1),(0,t_2),\dots,(0,t_s)$. Then

$$\begin{aligned}\fint\!\!\!\fint\limits_{Q_\rho^{\eta_{n+1}}}(1+|D\mathbf{w}|^p)\,dxd\tau&\le\frac{\rho^{\eta_n}}{\rho^{\eta_{n+1}}}\sum_{j=1}^{s}\ \fint\!\!\!\fint\limits_{[(0,t_j)+Q_\rho^{\eta_n}]}(1+|D\mathbf{w}|^p)\,dxd\tau\\&\le\gamma(\alpha_{n+1})\,\rho^{-\alpha_{n+1}p}.\end{aligned}$$

We may treat analogously the other points of $\mathcal{Q}_{\mathcal{R}/2}$ and the inductive inequality (5.2) follows. To prove the lemma it suffices to prove that $\{\alpha_n\}\to0$ as $n\to\infty$. The sequence $\{\alpha_n\}$ is deacreasing. We claim that $\{\alpha_n\}\to0$. Indeed if not,

$$\lim_{n\to\infty}\alpha_n=\alpha_o>0,$$

and the definitions of $\{\alpha_n\}$ and $\{\delta_n\}$ would imply

$$\bar{\delta}_n>\delta_o\equiv\frac{\alpha_o}{\alpha_o+N+2+\eta_o},\qquad \eta_o\equiv\alpha_o(p-2).$$

Therefore $\alpha_{n+1}\le\alpha_n(1-\delta_o)$. This in turn implies $\{\alpha_n\}\to0$.

Remark 5.1. The constant γ on the right hand side of (5.1) is '*stable*' as $p\searrow2$. This follows from the choice (5.3) of the parameter ν and Remark 3.1.

6. Estimating the local averages of w (the case $p>2$)

We return to cylinders bearing the natural parabolic geometry, i.e., $Q_\rho\equiv Q\left(\rho^2,\rho\right)$ and will work within the boxes

(6.1) $$[(x_o,t_o)+Q_\rho], \quad (x_o,t_o)\in \mathcal{Q}^+_{\mathcal{R}/2}, \quad 0<\rho\le \tfrac{1}{4}\mathcal{R}.$$

Since $\mathbf{w}$ vanishes for $x_N=0$, we regard it as defined in the whole $\mathcal{Q}_\mathcal{R}$, by an odd extension across $\{x_N=0\}$. Let

$$(\mathbf{w})_{o,\rho} \equiv \fint\!\!\fint_{[(x_o,t_o)+Q_\rho]} \mathbf{w}(x,t)dxd\tau$$

denote the integral average of $\mathbf{w}$ over $[(x_o,t_o)+Q_\rho]$. If (x_o,t_o) coincides with the origin, we let $(\mathbf{w})_{o,\rho}\equiv(\mathbf{w})_\rho$. Also let

$$(\mathbf{w})_{o,\rho}(t) \equiv \fint_{[x_o+K_\rho]} \mathbf{w}(x,t)dx, \quad t\in\left(t_o-\rho^2,t_o\right).$$

We observe that if $x_o\in\{x_N=0\}$, we have $(\mathbf{w})_{o,\rho}(t)=0$ for all $t\in(t_o-\rho^2,t_o)$, since $\mathbf{w}$ is odd across $\{x_N=0\}$, and in particular $(\mathbf{w})_{o,\rho}=0$.

LEMMA 6.1. *For every $\alpha\in(0,1)$ there exists a constant $\gamma=\gamma(\alpha,data)$, such that*

(6.2) $$\fint\!\!\fint_{[(x_o,t_o)+Q_\rho]} \left|\mathbf{w}-(\mathbf{w})_{o,\rho}\right|^p dxd\tau \le \gamma(\alpha)\rho^{p(1-\alpha)},$$

for all cylinders satisfying (6.1).

PROOF: We first observe that from Lemma 5.1 and its proof it follows that for every cylinder satisfying (6.1)

(6.3) $$\fint\!\!\fint_{[(x_o,t_o)+Q_\rho]} \left(1+|D\mathbf{w}|^p\right) dxd\tau \le \gamma(\alpha)\,\rho^{-\alpha p}.$$

If $x_o\in\{x_N=0\}$ by the Poincaré inequality and (6.3),

$$\fint\!\!\fint_{[(x_o,t_o)+Q_\rho]} \left|\mathbf{w}-(\mathbf{w})_{o,\rho}\right|^p dxd\tau \le \gamma(\alpha)\,\rho^{p(1-\alpha)}.$$

Consider next the case

(6.4) $$[(x_o,t_o)+Q_{\rho+\sigma\rho}]\subset \mathcal{Q}^+_{\mathcal{R}/2}, \quad \sigma\in(0,\tfrac{1}{2}) \text{ to be chosen.}$$

By a translation we may assume that (x_o,t_o) coincides with the origin. We have

$$\iint_{Q_\rho} \left|\mathbf{w} - (\mathbf{w})_\rho\right|^p dx d\tau \le \iint_{Q_\rho} \left|\mathbf{w} - (\mathbf{w})_\rho(\tau)\right|^p dx d\tau + \iint_{Q_\rho} \left|(\mathbf{w})_\rho(\tau) - (\mathbf{w})_\rho\right|^p dx d\tau \equiv I^{(1)} + I^{(2)}.$$

By the Poincaré inequality and (6.3)

$$I^{(1)} \le \gamma(\alpha)\, \rho^{p(1-\alpha)}.$$

Next

$$I^{(2)} = \iint_{Q_\rho} \left| \iint_{Q_\rho} [\mathbf{w}(x,t) - \mathbf{w}(x,\tau)]\, dx d\tau \right|^p dx dt. \tag{6.5}$$

We estimate the integrand on the right hand side of (6.5) by making use of the equation (2.5), over the cylinder $Q_{\rho+\sigma\rho}$. Let $x \to \zeta(x)$ be a non-negative piecewise smooth cutoff function in $K_{\rho+\sigma\rho}$ that equals one on K_ρ and such that $|D\zeta| \le 1/\sigma\rho$. In the weak formulation of (2.5), take ζ as a testing function and integrate over $K_{\rho+\sigma\rho} \times [\tau, t]$ to obtain

$$\begin{aligned} \left| \int_{K_{\rho+\sigma\rho}} \zeta\, [\mathbf{w}(x,t) - \mathbf{w}(x,\tau)]\, dx \right| &\le \sum_{i=1}^m \iint_{Q_{\rho+\sigma\rho}} \left| \mathbf{A}_i(x, \mathbf{w}, D\mathbf{w}) \cdot D\zeta + f_{i,\ell}\zeta_{x_\ell} + B_i \zeta \right| dx d\tau \\ &\le \frac{\gamma}{\sigma\rho} \iint_{Q_{\rho+\sigma\rho}} \left(1 + |D\mathbf{w}|^{p-1}\right) dx d\tau \\ &\le \frac{\gamma}{\sigma} \rho^{N+1} \left(\iint_{Q_{\rho+\sigma\rho}} (1 + |D\mathbf{w}|^p)\, dx d\tau \right)^{\frac{p-1}{p}}. \end{aligned}$$

By the properties of ζ and (6.3) with a suitable choice of α we conclude that

$$\left| \int_{K_\rho} [\mathbf{w}(x,t) - \mathbf{w}(x,\tau)]\, dx \right| \le \frac{\gamma}{\sigma} \rho^{N+(1-\alpha)} + \left| \int_{K_{\rho+\sigma\rho} \setminus K_\rho} [\mathbf{w}(x,t) - \mathbf{w}(x,\tau)]\, dx \right|.$$

Let $(\mathbf{w})_{\rho+\sigma\rho}$ denote the integral average of w over the $Q_{\rho+\sigma\rho}$, i.e.,

$$(\mathbf{w})_{\rho+\sigma\rho} \equiv \iint\limits_{Q_{\rho+\sigma\rho}} \mathbf{w}(x,\tau)dxd\tau.$$

Then

$$\Big| \int\limits_{K_{\rho+\sigma\rho}\backslash K_\rho} [\mathbf{w}(x,t) - \mathbf{w}(x,\tau)]\; dx \Big|$$

$$\le \left|K_{\rho+\sigma\rho}\backslash K_\rho\right|^{\frac{p-1}{p}} \left\{ \left(\int\limits_{K_{\rho+\sigma\rho}} \left|\mathbf{w}(x,t) - (\mathbf{w})_{\rho+\sigma\rho}\right|^p dx \right)^{\frac{1}{p}} \right.$$

$$\left. + \left(\int\limits_{K_{\rho+\sigma\rho}} \left|\mathbf{w}(x,\tau) - (\mathbf{w})_{\rho+\sigma\rho}\right|^p dx \right)^{\frac{1}{p}} \right\}.$$

Combining these estimates in (6.5) gives

$$I^{(2)} \le \frac{\gamma}{\sigma^p}\, \rho^{p(1-\alpha)} + \gamma\sigma^{p-1} \iint\limits_{Q_{\rho+\sigma\rho}} \left|\mathbf{w} - (\mathbf{w})_{\rho+\sigma\rho}\right|^p dxd\tau.$$

We conclude that for every $\alpha\in(0,1)$ there exists a constant $\gamma=\gamma(\alpha)$ such that for every $\sigma\in(0,1)$ and for every $\rho\in(0,\frac{1}{4}\mathcal{R})$

$$\iint\limits_{Q_\rho} \left|\mathbf{w} - (\mathbf{w})_\rho\right|^p dxd\tau \le \frac{\gamma(\alpha)}{\sigma^p}\, \rho^{(1-\alpha)p} \tag{6.6}$$

$$+ \gamma(\alpha)\sigma^{p-1} \iint\limits_{Q_{\rho+\sigma\rho}} \left|\mathbf{w} - (\mathbf{w})_{\rho+\sigma\rho}\right|^p dxd\tau.$$

This implies the lemma, in the case (6.4) holds, by the interpolation process of Lemma 4.3 of Chap. I. This process yields the choice of $\sigma\in(0,\frac{1}{2})$. Finally, having fixed $\sigma\in(0,\frac{1}{2})$, consider the case when

$$[(x_o,t_o) + Q_{\rho+\sigma\rho}] \cap \{x_N = 0\} \ne \emptyset.$$

Let $x_* \in\{x_N=0\}$ be defined as in (4.12) and observe that the box $[(x_*,t_o) + Q_{2\rho}]$ '*centered*' at (x_*,t_o) contains $[(x_o,t_o) + Q_\rho]$. Therefore, by the Poincaré inequality, since the average of $\mathbf{w}$ over $[(x_*,t_o) + Q_{2\rho}]$ is zero,

$$\iint\limits_{[(x_o,t_o)+Q_\rho]} \left|\mathbf{w} - (\mathbf{w})_{o,\rho}\right|^p dxd\tau \le \gamma \iint\limits_{[(x_*,t_o)+Q_{2\rho}]} |w|^p dxd\tau \le \gamma(\alpha)\, \rho^{p(1-\alpha)}.$$

6-(i). Proof of Theorem 1.1 (the case $p>2$*)*

The proof is a consequence of Lemma (6.1) and the averaging theory of Campanato-Morrey spaces [22,23,33,79]. It can also be proved directly, starting from (6.2), by arguments similar to those in §§2 and 3 of Chap. IX.

7. Comparing w and v (the case $\max\{1;\frac{2N}{N+2}\}<p<2$)

Assuming that $(x_o,t_o)\equiv(0,0)$, the functions **w** and **v** satisfy

$$
\begin{aligned}
(7.1)\qquad & \frac{\partial}{\partial t}(v_i-w_i)-\operatorname{div}\left(|D\mathbf{v}|^{p-2}Dv_i-|D\mathbf{w}|^{p-2}Dw_i\right)\\
&\quad=-\frac{\partial}{\partial x_\ell}\left(a_{\ell,k}(x,D\mathbf{u})u_{i,x_k}-|D\mathbf{u}|^{p-2}u_{i,x_k}\,\delta_{\ell,k}\right)\\
&\quad\quad-\operatorname{div}\left(|D\mathbf{u}|^{p-2}Du_i-|D\mathbf{w}|^{p-2}Dw_i\right)\\
&\quad\quad-B_i(x,t,\mathbf{w},D\mathbf{w}),\quad \text{in } {}^+Q_R^\eta\\
& v_i-w_i=0,\quad \text{on the parabolic boundary of } {}^+Q_R^\eta.
\end{aligned}
$$

The boxes Q_R^η are formally identical to those introduced in the degenerate case $p>2$. In the singular case we will take $\eta\in(-1,0)$. In writing (7.1) we have used the definitions (2.3) and (2.8). From (2.3)–(2.4) we derive the estimate

$$
\begin{aligned}
&\left|a_{\ell,k}(x,D\mathbf{u})u_{i,x_\ell}-|D\mathbf{u}|^{p-2}u_{i,x_\ell}\right|\le\gamma R^\lambda\,|D\mathbf{u}|^{p-1},\\
&i=1,2,\dots,m,\quad \ell,k=1,2,\dots,N.
\end{aligned}
$$

Moreover by Lemma 4.4 of Chap. I,

$$
\left||D\mathbf{u}|^{p-2}Du_i-|D\mathbf{w}|^{p-2}Dw_i\right|\le\gamma_1\left|D(\mathbf{u}-\mathbf{w})\right|^{p-1}\le\gamma,
$$

since the boundary data $\tilde{\mathbf{g}}$ are regular. In the weak formulation of (7.1) we take the testing functions v_i-w_i modulo a Steklov averaging process, integrate over ${}^+Q_R^\eta$ and add over $i=1,2,\dots,m$. Using the remarks above to estimate the corresponding terms on the right hand side gives

$$
\begin{aligned}
(7.2)\qquad & \iint_{{}^+Q_R^\eta}\left(\int_0^1\left|D\left(s\mathbf{v}+(1-s)\mathbf{w}\right)\right|^{p-2}ds\right)\left|D\mathbf{w}-D\mathbf{v}\right|^2dxd\tau\\
&\quad\le\gamma R^\lambda\iint_{{}^+Q_R^\eta}|D\mathbf{u}|^{p-1}\left|D\mathbf{w}-D\mathbf{v}\right|dxd\tau\\
&\quad\quad+\gamma\iint_{{}^+Q_R^\eta}\left|D\mathbf{w}-D\mathbf{v}\right|dxd\tau+\gamma\iint_{{}^+Q_R^\eta}|\mathbf{B}|\,|\mathbf{w}-\mathbf{v}|\,dxd\tau.
\end{aligned}
$$

In carrying the estimates below, we think of $\mathbf{v}$ and $\mathbf{w}$ as defined in the whole Q_R^η by an odd reflexion across $\{x_N=0\}$. By the Poincaré inequality and (2.7)

$$\leq \gamma \left(\iint_{Q_R^\eta} (1+|D\mathbf{w}|^p)\, dx d\tau \right)^{\frac{p-1}{p}}$$

$$\leq \gamma R \left(\iint_{Q_R^\eta} (1+|D\mathbf{w}|^p)\, dx d\tau \right)^{\frac{p-1}{p}} \left(\iint_{Q_R^\eta} |D\mathbf{v}-D\mathbf{w}|^p dx d\tau \right)^{\frac{1}{p}}.$$

Introduce the two sets

$$\mathcal{E}_1 \equiv \left\{(x,t)\in Q_R^\eta \mid |D\mathbf{w}-D\mathbf{v}| \geq |D\mathbf{w}|\right\},$$
$$\mathcal{E}_2 \equiv \left\{(x,t)\in Q_R^\eta \mid |D\mathbf{w}-D\mathbf{v}| < |D\mathbf{w}|\right\}.$$

On the set $\mathcal{E}_1$,

$$\int_0^1 |D(s\mathbf{v}+(1-s)\mathbf{w})|^{p-2} ds \geq \frac{1}{2} |D\mathbf{w}-D\mathbf{v}|^{p-2}.$$

Therefore from (7.2) it follows

$$\iint_{\mathcal{E}_1} |D\mathbf{w}-D\mathbf{v}|^p dx d\tau + \iint_{\mathcal{E}_2} |D\mathbf{w}|^{p-2} |D\mathbf{w}-D\mathbf{v}|^2 dx d\tau$$
$$\leq \gamma R^\lambda \left\{ \iint_{\mathcal{E}_1} |D\mathbf{u}|^{p-1} |D\mathbf{w}-D\mathbf{v}| dx d\tau \right.$$
$$+ \left. \iint_{\mathcal{E}_2} |D\mathbf{u}|^{p-1} |D\mathbf{w}-D\mathbf{v}| dx d\tau \right\}$$
$$+ \gamma \left\{ \iint_{\mathcal{E}_1} |D\mathbf{w}-D\mathbf{v}| dx d\tau + \iint_{\mathcal{E}_2} |D\mathbf{w}-D\mathbf{v}| dx d\tau \right\}$$
$$+ \gamma R \left(\iint_{Q_R^\eta} (1+|D\mathbf{w}|^p)\, dx d\tau \right)^{\frac{p-1}{p}}$$
$$\times \left\{ \iint_{\mathcal{E}_1} |D\mathbf{w}-D\mathbf{v}|^p dx d\tau + \iint_{\mathcal{E}_2} |D\mathbf{w}-D\mathbf{v}|^p dx d\tau \right\}^{1/p}.$$

In this inequality we absorb the integrals of $|D\mathbf{w}-\mathbf{v}|$ extended over $\mathcal{E}_1$ into the analogous term on the left hand side by means of Young's inequality. Using also the definition of $\mathcal{E}_2$ we arrive at

$$\iint_{\mathcal{E}_1} |D\mathbf{w} - D\mathbf{v}|^p dx d\tau + \iint_{\mathcal{E}_2} |D\mathbf{w}|^{p-2} |D\mathbf{w} - D\mathbf{v}|^2 dx d\tau \tag{7.3}$$
$$\leq \gamma R^{\lambda} \iint_{Q_R^{\eta}} (1 + |D\mathbf{w}|^p)\, dx d\tau + \gamma \iint_{Q_R^{\eta}} (1 + |D\mathbf{w}|)\, dx d\tau.$$

We estimate the right hand side of (7.3) by

$$\gamma\, R^{N+2+\eta+\lambda-\alpha p} \left\{ R^{\alpha p} \fint\!\!\fint_{[(x_o,t_o)+Q_R^{\eta}]} (1 + |D\mathbf{w}|^p)\, dt \right\}$$
$$+ \gamma\, R^{N+2+\eta-\alpha} \left\{ R^{\alpha p} \fint\!\!\fint_{[(x_o,t_o)+Q_R^{\eta}]} (1 + |D\mathbf{w}|^p)\, dt \right\}^{\frac{1}{p}}$$
$$\leq \gamma\, R^{N+2+\eta-\alpha p \lambda_o} \left[\mathcal{F}(\alpha) \wedge \mathcal{F}^{1/p}(\alpha) \right],$$

where

$$\lambda_o \equiv \max\left\{ \frac{1}{p};\, 1 - \frac{\lambda}{\alpha p} \right\} \in (0, 1)$$

and where as before we have set

$$\mathcal{F}(\alpha) \equiv \sup_{\substack{(x_o,t_o)\in \mathcal{Q}_{\mathcal{R}/2} \\ 0<\rho<\mathcal{R}/2}} \left\{ \rho^{\alpha p} \fint\!\!\fint_{[(x_o,t_o)+Q_{\rho}^{\eta}]} (1 + |D\mathbf{w}|^p)\, dt \right\}, \quad \alpha > 0.$$

Therefore

$$\iint_{\mathcal{E}_1} |D\mathbf{w} - D\mathbf{v}|^p dx d\tau + \iint_{\mathcal{E}_2} |D\mathbf{w}|^{p-2} |D\mathbf{w} - D\mathbf{v}|^2 dx d\tau \tag{7.4}$$
$$\leq R^{N+2+\eta-\alpha p \lambda_o} \left[\mathcal{F}(\alpha) \wedge \mathcal{F}^{1/p}(\alpha) \right].$$

Rewrite the integrand in the second integral on the left hand side of (7.4) as

$$|D\mathbf{w}|^{p-2} |D\mathbf{w} - D\mathbf{v}|^2 = |D\mathbf{w}|^{p-2} \left(|D\mathbf{w}|^2 - 2\langle D\mathbf{w}, D\mathbf{v}\rangle + |D\mathbf{v}|^2 \right).$$

Observe also that on the set $\mathcal{E}_2$, $|D\mathbf{v}| \leq 2|D\mathbf{w}|$, so that

$$|D\mathbf{w}|^{p-2} |D\mathbf{v}|^2 \leq 2^{2-p} |D\mathbf{v}|^p.$$

These remarks in (7.4) prove the following:

LEMMA 7.1. *There exits a constant* $\gamma=\gamma\,(data)$, *such that*

$$\text{for all } (x_o,t_o)\in\mathcal{Q}^+_{\mathcal{R}/2} \quad \text{and for all} \quad 0<\rho\le R\le\mathcal{R}/2,$$

$$\iint\limits_{[(x_o,t_o)+Q^\eta_\rho]}|D\mathbf{w}|^p dxd\tau \le \gamma R^{N+2+\eta-\alpha p\lambda_o}\max\left\{\mathcal{F}(\alpha);\mathcal{F}^{1/p}(\alpha)\right\} + \gamma\iint\limits_{[(x_o,t_o)+Q^\eta_\rho]}|D\mathbf{v}|^p dxd\tau, \tag{7.5}$$

$$\iint\limits_{[(x_o,t_o)+Q^\eta_R]}|D\mathbf{v}|^p dxd\tau \le \gamma R^{N+2+\eta-\alpha p\lambda_o}\max\left\{\mathcal{F}(\alpha);\mathcal{F}^{1/p}(\alpha)\right\} + \gamma\iint\limits_{[(x_o,t_o)+Q^\eta_R]}|D\mathbf{w}|^p dxd\tau. \tag{7.6}$$

To estimate the last integral on the right hand side of (7.5) we make use of Theorem 2.2, the definition of $\mathcal{F}(\alpha)$ and (7.6). Assuming that (x_o,t_o) coincides with the origin, we have for all $0<\rho\le\frac{1}{2}R\le\mathcal{R}/4$

$$\begin{aligned}\iint\limits_{Q^\eta_\rho}|D\mathbf{v}|^p dxd\tau &\le \rho^{N+2+\eta}\|D\mathbf{v}\|^p_{\infty,Q^\eta_\rho}\\ &\le \gamma\rho^{N+2+\eta}R^{p\frac{\eta}{2-p}}\\ &+\gamma\rho^{N+2+\eta}\left\{R^{-N\eta-\alpha p\lambda_o}\left[\mathcal{F}(\alpha)\wedge\mathcal{F}^{1/p}(\alpha)\right]\right.\\ &\left.+R^{-N\eta}\iint\limits_{Q^\eta_R}\!\!\!\!\!\!-\ (1+|D\mathbf{w}|^p)\,dxd\tau\right\}^{p/\nu_p},\end{aligned} \tag{7.7}$$

where $\nu_p=N(p-2)+2p>0$. Choose

$$\eta=\alpha(p-2),\qquad \alpha\in\left(0;\frac{2N}{N+2}\right]>0. \tag{7.8}$$

Then the first term on the right hand side of (7.7) is estimated above by

$$\gamma\,\rho^{N+2+\eta}R^{-\alpha p}.$$

Setting also

$$\mathcal{G}(\alpha)\equiv\max\left\{\mathcal{F}^{p/\nu_p}\,;\,\mathcal{F}^{1/\nu_p}\,;\,\mathcal{F}^{N(2-p)/\nu_p}\,;\,1\right\},$$

the second term on the right hand side of (7.7) is estimated by

$$\gamma\,\mathcal{G}(\alpha)\left(\frac{\rho}{R}\right)^{N+2+\eta}\iint_{Q_R^\eta}(1+|D\mathbf{w}|^p)\,dxd\tau$$
$$+\,\gamma\mathcal{G}(\alpha)\,\rho^{N+2+\eta}\,R^{-(N\eta+2\alpha p\lambda_o)p/\nu_p}.$$

Using the definition (7.8) of η we have

$$-(N\eta+2\alpha p\lambda_o)\frac{p}{\nu_p}=-\alpha+\frac{\alpha p^2}{\nu_p}(1-\lambda_o).$$

Since $\lambda_o\in(0,1)$ we estimate $R^{-(N\eta+\alpha p\lambda_o)p/\nu_p}\le R^{-\alpha\,p}$, and summarise:

LEMMA 7.1. *Let α and η be chosen as in (7.8). There exists a constant $\gamma=\gamma(data)$, independent of α,η,ρ,R, such that*

$$\textit{for all }\ (x_o,t_o)\in\mathcal{Q}^+_{\mathcal{R}/2}\qquad\textit{and for all}\quad 0<\rho\le R\le\mathcal{R}/2,$$

(7.9)
$$\iint_{[(x_o,t_o)+Q_\rho^\eta]}(1+|D\mathbf{w}|^p)dxd\tau$$
$$\le\gamma\,\mathcal{G}(\alpha)\left(\frac{\rho}{R}\right)^{N+2+\eta}\iint_{[(x_o,t_o)+Q_R^\eta]}(1+|D\mathbf{w}|^p)\,dxd\tau$$
$$+\,\gamma\,\mathcal{G}(\alpha)\,\rho^{N+2+\eta}R^{-\alpha\,p}.$$

8. Estimating the local average of $|D\mathbf{w}|$

LEMMA 8.1. *For every $\alpha\in(0,1)$ there exists a constant $\gamma=\gamma\,(\alpha,data)$, such that*

$$\textit{for all }\ (x_o,t_o)\in\mathcal{Q}^+_{\mathcal{R}/2}\qquad\textit{and for all}\quad 0<\rho\le R\le\mathcal{R}/2,$$

(8.1)
$$⨌_{[(x_o,t_o)+Q_\rho^\eta]}(1+|D\mathbf{w}|^p)\,dxd\tau\ \le\gamma(\alpha,data)\,\rho^{-\alpha p}.$$

PROOF: Define sequences $\alpha_o\equiv(N+2)/2$ and for $n=0,1,2\dots,$

$$\alpha_{n+1}=\alpha_n\left(1-\frac{N+2+\alpha_n(p-1)}{N+2+2\alpha_n(p-1)}\right),\qquad\eta_n=\alpha_n(p-2),$$
$$\delta_n=\alpha_n p\,\frac{\alpha_n(p-1)}{N+2+2\alpha_n(p-1)}.$$

It is apparent that $\{\alpha_n\}, \{\eta_n\}, \{\delta_n\} \to 0$ as $n \to \infty$. We will prove by induction that

$$\mathcal{F}(\alpha_n) \le \gamma_n \left(\alpha_n, \text{data}\right), \quad n = 0, 1, 2, \dots . \tag{8.2}$$

Since $|D\mathbf{w}| \in L^p(\Omega_T)$,

$$\iint\limits_{[(x_o,t_o)+Q_\rho^{\eta_o}]} \!\!\!\!\!\!\!\!\!\!-\;\; (1 + |D\mathbf{w}|^p)\, dx d\tau \le \gamma\, \rho^{-\frac{p(N+2)}{2}} \left(1 + \|D\mathbf{w}\|_{p,\Omega_T}\right)^p .$$

Therefore

$$\mathcal{F}(\alpha_o) \le \gamma_o \equiv \left(1 + \|D\mathbf{w}\|_{p,\Omega_T}\right)^p .$$

Assume now that (8.2) holds for some n and let us show that it continues to hold for $n+1$. We apply the iterative Lemma 3.1 with the choice of the parameters

$$\beta \equiv N + 2 + \eta_n, \quad \kappa = \alpha_n p, \quad \delta = \delta_n, \quad \nu = 0,$$

to the function

$$\varphi(\rho) = \iint\limits_{[(x_o,t_o)+Q_\rho^{\eta_n}]} (1 + |D\mathbf{w}|^p)\, dx d\tau,$$

which satisfies (7.9), to conclude that for all $(x_o, t_o) \in \mathcal{Q}^+_{\mathcal{R}/2}$ and for all $0 < \rho \le R \le \mathcal{R}/2$,

$$\iint\limits_{[(x_o,t_o)+Q_\rho^{\eta_n}]} (1 + |D\mathbf{w}|^p) \le \gamma \left(\frac{\rho}{R^q}\right)^{N+2+\eta_n-\alpha_n p+\delta_n} (\varphi(R) + 1) . \tag{8.3}$$

In particular, ρ being fixed, (8.3) must hold for radii ρ_* satisfying

$$\rho \equiv \rho_*^{\,1+\frac{\eta_n-\eta_{n+1}}{2+\eta_{n+1}}} .$$

Without loss of generality we may assume that

$$\rho_*^{\frac{\eta_n-\eta_{n+1}}{2+\eta_{n+1}}} \equiv \ell_* \quad \text{is an integer.}$$

Then we may regard the cube $[x_o + K_\rho]$ as the disjoint union, up to a set of measure zero, of $(\ell_*)^N$ cubes $[x_j + K_{\rho_*}]$ centered at points x_j of $[x_o + K_\rho]$. Similarly we regard the cylinders $\left[(x_o, t_o) + Q_\rho^{\eta_{n+1}}\right]$ as the disjoint union, up to a set of measure zero, of $(\ell_*)^N$ cylinders $\left[(x_j, t_o) + Q_{\rho_*}^{\eta_n}\right]$. We write (8.3) for each of these cylinders and add up for $j = 1, 2, \dots, \ell_*$ to obtain

$$\begin{aligned} \iint\limits_{[(x_o,t_o)+Q_\rho^{\eta_{n+1}}]} (1 + |D\mathbf{w}|^p) &\le \gamma\, (\ell_*)^N \rho_*^{N+2+\eta_n-\alpha_n p+\delta_n} \\ &\equiv \gamma \rho^{N+2+\eta_{n+1}} \rho_*^{\delta_n - \alpha_n p} . \end{aligned}$$

Therefore

$$\rho^{\alpha_{n+1}p} \iint\limits_{\left[(x_o,t_o)+Q_\rho^{\eta_{n+1}}\right]} (1+|D\mathbf{w}|^p)\, dxd\tau \le \gamma \quad \text{and} \quad \mathcal{F}(\alpha_{n+1},\eta_{n+1}) \le \gamma_{n+1}.$$

8-(I). PROOF OF THEOREM 1.1 (THE CASE $\max\left\{1; \frac{2N}{N+2}\right\} < p < 2$)

By a cube decomposition technique similar to the one outlined, Lemma 8.1 can be rephrased in terms of the parabolic cylinders $Q_\rho \equiv Q(\rho^2,\rho)$.

LEMMA 8.1′. *For every* $\alpha \in (0,1)$ *there exists a constant* $\gamma = \gamma(\alpha, data)$, *such that*

$$\text{for all } (x_o,t_o) \in \mathcal{Q}^+_{\mathcal{R}/2} \quad \text{and for all} \quad 0<\rho\le R\le \mathcal{R}/2,$$

(8.1)′
$$\iint\limits_{\left[(x_o,t_o)+Q_\rho\right]} (1+|D\mathbf{w}|^p)\, dxd\tau \le \gamma(\alpha, data)\, \rho^{-\alpha p}.$$

With this lemma at hand the proof is now concluded as in the degenerate case. First we may establish a version of Lemma 6.1 and then the Hölder continuity of $\mathbf{u}$ follows from the arguments of [22,23,32,79] or by those in §§2 and 3 of Chap. IX.

9. Bibliographical notes

The proof of Theorem 1.1 is in [27]. The iteration Lemma 3.1 is in the same spirit of similar results of Campanato [22,23]. The technique is indeed a *degenerate version* of [22,23]. Techniques of this type near the boundary appear in Giaquinta–Giusti [49]. The boundary behaviour of solutions of (1.1) is essentially not understood. In the case of a single equation some results appear in Lieberman [69] and Lin [72].

XI

Non-negative solutions in Σ_T. The case $p>2$

1. Introduction

Non-negative solutions of the heat equation in a strip $\Sigma_T \equiv \mathbf{R}^N \times (0,T)$ are somewhat special in the sense that they grow no faster than

$$e^{a|x|^2}, \qquad a < 1/4T, \qquad \text{as } |x| \to \infty. \tag{1.1}$$

Let μ be a $\sigma-$finite Borel measure in $\mathbf{R}^N$ with no sign restriction. We say that μ has the growth (1.1) if

$$\int_{\mathbf{R}^N} e^{-\frac{|x|^2}{4T}} |d\mu| < \infty, \tag{1.2}$$

where $|d\mu|$ is the variation of μ. Then the Cauchy problem

$$\begin{cases} u_t - \Delta u = 0, & \text{in } \Sigma_T, \\ u(\cdot, 0) = \mu, \end{cases} \tag{1.3}$$

is uniquely solvable within the class of functions satisfying (1.1). The *'initial measure'* is taken in the sense

$$\int_{\mathbf{R}^N} u(x,t)\varphi\, d\mu \longrightarrow \int_{\mathbf{R}^N} \varphi\, d\mu, \text{ as } t \searrow 0, \quad \forall \varphi \in C_o^\infty(\mathbf{R}^N). \tag{1.4}$$

Conversely every non-negative solution of the heat equation in Σ_T verifies (1.4) for some σ–finite non-negative Borel measure μ satisfying the growth condition (1.2). The measure μ is unique and it is called the *initial trace* of u. In turn the initial trace of u determines u uniquely. These are the basic elements of a classical theory developed by Tychonov [98], Tacklind [94] and Widder [105]. A perhaps rough summary of the theory is that the structure of all non-negative solutions of the heat equation is determined by the heat kernel

$$\Gamma(x,t)=\frac{1}{(4\pi t)^{N/2}}\,e^{-\frac{|x|^2}{4t}},\quad t>0.$$

Consider now non-negative local weak solutions in Σ_T of

$$\text{(1.5)}\qquad \begin{cases} u\in C_{loc}\left(0,T;L^2_{loc}(\mathbf{R}^N)\right)\cap L^p_{loc}\left(0,T;W^{1,p}_{loc}(\mathbf{R}^N)\right),\ p>2,\\ u_t-\operatorname{div}\left(|Du|^{p-2}Du\right)=0 \quad \text{in } \Sigma_T. \end{cases}$$

The analog of $\Gamma(x,t)$ for the degenerate p.d.e. (1.5) is the Barenblatt explicit solution

$$\text{(1.6)}\qquad \mathcal{B}(x,t)\equiv t^{-N/\lambda}\left\{1-\gamma_p\left(\frac{|x|}{t^{1/\lambda}}\right)^{\frac{p}{p-1}}\right\}_+^{\frac{p-1}{p-2}},\quad t>0,$$

$$\gamma_p\equiv\lambda^{-\frac{1}{p-1}}\frac{p-2}{p},\qquad \lambda=N(p-2)+p.$$

We call this a *'fundamental solution'* only in the sense that

$$\mathcal{B}(x,t)\longrightarrow(4\pi)^{N/2}\Gamma(x,t)\quad \text{pointwise in } \Sigma_T,\ \textit{as}\ p\searrow 2.$$

Solutions of (1.5) cannot be represented as convolutions of initial data with $\mathcal{B}(x,t)$. Nevertheless the sup-estimates of Chap. V and the global Harnack estimates of §7 of Chap. VI permit a precise characterisation of the class of non-negative solutions of (1.5) in the whole Σ_T, with no reference to possible initial data. Such a characterisation essentially says that all non-negative solutions of (1.5) behave as $t\searrow 0$ like the *'fundamental'* solution $\mathcal{B}(x,t)$, and as $|x|\to\infty$ they grow no faster than $|x|^{p/(p-2)}$. For these solutions we will establish the existence of initial traces and prove their uniqueness when the initial datum is taken in the sense of $L^1_{loc}(\mathbf{R}^N)$.

2. Behaviour of non-negative solutions as $|x|\to\infty$ and as $t\searrow 0$

Let u be a non-negative local weak solution of (1.5) in Σ_T. For $\varepsilon\in(0,T)$ and $r>0$ set

$$|||u|||_{r,T-\varepsilon}\equiv\sup_{0<\tau\le T-\varepsilon}\ \sup_{\rho>r}\int_{K_\rho}\frac{u(x,\tau)}{\rho^{\lambda/(p-2)}}\,dx,\quad \lambda=N(p-2)+p.$$

THEOREM 2.1. *There exists a constant $\gamma=\gamma(N,p)$ such that for all $\varepsilon\in(0,T)$,*

$$\|\|u\|\|_{r,T-\varepsilon}\leq\gamma\,\varepsilon^{-\frac{1}{p-2}}\left\{1+\left(\frac{T}{r^p}\right)^{\frac{1}{p-2}}u(0,T-\varepsilon)\right\}^{\lambda/p}. \tag{2.1}$$

Moreover for all $t\in(0,T-\varepsilon)$, and all $\rho\geq r$,

$$\|u(\cdot,t)\|_{\infty,K_\rho}\leq\gamma\,\frac{\rho^{p/(p-2)}}{t^{N/\lambda}}\,\|\|u\|\|_{r,T-\varepsilon}^{p/\lambda}, \tag{2.2}$$

$$\int_0^t\int_{K_\rho}|Du|^{p-1}dxd\tau\leq\gamma\,t^{1/\lambda}\rho^{1+\frac{\lambda}{p-2}}\,\|\|u\|\|_{r,T-\varepsilon}^{1+\frac{p-2}{\lambda}}, \tag{2.3}$$

$$\|Du(\cdot,t)\|_{\infty,K_\rho}\leq\gamma\,\frac{\rho^{2/(p-2)}}{t^{(N+1)/\lambda}}\,\|\|u\|\|_{r,T-\varepsilon}^{2/\lambda}. \tag{2.4}$$

Moreover $(x,t)\to Du(x,t)$ is Hölder continuous in $K_\rho\times(\varepsilon,T-\varepsilon)$ with Hölder constants and exponent depending only upon $N,p,\gamma,\rho,\varepsilon$ and $\|\|u\|\|_{r,\varepsilon}$.

Remark 2.1. The functional dependence of these estimates is optimal as it can be verified for the explicit solution $\mathcal{B}(x,t)$.

Remark 2.2. The estimates (2.2)-(2.4) hold for solutions of variable sign, provided we assume (2.1).

PROOF OF THEOREM 1.1: The estimate (2.1) is the content of Corollary 7.1 of Chap. VI, whereas (2.2) follows from Theorem 4.5 of Chap. V. The gradient bound (2.3) is Lemma 9.1 of Chap. V and Remark 9.2. Inequalities (2.2)-(2.3) hold for a time interval

$$0\leq t\leq\gamma_*\|\|u\|\|_{r,T-\varepsilon}^{2-p}, \tag{2.5}$$

for a constant $\gamma_*=\gamma_*(N,p)$. In view of (2.1) they can be considered valid for all $t\in(0,T-\varepsilon)$. Indeed working within Σ_T, we may state them for every substrip

$$\mathbf{R}^N\times[t_1,t_2],\quad 0\leq t_1<t_2\leq T-\varepsilon,\quad t_2-t_1\leq\gamma_*\|\|u\|\|_{r,T-\varepsilon}.$$

In the proof of (2.4) we will work in the time interval (2.5). We begin with a *qualitative* information.

LEMMA 2.1. *For every $\varepsilon\in(0,T)$ and for every $r>0$ the quantities*

$$\sup_{\rho>r}\frac{\|Du(\cdot,t)\|_{\infty,K_\rho}}{\rho^{2/(p-2)}}$$

are finite for all $0<t\leq T-\varepsilon$.

PROOF: By the interpolation Theorem 5.1′ of Chap. VIII with $\varepsilon=1$, $\sigma=1/2$ and $\theta=\frac{1}{2}t$ we deduce

$$\|Du(\cdot,\tau)\|_{\infty,K_\rho} \le \gamma\rho^{-(N+2)} \int_0^t\int_{K_{2\rho}} |Du|^{p-1} dx d\tau + \left(\frac{\rho^2}{t}\right)^{\frac{1}{p-2}},$$

for all $\tau\in(\frac{3}{4}t,t)$. Estimating the right hand side by (2.3) we obtain

$$\frac{\|Du(\cdot,\tau)\|_{\infty,K_\rho}}{\rho^{2/(p-2)}} \le \gamma\left(t\,|\!|\!|u|\!|\!|_{r,T-\varepsilon}^{p-2}\right)^{1/\lambda} |\!|\!|u|\!|\!|_{r,T-\varepsilon} + t^{-\frac{1}{p-2}}.$$

Next we will turn such information into the quantitative estimate (2.4).

3. Proof of (2.4)

Let $t>0$ and $\rho>r$ be fixed and consider the box $\mathcal{Q}_o\equiv K_{2\rho}\times[\frac{1}{2}t,t]\subset\Sigma_T$. Let the radii $\{\rho_n\}$ and time levels $\{t_n\}$ be defined by

$$\rho_n\equiv\rho+2^{-n}\rho,\quad t_n\equiv\tfrac{1}{4}(1+2^{-n})t,\qquad n=0,1,2,\dots$$

and introduce the corresponding family of nested shrinking cylinders

$$\mathcal{Q}_n\equiv K_{\rho_n}\times\{t_n,t\},$$

with *vertex* at $(0,t)$. We will estimate the quantity $\|Du(\cdot,t)\|_{\infty,K_\rho}$, by using the techniques developed in Chap. VIII. The starting point is the the iterative inequality (5.4) in that chapter, which we rewrite here in the context of the cubes $\mathcal{Q}_n$ as

$$Y_{n+1}\le\gamma b^n k^{-\frac{\lambda_2}{N+2}}\mathcal{H}\,Y_n^{1+\frac{2}{N+2}},\quad \lambda_2\equiv N(p-2)+2p,$$

where

$$Y_n\equiv\iint_{\mathcal{Q}_n}(|Du|-k_n)_+^p\,dxd\tau,\qquad k_n=k-2^{-n}k,$$

(3.1) $$\mathcal{H}\equiv\left\{\sup_{\mathcal{Q}_o}|Du|^{p-2}\rho^{-2}+t^{-1}\right\},$$

and k is a positive number to be chosen. By Lemma 4.1 of Chap. I, the sequence $\{Y_n\}$ tends to zero as $n\to\infty$ if k is chosen to satisfy

$$k^{\lambda_2/2}\equiv\left(\gamma b^{\frac{N+2}{2}}\right)^{\frac{N+2}{2}}\mathcal{H}^{\frac{N+2}{2}}\int_{t/2}^t\int_{K_{2\rho}}|Du|^p dxd\tau.$$

We conclude that there exists a constant $\gamma=\gamma(N,p)$ such that for all $\frac{1}{4}t\le\tau\le t$,

(3.2) $$\|Du(\cdot,\tau)\|_{\infty,K_\rho}\le\gamma\mathcal{H}^{\frac{N+2}{\lambda_2}}\left(\int_{t/2}^t\int_{K_{2\rho}}|Du|^p dxd\tau\right)^{2/\lambda_2}.$$

To proceed we introduce the non-decreasing function of t

$$\Phi(t) \equiv \sup_{0<\tau\le t} \tau^{\frac{N+1}{\lambda}} \sup_{\rho>r} \frac{\|Du(\cdot,\tau)\|_{\infty,K_\rho}}{\rho^{2/(p-2)}}, \tag{3.3}$$

By Lemma 2.1 and (2.2) this quantity is well defined. In the estimates below we write $\Phi\equiv\Phi(t)$ if the dependence upon t is unambiguous. We estimate the quantity $\mathcal{H}$ introduced in (3.1) by

$$\mathcal{H} \le \Phi^{p-2}(t)\, t^{-\frac{(N+1)(p-2)}{\lambda}} + t^{-1},$$

and deduce from (3.2) that for all $\frac{1}{4}t \le \tau \le t$

$$\|Du(\cdot,\tau)\|_{\infty,K_\rho} \le \gamma \left[\Phi(t)\, t^{-\frac{N+1}{\lambda}}\right]^{\frac{(N+2)(p-2)}{\lambda_2}} \left(\int_{t/2}^{t} \int_{K_{2\rho}} |Du|^p dx d\tau \right)^{2/\lambda_2}$$
$$+ \gamma t^{-\frac{N+2}{\lambda_2}} \left(\int_{t/2}^{t} \int_{K_{2\rho}} |Du|^p dx d\tau \right)^{2/\lambda_2}$$

and

$$t^{\frac{N+1}{\lambda}} \frac{\|Du(\cdot,t)\|_{\infty,K_\rho}}{\rho^{2/(p-2)}} \tag{3.4}$$
$$\le \gamma \Phi^{\frac{(N+2)(p-2)}{\lambda_2}} t^{\frac{N+1}{\lambda}\frac{4}{\lambda_2}} \left(\int_{t/2}^{t} \rho^{-\frac{\lambda_2}{p-2}} \int_{K_{2\rho}} |Du(x,\tau)|^p dx d\tau \right)^{2/\lambda_2}$$
$$+ \gamma \left(t^{N/\lambda} \int_{t/2}^{t} \rho^{-\frac{\lambda_2}{p-2}} \int_{K_{2\rho}} |Du(x,\tau)|^p dx d\tau \right)^{2/\lambda_2}$$
$$\equiv G_1(t) + G_2(t).$$

Estimating $G_1(t)$ we have

$$G_1(t) \le \gamma \Phi^{\frac{(N+2)(p-2)}{\lambda_2}} \left\{ \int_{t/2}^{t} \tau^{-\frac{(N+1)(p-2)}{\lambda}} \tau^{\frac{N+1}{\lambda} p} \right.$$

$$\left. \times \left(\sup_{\rho>r} \frac{\|Du(\cdot,\tau)\|_{\infty,K_\rho}}{\rho^{2/(p-2)}} \right)^p d\tau \right\}^{2/\lambda_2}$$

$$\le \gamma \Phi^{\frac{(N+2)(p-2)}{\lambda_2}+\frac{2}{\lambda_2}} \left(\int_{t/2}^{t} \tau^{-\frac{(N+1)(p-2)}{\lambda}} \Phi^{p-1}(\tau)\, d\tau \right)^{2/\lambda_2}$$

$$\le \frac{1}{4}\Phi(t) + \gamma \int_0^t \tau^{-\frac{(N+1)(p-2)}{\lambda}} \Phi^{p-1}(\tau)\, d\tau.$$

To estimate $G_2(t)$ we refer back to the p.d.e. in (1.5). Let ζ be a non-negative piecewise smooth cutoff function in $K_{4\rho}\times\{\frac{1}{4}t,t\}$ that equals one on $K_{2\rho}\times\{\frac{1}{2}t,t\}$ and such that $|D\zeta| \le 2/\rho$ and $\zeta_t \le 4/t$. Taking $u\zeta^p$ in the weak formulation of (1.5) we obtain

$$\int_{t/2}^{t}\int_{K_{2\rho}} |Du|^p dx d\tau \le \frac{\gamma}{\rho^p}\int_{t/4}^{t}\int_{K_{4\rho}} u^p dx d\tau + \frac{\gamma}{t}\int_{t/4}^{t}\int_{K_{4\rho}} u^2 dx d\tau.$$

In estimating $G_2(t)$ we use the estimation (2.2) and the range (2.5) of t.

$$G_2(t) \le \gamma \left\{ t^{N/\lambda} \sup_{\rho>r} \int_{t/2}^{t} \rho^{-\frac{\lambda_2}{p-2}} \int_{K_{2\rho}} |Du|^p dx d\tau \right\}^{2/\lambda_2}$$

$$\le \gamma \left\{ \int_{t/4}^{t} \tau^{N/\lambda} \left(\sup_{\rho>r} \frac{\|u(\cdot,\tau)\|_{\infty,K_\rho}}{\rho^{p/(p-2)}} \right)^{p-1} \left(\sup_{\rho>r} \int_{K_\rho} \frac{u(x,\tau)}{\rho^{\lambda/(p-2)}}\, dx \right) \right\}^{2/\lambda_2}$$

$$+ \gamma \left\{ \int_{t/4}^{t} \tau^{\frac{N}{\lambda}-1} \left(\sup_{\rho>r} \frac{\|u(\cdot,\tau)\|_{\infty,K_\rho}}{\rho^{p/(p-2)}} \right) \left(\sup_{\rho>r} \int_{K_\rho} \frac{u(x,\tau)}{\rho^{\lambda/(p-2)}}\, dx \right) \right\}^{2/\lambda_2}$$

$$\le \gamma \left\{ t^{\frac{N}{\lambda}+1} t^{-\frac{N}{\lambda}(p-1)} |||u|||_{r,T-\varepsilon}^{1+(p-1)p/\lambda} \right\}^{2/\lambda_2} + \gamma \left\{ |||u|||_{r,T-\varepsilon}^{1+p/\lambda} \right\}^{2/\lambda_2}$$

$$\le \gamma |||u|||_{r,T-\varepsilon}^{2/\lambda} \left\{ 1 + t\, |||u|||_{r,T-\varepsilon}^{p-2} \right\}^{\frac{2}{\lambda_2}\frac{p}{\lambda}}$$

$$\le \gamma |||u|||_{r,T-\varepsilon}^{2/\lambda}.$$

Combining these estimates in (3.4) gives for all $0<t\le\gamma_* |||u|||_{r,T-\varepsilon}^{2-p}$,

$$\Phi(t) \le \gamma \int_0^t \tau^{-\frac{(N+1)(p-2)}{\lambda}} \Phi^{p-1}(\tau)\, d\tau + \gamma |\!|\!| u |\!|\!|_{r,T-\varepsilon}^{2/\lambda},$$

for a constant $\gamma = \gamma(N,p)$. It follows that $\Phi(\cdot)$ is majorised by the solution of

$$\begin{cases} V'(t) \le \gamma t^{-\frac{(N+1)(p-2)}{\lambda}} V^{p-1}(t), \\ V(0) = \gamma |\!|\!| u |\!|\!|_{r,T-\varepsilon}^{2/\lambda}, \ 0 < t \le \gamma_* |\!|\!| u |\!|\!|_{r,T-\varepsilon}^{2/\lambda}. \end{cases}$$

Solving this explicitly gives

$$\Phi(t) \le \gamma |\!|\!| u |\!|\!|_{r,T-\varepsilon}^{2/\lambda} \left\{ 1 - \gamma \left(t\, |\!|\!| u |\!|\!|_{r,T-\varepsilon}^{p-2} \right)^{2/\lambda} \right\}^{-1/(p-2)}.$$

Therefore choosing t so small that

$$\left\{ 1 - \gamma \left(t\, |\!|\!| u |\!|\!|_{r,T-\varepsilon}^{p-2} \right)^{2/\lambda} \right\}^{-1/(p-2)} \le 2,$$

we will have

$$t^{\frac{N+1}{\lambda}} \frac{\| Du(\cdot,t) \|_{\infty,K_\rho}}{\rho^{2/(p-2)}} \le 2\gamma |\!|\!| u |\!|\!|_{r,T-\varepsilon}^{2/\lambda}$$

for all such t and all $\rho > r$.

4. Initial traces

THEOREM 4.1. *Let u be a non-negative local weak solution of (1.5) in Σ_T. There exists a unique Radon measure μ such that*

$$\lim_{t \searrow 0} \int_{\mathbf{R}^N} u(x,t)\varphi\, dx = \int_{\mathbf{R}^N} \varphi\, d\mu, \quad \forall \varphi \in C_o^\infty(\mathbf{R}^N). \tag{4.1}$$

Moreover, as $|x| \to \infty$, μ 'grows' at most as $|x|^{p/(p-2)}$. Precisely,

$$\sup_{\rho > r} \fint_{K_\rho} \frac{d\mu}{\rho^{p/(p-2)}} < \infty, \qquad \forall r > 0. \tag{4.2}$$

PROOF: The existence of a Radon measure μ satisfying (4.1)-(4.2) follows from the global Harnack estimates of §7 of Chap. VI. Indeed by Corollary 7.1 of that Chapter, for every cube $[x_o + K_\rho] \subset \mathbf{R}^N$ and all $\varphi \in C_o^\infty(K_\rho)$,

$$\left| \int_{K_\rho} u(x,t)\varphi(x) dx \right| \le \gamma\, (N, p, \rho, T, u(x_o, T-\varepsilon))\ \|\varphi\|_{\infty,K_\rho},$$

for all $0<t\le T-\varepsilon$ and all $\varepsilon\in(0,T)$. Therefore $\{u(\cdot,t)\}_{0<t<T-\varepsilon}$ is a net of equibounded linear operators in $C_o^\infty(\mathbf{R}^N)$, and for a subnet, indexed with t',

$$\lim_{t'\searrow 0}\int_{\mathbf{R}^N} u(x,t')\varphi\,dx=\int_{\mathbf{R}^N}\varphi\,d\mu,\qquad \forall\varphi\in C_o^\infty(\mathbf{R}^N),$$

for a Radon measure μ. The uniqueness of such a measure is a consequence of the following:

LEMMA 4.1. *Let u be a non-negative local weak solution of (1.5) in Σ_T. Then*

$$\forall\rho>0,\quad \forall\sigma\in(0,1),\quad \forall 0<\tau<t<T/2,$$

$$\int_{K_{(1+\sigma)\rho}} u(x,t)dx\ge\int_{K_\rho} u(x,\tau)dx-\frac{\gamma}{\sigma}(t-\tau)^{1/\lambda}\rho^{\frac{\lambda}{p-2}}\,|||u|||_{r,T-\varepsilon}^{1+\frac{p-2}{\lambda}}. \tag{4.3}$$

PROOF: Fix $0<\tau<t$ and $\sigma\in(0,1)$, and let $x\to\zeta(x)$ be a non-negative piecewise smooth cutoff function in $K_{(1+\sigma)\rho}$ that equals one on K_ρ and such that $|D\zeta|\le 2\rho/\sigma$. In the weak formulation of (1.5) take ζ as a testing function. Integrating over (τ,t) gives

$$\int_{K_{(1+\sigma)\rho}} u(x,t)\,dx\ge\int_{K_\rho} u(x,\tau)\,dx-\frac{2}{\sigma\rho}\int_\tau^t\int_{K_{(1+\sigma)\rho}}|Du|^{p-1}\,dx\,ds.$$

To prove (4.3) we estimate the right hand side of this inequality by (2.3).

We now prove the uniqueness part of Theorem 4.1. Suppose that out of the net $\{u(x,t)\}_{0<t<T-\varepsilon}$ we may select two subnets indexed with τ' and t' such that

$$\lim_{t'\searrow 0}\int_{\mathbf{R}^N} u(x,t')\varphi\,dx=\int_{\mathbf{R}^N}\varphi\,d\mu,\qquad \lim_{\tau'\searrow 0}\int_{\mathbf{R}^N} u(x,\tau')\varphi\,dx=\int_{\mathbf{R}^N}\varphi\,d\nu,$$

for all $\varphi\in C_o^\infty(\mathbf{R}^N)$ and $\mu\ne\nu$. Then we let $\tau\to 0$ along τ' in (4.2) and then let $t\to 0$ along t'. This gives

$$\int_{K_{(1+\sigma)\rho}} d\mu\ge\int_{K_\rho} d\nu.$$

Interchanging the role of μ and ν proves the Theorem since $\sigma\in(0,1)$ is arbitrary.

5. Estimating $|Du|^{p-1}$ in Σ_T

Local integral estimates of $|Du|^{p-1}$ are crucial both in the global Harnack estimate of §7 Chap. VI and in the theory of initial traces. The inequality (2.3) of Theorem

2.1 is *local* but holds for all $\rho > r$. Therefore it implies some control on the behaviour of $|Du|$ as $|x| \to \infty$. This behaviour can be given an integral form, by means of the weights

$$A_\alpha(x) \equiv (1+|x|^p)^{-\alpha}, \tag{5.1}$$

where α is a positive number satisfying

$$\alpha p = \frac{\lambda}{p-2} + \sigma, \qquad \text{for some } \sigma > 0. \tag{5.2}$$

THEOREM 5.1. *Let u be a non-negative local weak solution of (1.5) in Σ_T. Then for every $\sigma > 0$, there exists a constant $\gamma = \gamma(N,p,\sigma)$ such that for all $r > 0$ and all $\varepsilon \in (0,T)$,*

$$\sup_{0<t\le T-\varepsilon} \int_{\mathbf{R}^N} u(x,t) A_\alpha(x)\,dx \le \gamma \|\!|u|\!\|_{r,T-\varepsilon}, \tag{5.3}$$

$$\int_0^t \int_{\mathbf{R}^N} |Du|^{p-1} A_\alpha(x)\,dx d\tau \le \gamma\, t^{1/\lambda}\, \|\!|u|\!\|_{r,T-\varepsilon}^{1+\frac{p-2}{\lambda}}. \tag{5.4}$$

Remark 5.1. *The constant $\gamma(N,p,\sigma) \nearrow \infty$ as $\sigma \searrow 0$.*

PROOF OF (5.3): Without loss of generality we may assume that $r=1$. Then for all $0<t\le T-\varepsilon$,

$$\begin{aligned}\int_{\mathbf{R}^N} u(x,t)A_\alpha(x)\,dx &\le \int_{\{|x|<1\}} u(x,t)A_\alpha(x)\,dx + \sum_{n=0}^{\infty} \int_{\{2^n<|x|<2^{n+1}\}} u(x,t)A_\alpha(x)\,dx \\ &\le \|\!|u|\!\|_{r,T-\varepsilon} + 2^{\frac{\lambda}{p-2}} \sum_{n=0}^{\infty} 2^{-\sigma n}\, \|\!|u|\!\|_{r,T-\varepsilon}.\end{aligned}$$

PROOF OF (5.4): It will suffice to establish the estimate for t in the interval (2.5). We will use this fact with no further mention. First we observe that the inequality

$$u(x,t)A_{1/(p-2)} \le \gamma t^{-\frac{N}{\lambda}} \|\!|u|\!\|_{r,T-\varepsilon}^{p/\lambda} \tag{5.5}$$

holds for all $x \in \mathbf{R}^N$ and all $0 < t \le T-\varepsilon$. This is obvious if $|x| \le r$ with the constant γ depending also upon r. If $|x| > r$, we apply (2.2) to the cube $K_{2|x|}$.

Let $\eta \in (0, T-\varepsilon)$ and in the weak formulation of (1.5), take the testing function

$$(t-\eta)_+^{1/p} u^{1-\frac{2}{p}} \left(A_{\alpha+\frac{1}{p}}^{1/p} \zeta\right)^p,$$

where $x \to \zeta(x)$ is the usual cutoff function in K_ρ. After a Steklov averaging process and standard calculations, we obtain

$$(5.6)\qquad \int_\eta^t (\tau-\eta)^{1/p}\int_{K_\rho}\frac{|Du|^p}{u^{2/p}}A_{\alpha+\frac{1}{p}}\zeta^p dx d\tau$$

$$\le \gamma\int_\eta^t(\tau-\eta)^{1/p}\int_{K_\rho}u^{\frac{p-2}{p}}u^{p-1}\left|D\left(A^{1/p}_{\alpha+\frac{1}{p}}\zeta\right)\right|^p dxd\tau$$

$$+\gamma\int_\eta^t(\tau-\eta)^{\frac{1}{p}-1}\int_{K_\rho}u^{\frac{p-2}{p}}A_{1/p}\,uA_\alpha dxd\tau = J_\rho^{(1)}+J_\rho^{(2)}.$$

As for $J_\rho^{(2)}$ we have

$$J_\rho^{(2)}\le\gamma\int_\eta^t(\tau-\eta)^{\frac{1}{\lambda}-1}\int_{K_\rho}\tau^{-\frac{N(p-2)}{p\lambda}}\frac{|u(x,\tau)|^{\frac{p-2}{p}}}{(1+|x|^p)^{1/p}}u(x,\tau)A_\alpha(x)dxd\tau,$$

so that by (5.5) and (5.3),

$$J_\rho^{(2)}\le\gamma(t-\eta)_+^{1/\lambda}|||u|||_{r,T-\varepsilon}^{1+\frac{p-2}{\lambda}}.$$

We estimate $J_\rho^{(1)}$:

$$J_\rho^{(1)}\le\gamma\int_\eta^t(\tau-\eta)^{\frac{1}{p}}\int_{K_\rho}u^{\frac{p-2}{p}}u^{p-1}A_{\alpha+\frac{1}{p}}|D\zeta|^p dxd\tau$$

$$+\gamma\int_\eta^t(\tau-\eta)^{1/p}\int_{K_\rho}u^{\frac{p-2}{p}}u^{p-1}|DA^{1/p}_{\alpha+\frac{1}{p}}|^p dxd\tau$$

$$=J_\rho^{(1,1)}+J_\rho^{(1,2)}.$$

Since

$$|DA^{1/p}_{\alpha+\frac{1}{p}}|\le\gamma|A_{\frac{\alpha}{p}+\frac{1}{p^2}+\frac{1}{p}}|^p\le\gamma A_\alpha\,A_{1/p}\,A_1,$$

by (5.5), (5.3) and the range (2.5) of t,

$$J_\rho^{(1,2)}\le\gamma\int_\eta^t(\tau-\eta)^{\frac{1}{p}}\int_{K_\rho}\left(u^{\frac{p-2}{p}}A_{\frac{1}{p}}\right)\left(u^{p-2}A_1\right)u(x,\tau)A_\alpha(x)\,dxd\tau$$

$$\le\gamma(t-\eta)_+^{1/\lambda}|||u|||_{r,T-\varepsilon}^{1+\frac{p-2}{\lambda}}.$$

As for $J_\rho^{(1,1)}$, since $|D\zeta|\le 2/\rho$, again by (5.5) and (5.3)

$$J_\rho^{(1,1)}\le\gamma(t-\eta)_+^{1/\lambda}|||u|||_{r,T-\varepsilon}^{1+\frac{p-2}{\lambda}}.$$

Combining these estimates in (5.6),

$$\int_\eta^t\int_{K_\rho}(\tau-\eta)^{\frac{1}{p}}\frac{|Du|^p}{u^{2/p}}\,A_{\alpha+\frac{1}{p}}\,dxd\tau\le\gamma(t-\eta)^{1/\lambda}|||u|||_{r,T-\varepsilon}^{1+\frac{p-2}{\lambda}},\tag{5.7}$$

where we have changed ρ into 2ρ. Next, for all $\eta\le t\le T-\varepsilon$

$$\begin{aligned}\int_\eta^t\int_{K_\rho}|Du|^{p-1}A_{\alpha+\frac{1}{p}}dxd\tau&=\int_\eta^t\int_{K_\rho}\left\{(\tau-\eta)^{(p-1)/p^2}\frac{|Du|^{p-1}}{u^{2(p-1)/p^2}}A_{\alpha+\frac{1}{p}}^{\frac{p-1}{p}}\right\}\\&\quad\times\left\{(\tau-\eta)^{-(p-1)/p^2}u^{2(p-1)/p^2}A_{\alpha+\frac{1}{p}}^{\frac{1}{p}}\right\}dxd\tau\\&\le\left(\int_\eta^t\int_{K_\rho}(t-\eta)^{\frac{1}{p}}\frac{|Du|^p}{u^{2/p}}A_{\alpha+\frac{1}{p}}dxd\tau\right)^{(p-1)/p}\\&\quad\times\left(\int_\eta^t\int_{K_\rho}(\tau-\eta)^{-\frac{p-1}{p}}u^{\frac{p-2}{p}}A_{\frac{1}{p}}uA_\alpha dxd\tau\right)^{1/p}\\&\le\gamma(t-\eta)^{\frac{1}{\lambda}}|||u|||_{r,T-\varepsilon}^{1+\frac{p-2}{\lambda}}.\end{aligned}$$

6. Uniqueness for data in $L^1_{loc}(\mathbf{R}^N)$

THEOREM 6.1. *Let u and v be two non-negative local weak solutions of (1.5) in Σ_T, satisfying*

$$\lim_{t\searrow 0}[u(\cdot,t)-v(\cdot,t)]=0,\quad in\ L^1_{loc}(\mathbf{R}^N).$$

Then $u\equiv v$ in Σ_T.

PROOF: Fix $r>0$ and some $\varepsilon\in(0,T)$ and set

$$|||u|||_{r,T-\varepsilon}+|||v|||_{r,T-\varepsilon}\equiv\Lambda.$$

Then and u and v satisfy all the estimates of Theorem 2.1 with the quantities $|||u,v|||_{r,T-\varepsilon}$ replaced by Λ within the strip Σ_{T_o}, where

$$0<T_o=\min\{T;\gamma_*\Lambda^{-(p-2)}\},$$

and γ_* is the constant appearing in (2.5). It will suffice to prove uniqueness within the strip Σ_{T_o}. The difference $w=u-v$ satisfies

$$w_t - (a^{i,j}(x,t)w_{x_i})_{x_j} = 0, \qquad \text{in } \Sigma_{T_o} \tag{6.1}$$

where

$$a^{i,j}(x,t) = \left(\int_0^1 |D(su+(1-s)v)|^{p-2} ds \right) \delta_{ij}$$
$$+ (p-2) \int_0^1 |D(su+(1-s)v)|^{p-4}$$
$$\times (su+(1-s)v)_{x_i}(su+(1-s)v)_{x_j} ds.$$

The matrix $(a^{i,j})$ is positive semi-definite and for all $\xi \in \mathbf{R}^N$ and $(x,t) \in \Sigma_{T_o}$

$$\begin{cases} a_o(x,t)|\xi|^2 \le a^{i,j}(x,t)\xi_i\xi_j \le (p-1)a_o(x,t)|\xi|^2, \\ a_o(x,t) = \int_0^1 |D(su+(1-s)v)|^{p-2} ds, \ (x,t) \in \Sigma_{T_o}. \end{cases} \tag{6.2}$$

Let $A_\alpha(x)$ be the weight introduced in (5.1) with α satisfying (5.2). In the arguments below, γ denotes a positive constant that can be determined a priori only in terms of N, p, σ and Λ.

6-(i). Auxiliary lemmas

LEMMA 6.1. *There exists a constant $\gamma = \gamma(N,p,\sigma,\Lambda)$ such that if $w(\cdot,t) \to 0$ in $L^1_{loc}(\mathbf{R}^N)$ as $t \searrow 0$, then*

$$\int_{\mathbf{R}^N} |w(x,t)| A_\alpha(x)\, dx \le \gamma t^{1/\lambda}, \qquad 0 < t < T_o.$$

PROOF: The functions $w^\pm$ are both weak subsolutions of (6.1), i.e.,

$$w_t^\pm - \left(a^{i,j}(x,t) w_{x_i}^\pm\right)_{x_j} \le 0 \ \text{ weakly in } \Sigma_{T_o}.$$

By working separately with w^+ and w^- we may assume that w is a non-negative subsolution of (6.1). In the weak formulation of (6.1) take the test function $x \to A_\alpha(x)\zeta(x)$, where ζ is the usual cutoff function in K_ρ. Using the assumptions of the lemma we deduce

$$\int_{K_\rho} |w(x,t)| A_\alpha(x)\zeta\, dx \le \gamma \int_0^t \int_{K_\rho} \left(|Dv| + |Dv|\right)^{p-1} \left|DA_\alpha \zeta\right| dx d\tau$$
$$\le \gamma \int_0^t \int_{K_\rho} \left(|Dv| + |Dv|\right)^{p-1} A_\alpha \left|D\zeta\right| dx d\tau$$
$$+ \gamma \int_0^t \int_{K_\rho} \left(|Dv| + |Dv|\right)^{p-1} \left|DA_\alpha\right| dx d\tau.$$

In the last integral, $|DA_\alpha|\le\gamma A_{\alpha+1/p}$ and in the first integral, since $|D\zeta|=0$ on $K_{\rho/2}$ we have $A_\alpha|D\zeta|\le\gamma A_{\alpha+1/p}$ for $\rho>1$. Therefore letting $\rho\to\infty$

$$\int_{\mathbf{R}^N}|w(x,t)|A_\alpha(x)\,dx\le\gamma\int_0^t\int_{\mathbf{R}^N}\left(|Dv|+|Dv|\right)^{p-1}A_{\alpha+1/p}\,dxd\tau,$$

and the conclusion follows from Theorem 5.1.

LEMMA 6.2. *If $w(\cdot,t)\to 0$ in $L^1_{loc}(\mathbf{R}^N)$ as $t\to 0$, then*

$$w(\cdot,t)\longrightarrow 0\quad in\ L^{1+\eta}_{loc}(\mathbf{R}^N)\ \ as\ t\to 0,\quad \forall\eta\in\left(0,\frac{1}{N}\right).$$

PROOF: Let $\eta\in(0,\frac{1}{N})$ be fixed. Then $\forall t\in(0,T_o)$

$$\int_{\mathbf{R}^N}|w(x,t)|^{1+\eta}A_{\alpha+\eta/(p-2)}(x)dx$$
$$\le\int_{\mathbf{R}^N}|w(x,t)|^{\eta}A_{\eta/(p-2)}(x)|w(x,t)|A_\alpha(x)dx.$$

By (5.5), $|w(x,t)|^\eta A_{\eta/(p-2)}(x)\le\gamma t^{-\frac{N}{\lambda}\eta}$, so that by Lemma 6.1,

$$\int_{\mathbf{R}^N}|w(x,t)|^{1+\eta}A_{\alpha+\eta/(p-2)}(x)dx\le\gamma t^{-\frac{N\eta}{\lambda}}\int_{\mathbf{R}^N}|w(x,t)|A_\alpha(x)\,dx$$
$$\le\gamma t^{\frac{1}{\lambda}(1-N\eta)}.$$

6–(ii). *Proof of Theorem 6.1*

In (6.1) we may assume, by working separately with w^+ and w^-, that $w\ge 0$. In its weak formulation we take the testing functions

$$(w+\delta)^\eta\left(A_\alpha^{\frac{1}{2}}\zeta\right)^2,\quad\eta\in\left(0,\frac{1}{N}\right),\quad\delta\in(0,T_o).$$

Integrating over $K_\rho\times(\delta,t)$, $0<\delta<t\le T_o$, we obtain

$$\text{(6.3)} \qquad \frac{1}{1+\eta}\int_{K_\rho}(w+\delta)^{1+\eta}A_\alpha\zeta^2dx$$

$$+\eta\int_\delta^t\int_{K_\rho} a_o(x,\tau)\frac{|Dw|^2}{(w+\delta)^{1-\eta}}\left(A_\alpha^{\frac{1}{2}}\zeta\right)^2dxd\tau$$

$$\leq\frac{1}{1+\eta}\int_{K_\rho\times\{\delta\}}(w+\delta)^{1+\eta}A_\alpha\zeta^2dx$$

$$+\gamma\int_\delta^t\int_{K_\rho} a_o(x,\tau)\frac{|Dw|}{(w+\delta)^{\frac{1-\eta}{2}}}(w+\delta)^{\frac{1+\eta}{2}}$$

$$\times\left(A_\alpha^{\frac{1}{2}}\zeta\right)\left|D\left(A_\alpha^{\frac{1}{2}}\zeta\right)\right|dxd\tau,$$

where $a_o(x,t)$ has been defined in (6.2). By the Schwartz inequality the last integral is majorized by

$$\frac{\eta}{2}\int_\delta^t\int_{K_\rho} a_o(x,\tau)\frac{|Dw|^2}{(w+\delta)^{1-\eta}}\left(A_\alpha^{\frac{1}{2}}\zeta\right)^2dxd\tau$$

$$+\gamma(\eta)\int_\delta^t\int_{K_\rho} a_o(x,\tau)(w+\delta)^{1+\eta}\left(A_\alpha|D\zeta|^2+\left|DA_\alpha^{\frac{1}{2}}\right|^2\right)dxd\tau.$$

We absorb the integral involving $|Dw|^2$ on the left-hand side of (6.3) and discard the resulting non-negative term. Finally, we observe that by the definition of A_α and the structure of ζ we have

$$A_\alpha|D\zeta|^2+\left|DA_\alpha^{\frac{1}{2}}\right|^2\leq\gamma A_\alpha(x)\,A_{\frac{2}{p}}(x).$$

Carrying these remarks in (6.3) gives

$$\text{(6.4)} \qquad \int_{K_\rho\times\{t\}}(w+\delta)^{1+\eta}A_\alpha\zeta^2dx\leq\int_{K_\rho\times\{t\}}(w+\delta)^{1+\eta}A_\alpha(x)\,dx$$

$$+\gamma\int_\delta^t\int_{K_\rho} a_o(x,\tau)A_{\frac{2}{p}}(x)(w+\delta)^{1+\eta}A_\alpha(x)\,dxd\tau.$$

Next by (6.2) and (2.4)

$$a_o(x,\tau)A_{\frac{2}{p}}(x)\leq\gamma\frac{|x|^2}{(1+|x|^p)^{2/p}}\,\Lambda^{\frac{2}{\lambda}(p-2)}\tau^{-\frac{(N+1)}{\lambda}(p-2)}.$$

Substitute this last estimate in (6.4) and let $\delta\to0$ for $\rho\geq1$ fixed so that by Lemma 6.2

$$\int_{K_\rho\times\{\delta\}} (w+\delta)^{1+\eta} A_\alpha(x)\, dx \longrightarrow 0 \text{ as } \delta\to 0.$$

Then we let $\rho\to\infty$. The net result is

$$\int_{\mathbf{R}^N} |w(x,t)|^{1+\eta} A_\alpha(x)\, dx$$
$$\leq \gamma \int_0^t \tau^{-\frac{(N+1)}{\lambda}(p-2)} \int_{\mathbf{R}^N} |w(x,\tau)|^{1+\eta} A_\alpha(x)\, dx d\tau.$$

Since $\tau^{-\frac{(N+1)}{\lambda}(p-2)} \in L^1(0,t)$, this implies

$$t\to \int_{\mathbf{R}^N} |w(x,t)|^{1+\eta} A_\alpha(x) dx \equiv 0,$$

by Gronwall's lemma, provided

$$t\to \int_{\mathbf{R}^N} |w(x,t)|^{1+\eta} A_\alpha(x) dx \in L^\infty(0,T_o).$$

Now the parameter α in the calculations above is arbitrary and only restricted by (5.2). If α is replaced by $\alpha+\eta/(p-2)$, then Lemma 6.2 and its proof ensure the $L^\infty(0,T_o)$ requirement and the theorem follows.

Remark 6.1. For non-negative solutions u and v of (1.5) in Σ_T, the quantities

(6.5) $$|\!|\!|u|\!|\!|_{r,T-\varepsilon}, \quad |\!|\!|v|\!|\!|_{r,T-\varepsilon} \quad \text{are finite.}$$

The proof of Theorem 6.1 uses only this information. Indeed by Remark 2.2 such a *growth condition* implies all the estimates of Theorem 2.1. We conclude that the uniqueness theorem for initial data taken in the sense of $L^1_{loc}(\mathbf{R}^N)$ holds for solutions of variable sign provided (6.5) holds.

7. Solving the Cauchy problem

Consider the Cauchy problem

(7.1) $$\begin{cases} u\in C\left(0,T;L^1_{loc}(\mathbf{R}^N)\right)\cap L^p_{loc}\left(0,T;W^{1,p}_{loc}(\mathbf{R}^N)\right), & p>2,\\ u_t-\operatorname{div}\left(|Du|^{p-2}Du\right)=0 \text{ in } \Sigma_T, \text{ for some } T>0\\ u(\cdot,0)=u_o\in L^1_{loc}(\mathbf{R}^N). \end{cases}$$

As indicated in the first of (7.1) the initial datum is taken in the sense of $L^1_{loc}(\mathbf{R}^N)$. By Theorem 6.1 and Remark 6.1 there is at most one solution to (7.1) within the class of functions u satisfying

(7.2) $$|||u|||_{r,T-\varepsilon} < \infty \quad \text{for some } \varepsilon \in (0,T).$$

Existence of a solution satisfying (7.2) can be established if the initial datum u_o satisfies the *growth condition*

(7.3) $$|||u_o|||_r \equiv \sup_{\rho>r} \int_{K_\rho} \frac{|u_o(x)|}{\rho^{\lambda/(p-2)}}\, dx < \infty, \quad \text{for some } r>0.$$

Since $u_o \in L^1_{loc}(\mathbf{R}^N)$ if $|||u_o|||_r$ is finite for some $r>0$, it is finite for all $r>0$.

THEOREM 7.1. *Let u_o satisfy (7.3) for some $r>0$. There exists a constant $\gamma_* = \gamma_*(N,p)$ such that defining*

(7.4) $$T = \gamma_* \lim_{r\to\infty} |||u_o|||_r^{2-p}$$

there exists a unique solution u to (7.1) in Σ_T. Moreover u satisfies (7.2) for all $\varepsilon \in (0,T)$ and the estimates (2.2)-(2.4) of Theorem 2.1.

Remark 7.1. This is an existence theorem *local in time* and the largest existence time is estimated by (7.4). The functional dependence in (7.4) is optimal as shown by the following explicit solution.

$$\mathcal{D}(x,t) = \left\{ A\left(\frac{T}{T-t}\right)^{\frac{N(p-2)}{\lambda(p-1)}} + \left(\frac{p-2}{p}\right)\lambda^{-\frac{1}{p-1}} \left(\frac{|x|^p}{T-t}\right)^{\frac{1}{p-1}} \right\}^{\frac{p-1}{p-2}},$$

where A and T are two positive parameters. By direct calculation we have

$$\lim_{r\to\infty} |||\mathcal{D}(\cdot,0)|||_r = \frac{\omega_N}{N}\left(\frac{p-2}{p}\right)^{\frac{p-1}{p-2}} (\lambda T)^{-\frac{1}{p-2}},$$

where ω_N is the area of the unit sphere in $\mathbf{R}^N$. Therefore $\mathcal{D}(x,t)$ exists up to the blow-up time

$$T = \gamma_* \left\{ \lim_{r\to\infty} |||\mathcal{D}(\cdot,0)|||_r \right\}^{-(p-2)},$$

where

$$\gamma_* = \lambda^{-\frac{1}{p-2}} \left(\frac{\omega_N}{N}\right)^{p-2} \left(\frac{p-2}{p}\right)^{p-1}.$$

For $n=1,2,\ldots$, consider the sequence of *truncated* initial data

$$u_{o,n}(x) \equiv \begin{cases} \max\{-n;\, \min\{u_o(x);n\}\}, & \text{for } |x|<n \\ 0, & \text{for } |x| \ge n. \end{cases}$$

It is apparent that for all $n=1,2,\ldots$,

(7.5) $$|||u_{o,n}|||_r \le |||u_o|||_r .$$

Consider also the family of approximating problems

$$(7.1)_n \qquad \begin{cases} u_{n,t} - \operatorname{div}|Du_n|^{p-2}Du_n = 0, & \text{in } \mathbf{R}^N \times \mathbf{R}^+ \\ u_n(\cdot,0) = u_{o,n}. \end{cases}$$

Since $u_{o,n}$ are compactly supported in $\mathbf{R}^N$, $(7.1)_n$ can be uniquely solved as indicated in §12 of Chap. VI. By the maximum principle the solutions u_n are bounded by n. Therefore the quantities

$$|||u_n|||_{r,t} \equiv \sup_{0<\tau\le t}\ \sup_{\rho>r} \int_{K_\rho} \frac{u_n(x,\tau)}{\rho^{\lambda/(p-2)}}\,dx$$

are finite for all $r,t>0$. It follows that the sequence $\{u_n\}$ satisfies (2.2)-(2.4) of Theorem 2.1. We will turn such n-dependent information into a quantitative sup-estimate of $\{u_n\}$ independent of n. Let $x\to\zeta(x)$ be the standard cutoff function in $K_{2\rho}$. Then $(7.1)_n$ implies

$$\int_{K_\rho} u_n(x,t)\,dx \le \int_{K_{2\rho}} u_{o,n}dx + \frac{2}{\rho}\int_0^t\int_{K_{4\rho}} |Du_n|^{p-1}dxd\tau.$$

We divide by $\rho^{\lambda/(p-2)}$ and take the supremum over all $\rho>r$. Taking into account (7.5) and (2.3) this gives

$$(7.6) \qquad |||u_n|||_{r,t} \le \gamma_o\,|||u_o|||_r + \gamma_1\left(t\,|||u_n|||_{r,t}^{p-2}\right)^{1/\lambda} |||u_n|||_{r,t},$$

for two constants $\gamma_i=\gamma_i(N)$, $i=0,1$. Let t_n be defined by

$$\gamma_1\left(t_n\,|||u_n|||_{r,t}^{p-2}\right)^{1/\lambda} = \frac{1}{2}.$$

Then from (7.6) for all $t\in(0,t_n)$

$$|||u_n|||_{r,t} \le 2\gamma_o\,|||u_o|||_r.$$

This implies that $t_n\ge T_r$ for all $n=1,2,\ldots,$, where T_r is defined by

$$\gamma_1\left\{T_r\,(2\gamma_o|||u_o|||_r)^{p-2}\right\} = \frac{1}{2}.$$

We summarise:

LEMMA 7.1. *Let $\{u_n\}$ be the sequence of the approximating solutions $(7.1)_n$. There exists a constants $\gamma=\gamma(N,p)$ and $\gamma_*=\gamma_*(N,p)$ independent of n, such that*

$$(7.7) \qquad |||u_n|||_{r,t} \le \gamma|||u_o|||_r, \quad \textit{for all } 0<t\le T_r,$$

where

$$(7.8) \qquad T_r \equiv \gamma_*|||u_o|||_r^{2-p}.$$

Given such an estimate, the Cauchy problem (7.1) can be solved by a standard limiting process. Indeed by Theorem 2.1 the sequences

$$\{u_n\}_{n\in\mathbf{N}}, \qquad \left\{\frac{\partial}{\partial x_i}u_n\right\}_{n\in\mathbf{N}}, \qquad i=1,2,\ldots,N,$$

are locally equibounded and equi−Hölder continuous in $\mathbf{R}^N\times(0,T_r)$. This gives the existence of a unique solution in Σ_{T_r}. The largest time of existence can be calculated from (7.8) by letting $r\to\infty$. In particular the solution to (7.1) is global in time if

$$\lim_{r\to\infty}\sup_{\rho>r}\int_{K_\rho}\frac{u_o(x)}{\rho^{\lambda/(p-2)}}\,dx=0.$$

This occurs for example if $u_o\in L^1(\mathbf{R}^N)$.

8. Bibliographical notes

Theorem 2.1 is taken from [41]. A weaker version of (2.2) in 1−space dimension is due to Kalashnikov [58]. It is remarkable that in (2.4) one can also control the behaviour of the space-gradient $|Du|$ as $|x|\to\infty$. Since $|Du|^2$ is a non-negative subsolution of a porous medium−type equation (see (1.8) of Chap. IX) the same techniques yield a version of (2.2) for such degenerate p.d.e. The analog of (2.2) for the porous medium equation is due to Bénilan−Crandall−Pierre [10] in the context of an existence theorem. A rather general version is in [4]. Perhaps the most relevant estimate of Theorem 2.1 is the integral gradient bound (2.3) proved in [41]. A version of such a local bound, for the porous medium equation is in [4] and reads

$$\int_{K_\rho}\left|Du^m\right|dxd\tau\le\gamma\,t^{1/\kappa}\rho^{1+\frac{\kappa}{m-1}}|||u|||_{r,T-\varepsilon}^{1+\frac{m-1}{\kappa}},\qquad \kappa=N(m-1)+2,$$

where $\gamma=\gamma(N,m)$ and

$$|||u|||_{r,T-\varepsilon}\equiv\sup_{0<t\le T-\varepsilon}\sup_{\rho>r}\int_{K_\rho}\frac{u(x,t)}{\rho^{\kappa/(m-1)}}\,dx.$$

The estimate holds for *small* time intervals and for general non-linearities. We refer to [4] for details. There is no analog of (2.4) for the porous medium equation. Theorems 4.1 is taken from [41]. The analog for the porous medium equations is in [6] and for general non-linearities [4]. It would be desirable to have a version of the uniqueness Theorem 6.1 for initial data measures. This would parallel the analogous theory for the heat equation.

XII

Non-negative solutions in Σ_T. The case $1<p<2$

1. Introduction

We will investigate the structure of non-negative solutions in the strip Σ_T of the singular p.d.e.

$$u_t - \operatorname{div} |Du|^{p-2} Du = 0, \qquad 1<p<2. \tag{1.1}$$

A striking feature of these singular equations is that, unlike the degenerate case $p>2$, non-negative solutions of (1.1) are not restricted by any *'growth condition'* as $|x|\to\infty$. Nevertheless they have initial traces that are Radon measures. Moreover they are *unique* whenever the initial traces are in $L^1_{loc}(\mathbf{R}^N)$. Accordingly, the Cauchy problem for (1.1) associated with an initial datum

$$u_o \in L^1_{loc}(\mathbf{R}^N), \qquad u_o \geq 0, \tag{1.2}$$

is uniquely solvable, regardless of the behaviour of $x \to u_o(x)$ as $|x|\to\infty$.

The case $1<p<2$ is noticeably different from the case $p>2$, both in terms of results and techniques. The main difference stems from the fact that, unlike the degenerate case, solutions of (1.1) are not, in general, locally bounded. In a precise way, if

$$u_o \in L^r_{loc}(\mathbf{R}^N), \quad r \geq 1, \quad \text{and} \quad p > \frac{2N}{N+r}, \tag{1.3}$$

then the solution u of (1.1)-(1.2) belongs to $L^\infty_{loc}(S_T), \forall t > 0$. This is the content of Theorem 5.1 of Chap. V. In §13 we will give a counterexample that shows that if u_o violates (1.3), then $u \notin L^\infty_{loc}(\Sigma_T)$. The basic *formal* energy estimate for (1.1) is

$$\forall 0 < s < t \leq T, \quad \forall K_\rho$$

$$\text{(1.4)} \qquad \sup_{s \leq \tau \leq t} \int_{K_\rho} u^2(x,\tau)\, dx + \int_s^t \int_{K_\rho} |Du|^p dx d\tau$$

$$\leq \frac{\gamma}{\rho^p} \int_s^t \int_{K_{2\rho}} u^p dx d\tau + \frac{\gamma}{(t-s)} \int_s^t \int_{K_{2\rho}} u^2 dx d\tau.$$

Thus if $u \in L^2_{loc}(\Sigma_T)$, the left hand side of (1.4) is finite and $|Du| \in L^p_{loc}(\Sigma_T)$. However if $u_o \in L^1_{loc}(\mathbf{R}^N)$, there is no a priori information to guarantee that

$$\text{(1.5)} \qquad |Du| \in L^p_{loc}(\Sigma_T).$$

We have spoken of solutions of (1.1); however if (1.5) fails, one of the main problems is to make precise what it is meant by solution. Thus the starting point of the theory is to give a precise meaning to Du to make sense out of (1.1). The previous remarks suggest that $|Du|$ might fail to be in $L^p_{loc}(\Sigma_T)$, roughly speaking at those points where u is unbounded. Motivated by these remarks, we have given a novel formulation of non-negative weak solutions. Such solutions are *'regular'* in the sense that the truncations

$$\text{(1.6)} \qquad \forall k > 0, \qquad u_k = \min\{u, k\},$$

satisfy

$$\text{(1.7)} \qquad |Du_k| \in L^p_{loc}(\Sigma_T), \qquad \frac{\partial}{\partial t} u_k \in L^1_{loc}(\Sigma_T).$$

Then (1.1) can be interpreted weakly against testing functions that vanish 'whenever u is large'. A suitable choice of such testing functions is

$$(\varphi - u)_+ \equiv \max\{(\varphi - u); 0\}, \quad \varphi \in C^\infty_o(\Sigma_T); \Sigma_T.$$

The notion is introduced and discussed §§2 and 3. We prove that these solutions coincide with the distributional ones if (1.5) holds and that the truncations u_k are distributional super-solutions of (1.1) $\forall k > 0$. We derive a spectrum of properties of such local weak solutions, regardless of their initial datum. In particular we investigate the behaviour of Du_k as $k \to \infty$. A relevant fact is the estimate

$$\text{(1.8)} \qquad \int_s^t \int_{K_\rho} |Du|^{p-1} dx d\tau \leq \gamma \sup_{s \leq \tau \leq t} \int_{K_{2\rho}} u(x,\tau)\, dx + \gamma \left(\frac{t-s}{\rho^\lambda} \right)^{\frac{1}{2-p}},$$

$$\forall 0 < s < t \leq T, \quad \forall K_{2\rho},$$

where $\lambda = N(p-2)+p$ and $\gamma=\gamma(N,p)$. We remark that in Chap. XI an estimate of the local integral norm of $|Du|^{p-1}$ was crucial to establish the existence of initial traces. In the singular case $1<p<2$ it is precisely (1.8) that permits one to prove an integral Harnack-type inequality, which in turns implies the existence of initial traces. The estimate (1.8) is essential also for the solvability of the Cauchy problem. A solution to (1.1)-(1.2) is constructed by using the increasing sequence $\{u_{o,n}\}$ of approximating initial data

$$u_{o,n} = \min\{u_o\,;\, n\}, \qquad n = 1, 2, \ldots, \tag{1.9}$$

and solving the approximating problems

$$\begin{cases} u_n \in C\left(0,T; L^2_{loc}(\mathbf{R}^N)\right) \cap L^p\left(0,T; W^{1,p}_{loc}(\mathbf{R}^N)\right), \\ u_{n,t} - \operatorname{div} |Du_n|^{p-2} Du_n = 0, \ \text{ in } \Sigma_T, \\ u_n(\cdot, 0) = u_{o,n}, \ \text{ in the sense of } L^1_{loc}(\mathbf{R}^N). \end{cases} \tag{1.10}$$

The comparison principle and (1.8) yield the $L^1_{loc}(\Sigma_T)$ convergence of the approximating solutions $\{u_n\}$. A one-sided bound on $u_{o,n}$ and hence on u_n is crucial to this process in view of the regularising effect of Proposition 6.1 of Chap. VI.

In §5 we show uniqueness of weak solutions if they take their initial datum in the sense of $L^1_{loc}(\mathbf{R}^N)$. Namely, if u and v solve (1.1) weakly and if

$$t \to (u-v)(t) \to 0, \ \text{ in } L^1_{loc}(\mathbf{R}^N) \text{ as } t \searrow 0,$$

then the difference $w = u - v$ satisfies

$$\int_{K_\rho} |w|^q(t)dx \le \gamma(q) \left(\frac{t}{\rho^{\frac{N(p-2)+pq}{q}}} \right)^{\frac{q}{2-p}}, \quad \forall q \ge 1, \forall t > 0, \forall \rho > 0 \tag{1.11}$$

for a constant $\gamma = \gamma(N,p,q)$. The theorem follows by letting $\rho \to \infty$ after we choose q so large that $N(p-2)+pq > 0$.

If, in (1.3), $r=1$ and $p > \frac{2N}{N+1}$, the existence and uniqueness theory remains valid if $u_o \in L^1_{loc}(\mathbf{R}^N)$ with *no sign restriction*. Indeed in such a case the sequences

$$\{u_n\} \quad \text{and} \quad \left\{ \frac{\partial}{\partial x_i} u_n \right\}, \quad i = 1, 2, \ldots, N,$$

are locally equibounded and equi-Hölder continuous in Σ_T.

If $1 < p < \frac{2N}{N+2}$, the singular equation (1.1) is not fully understood. For example it would be of interest to investigate questions of existence and uniqueness for the Cauchy problem (1.1)-(1.2) if the initial datum is a measure μ. Finally, we notice that all the results of this chapter hold true for equations of the type

$$u_t - \sum_{i=1}^N (|u_{x_i}|^{p-2} u_{x_i})_{x_i} = 0 \qquad \text{in } \Sigma_T.$$

2. Weak solutions

A measurable function $u:\Sigma_T\to\mathbf{R}^+$ is a *local* weak solution of (1.1) in Σ_T if

$$u\in C(0,T:L^1_{loc}(\mathbf{R}^N)),\ \ |Du_k|\in L^p_{loc}(\Sigma_T),\ \ \frac{\partial}{\partial t}u_k\in L^1_{loc}(\Sigma_T) \tag{2.1}$$

for all $k>0$ and $\forall\varphi\in C_o^\infty(\Sigma_T)$,

$$\iint_{\Sigma_T}\{u_t(\varphi-u)_+ + |Du|^{p-2}DuD(\varphi-u)_+\}\,dxd\tau=0. \tag{2.2}$$

Introduce the spaces

$$X_{loc}(\Sigma_T)\equiv L^p_{loc}(0,T;W^{1,p}_{loc}(\mathbf{R}^N))\cap L^\infty_{loc}(\Sigma_T), \tag{2.3}$$

$$\overset{\circ}{X}_{loc}(\Sigma_T)\equiv\left\{\begin{array}{c}\varphi\in X_{loc}(\Sigma_T)\mid\varphi(x,t)=0,\quad\forall|x|>\rho\\ \forall t\in(0,T),\ \text{ for some }\rho>0\end{array}\right\}. \tag{2.4}$$

By density, (2.2) holds for all $\varphi\in\overset{\circ}{X}_{loc}(\Sigma_T)$. We denote with $\mathcal{S}$ the set of all non-negative local weak solutions of (1.1) in Σ_T.

LEMMA 2.1. *Let* $u\in\mathcal{S}$. *Then* $\forall\psi\in X_{loc}(\Sigma_T)$ *and* $\forall\eta\in C_o^\infty(\Sigma_T)$,

$$\iint_{\Sigma_T}\{u_t(\psi-u)_+\eta+|Du|^{p-2}DuD[(\psi-u)_+\eta]\}\,dxd\tau=0. \tag{2.5}$$

PROOF: Let $\mathcal{K}\subset\mathcal{K}'$ be compact subsets of Σ_T such that dist $(\partial\mathcal{K},\partial\mathcal{K}')=d>0$ and let $\zeta\in C_o^\infty(\mathcal{K}')$ be such that $0\le\zeta\le1$ and $\zeta\equiv1$ on $\mathcal{K}$. Choose $\psi\in X_{loc}(\Sigma_T)$ and in (2.2) take

$$\varphi=(\psi-u)_+\eta+u_k\zeta$$

where

$$\eta\in C_o^\infty(\mathcal{K})\quad\text{ and }\quad k=\|\psi\|_{\infty,\mathcal{K}'}. \tag{2.6}$$

We have a.e. in $\mathcal{K}'\backslash\mathcal{K}$

$$\begin{aligned}(\varphi-u)_+&=((\psi-u)_+\eta+u_k\zeta-u)_+\\&=(u_k\zeta-u)_+=0.\end{aligned}$$

Moreover

$$(\varphi-u)_+=((\psi-u)_+\eta+u_k-u)_+,\qquad\text{a.e. in }\mathcal{K}.$$

This vanishes unless $u<\psi$. In such a case, $u_k=u$ and

$$(\varphi-u)_+=(\psi-u)_+\eta\qquad\text{a.e. in }\mathcal{K}.$$

We conclude that this holds a.e. in Σ_T and (2.5) follows.

Let $\sigma\in(0,1)$ and let $x\to\zeta(x)$ denote the standard cutoff function in K_ρ that equals one on $K_{\sigma\rho}$, $\sigma\in(0,1)$. By density, (2.5) implies

$$\text{(2.7)}\qquad \forall\psi\in X_{loc}(\Sigma_T),\quad \forall\, 0<s<t\le T,$$
$$\int_s^t\int_{\mathbf{R}^N}\{u_t(\psi-u)_+\zeta^p+|Du|^{p-2}DuD[(\psi-u)_+\zeta^p]\}\,dxd\tau=0.$$

Conversely, if $\psi\in C_o^\infty(\Sigma_T)$, we may write (2.7) for $s<t$ such that $\operatorname{supp}\{\psi\}\subset\mathbf{R}^N\times(s,t)$. By taking ζ so that $\rho>2\operatorname{diam}(\operatorname{supp}\{\psi\})$, we obtain (2.2). We conclude that the formulations (2.2), (2.5) and (2.7) are equivalent.

LEMMA 2.2. *Let $u\in\mathcal{S}$ satisfy*

$$|Du|\in L^p_{loc}(\Sigma_T),\qquad u_t\in L^1_{loc}(\Sigma_T).$$

Then

$$u_t-\operatorname{div}|Du|^{p-2}Du=0\qquad in\ \mathcal{D}'(\Sigma_T).$$

PROOF: In (2.5) take $\psi=u_n+1\in X_{loc}(\Sigma_T)$, $n\in\mathbf{N}$. We obtain $\forall\eta\in C_o^\infty(\Sigma_T)$

$$\iint_{\Sigma_T}\{u_t\eta+|Du|^{p-2}DuD\eta\}(u_n-u+1)_+dxd\tau=\iint_{\Sigma_T\cap[n<u<n+1]}|Du|^p\eta\,dxd\tau.$$

Since $|Du|\in L^p_{loc}(\Sigma_T)$, the right-hand side tends to zero as $n\to\infty$. The left-hand side converges to

$$\iint_{\Sigma_T}\{u_t\eta+|Du|^{p-2}DuD\eta\}\ dxd\tau=0.$$

LEMMA 2.3. *Let $u\in\mathcal{S}$. Then for all $k>0$, u_k is a distributional super-solution of (1.1) in Σ_T.*

PROOF: Fix $k>0$ and $\alpha,\varepsilon\in(0,1)$, and in (2.5) take

$$\psi=u_k+[(k-u)_++\varepsilon]^\alpha\in X_{loc}(\Sigma_T)$$

to obtain $\forall\eta\in C_o^\infty(\Sigma_T)$, $\eta\ge0$

$$\iint_{\Sigma_T}\{u_t\eta+|Du|^{p-2}DuD\eta\}(\psi-u)_+dxd\tau=\iint_{\Sigma_T\cap[k<u<\psi]}|Du|^p\eta\,dxd\tau$$
$$+\alpha\iint_{\Sigma_T}|Du_k|^p[(k-u)_++\varepsilon]^{\alpha-1}\eta\,dxd\tau\ge0.$$

First we let $\varepsilon\to0$ as $\alpha\in(0,1)$ remains fixed. Since

$$(\psi - u)_+ \to (k-u)_+^\alpha \qquad \text{a.e. } \Sigma_T,$$

we deduce

$$\iint_{\Sigma_T} \{u_t\eta + |Du|^{p-2} DuD\eta\}(k-u)_+^\alpha dx d\tau \geq 0, \qquad \forall \alpha \in (0,1).$$

Now letting $\alpha \to 0$ gives for every non-negative $\eta \in C_o^\infty(\Sigma_T)$

$$\iint_{\Sigma_T} \left\{\frac{\partial}{\partial t} u_k\eta + |Du_k|^{p-2} Du_k \cdot D\eta\right\} dx d\tau \geq 0. \tag{2.8}$$

The next proposition permits a large class of testing functions in (2.5). If $k_o > 0$, let $\mathcal{F}(k_o)$ denote the set of all the Lipschitz–continuous functions $f : \mathbf{R}^+ \to \mathbf{R}$ such that $f(k) = 0$, $\forall k > k_o$, and set

$$\mathcal{F} = \bigcup_{k_o \in \mathbf{R}^+} \mathcal{F}(k_o).$$

PROPOSITION 2.1. *Let $u \in \mathcal{S}$. Then $\forall f \in \mathcal{F}$ and $\forall \eta \in C_o^\infty(\Sigma_T)$,*

$$\iint_{\Sigma_T} \{u_t f(u)\eta + |Du|^{p-2} Du \cdot D(f(u)\eta)\} dx d\tau = 0.$$

PROOF: Assume first that $f \in C^2(0,\infty)$. Write (2.5) for $\psi = k$, multiply it by $-f''(k)$ and integrate in dk over $(0,\infty)$. By interchanging the order of integration with the aid of Fubini's theorem we obtain

$$\iint_{\Sigma_T} \left\{ u_t\eta \int_u^\infty f''(k)(k-u)dk \right.$$

$$\left. + |Du|^{p-2} Du \cdot D\left[\eta \int_u^\infty f''(k)(k-u)dk\right]\right\} dx d\tau = 0.$$

Since

$$\int_u^\infty f''(k)(k-u)\, dk = f(u),$$

the assertion follows for $f \in C^2(0,\infty)$. The general case is proved by approximation.

3. Estimating $|Du|$

LEMMA 3.1. *There exists a constant* $\gamma=\gamma(N,p)$ *such that*

$$\forall k>0,\quad \forall\rho>0,\quad \forall\, 0<s<t\le T,\quad \forall u\in\mathcal{S}$$

$$\int_s^t\int_{\mathbf{R}^N}|Du_k|^p dxd\tau\le\gamma\, k^p|K_\rho|\left(k^{2-p}+\frac{t-s}{\rho^p}\right).$$

PROOF: Let ζ be the standard cutoff function in $K_{2\rho}$. Then from (2.7) with $\psi=k$

$$\begin{aligned}\int_s^t\int_{\mathbf{R}^N}|Du_k|^p\zeta^p dxd\tau &\le p\int_s^t\int_{\mathbf{R}^N}|Du_k|^{p-1}\zeta^{p-1}(k-u)_+|D\zeta|dxd\tau\\ &\quad+\frac{1}{2}\int_s^t\int_{\mathbf{R}^N}\frac{\partial}{\partial\tau}(k-u)_+^2\zeta^p dxd\tau\\ &\le\frac{p-1}{p}\int_s^t\int_{\mathbf{R}^N}|Du_k|^p\zeta^p dxd\tau\\ &\quad+p^{p-1}\int_s^t\int_{\mathbf{R}^N}(k-u)_+^p|D\zeta|^p dxd\tau\\ &\quad+\frac{1}{2}\int_{\mathbf{R}^N\times\{t\}}(k-u)_+^2\zeta^p dx.\end{aligned}$$

For all $0<s<t\le T$ and all $\rho>0$ set

$$M_{s,t}(\rho)=\sup_{\tau\in(s,t)}\fint_{K_\rho}u(x,\tau)\,dx. \tag{3.1}$$

LEMMA 3.2. *Let* $u\in\mathcal{S}$. *Then* $\forall\alpha\in(0,p-1)$

$$\left|Du^{\frac{p-1-\alpha}{p}}\right|\in L^p_{loc}(\Sigma_T),$$

and there exists a constant $\gamma=\gamma(N,p)$ *such that* $\forall\, 0<s<t\le T$ *and for all* $\rho>0$

$$\int_s^t\fint_{K_\rho}|Du^{\frac{p-1-\alpha}{p}}|^p dxd\tau\le\frac{\gamma}{\alpha^p}\left[M_{s,t}(2\rho)+\left(\frac{t-s}{\rho^p}\right)^{\frac{1}{2-p}}\right]^{1-\alpha}. \tag{3.2}$$

PROOF: Fix $k>0$ and $\varepsilon\in(0,1)$, and in (2.8) take $\eta=\zeta^p\psi^{-\alpha}$, where ζ is the standard cutoff function in $K_{2\rho}$ and

$$\psi = \begin{cases} u_k, & u > \varepsilon \\ \varepsilon, & u \le \varepsilon. \end{cases}$$

We obtain

$$\begin{aligned}
&\alpha \int_s^t\!\!\int_{\mathbf{R}^N} |Du|^p u^{-\alpha-1} \zeta^p \chi[\varepsilon < u < k]\, dx d\tau \\
&\le p \int_s^t\!\!\int_{\mathbf{R}^N} |Du|^{p-1} u^{-\alpha} \zeta^{p-1} |D\zeta| \chi[\varepsilon < u < k]\, dx d\tau \\
&+ p \int_s^t\!\!\int_{\mathbf{R}^N} |Du_\varepsilon|^{p-1} \varepsilon^{-\alpha} \zeta^{p-1} |D\zeta|\, dx d\tau \\
&+ \frac{1}{1-\alpha} \int_s^t\!\!\int_{\mathbf{R}^N} \frac{\partial}{\partial \tau} \psi^{1-\alpha} \zeta^p\, dx d\tau + \int_s^t\!\!\int_{\mathbf{R}^N} \frac{\partial}{\partial \tau} u_\varepsilon \varepsilon^{-\alpha} \zeta^p dx d\tau.
\end{aligned} \tag{3.3}$$

By Young's inequality, the first integral on the right-hand side is majorised by

$$\frac{\alpha}{2} \int_s^t\!\!\int_{\mathbf{R}^N} |Du|^p u^{-\alpha-1} \zeta^p \chi[\varepsilon < u < k] dx d\tau + \left(\frac{2}{\alpha}\right)^{p-1} p^p \int_s^t\!\!\int_{\mathbf{R}^N} u^{p-1-\alpha} |D\zeta|^p dx d\tau.$$

By virtue of Lemma 3.1 the second integral tends to zero as $\varepsilon \to 0$ at the rate of $\varepsilon^{p-1-\alpha}$. Combining these calculations we deduce

$$\begin{aligned}
&\alpha \int_s^t\!\!\int_{K_\rho} |Du|^p u^{-\alpha-1} \chi[\varepsilon < u < k]\, dx d\tau \\
&\le O(\varepsilon^{p-1-\alpha}) + \frac{\gamma}{\alpha^{p-1}} \left\{ \left(\sup_{\tau \in (s,t)} \int_{K_{2\rho}} u(x,\tau) dx \right)^{1-\alpha} (2\rho)^{\alpha N} \right. \\
&\left. + \left(\frac{t-s}{\rho^p}\right) \left(\sup_{\tau \in (s,t)} \int_{K_{2\rho}} u(x,\tau) dx \right)^{p-1-\alpha} (2\rho)^{N(2-p+\alpha)} \right\} \\
&\le \frac{\gamma}{\alpha^{p-1}} \left\{ [M_{s,t}(2\rho)]^{1-\alpha} + \left(\frac{t-s}{\rho^p}\right) [M_{s,t}(2\rho)]^{p-1-\alpha} \right\} \rho^N \\
&+ O(\varepsilon^{p-1-\alpha}).
\end{aligned} \tag{3.4}$$

If

$$\left(\frac{t-s}{\rho^p}\right) \le [M_{s,t}(2\rho)]^{2-p},$$

the quantity in braces on the rightmost side of (3.4) is majorised by $[M_{s,t}(2\rho)]^{1-\alpha}$. Otherwise it is majorised by

$$[M_{s,t}(2\rho)]^{1-\alpha} + \left(\frac{t-s}{\rho^p}\right)^{\frac{1-\alpha}{2-p}}.$$

In either case

$$\int_s^t\int_{K_\rho} |Du|^p u^{-(\alpha+1)} \chi[\varepsilon<u<k]\,dxd\tau \tag{3.5}$$

$$\le O(\varepsilon^{p-(\alpha+1)}) + \frac{\gamma}{\alpha^p}\rho^N \left\{ M_{s,t}(2\rho) + \left(\frac{t-s}{\rho^p}\right)^{\frac{1}{2-p}} \right\}^{1-\alpha},$$

and the lemma follows by letting first $\varepsilon\to 0$ and then $k\to\infty$.

Estimate (3.2) deteriorates as $\alpha\to 0$. The next lemma gives some information for the case $\alpha=0$.

LEMMA 3.3. *Let $u\in\mathcal{S}$. There exists $\gamma=\gamma(N,p)$ such that*

$$\forall 0<s<t\le T, \quad \forall\rho>0, \quad \forall n\ge 1$$

$$\int_s^t \fint_{K_\rho} |Du^{\frac{p-1}{p}}|^p \chi[n<u<n+1]\,dxd\tau$$

$$\le \gamma \ln\left(1+\frac{1}{n}\right)\left[M_{s,t}(2\rho) + \left(\frac{t-s}{\rho^p}\right)^{\frac{1}{2-p}} \right].$$

PROOF: In (2.8) we take $\eta=\zeta\psi$, where ζ is the standard cutoff function in $K_{2\rho}$ and $\psi=\ln^+\left(\frac{n+1}{u^{(n)}}\right)$. Here

$$u^{(n)} = \begin{cases} n, & \text{if } 0<u\le n \\ u, & \text{if } u>n. \end{cases}$$

We get

$$\int_s^t\int_{K_\rho} |Du|^p u^{-1}\chi[n<u<n+1]\,dxd\tau \tag{3.6}$$

$$\le \int_s^t\int_{K_{2\rho}} \frac{\partial}{\partial\tau} u_{n+1} \ln^+\left(\frac{n+1}{u^{(n)}}\right)\zeta^p dxd\tau$$

$$+\frac{p}{\rho}\int_s^t\int_{K_{2\rho}} |Du|^{p-1} \ln^+\left(\frac{n+1}{u^{(n)}}\right) dxd\tau = I_n^{(1)} + \frac{p}{\rho} I_n^{(2)}.$$

Setting, for simplicity of notation,

$$A = K_{2\rho} \times (s,t),$$

we have

$$I_n^{(2)} \leq \ln\left(1+\frac{1}{n}\right) \iint\limits_{A\cap[u<n+1]} |Du|^{p-1} u^{-(\alpha+1)\frac{(p-1)}{p}} u^{(\alpha+1)\frac{(p-1)}{p}}\, dx d\tau$$

$$\leq \gamma \ln\left(1+\frac{1}{n}\right)\left(\iint\limits_{A} |Du^{\frac{p-(\alpha+1)}{p}}|^p dx d\tau\right)^{\frac{p-1}{p}} \left(\iint\limits_{A} u^{(\alpha+1)(p-1)} dx d\tau\right)^{\frac{1}{p}}.$$

If $\alpha \in (0, p-1)$ is so small that $(\alpha+1)(p-1) \leq 1$, both integrals in parentheses are finite. Taking Lemma 3.2 into account in estimating the first integral we have

$$\frac{p}{\rho} I_n^{(2)} \leq \gamma \rho^N \ln\left(1+\frac{1}{n}\right),$$

$$\left[M_{s,t}(2\rho) + \left(\frac{t-s}{\rho^p}\right)^{\frac{1}{2-p}}\right]^{(1-\alpha)\frac{(p-1)}{p}} \left(\frac{1}{\rho^p}\int_s^t \fint_{K_{2\rho}} u^{(\alpha+1)(p-1)} dx d\tau\right)^{\frac{1}{p}}.$$

The last integral above is estimated by

$$\left(\frac{1}{\rho^p}\int_s^t \fint_{K_{2\rho}} u^{(\alpha+1)(p-1)}(x,\tau)\, dx d\tau\right)^{\frac{1}{p}}$$

$$\leq \left(\frac{t-s}{\rho^p}\right)^{\frac{1}{p}} [M_{s,t}(2\rho)]^{(\alpha+1)\frac{(p-1)}{p}}$$

$$\leq \left[M_{s,t}(2\rho) + \left(\frac{t-s}{\rho^p}\right)^{\frac{1}{2-p}}\right]^{(\alpha+1)\frac{(p-1)}{p}+\frac{2-p}{p}}.$$

Therefore

$$\frac{p}{\rho} I_n^{(2)} \leq \gamma \rho^N \ln\left(1+\frac{1}{n}\right)\left[M_{s,t}(2\rho) + \left(\frac{t-s}{\rho^p}\right)^{\frac{1}{2-p}}\right].$$

As for $I_n^{(1)}$ we write

$$I_n^{(1)} = \iint\limits_{A\cap[u<n]} u_t \ln\left(1+\frac{1}{n}\right)\zeta^p dx d\tau + \iint\limits_{A\cap[u>n]} u_t \ln^+\left(\frac{n+1}{u}\right)\zeta^p dx d\tau$$

$$= \iint\limits_{A} \frac{\partial}{\partial \tau} u_n \ln\left(1+\frac{1}{n}\right)\zeta^p dx d\tau + \iint\limits_{A} \frac{\partial}{\partial \tau} u^{(n)} \ln^+\left(\frac{n+1}{u}\right)\zeta^p dx d\tau$$

$$\leq \gamma \rho^N \ln\left(1+\frac{1}{n}\right) M_{s,t}(2\rho) + \iint\limits_{A} \frac{\partial}{\partial \tau}\left(\int_n^u \ln^+\left(\frac{n+1}{\xi}\right) d\xi\right)_+ \zeta^p dx d\tau.$$

COROLLARY 3.1. *Let $u \in S$ and define*

$$(x,t) \to z(x,t) = \int_e^{u(x,t)} (\xi \ln^{1+\varepsilon} \xi)^{-\frac{1}{p}} dx, \qquad \varepsilon \in (0, p-1).$$

Then $|Dz| \in L^p_{loc}(\Sigma_T)$ and there exists $\gamma = \gamma(N,p)$ such that $\forall\, 0<s<t\leq T$ and $\forall \rho>0$,

$$\int_s^t \fint_{K_\rho} |Dz|^p dx d\tau \leq \gamma \varepsilon^{-1} \left[M_{s,t}(2\rho) + \left(\frac{t-s}{\rho^p} \right)^{\frac{1}{2-p}} \right]. \tag{3.7}$$

PROOF: Divide both sides of the inequality of Lemma 3.3 by $\ln^{1+\varepsilon} n$, and add over all $n = 2, 3, \ldots$

The estimate (3.7) deteriorates as $\varepsilon \to 0$. The following corollary gives some information in the case $\varepsilon = 0$.

COROLLARY 3.2. *Let $u \in S$. Then $\forall\, 0<s<t\leq T$, and for all $C>1$,*

$$\overline{\lim_{k\to\infty}} \int_s^t \int_{K_\rho} |Du|^p \frac{1}{u \ln u} \chi[k<u<Ck]\, dx d\tau = 0.$$

PROOF: Without loss of generality we may assume that k and Ck are positive integers. Divide both sides of the inequality of Lemma 3.3 by $\ln n$ and add for $n = k, k+1, \ldots, Ck$. This gives

$$\int_s^t \int_{K_\rho} |Du|^p (u \ln u)^{-1} \chi[k<u<Ck]\, dx d\tau$$

$$\leq \gamma (\ln \ln Ck - \ln \ln k) \left[M_{s,t}(2\rho) + \left(\frac{t-s}{\rho^p} \right)^{\frac{1}{2-p}} \right]$$

$$= \gamma \ln \left(1 + \frac{\ln C}{\ln k} \right) \left[M_{s,t}(2\rho) + \left(\frac{t-s}{\rho^p} \right)^{\frac{1}{2-p}} \right].$$

4. The weak Harnack inequality and initial traces

In the definition of local weak solutions of (2.1) in Σ_T, no reference has been made to initial data. We will show that each $u \in S$ has a *unique* non-negative σ–finite Borel measure μ as the initial trace. The existence of such a trace will be a consequence of the following weak Harnack-type estimate.

THEOREM 4.1. *Let $u \in \mathcal{S}$. There exists $\gamma = \gamma(N,p)$, such that $\forall\, 0 < s < t \le T$ and $\forall \rho > 0$*

$$\sup_{\tau\in(s,t)} \int_{K_\rho} u(x,\tau)dx \le \gamma \int_{K_{2\rho}} u(x,t)dx + \gamma \left(\frac{t-s}{\rho^\lambda}\right)^{\frac{1}{2-p}}, \tag{4.1}$$

$$\lambda = N(p-2) + p.$$

The uniqueness of the initial trace μ relies on the next gradient estimates.

LEMMA 4.1. *Let $u \in \mathcal{S}$. There exists a constant $\gamma = \gamma(N,p)$ such that*

$$\forall\, 0<s<t\le T, \quad \forall \rho > 0, \quad \forall \sigma \in (0,1), \quad \forall \nu > 0,$$

$$\frac{1}{\rho} \int_s^t \int_{K_\rho} |Du|^{p-1} dx d\tau \le \gamma \left(\frac{t-s}{\rho^\lambda}\right)^{\frac{1}{2-p}} + \gamma \left(\frac{t-s}{\rho^\lambda}\right)^{\frac{1}{p}} \left\{ \sup_{s<\tau<t} \int_{K_{2\rho}} u(x,\tau)\, dx \right\}^{\frac{2(p-1)}{p}}. \tag{4.2}$$

Moreover

$$\frac{1}{\rho} \int_s^t \int_{K_\rho} |Du|^{p-1} dx d\tau \le \gamma \sup_{s<\tau<t} \int_{K_{2\rho}} u(x,\tau)\, dx + \gamma \left(\frac{t-s}{\rho^\lambda}\right)^{\frac{1}{2-p}}. \tag{4.3}$$

PROOF: The proof is the same as that of Propositions 4.1 and 4.2 of Chap. VII. The only difference is that instead of working with the solution u we work with the truncations u_k and use the fact that these are supersolutions. In (2.8) we take the testing functions

$$\psi = (t-\tau)^{\frac{1}{p}} (u_k + \nu)^{1-\frac{2}{p}} \in X_{loc}(\Sigma_T),$$

where $\nu > 0$ is arbitrary. We proceed as in Chap. VII and then let $k \to \infty$.

THEOREM 4.2. *Every $u \in \mathcal{S}$ has a unique Radon measure μ as initial trace at $t = 0$.*

PROOF: From Theorem 4.1 it follows that $\forall \eta \in C_o^\infty(\mathbf{R}^N)$, the net

$$\left\{ \int_{\mathbf{R}^N} u(\tau)\eta\, dx \right\}_{\tau\in(0,t)}$$

is equibounded, with bound depending only upon $\|\eta\|_{\infty,\mathbf{R}^N}$. A subnet indexed with $\{\tau'\}$ converges to a Radon measure μ, in the sense of the measures, i.e.,

$$\int_{\mathbf{R}^N} u(\tau')\eta\, dx \quad \longrightarrow \quad \int_{\mathbf{R}^N} \eta\, d\mu, \qquad \forall \eta \in C_0(\mathbf{R}^N), \ \text{ as } \tau' \searrow 0.$$

Suppose now that there exist another subnet, indexed with $\{\tau''\}$ and a Radon measure $\hat{\mu}$, such that

$$\int_{\mathbf{R}^N} u(\tau'')\eta\, dx \quad \longrightarrow \quad \int_{\mathbf{R}^N} \eta\, d\hat{\mu}, \qquad \forall \eta \in C_0(\mathbf{R}^N), \ \text{ as } \tau'' \searrow 0.$$

We will prove that $\mu \equiv \hat{\mu}$. Let $\sigma \in (0,1)$ and write (2.8) with $\psi \equiv 1$ and ζ the standard cutoff function in $K_{(1+\sigma)\rho}$. Letting $k \to \infty$, standard calculations give $\forall\, 0<s<t\leq T$

$$\int_{K_\rho} u(s)\, dx \leq \int_{K_{(1+\sigma)\rho}} u(t)\, dx + \frac{1}{\sigma\rho} \int_s^t\!\!\int_{K_{2\rho}} |Du|^{p-1}\, dx d\tau. \tag{4.4}$$

We estimate the last term by using (4.2) and let $s \searrow 0$ along τ' while $t>0$ remains fixed. Then we let $t \searrow 0$ along the net τ'' to get

$$\int_{K_\rho} d\mu \leq \int_{K_{(1+\sigma)\rho}} d\hat{\mu}.$$

Since $\sigma \in (0,1)$ is arbitrary, interchanging the role of μ and $\hat{\mu}$ proves the theorem.

5. The uniqueness theorem

Let $\mathcal{S}^*$ denote the subclass of $\mathcal{S}$ of those non-negative local weak solutions of (1.1) in Σ_T, satisfying

$$\begin{aligned} &\frac{\partial}{\partial t} u_k(x,t) \leq \gamma u_k(x,t), \qquad \text{a.e. } (x,t) \in \Sigma_T \\ &\text{for some } \gamma = \gamma(N,p,t), \qquad \forall k \in \mathbf{R}^+, \end{aligned} \tag{5.1}$$

$$\overline{\lim_{k\to\infty}} \iint_{\mathcal{K}\cap[k<u<Ck]} |Du|^p \frac{1}{u}\, dx d\tau = 0, \tag{5.2}$$

for every compact subset $\mathcal{K} \subset \Sigma_T$ and for all $C>1$. In section §§8-12 we will construct solutions of the Cauchy problem (1.1)-(1.2) that satisfy both (5.1) and (5.2); therefore $\mathcal{S}^*$ is not empty. Corollary 3.2 suggests that (5.2) is *almost* satisfied by all solutions in $\mathcal{S}$. It would be of interest to know whether the inclusion $\mathcal{S}^* \subset \mathcal{S}$ is strict.

THEOREM 5.1. *Let $u_1, u_2 \in \mathcal{S}^*$ satisfy*

$$(u_1 - u_2)(t) \to 0 \quad in\ L^1_{loc}(\mathbf{R}^N)\ as\ t \to 0.$$

Then $u_1 = u_2$ a.e. in Σ_T.

5-(i). Preliminaries

LEMMA 5.1. *Let $u \in s^*$. Then for all $0 < s < t \leq T, \forall \rho > 0, \forall C > 1$,*

$$\overline{\lim_{k\to\infty}} \int_s^t\!\!\int_{K_\rho} |u_t| \chi[k < u < Ck]\, dx d\tau = 0.$$

PROOF: Consider (2.8) written for u_k replaced by u_{Ck}, against testing functions

$$\eta = \zeta \ \ln\left(k/2w_{k,C}\right)$$

where $x \to \zeta(x)$ is the standard cutoff function in $K_{2\rho}$, and

$$w_{k,C} \equiv \begin{cases} \frac{1}{2}k, & 0 \leq u \leq \frac{1}{2}k \\ u, & \frac{1}{2}k < u < Ck \\ Ck, & u \geq Ck. \end{cases}$$

It follows from these definitions that $\eta \leq 0$ a.e. in Σ_T and $\eta = 0$ a.e. on the set $[0 < u \leq \frac{1}{2}k]$. By calculation from (2.8) we obtain

$$(5.3) \quad \int_s^t\!\!\int_{K_{2\rho}} \frac{\partial}{\partial t} u_{Ck} \eta \, dx d\tau \leq \int_s^t\!\!\int_{K_{2\rho}} |Du|^p \frac{1}{u} \chi[k/2 < u < Ck] dx d\tau$$
$$+ \ln 2C \int_s^t\!\!\int_{K_{2\rho}} |Du_{Ck}|^{p-1} \chi[u > k/2] |D\zeta| \, dx d\tau.$$

The first integral on the right hand side of (5.3) tends to zero as $k \to \infty$ by virtue of (5.2). We estimate the second integral, formally, by

$$\frac{\ln 2C}{\rho}\iint_{{}_sK_{2\rho}}^{t}|Du_{Ck}|^{p-1}\chi[u>k/2]\,dxd\tau$$

$$=\frac{\ln 2C}{\rho}\iint_{{}_sK_{2\rho}}^{t}|Du_{Ck}|^{p-1}u^{-\frac{(\alpha+1)(p-1)}{p}}u^{\frac{(\alpha+1)(p-1)}{p}}\chi[u>k/2]dxd\tau$$

$$\leq\frac{\ln 2C}{\rho}\left(\frac{p}{p-1-\alpha}\right)^{p-1}\left(\iint_{{}_sK_{2\rho}}^{t}|Du^{\frac{p-1-\alpha}{p}}|^{p}dxd\tau\right)^{\frac{p-1}{p}}$$

$$\times\left(\iint_{{}_sK_{2\rho}}^{t}u^{(\alpha+1)(p-1)}\chi[u>k/2]dxd\tau\right)^{\frac{1}{p}}.$$

If we choose $\alpha\in(0,p-1)$ so small that $(\alpha+1)(p-1)\leq 1$, the estimate is rigorous and the last term in the right hand side of (5.3) tends to zero as $k\to\infty$, since $u\in L^1_{loc}(\Sigma_T)$. These remarks in (5.3) give

$$\iint_{{}_sK_{2\rho}}^{t}u_t\ln\left(\frac{k}{2w_{k,C}}\right)\zeta\,\chi[u_t<0]\,\chi\,[(k/2)<u<Ck]\,dxd\tau$$

$$\leq\iint_{{}_sK_{2\rho}}^{t}u_t|\ln\frac{k}{2w_{k,C}}|\,\zeta\,\chi[u_t\geq 0]\chi[u>k/2]\,dxd\tau+O\left(\frac{1}{k}\right).$$

In view of the definition of $w_{k,C}$ this gives in turn

$$\iint_{{}_sK_{2\rho}}^{t}|u_t|\chi[k<u<Ck]dxd\tau\leq\gamma\iint_{{}_sK_{2\rho}}^{t}u_t\chi[u_t>0]\chi[u>k/2]dxd\tau+O\left(\frac{1}{k}\right).$$

The last integral is estimated by means of (5.1) and the lemma follows.

Remark 5.1. The assertion of the lemma is trivial if $u_t\in L^1_{loc}(\Sigma_T)$.

We give next a weak formulation for the difference of two solutions u_1,u_2. First we recall that, by Lemma 2.3, the truncated function

$$u_{2,k}\equiv\begin{cases}u_2, & \text{if } 0<u_2<k\\ k, & \text{if } \quad u_2\geq k\end{cases}$$

is a distributional supersolution of (1.1), $\forall k>0$. We write (2.8) for $u_{2,k}$ against the testing functions

$$\eta=(\psi-u_1)_+\zeta^p,\qquad \forall\psi\in X_{loc}(\Sigma_T)$$

where ζ is a non-negative piecewise smooth cutoff function in $K_{(1+\sigma)\rho}$, $\sigma \in (0,1)$, such that

$$\zeta \equiv 1 \quad \text{on } K_\rho \quad \text{and} \quad |D\zeta| \le 1/\sigma\rho. \tag{5.4}$$

In view of the definition of $X_{loc}(\Sigma_T)$ and the regularity properties (2.1) of $u_i, i = 1, 2$, such a choice of testing function is admissible, modulo a density argument. On the other hand the weak formulation (2.7) of u_1 holds against the same testing functions. Therefore setting

$$w \equiv u_1 - u_2, \qquad w_{(k)} \equiv u_1 - u_{2,k}, \qquad k \in \mathbf{R}^+,$$

we obtain by difference the weak formulation

$$\iint_{s\,K_{(1+\sigma)\rho}}^{t} \left\{ \frac{\partial}{\partial t} w_{(k)} (\psi - u_1)_+ \zeta^p + \mathbf{J}_k D(\psi - u_1)_+ \zeta^p \right\} dx d\tau \tag{5.5}$$
$$\le -p \iint_{s\,K_{(1+\sigma)\rho}}^{t} \mathbf{J}_k (\psi - u_1)_+ \zeta^{p-1} D\zeta dx d\tau \quad \forall \psi \in X_{loc}(\Sigma_T),$$

where

$$\begin{aligned}
\mathbf{J}_k &\equiv |Du_1|^{p-2} Du_1 - |Du_{2,k}|^{p-2} Du_{2,k} \\
&= \int_0^1 \frac{d}{d\xi} \left\{ \left| D\left(\xi u_1 + (1-\xi) u_{2,k}\right) \right|^{p-2} D\left(\xi u_1 + (1-\xi) u_{2,k}\right) \right\} d\xi \\
&= \left(\int_0^1 |D(\xi u_1 + (1-\xi) u_{2,k})|^{p-2} d\xi \right) Dw_{(k)} \\
&+ (p-2) \Bigg(\int_0^1 |D(\xi u_1 + (1-\xi) u_{2,k})|^{p-4} \\
&\times D(\xi u_1 + (1-\xi) u_{2,k}) (\xi u_1 + (1-\xi) u_{2,k})_{x_j} d\xi \Bigg) w_{(k),x_j}.
\end{aligned}$$

Set also

$$A_o \equiv \int_0^1 |D(\xi u_1 + (1-\xi) u_{2,k})|^{p-2} d\xi.$$

LEMMA 5.2. $A_o \le \frac{2}{p-1} |Dw_{(k)}|^{p-2}$.

PROOF: If $|Du_{2,k}| \ge |Dw_{(k)}|$, we have

$$\begin{aligned}|D(\xi u_1+(1-\xi)u_{2,k})| &= |Du_{2,k}+\xi Dw_{(k)}|\\ &\ge \Big|\,|Du_{2,k}|-\xi|Dw_{(k)}|\,\Big|\\ &\ge (1-\xi)|Dw_{(k)}|.\end{aligned}$$

Therefore

$$\begin{aligned}A_o &\le \left(\int_0^1 (1-\xi)^{p-2}d\xi\right)|Dw_{(k)}|^{p-2}\\ &= \frac{1}{p-1}|Dw_{(k)}|^{p-2}.\end{aligned}$$

If $|Du_{2,k}|<|Dw_{(k)}|$,

$$\begin{aligned}\int_0^1 |Du_{2,k}+\xi Dw_{(k)}|^{p-2}d\xi &\le \int_0^1 \Big|\,|Du_{2,k}|-\xi|Dw_{(k)}|\,\Big|^{p-2}d\xi\\ \le \frac{1}{(p-1)|Dw_{(k)}|}&\left\{-\int_0^{\xi_o}\frac{d}{d\xi}\left(|Du_{2,k}|-\xi|Dw_{(k)}|\right)^{p-1}d\xi\right.\\ &\left.+\int_{\xi_o}^1\frac{d}{d\xi}\left(\xi|Dw_{(k)}|-|Du_{2,k}|\right)^{p-1}d\xi\right\},\end{aligned}$$

where $\xi_o\in(0,1)$ is defined by

$$\xi_o\equiv\frac{|Du_{2,k}|}{|Dw_{(k)}|}\in(0,1).$$

From the definitions set forth and Lemma 5.2 we have

$$\begin{cases}\mathbf{J}_k Dw_{(k)}\ge(p-1)A_o|Dw_{(k)}|^2,\\ |\mathbf{J}_k|\le A_0|Dw_{(k)}|\le\frac{2}{p-1}|Dw_{(k)}|^{p-1}.\end{cases}\tag{5.6}$$

In what follows we will use these inequalities without specific mention.

6. An auxiliary proposition

PROPOSITION 6.1. *Let $u_i\in S^*$, $i=1,2$, satisfy*

$$w(t)\equiv(u_1-u_2)(t)\to 0\quad in\ L^1_{loc}(\mathbf{R}^N)\ as\ t\to 0.$$

Then $w \in L^\infty\left(0,T; L^q_{loc}(\mathbf{R}^N)\right), \forall q \in [1,\infty)$. *Moreover* $\forall q \geq 1$ *there exists a constant* $\gamma = \gamma(N,p,q)$, *such that*

$$\int_{K_\rho} |w(t)|^q dx \leq \frac{\gamma}{(\sigma\rho)^p} \int_0^t \int_{K_{(1+\sigma)\rho}} |w|^{q+(p-2)} dx d\tau, \tag{6.1}$$

for all $\rho > 0$ *and for all* $\sigma \in (0,1)$.

The proof is based on an iteration procedure and uses recursive inequalities obtained from (5.5) with suitable choices of testing functions ψ.

6-(i). Testing functions in (5.5)

For $h > 0$, set

$$w^+_{(k),h} \equiv (u_1 - u_{2,k})^+_h = \begin{cases} 0 & \text{if } w_{(k)} \leq 0 \\ w^+_{(k)} & \text{if } w_{(k)} < h \\ h & \text{if } w_{(k)} \geq h \end{cases} \tag{6.2}$$

and in (5.5) consider the testing function

$$\psi \equiv u_{1,1/\varepsilon} + \frac{1}{\varepsilon}\left(w^+_{(k),n} + \varepsilon\right)^a \left(w^+_{(k),m} + \varepsilon\right)^b \in X_{loc}(\Sigma_T), \tag{6.3}$$

where

$$\varepsilon \in (0,1), \quad a,b > 0, \quad n,m \in \mathbf{N}; \quad n > m+1.$$

We obtain

$$\begin{aligned} &\int_{\mathbf{R}^N \times\{t\}} w_{(k)}(\psi - u_1)_+ \zeta^p dx - \int_{\mathbf{R}^N \times\{s\}} w_{(k)}(\psi - u_1)_+ \zeta^p dx \\ &- \int_s^t \int_{\mathbf{R}^N} w_{(k)} \frac{\partial}{\partial \tau}(\psi - u_1)_+ \zeta^p dx d\tau + \int_s^t \int_{\mathbf{R}^N} \mathbf{J}_k D(\psi - u_1)_+ \zeta^p dx d\tau \\ &\leq -p \int_s^t \int_{\mathbf{R}^N} \mathbf{J}_k (\psi - u_1)_+ \zeta^{p-1} D\zeta dx d\tau. \end{aligned} \tag{6.4}$$

In using ψ as a testing function in (6.4) we keep in mind that the truncated functions $u_{i,h}, i = 1,2, \forall h > 0$, are *regular* in the sense of (2.1). In particular the first two integrals on the left hand side of (6.4) are well defined $\forall\, 0 < s < t \leq T$. We will eliminate the parameters ε, k, s, n, m by letting $\varepsilon \to 0$, $k \to \infty$, $s \to 0$, $n, m \to \infty$ in the indicated order.

6-(ii). The limit as $\varepsilon \to 0$

We multiply both sides of (6.4) by ε and let $\varepsilon \to 0$, while k, s, n, m remain fixed. From the definition (6.3) of ψ it follows that $\forall \tau \in (0, T]$ the net $[w_{(k)}(\varepsilon\psi - \varepsilon u_1)_+](\cdot, \tau)$, is equibounded in $L^1_{loc}(\mathbf{R}^N)$. Moreover it converges to

$$\left[w_{(k)}\left(w^+_{(k),n}\right)^a\left(w^+_{(k),m}\right)^b\right](\cdot,\tau) \quad \text{a.e. } K_{2\rho},$$

and it is majorised a.e. in $\mathbf{R}^N$ by

$$w_{(k)}\left(w^+_{(k),n}+1\right)^a\left(w^+_{(k),m}+1\right)^b(\cdot,\tau)\in L^1_{loc}(\mathbf{R}^N).$$

Therefore for all $0<\tau\le T$, as $\varepsilon\to 0$

$$\int\limits_{\mathbf{R}^N\times\{\tau\}} w_{(k)}(\psi-u_1)_+\zeta^p dx \longrightarrow \int\limits_{\mathbf{R}^N\times\{\tau\}} w_{(k)}\left(w^+_{(k),n}\right)^a\left(w^+_{(k),m}\right)^b\zeta^p dx. \tag{6.5}$$

This determines the limit for the first two terms on the left hand side of (6.4). To examine the remaining terms we let $\overline{u}_i, i=1,2$, be arbitrarily selected but fixed representatives out of the equivalence classes u_i, define $\overline{w}, \overline{w}_{(k)}$ accordingly, and let

$$\mathcal{F}_\varepsilon \equiv \left\{\begin{array}{c}(x,\tau)\in\Sigma_T, \text{ such that}\\ \frac{1}{\varepsilon}\le\overline{u}_1(x,\tau)\le\frac{1}{\varepsilon}+\frac{1}{\varepsilon}\left(\overline{w}^+_{(k),n}(x,\tau)+\varepsilon\right)^a\left(\overline{w}^+_{(k),m}(x,\tau)+\varepsilon\right)^b\end{array}\right\},$$

$$\mathcal{E}_\varepsilon\equiv\left\{(x,\tau)\mid \overline{u}_1(x,\tau)<1/\varepsilon\right\}, \qquad \mathcal{G}_\varepsilon\equiv\mathcal{E}_\varepsilon\cup\mathcal{F}_\varepsilon.$$

Next

$$\begin{aligned}
L_\varepsilon &\equiv -\varepsilon\int_s^t\int_{\mathbf{R}^N} w_{(k)}\frac{\partial}{\partial\tau}(\psi-u_1)_+\zeta^p dxd\tau\\
&= -a\int_s^t\int_{\mathbf{R}^N} w^+_{(k),n}\left(w^+_{(k),n}+\varepsilon\right)^{a-1}\left(w^+_{(k),m}+\varepsilon\right)^b\frac{\partial}{\partial\tau}w^+_{(k),n}\zeta^p\chi(\mathcal{G}_\varepsilon)dxd\tau\\
&\quad -b\int_s^t\int_{\mathbf{R}^N} w^+_{(k),m}\left(w^+_{(k),n}+\varepsilon\right)^a\left(w^+_{(k),m}+\varepsilon\right)^{b-1}\frac{\partial}{\partial\tau}w^+_{(k),m}\zeta^p\chi(\mathcal{G}_\varepsilon)dxd\tau\\
&\quad -\int_s^t\int_{\mathbf{R}^N} w_{(k)}(1-\varepsilon u_1)_t\chi(\mathcal{F}_\varepsilon)\zeta^p dxd\tau\\
&\equiv L^{(1)}_\varepsilon+L^{(2)}_\varepsilon+L^{(3)}_\varepsilon\,.
\end{aligned}$$

We claim that $L^{(3)}_\varepsilon\to 0$ as $\varepsilon\to 0$. Indeed

$$\left|L_\varepsilon^{(3)}\right| \le \iint\limits_{s\,K_{2\rho}}^{t} \varepsilon |w_{(k)}| \left|\frac{\partial}{\partial\tau}u_1\right| \chi(\mathcal{F}_\varepsilon)\, dx d\tau.$$

On the set $\mathcal{F}_\varepsilon$ we have

$$\frac{1}{\varepsilon} \le u_1 \le \frac{1}{\varepsilon} + \frac{1}{\varepsilon}(n+1)^{a+b} \equiv \frac{\gamma}{\varepsilon},$$

$$\varepsilon |w_{(k)}| \le \gamma \qquad \text{a.e.} \quad \mathcal{F}_\varepsilon.$$

Therefore

$$\left|L_\varepsilon^{(3)}\right| \le \gamma \iint\limits_{s\,K_{2\rho}}^{t} \left|\frac{\partial}{\partial\tau}u_1\right| \chi\left[\frac{1}{\varepsilon} \le u_1 \le \frac{\gamma}{\varepsilon}\right] dx d\tau \tag{6.6}$$

and the assertion follows from Lemma 5.1.

Since k, n, m are fixed, the integrands in $L_\varepsilon^{(i)}, i=1,2$, are in $L^1_{loc}(\Sigma_T)$ uniformly in ε. Moreover they have a.e. limits that are in $L^1_{loc}(\Sigma_T)$ and their absolute value is majorised almost everywhere in Σ_T, uniformly in ε, by functions in $L^1_{loc}(\Sigma_T)$. Therefore as $\varepsilon \to 0$

$$L_\varepsilon^{(1)} + L_\varepsilon^{(2)} \longrightarrow \mathcal{L} \tag{6.7}$$

$$\equiv -\frac{a}{a+1} \iint\limits_{s\,\mathbf{R}^N}^{t} \frac{\partial}{\partial\tau}\left(w^+_{(k),n}\right)^{a+1} \left(w^+_{(k),m}\right)^{b} \zeta^p dx d\tau$$

$$- \frac{b}{b+1} \iint\limits_{s\,\mathbf{R}^N}^{t} \left(w^+_{(k),n}\right)^{a} \frac{\partial}{\partial\tau}\left(w^+_{(k),m}\right)^{b+1} \zeta^p dx d\tau.$$

Since $n > m+1$,

$$\begin{aligned}\left(w^+_{(k),n}\right)^{a} \frac{\partial}{\partial\tau}\left(w^+_{(k),m}\right)^{b+1} &= \left(w^+_{(k),m}\right)^{a} \frac{\partial}{\partial\tau}\left(w^+_{(k),m}\right)^{b+1} \\ &= \frac{b+1}{a+b+1}\frac{\partial}{\partial\tau}\left(w^+_{(k),m}\right)^{a+b+1}, \qquad \text{a.e. } \Sigma_T,\end{aligned}$$

We obtain from (6.7)

$$\begin{aligned}\mathcal{L} \equiv &-\frac{a}{a+1} \int\limits_{\mathbf{R}^N\times\{t\}} \left(w^+_{(k),n}\right)^{a+1} \left(w^+_{(k),m}\right)^{b} \zeta^p dx \\ &- \frac{b}{(a+1)(a+b+1)} \int\limits_{\mathbf{R}^N\times\{t\}} \left(w^+_{(k),m}\right)^{a+b+1} \zeta^p dx \\ &+ \frac{a}{a+1} \int\limits_{\mathbf{R}^N\times\{s\}} \left(w^+_{(k),n}\right)^{a+1} \left(w^+_{(k),m}\right)^{b} \zeta^p dx \\ &+ \frac{b}{(a+1)(a+b+1)} \int\limits_{\mathbf{R}^N\times\{s\}} \left(w^+_{(k),m}\right)^{a+b+1} \zeta^p dx.\end{aligned} \tag{6.7'}$$

We combine this with (6.5) and conclude that the sum of the first three terms on the left hand side of (6.4) has a limit, as $\varepsilon \to 0$, that is minorised by

$$\text{(6.8)}\qquad \frac{1}{a+b+1}\int_{\mathbf{R}^N\times\{t\}}\left(w^+_{(k),m}\right)^{a+b+1}\zeta^p dx - \frac{1}{a+b+1}\int_{\mathbf{R}^N\times\{s\}} w^+_{(k)}\left(w^+_{(k),n}\right)^{a+b}\zeta^p dx.$$

We turn to estimate below the lim-inf as $\varepsilon \to 0$ of the last integral on the left hand side of (6.4):

$$\begin{aligned}
&\varepsilon\int_s^t\!\!\int_{\mathbf{R}^N}\mathbf{J}_k D(\psi-u_1)_+\zeta^p dxd\tau\\
&= a\int_s^t\!\!\int_{\mathbf{R}^N}\mathbf{J}_k Dw^+_{(k),n}\left(w^+_{(k),n}+\varepsilon\right)^{a-1}\\
&\qquad\times\left(w^+_{(k),m}+\varepsilon\right)^{b}\zeta^p\chi(\mathcal{G}_\varepsilon)dxd\tau\\
&+b\int_s^t\!\!\int_{\mathbf{R}^N}\mathbf{J}_k Dw^+_{(k),m}\left(w^+_{(k),n}+\varepsilon\right)^{a}\\
\text{(6.9)}\qquad&\qquad\times\left(w^+_{(k),m}+\varepsilon\right)^{b-1}\zeta^p\chi(\mathcal{G}_\varepsilon)dxd\tau\\
&+\int_s^t\!\!\int_{\mathbf{R}^N}\mathbf{J}_k D(1-\varepsilon u_1)\zeta^p\chi(\mathcal{F}_\varepsilon)dxd\tau\\
&\ge a(p-1)\int_s^t\!\!\int_{\mathbf{R}^N}A_o|Dw^+_{(k),n}|^2\left(w^+_{(k),n}+\varepsilon\right)^{a-1}\\
&\qquad\times\left(w^+_{(k),m}+\varepsilon\right)^{b}\zeta^p\chi(\mathcal{G}_\varepsilon)dxd\tau\\
&-\frac{4\varepsilon}{p-1}\int_s^t\!\!\int_{\mathbf{R}^N}A_o|Dw_{(k)}||Du_1|\zeta^p\chi(\mathcal{F}_\varepsilon)dxd\tau\\
&\equiv H^{(1)}_\varepsilon + H^{(2)}_\varepsilon.
\end{aligned}$$

By weak lower semicontinuity

$$\begin{aligned}
\text{(6.10)}\quad &\liminf_{\varepsilon\to 0} H^{(1)}_\varepsilon\\
&\ge a(p-1)\int_s^t\!\!\int_{\mathbf{R}^N}A_o|Dw^+_{(k),n}|^2\left(w^+_{(k),n}\right)^{a-1}\left(w^+_{(k),m}\right)^{b}\zeta^p dxd\tau.
\end{aligned}$$

We claim that $H_\varepsilon^{(2)} \to 0$ as $\varepsilon \to 0$. Using Lemma 5.2 we have

$$\begin{aligned}
\left|H_\varepsilon^{(2)}\right| &\le \varepsilon C(N,p) \iint_{s\,K_{2\rho}}^{t} \left|Du_1 - Du_{2,k}\right|^{p-1} |Du_1| \chi(\mathcal{F}_\varepsilon)\, dx d\tau \\
&\le C\varepsilon \iint_{s\,K_{2\rho}}^{t} |Du_1|^p \chi(\mathcal{F}_\varepsilon) dx d\tau + C\varepsilon \iint_{s\,K_{2\rho}}^{t} |Du_{2,k}|^p \chi(\mathcal{F}_\varepsilon)\, dx d\tau \\
&\equiv H_{\varepsilon,1}^{(2)} + H_{\varepsilon,2}^{(2)}.
\end{aligned}$$

Since $|Du_{2,k}| \in L^p_{loc}(\Sigma_T)$ the second term tends to zero as $\varepsilon \to 0$. As for $H_{\varepsilon,1}^{(2)}$ write

$$H_{\varepsilon,1}^{(2)} \le C \iint_{s\,K_{2\rho}}^{t} |Du_1|^p \frac{1}{u_1} \varepsilon u_1 \, \chi\left(\frac{1}{\varepsilon} \le u_1 < \frac{\gamma}{\varepsilon}\right) dx d\tau$$

where $\gamma = 1 + (n+1)^a (m+1)^b$. This implies, since $u_1 \in \mathcal{S}^*$

$$H_{\varepsilon,1}^{(2)} \le C(p,n,m) \iint_{s\,K_{2\rho}}^{t} |Du_1^{\frac{p-1}{p}}|^p \chi\left(\frac{1}{\varepsilon} \le u_1 \le \frac{\gamma}{\varepsilon}\right) dx d\tau \longrightarrow 0 \text{ as } \varepsilon \to 0.$$

We finally estimate above the lim-sup as $\varepsilon \to 0$ of the integral on the right-hand side of (6.4). Using the definition (6.3) of ψ and (5.6)

$$\begin{aligned}
&\left| p \iint_{s\,\mathbf{R}^N}^{t} \mathbf{J}_k (\psi - u_1)_+ \zeta^{p-1} D\zeta \, dx d\tau \right| \\
&\le \gamma \iint_{s\,\mathbf{R}^N}^{t} A_o |Dw_{(k)}| \left(w_{(k),n}^+ + \varepsilon\right)^a \\
&\qquad\qquad \times \left(w_{(k),m}^+ + \varepsilon\right)^b \zeta^{p-1} |D\zeta| \chi(\mathcal{G}_\varepsilon) dx d\tau \\
&+ \gamma \iint_{s\,\mathbf{R}^N}^{t} A_o |Dw_{(k)}| \varepsilon u_1 \chi(\mathcal{F}_\varepsilon) \zeta^{p-1} |D\zeta| dx d\tau.
\end{aligned} \tag{6.11}$$

The last integral tends to zero as $\varepsilon \to 0$. Indeed it can be majorised by

$$
\begin{aligned}
(6.12) \qquad & \gamma \int_s^t\!\!\int_{K_{2\rho}} |Dw_{(k)}|^{p-1}\chi(\mathcal{F}_\varepsilon)\,dxd\tau \\
& \le \gamma \int_s^t\!\!\int_{K_{2\rho}} |Du_1|^{p-1}\chi[u_1 \ge \tfrac{1}{\varepsilon}]dxd\tau \\
& + \gamma \int_s^t\!\!\int_{K_{2\rho}} |Du_{2,k}|^{p-1}\chi[u_1 \ge \tfrac{1}{\varepsilon}]dxd\tau,
\end{aligned}
$$

for a constant $\gamma = \gamma(p, n, m, a, b)$. The second integral on the right hand side of (6.12) tends to zero as $\varepsilon \to 0$, since $u_1 \in L^1_{loc}(\Sigma_T)$. As for the first integral, let $\alpha \in (0, p-1)$ be so small that $(\alpha+1)(p-1) < 1$. Then

$$
\begin{aligned}
& \overline{\gamma} \int_s^t\!\!\int_{K_{2\rho}} |Du_1|^{p-1}\chi[u_1 > \tfrac{1}{\varepsilon}]dxd\tau \\
&= \overline{\gamma} \int_s^t\!\!\int_{K_{2\rho}} |Du_1|^{p-1} u_1^{-\frac{(\alpha+1)(p-1)}{p}} u_1^{\frac{(\alpha+1)(p-1)}{p}} \chi[u_1 > \tfrac{1}{\varepsilon}]dxd\tau \\
&= \overline{\overline{\gamma}} \int_s^t\!\!\int_{K_{2\rho}} |Du_1^{\frac{p-1-\alpha}{p}}|^{p-1} u_1^{\frac{(\alpha+1)(p-1)}{p}} \chi[u_1 > \tfrac{1}{\varepsilon}]dxd\tau \\
&\le \overline{\overline{\gamma}} \| Du_1^{\frac{p-1-\alpha}{p}} \|^{p-1}_{p, K_{2\rho}\times(s,t)} \left(\int_s^t\!\!\int_{K_{2\rho}} u_1^{(\alpha+1)(p-1)} \chi[u_1 > \tfrac{1}{\varepsilon}]dxd\tau \right)^{\frac{1}{p}} \\
&\qquad \longrightarrow 0 \quad \text{as } \varepsilon \to 0.
\end{aligned}
$$

We examine the lim-sup as $\varepsilon \to 0$ of the first integral on the right-hand side of (6.11). The numbers $k \in \mathbf{R}^+, n \in \mathbf{N}$ being fixed, if ε is small enough, we have the inclusion

$$
\{(x,\tau) \in \Sigma_T \,|\, 0 \le \overline{w}_{(k)}(x,\tau) \le n\} \subset \mathcal{G}_\varepsilon .
$$

Moreover since $w^+_{(k),n} \in L^p_{loc}\left(0, T; W^{1,p}_{loc}(\mathbf{R}^N)\right)$

$$
|Dw^+_{(k),n}| = 0 \quad \text{a.e. on } \left\{(x,\tau) \in \Sigma_T | \overline{w}^+_{(k)}(x,\tau) > n\right\}.
$$

We write

$$
\begin{aligned}
&\int_s^t\!\!\int_{\mathbf{R}^N} A_o|Dw_{(k)}|\left(w^+_{(k),n}+\varepsilon\right)^a\left(w^+_{(k),m}+\varepsilon\right)^b\zeta^{p-1}|D\zeta|\chi(\mathcal{G}_\varepsilon)dxd\tau\\
&=\int_s^t\!\!\int_{\mathbf{R}^N} A_o|Dw^+_{(k),n}|\left(w^+_{(k),n}+\varepsilon\right)^a\left(w^+_{(k),m}+\varepsilon\right)^b\zeta^{p-1}|D\zeta|dxd\tau\\
&+\int_s^t\!\!\int_{\mathbf{R}^N} A_o|Dw^+_{(k)}|(n+\varepsilon)^a(m+\varepsilon)^b\zeta^{p-1}|D\zeta|\chi[w^+_{(k)}>n]\chi(\mathcal{G}_\varepsilon)dxd\tau\\
&+\int_s^t\!\!\int_{\mathbf{R}^N} A_o|Dw^+_{(k)}|\varepsilon^{a+b}\zeta^{p-1}|D\zeta|\chi(\mathcal{G}_\varepsilon)dxd\tau=K^{(1)}_\varepsilon+K^{(2)}_\varepsilon+K^{(3)}_\varepsilon.
\end{aligned}
\tag{6.13}
$$

As for $K^{(1)}_\varepsilon$ the integrand tends to

$$A_o|Dw^+_{(k),n}|(w^+_{(k),n})^a(w^+_{(k),m})^b\zeta^{p-1}|D\zeta| \quad \text{a.e. } K_{2\rho}\times(s,t),$$

in a decreasing way. Therefore

$$K^{(1)}_\varepsilon \longrightarrow \int_s^t\!\!\int_{\mathbf{R}^N} A_o|Dw^+_{(k),n}|(w^+_{(k),n})^a(w^+_{(k),m})^b\zeta^{p-1}|D\zeta|dxd\tau.$$

The last integral tends to zero as $\varepsilon\to 0$. Indeed

$$K^{(3)}_\varepsilon \le C(p)\frac{\varepsilon^{a+b}}{\sigma\rho}\int_s^t\!\!\int_{K_{2\rho}}\left(|Du_{1,k}|^{p-1}+|Du_{2,k}|^{p-1}\right)dxd\tau \longrightarrow 0 \ \text{ as } \varepsilon\to 0.$$

The operation $Dw^+_{(k)}$ coincides with the weak derivative of $w^+_{(k)}$ only on those sets $\mathcal{A}^\ell$ where $w^+_{(k)}$ is bounded by a positive constant ℓ, i.e.,

$$Dw^+_{(k)}\chi\left(\mathcal{A}^\ell\right)\equiv Dw^+_{(k),\ell}.$$

Since $Dw^+_{(k)}$ is not well defined a.e. in the whole strip Σ_T we estimate $K^{(2)}_\varepsilon$ as follows:

$$
\begin{aligned}
K^{(2)}_\varepsilon &\le \gamma\frac{(m+1)^b}{\sigma\rho}\int_s^t\!\!\int_{K_{2\rho}}\left|Du_1-Du_{2,k}\right|^{p-1}u_1^a\chi[u_1>n+u_{2,k}]\chi(\mathcal{G}_\varepsilon)\,dxd\tau\\
&\le \gamma\frac{(m+1)^b}{\sigma\rho}\int_s^t\!\!\int_{K_{2\rho}}|Du_1|^{p-1}u_1^a\chi[u_1>n]\chi(\mathcal{G}_\varepsilon)\,dxd\tau\\
&+\gamma\frac{(m+1)^b}{\sigma\rho}\int_s^t\!\!\int_{K_{2\rho}}|Du_{2,k}|^{p-1}u_1^a\chi[u_1>n+u_{2,k}]\chi(\mathcal{G}_\varepsilon)\,dxd\tau.
\end{aligned}
$$

If $\alpha\in(0,p-1)$, write

$$\int_s^t\int_{K_{2\rho}}|Du_1|^{p-1}u_1^a\chi[u_1>n]\chi(\mathcal{G}_\varepsilon)dxd\tau$$

$$=\int_s^t\int_{K_{2\rho}}|Du_1|^{p-1}u_1^{-\frac{(\alpha+1)(p-1)}{p}}u_1^{\frac{(\alpha+1)(p-1)+ap}{p}}\chi[u_1>n]\chi(\mathcal{G}_\varepsilon)dxd\tau$$

$$\leq\gamma\left(\int_s^t\int_{K_{2\rho}}|Du_1^{\frac{p-1-\alpha}{p}}|^p dxd\tau\right)^{\frac{p-1}{p}}\left(\int_s^t\int_{K_{2\rho}}u_1^{(\alpha+1)(p-1)+ap}\chi[u_1>n]dxd\tau\right)^{\frac{1}{p}}.$$

Choose α and $a>0$ so small that

$$(\alpha+1)(p-1)+ap\leq 1. \tag{6.14}$$

Then $u_1^{(\alpha+1)(p-1)+ap}\in L^1_{loc}(\Sigma_T)$ and $\forall\varepsilon\in(0,1)$

$$\int_s^t\int_{K_{2\rho}}|Du_1|^{p-1}u_1^a\chi[u_1>n]\chi(\mathcal{G}_\varepsilon)\,dxd\tau\leq O\left(\frac{1}{n}\right).$$

Analogously

$$\int_s^t\int_{K_{2\rho}}|Du_{2,k}|^{p-1}u_1^a\chi[u_1>n+u_{2,k}]\chi(\mathcal{G}_\varepsilon)dxd\tau$$

$$\leq\gamma\left(\int_s^t\int_{K_{2\rho}}|Du_2^{\frac{p-1-\alpha}{p}}|^p dxd\tau\right)^{\frac{p-1}{p}}$$

$$\times\left(\int_s^t\int_{K_{2\rho}}u_1^{ap}u_2^{(\alpha+1)(p-1)}\chi[u_1>n+u_{2,k}]dxd\tau\right)^{\frac{p-1}{p}}$$

$$\leq\gamma\left(\int_s^t\int_{K_{2\rho}}|Du_2^{\frac{p-1-\alpha}{p}}|^p dxd\tau\right)^{\frac{p-1}{p}}$$

$$\times\left(\int_s^t\int_{K_{2\rho}}u_1^{(\alpha+1)(p-1)+ap}\chi[u_1>n]dxd\tau\right)^{\frac{1}{p}}$$

$$=O\left(\frac{1}{n}\right).$$

We conclude that

$$\overline{\lim_{\varepsilon\to 0}} K_\varepsilon^{(2)} \le \gamma(m+1)^b O(\frac{1}{n}), \qquad \text{uniformly for all } k\in\mathbf{R}^+, \tag{6.15}$$

provided α and $a>0$ are chosen so that (6.14) holds. Combining these estimates and limiting processes as parts of (6.4) we obtain

$$\begin{aligned}
&\frac{1}{a+b+1}\int_{\mathbf{R}^N\times\{t\}}\left(w^+_{(k),m}\right)^{a+b+1}\zeta^p\,dx \\
&+a(p-1)\int_s^t\int_{\mathbf{R}^N} A_o|Dw^+_{(k),n}|^2\left(w^+_{(k),n}\right)^{a-1}\left(w^+_{(k),m}\right)^b\zeta^p dx d\tau \\
&\le \frac{1}{a+b+1}\int_{\mathbf{R}^N\times\{s\}} w_k^+\left(w^+_{(k),n}\right)^{a+b}\zeta^p dx \\
&+\frac{C}{\sigma\rho}\int_s^t\int_{\mathbf{R}^N} A_o|Dw^+_{(k),n}|\left(w^+_{(k),n}\right)^{a}\left(w^+_{(k),m}\right)^b\zeta^{p-1} dx d\tau \\
&+\gamma(m+1)^b O\left(\frac{1}{n}\right).
\end{aligned} \tag{6.16}$$

6-(iii). The limits as $k\to\infty$ and $s\to 0$

If $n\in\mathbf{N}$ and $k>0$ are fixed, we let $\overline{w}^+_{(k),n}$ and $\overline{D}w^+_{(k),n}$ be arbitrarily selected but fixed representatives out of the equivalence classes $w^+_{(k),n}$ and $Dw^+_{(k),n}$ and introduce the sets

$$\mathcal{E}_1 \equiv \left\{(x,\tau)\in\Sigma_T\middle|\overline{w}^+_{(k),n}(x,\tau)\le\frac{a(p-1)\sigma\rho}{4C}\zeta(x)|\overline{D}w^+_{(k),n}|(x,\tau)\right\},$$

$$\mathcal{E}_2 \equiv \left\{(x,\tau)\in\Sigma_T\middle||\overline{D}w^+_{(k),n}|(x,\tau)\le\frac{4C}{a(p-1)\sigma\rho\zeta(x)}\overline{w}^+_{(k),n}(x,\tau)\right\},$$

where C is the constant appearing in the last integral on the right hand side of (6.16). This integral is estimated as follows:

$$\begin{aligned}
\frac{C}{\sigma\rho}&\int_s^t\!\!\int_{\mathbf{R}^N} A_o|Dw^+_{(k),n}|\left(w^+_{(k),n}\right)^a\left(w^+_{(k),m}\right)^b\zeta^{p-1}dxd\tau\\
&\le \frac{C}{\sigma\rho}\int_s^t\!\!\int_{\mathbf{R}^N} A_o|Dw^+_{(k),n}|\left(w^+_{(k),n}\right)^a\left(w^+_{(k),m}\right)^b\zeta^{p-1}\chi(\mathcal{E}_1)dxd\tau\\
&+\frac{2C}{(p-1)\sigma\rho}\int_s^t\!\!\int_{\mathbf{R}^N} |Dw^+_{(k),n}|^{p-1}\left(w^+_{(k),n}\right)^a\left(w^+_{(k),m}\right)^b\zeta^{p-1}\chi(\mathcal{E}_2)dxd\tau\\
&\le \frac{a(p-1)}{2}\int_s^t\!\!\int_{\mathbf{R}^N} A_o|Dw^+_{(k),n}|^2\left(w^+_{(k),n}\right)^{a-1}\left(w^+_{(k),m}\right)^b\zeta^{p}dxd\tau\\
&+\frac{4^pC^p}{a^{p-1}(p-1)^p(\sigma\rho)^p}\int_s^t\!\!\int_{\mathbf{R}^N}\left(w^+_{(k),n}\right)^{p-1+a}\left(w^+_{(k),m}\right)^b dxd\tau.
\end{aligned}$$

We carry this estimate in (6.16), move the integral involving $\left|Dw^+_{(k),n}\right|^2$ on the left hand side and discard the resulting non-negative term to obtain

$$\begin{aligned}
\frac{1}{a+b+1}&\int_{\mathbf{R}^N\times\{t\}}\left(w^+_{(k),m}\right)^{a+b+1}\zeta^p dx\\
&\le \frac{1}{a+b+1}\int_{\mathbf{R}^N\times\{s\}} w^+_k\left(w^+_{(k),n}\right)^{a+b}\zeta^p dx\\
&+\frac{\gamma(p)}{(\sigma\rho)^p}\int_s^t\!\!\int_{K_{(1+\sigma)\rho}}\left(w^+_{(k),n}\right)^{p-1+a}\left(w^+_{(k),m}\right)^b dxd\tau\\
&+\gamma(m+1)^bO\left(\frac{1}{n}\right).
\end{aligned}\tag{6.17}$$

We let now $k\to\infty$ while $s>0$, $n,m\in\mathbf{N}$ remain fixed. Since $w^+_{(k)}\to w^+$ in a decreasing way we may pass to the limit under the integrals in (6.17) and obtain the same integral inequality written for w^+. In particular the first integral on the right hand side takes the form

$$\frac{1}{a+b+1}\int_{\mathbf{R}^N\times\{s\}} w^+(w^+_n)^{a+b}\zeta^p dx.\tag{6.18}$$

Now letting $s\to 0$, the integral in (6.18) tends to zero since it can be majorised by

$$\frac{n^{a+b}}{a+b+1}\int_{\mathbf{R}^N\times\{s\}} w^+\zeta^p dx\longrightarrow 0\quad\text{as } s\to 0.$$

These limiting processes yield

$$\int\limits_{\mathbf{R}^N\times\{t\}} \left(w_m^+\right)^{a+b+1} \zeta^p dx \le \frac{\gamma(p)(a+b+1)}{(\sigma\rho)^p} \int\limits_0^t \int\limits_{K_{(1+\sigma)\rho}} \left(w_n^+\right)^{p-1+a} \left(w_m^+\right)^b dx d\tau + \gamma(a+b+1)(m+1)^b O\left(\frac{1}{n}\right). \tag{6.19}$$

6-(iv). *Proof of Proposition 6.1*

We let $n \to \infty$ in (6.19), while $m \in \mathbf{N}$ remains fixed. The integrand in the last integral tends to $(w^+)^{p-1+a}(w_m^+)^b$ a.e. in $K_{(1+\sigma)\rho} \times (0,t)$ in an increasing fashion. Moreover if a is so small that

$$p - 1 + a \in (0,1), \tag{6.20}$$

it is dominated, uniformly in n, by the function

$$(w^+)^{p-1+a}(w_m^+)^b \in L^1_{loc}(\Sigma_T).$$

The limit process gives

$$\int\limits_{\mathbf{R}^N\times\{t\}} \left(w_m^+\right)^{a+b+1} \zeta^p dx \le \frac{\gamma(p)(a+b+1)}{(\sigma\rho)^p} \int\limits_0^t \int\limits_{K_{(1+\sigma)\rho}} (w^+)^{p-1+a}(w_m^+)^b dx d\tau. \tag{6.21}$$

This inequality holds true $\forall m \in \mathbf{N}, \forall b \ge 0, \forall \sigma \in (0,1), \forall \rho > 0$. The positive number a is fixed, satisfying the restrictions (6.14) and (6.20). The sequence $\{w_m^+\}$ increases to w^+ a.e. in Σ_T. Therefore as $m \to \infty$, we may pass to the limit under the integrals in (6.21) for those $b \ge 0$ for which

$$(w^+)^{p-1+a+b} \in L^1_{loc}(\Sigma_T).$$

If $b_i \ge 0$ is one such b, letting $m \to \infty$ we find that

$$(w^+)^{a+b_i+1} \in L^1_{loc}(\Sigma_T),$$

which implies that

$$(w^+)^{p-1+a+b_{i+1}} \in L^1_{loc}(\Sigma_T), \quad b_{i+1} = b_i + 2 - p > b_i.$$

Let $b_o \ge 0$ be defined by $p - 1 + a + b_o = 1$. Then the previous remarks show that

$$(w^+)^{p-1+a+b_o+i(2-p)} \equiv (w^+)^{1+i(2-p)} \in L^1_{loc}(\Sigma_T), \quad i = 0,1,2,\ldots .$$

Interchanging the role of u_1 and u_2 proves the Proposition.

7. Proof of the uniqueness theorem

From (6.1) by Hölder's inequality, since $p\in(1,2)$

$$\int_{K_\rho} |w(t)|^q dx \le \frac{\gamma}{(\sigma\rho)^p} \int_0^t \left(\int_{K_{(1+\sigma)\rho}} |w(\tau)|^q dx \right)^{\frac{q+(p-2)}{q}} \rho^{\frac{N(2-p)}{q}} d\tau \tag{7.1}$$

$$\le \frac{\gamma t}{\sigma^p} \left(\sup_{0<\tau<t} \int_{K_{(1+\sigma)\rho}} |w(\tau)|^q dx \right)^{1-\frac{2-p}{q}} \rho^{\frac{N(2-p)}{q}-p}.$$

Let $\rho>0$ be fixed and for $n=1,2,\dots$ define

$$\rho_n = \left(\sum_{i=0}^{n} 2^{-i} \right) \rho, \quad K_n = K_{\rho_n}, \quad \sigma_n = 2^{-(n+1)},$$

$$Y_n = \sup_{0<\tau\le t} \int_{K_n} |w(\tau)|^q dx.$$

Rewrite (7.1) over K_n and K_{n+1} to obtain

$$Y_n \le \gamma 2^{np} \left(\frac{t}{\rho^{\frac{N(p-2)+pq}{p}}} \right) Y_{n+1}^{1-\frac{2-p}{q}}. \tag{7.2}$$

By the interpolation Lemma 4.3 of Chap. I we conclude that for every $q\in[1,\infty)$ there exists a constant $\gamma=\gamma(N,p,q)$, independent of ρ, such that for all $t\in(0,T)$

$$\int_{K_\rho} |w(t)|^q dx \le \gamma \left(\frac{t}{\rho^{\frac{N(p-2)+pq}{q}}} \right)^{\frac{q}{2-p}} \tag{7.3}$$

To prove the theorem we choose q so large that

$$\frac{N(p-2)+pq}{q} > 0$$

and then, such a q being fixed, we let $\rho\to\infty$ in (7.3).

8. Solving the Cauchy problem

We will establish the existence of a unique non-negative solution to the Cauchy problem (1.1)-(1.2) where u_o is non-negative and merely in $L^1_{loc}(\mathbf{R}^N)$. For $n=1,2,\dots$ consider the sequence of approximating problems

$$(8.1)_n \qquad \begin{cases} u_n \in C\left(0,T; L^2_{loc}(\mathbf{R}^N)\right) \cap L^p\left(0,T; W^{1,p}_{loc}(\mathbf{R}^N)\right) \\ \frac{\partial}{\partial \tau} u_n - \operatorname{div} |Du_n|^{p-2} Du_n = 0, \quad \text{in } \Sigma_T \\ u_n(x,0) = u_{o,n} \equiv \begin{cases} \min\{u_o; n\} & \text{for } |x| < n \\ 0 & \text{for } |x| \geq n. \end{cases} \end{cases}$$

The initial data are bounded and compactly supported in $\mathbf{R}^N$. Therefore the unique solvability of $(8.1)_n$ can be established as indicated in §12 of Chap. VI. Since the initial data $\{u_{o,n}\}_{n\in\mathbf{N}}$ form an increasing sequence of functions in $L^1_{loc}(\mathbf{R}^N)$ we have

$$(8.2) \qquad \int_{K_\rho} u_{o,n} dx \leq \int_{K_\rho} u_o dx \qquad \forall \rho > 0.$$

The solution of (1.1)-(1.2) will be constructed as the limit of the sequence $\{u_n\}_{n\in\mathbf{N}}$ in a suitable topology. For this we establish first some basic compactness of $\{u_n\}_{n\in\mathbf{N}}$.

LEMMA 8.1. *There exists a constant $\gamma = \gamma(N,p)$ independent of n such that for all $t, \rho > 0$*

$$(8.3) \qquad \sup_{0<\tau\leq t} \int_{K_\rho} u_n(x,\tau)\, dx \leq \gamma \left\{ \int_{K_{2\rho}} u_o\, dx + \left(\frac{t}{\rho^\lambda}\right)^{\frac{1}{2-p}} \right\},$$

$$(8.4) \qquad \frac{1}{\rho} \int_0^t \int_{K_\rho} |Du_n|^{p-1} dx d\tau \leq \gamma \left(\frac{t}{\rho^\lambda}\right)^{\frac{1}{p}} \left\{ \int_{K_{2\rho}} u_o\, dx + \left(\frac{t}{\rho^\lambda}\right)^{\frac{1}{2-p}} \right\}^{\frac{2(p-1)}{p}}.$$

Moreover for all $\alpha \in (0, p-1)$,

$$(8.5) \qquad \int_0^t \int_{K_\rho} \left|Du_n^{\frac{p-1-\alpha}{p}}\right|^p dx d\tau \leq \frac{\gamma \rho^{\alpha N}}{\alpha^p} \left\{ \int_{K_{2\rho}} u_o\, dx + \left(\frac{t}{\rho^\lambda}\right)^{\frac{1}{2-p}} \right\}^{1-\alpha}.$$

PROOF: The L^1_{loc}–estimate follows from (4.1) with $s = 0$, and the gradient estimate (8.4) is a consequence of (4.2) with $s = 0$. Finally (8.5) is the content of Lemma 3.2.

9. Compactness in the space variables

LEMMA 9.1. *Let $\alpha \in (0, p-1)$ be so small that $(\alpha+1)(p-1) < 1$. There exists a constant $\gamma = \gamma(N, p, \alpha)$ such that*

$$\forall 0 < t \le T, \quad \forall k, \quad \forall C > 1, \quad \forall n = 1, 2, \dots,$$

(9.1)
$$\int_0^t \int_{K_\rho} |Du_n|^p u_n^{-1} \chi[k < u_n < Ck] \, dx d\tau$$

$$\le \gamma k^{-\left(\frac{1-(\alpha+1)(p-1)}{p}\right)} \rho^{N\alpha \frac{p-1}{p}} \left(\frac{t}{\rho^p}\right)^{\frac{1}{p}} \left\{ \int_{K_{2\rho}} u_o \, dx + \left(\frac{t}{\rho^\lambda}\right)^{\frac{1}{2-p}} \right\}^{\frac{p-\alpha(p-1)}{p}}$$

$$+ \ln C \int_{K_{2\rho}} u_o \chi[u_0 > k] \, dx.$$

The constant $\gamma(\alpha) \nearrow \infty$ *as either* $\alpha \searrow 0$ *or* $\alpha \nearrow p - 1$.

PROOF: We drop the subscript n for simplicity of notation. If $C > 1$ is fixed, let

$$u_{Ck}^{(k)} \equiv \begin{cases} k & \text{if } 0 < u \le k \\ u & \text{if } k < u < Ck \\ Ck & \text{if } u \ge Ck \end{cases}$$

and in the weak formulation of $(8.1)_n$, take the testing function

$$\left(\ln \frac{u_{Ck}^{(k)}}{k} \right) \zeta(x),$$

where $x \to \zeta(x)$ is the standard cutoff function in $K_{2\rho}$ that equals one on K_ρ. We obtain

$$\int_0^t \int_{K_\rho} |Du|^p \frac{1}{u} \chi[k < u < Ck] dx d\tau \le \frac{2\gamma}{\rho} \int_0^t \int_{K_{2\rho}} |Du|^{p-1} \chi[u > k] dx d\tau$$

$$- \int_0^t \int_{K_{2\rho}} \frac{\partial}{\partial \tau} \left(\int_k^u \ln \frac{\min\{\xi; Ck\}}{k} d\xi \right)_+ \zeta(x) dx d\tau \equiv G_k^{(1)} + G_k^{(2)}.$$

Let α be any positive number satisfying

$$\alpha \in (0, p - 1) \quad \text{and} \quad (\alpha + 1)(p - 1) < 1.$$

Then by virtue of (8.5)

$$G_k^{(1)} \le \frac{2\gamma}{\rho}\int_0^t\int_{K_{2\rho}} |Du|^{p-1} u^{-\frac{(\alpha+1)(p-1)}{p}} u^{\frac{(\alpha+1)(p-1)}{p}} \chi[u>k]\,dx d\tau$$

$$= \frac{\gamma(\alpha,p)}{\rho}\int_0^t\int_{K_{2\rho}} \left|Du^{\frac{p-1-\alpha}{p}}\right|^{p-1} u^{\frac{(\alpha+1)(p-1)}{p}} \chi[u>k]\,dx d\tau$$

$$\le \frac{\gamma(\alpha,p)}{\rho}\left\{\int_0^t\int_{K_{2\rho}} \left|Du^{\frac{p-1-\alpha}{p}}\right|^{p} dx d\tau\right\}^{\frac{p-1}{p}}$$

$$\times\left\{\int_0^t\int_{K_{2\rho}} u^{(\alpha+1)(p-1)} \chi[u>k]\,dx d\tau\right\}^{\frac{1}{p}}$$

$$\le \frac{\gamma(\alpha,p)\rho^{N\alpha\frac{p-1}{p}}}{\rho}\left\{\int_{K_{4\rho}} u_o\,dx + \left(\frac{t}{\rho^\lambda}\right)^{\frac{1}{2-p}}\right\}^{\frac{p-1}{p}(1-\alpha)}$$

$$\times\left\{\int_0^t\int_{K_{2\rho}} u^{(\alpha+1)(p-1)} \chi[u>k]\,dx d\tau\right\}^{\frac{1}{p}}$$

$$\le \frac{\gamma(\alpha,p)\,t^{1/p}}{\rho}\left\{\int_{K_{4\rho}} u_o\,dx + \left(\frac{t}{\rho^\lambda}\right)^{\frac{1}{2-p}}\right\}^{\frac{p-1}{p}(1-\alpha)}$$

$$\times\left\{\sup_{0<\tau<t}\int_{K_{2\rho}} u(x,\tau)dx\right\}^{\frac{(\alpha+1)(p-1)}{p}}$$

$$\times\left\{\sup_{0<\tau<t}\int_{K_{2\rho}} \chi[u>k]\,dx\right\}^{\frac{1-(\alpha+1)(p-1)}{p}}$$

$$\le \gamma(\alpha,p)\left(\frac{t}{\rho^p}\right)^{\frac{1}{p}}\left\{\int_{K_{4\rho}} u_o\,dx + \left(\frac{t}{\rho^\lambda}\right)^{\frac{1}{2-p}}\right\}^{\frac{2(p-1)}{p}}$$

$$\left\{\sup_{0<\tau<t}\int_{K_{2\rho}} \chi[u>k]\,dx\right\}^{\frac{1-(\alpha+1)(p-1)}{p}}.$$

The last step follows by use of (8.3). Using it again we obtain

$$\sup_{0<\tau<t} \int_{K_{2\rho}} \chi[u>k]\,dx \le \frac{\gamma}{k}\left\{\int_{K_{4\rho}} u_o\,dx + \left(\frac{t}{\rho^\lambda}\right)^{\frac{1}{2-p}}\right\}.$$

Therefore

$$G_k^{(1)} \le \gamma k^{-\frac{1-(\alpha+1)(p-1)}{p}} \left(\frac{t}{\rho^p}\right)^{\frac{1}{p}} \left\{\int_{K_{4\rho}} u_o\,dx + \left(\frac{t}{\rho^\lambda}\right)^{\frac{1}{2-p}}\right\}^{\frac{p-\alpha(p-1)}{p}}.$$

As for $G_k^{(2)}$ it is estimated above by

$$\int_{K_{2\rho}} \left(\int_k^{u_0} \ln\frac{\min\{\xi; Ck\}}{k}\,d\xi\right)_+ dx \le \ln C \int_{K_{2\rho}} u_o\chi[u_o>k]\,dx.$$

10. Compactness in the t variable

LEMMA 10.1. *Let $\alpha\in(0,p-1)$. There exists a constant $\gamma=\gamma(N,p,\alpha)$ such that*

$$\forall\, 0<s<t\le T,\quad \forall\theta\ge\alpha+1,\quad \forall n=1,2,\ldots,$$

$$\int_0^t\!\!\int_{K_\rho} (u_n+1)^{-\theta}u_{n,t}^2\,dxd\tau \le \gamma\,\rho^N\left(\frac{t}{\rho^2}\right)^{\frac{p}{2-p}} \tag{10.1}$$

$$+\frac{\gamma\rho^{-\alpha N}}{s}\left\{\int_{K_{2\rho}} u_o\,dx + \left(\frac{t}{\rho^\lambda}\right)^{\frac{1}{2-p}}\right\}^{1-\alpha}.$$

The constant $\gamma(N,p,\alpha)\nearrow\infty$ as either $\alpha\searrow 0$ or $\alpha\nearrow(p-1)$.

PROOF: Let $0<s<t\le T$ and $\rho>0$ be fixed. Consider the cylinders

$$Q_o\equiv K_\rho\times(s,t),\qquad Q_1\equiv B_{\frac{3}{2}\rho}\times\left(\frac{s}{2},t\right),$$

and let $(x,\tau)\to\zeta(x,\tau)$ be a non-negative piecewise smooth cutoff function in Q_1 which equals one on Q_o and such that $|D\zeta|\le 2/\rho$ and $\zeta_t\le 2/s$. At first we will proceed formally. The calculations below will be made rigorous later. In the weak formulation of $(8.1)_n$, take the testing function

$$u_{n,t}\,(u_n+1)^{-\theta}\zeta^2,$$

and integrate by parts over Q_1. Dropping the subscript n, we obtain

$$\begin{aligned}
\iint_{Q_1}(u+1)^{-\theta}u_t^2\zeta^2dxd\tau &= -\iint_{Q_1}|Du|^{p-2}DuD(u_t(u+1)^{-\theta}\zeta^2)dxd\tau\\
&= -\frac{1}{p}\iint_{Q_1}\frac{\partial}{\partial\tau}|Du|^p(u+1)^{-\theta}\zeta^2dxd\tau\\
&+\theta\iint_{Q_1}|Du|^p(u+1)^{-\theta-1}u_t\zeta^2dxd\tau\\
&-2\iint_{Q_1}|Du|^{p-2}Duu_t(u+1)^{-\theta}\zeta D\zeta\,dxd\tau\\
&\le\frac{\theta}{p}(p-1)\iint_{Q_1}|Du|^p(u+1)^{-\theta-1}u_t\zeta^2dxd\tau\\
&+\frac{2}{p}\iint_{Q_1}|Du|^p(u+1)^{-\theta}\zeta\zeta_\tau dxd\tau\\
&+\frac{4}{\rho}\iint_{Q_1}|Du|^{p-1}(u+1)^{-\frac{\theta}{2}}\left((u+1)^{-\theta}u_t^2\zeta^2\right)^{\frac{1}{2}}dxd\tau\\
&=\mathcal{R}^{(1)}+\mathcal{R}^{(2)}+\mathcal{R}^{(3)}.
\end{aligned}$$

In estimating $\mathcal{R}^{(1)}$ we use the regularising effect of Proposition 6.1 of Chap. VI,

$$u_t\le\frac{1}{2-p}\frac{u}{t}. \tag{10.2}$$

Then,

$$\mathcal{R}^{(1)}\le\frac{\theta(p-1)}{p(2-p)}\frac{1}{s}\iint_{Q_1}|Du|^p(u+1)^{-\theta}\zeta^2dxd\tau.$$

By Young's inequality

$$\mathcal{R}^{(3)}\le\frac{1}{2}\iint_{Q_o}(u+1)^{-\theta}u_t^2\zeta^2dxd\tau+\frac{8}{\rho^2}\iint_{Q_o}|Du|^{2(p-1)}(u+1)^{-\theta}dxd\tau.$$

Since $1<p<2$, this last integral is majorised by

$$\frac{1}{s}\iint_{Q_1}|Du|^p(u+1)^{-\theta}dxd\tau+\gamma(p)\left(\frac{t}{\rho^2}\right)^{\frac{p}{2-p}}\rho^N.$$

Combining these estimates we find that

$$\iint\limits_{Q_o}(u+1)^{-\theta}u_t^2 dx d\tau \le \gamma\,\rho^N\left(\frac{t}{\rho^2}\right)^{\frac{p}{2-p}} + \frac{\gamma}{s}\iint\limits_{Q_o}|Du|^p(u+1)^{-\theta}dx d\tau.$$

By (8.5), if $\alpha\in(0,p-1)$, this is estimated by

$$\begin{aligned}
&\frac{\gamma}{s}\iint\limits_{Q_1}|Du|^p u^{-(\alpha+1)}(u+1)^{-[\theta-(\alpha+1)]}dx d\tau\\
&\qquad\le \frac{\gamma(\alpha,p)\rho^{\alpha N}}{s}\iint\limits_{Q_1}|Du^{\frac{p-1-\alpha}{p}}|^p dx d\tau\\
&\qquad\le \frac{\gamma(\alpha,p)}{s}\left\{\int\limits_{K_{2\rho}}u_o\,dx + \left(\frac{t}{\rho^\lambda}\right)^{\frac{1}{2-p}}\right\}^{1-\alpha},
\end{aligned}$$

and the lemma follows by formal calculations. The calculations are formal since $u_{n,t}\,(u_n+1)^{-\theta}\,\zeta^2$, need not be an admissible testing functions in $(8.1)_n$. The arguments would be rigorous if

$$|Du_{n,t}| \in L^2_{loc}(\Sigma_T)\,. \tag{10.3}$$

Indeed, if so, we may take in the weak formulation $(8.1)_n$ the testing function

$$\frac{u_n(t+h)-u_n(t)}{h}\,(u_n+1)^{-\theta}\,\zeta^2,\quad h\in(0,T-\tfrac{1}{2}s)\,,\quad \tfrac{1}{2}s\le t<T-h.$$

The limit as $h\to 0$ is justified and we may proceed as before.

10-(i). Approximating estimates

Therefore to prove the lemma it suffices to establish (10.1) for a sequence of approximating solutions satisfying (10.3). The unique solution of $(8.1)_n$, can be approximated by the solutions of

$$\begin{cases} v_{n,j}\in C\left(0,T;L^2(B_j)\right)\cap L^p\left(0,T;W_o^{1,p}(B_j)\right),\\ j=n+1,n+2,\dots,\\ \frac{\partial}{\partial\tau}v_{n,j}-\operatorname{div}\left(|Dv_{n,j}|^{p-2}Dv_{n,j}\right)=0 \ \text{ in } \ B_j\times(0,T),\\ v_{n,j}(\cdot,t)\,\big|_{|x|=j}=0,\ \ v_{n,j}(\cdot,0)=u_{o,n,j}, \end{cases}\tag{10.4}$$

where B_j is the ball of radius j about the origin and $\{u_{o,n,j}\}_{j=n+1}^{\infty}$, is a sequence of functions in $C_o^\infty(B_{n+1})$, such that

$$u_{o,n,j}\longrightarrow u_{o,n}\ \text{ in }\ L^2_{loc}(B_{n+1})\,,$$

and

$$\int_{K_\rho} u_{o,n,j}dx \leq 2\int_{K_\rho} u_o dx \quad \forall \rho > 0.$$

As indicated in §12 of Chap. VI,

$$v_{n,j},\ \frac{\partial}{\partial x_\ell}v_{n,j} \longrightarrow u_n,\ \frac{\partial}{\partial x_\ell}u_n,\ \text{ in } C^\alpha_{loc}(\Sigma_T)$$
$$\forall \ell = 1,2,\dots,N,$$

for some $\alpha \in (0,1)$. The unique solvability of (10.4) can be established by a Galerkin procedure. Such a method also yields

$$\int_0^T\int_{B_j}\left|\frac{\partial}{\partial t}v_{n,j}\right|^2 dxd\tau \leq \frac{1}{p}\int_{B_{n+1}}|Du_{o,n,j}|^p\,dx. \tag{10.5}$$

To establish (10.1) for $v_{n,j}$ is suffices to show that

$$\left|D\frac{\partial}{\partial t}v_{n,j}\right| \in L^2_{loc}(B_j). \tag{10.6}$$

In the remarks below we drop the subscript n,j and write $v = v_{n,j}$. We write (10.4) for the time levels $t+h$ and t and set

$$w = \frac{v(t+h)-v(t)}{h},\quad h\in(0,T-\tfrac{1}{2}s),\ \tfrac{1}{2}s \leq t < T-h.$$

By difference

$$w_t - h^{-1}\operatorname{div}\mathbf{J}_h = 0 \quad \text{in } B_j\times(0,T-h), \tag{10.7}$$

where

$$\mathbf{J}_h \equiv |Dv(t+h)|^{p-2}Dv(t+h) - |Dv(t)|^{p-2}Dv(t).$$

In the weak formulation of (10.7) take the testing function $w\left(t-\frac{s}{2}\right)_+$ which vanishes on $|x|=j$ and for $t\leq\frac{s}{2}$. This gives

$$\int_{\frac{s}{2}}^{T-h}\int_{B_j}\left(t-\frac{s}{2}\right)_+ A_{o,j}|Dw|^2dxd\tau \leq \gamma\int_0^{T-h}\int_{B_j}\left|\frac{\partial}{\partial\tau}\fint_t^{t+h}v(x,\tau)d\tau\right|^2 dxd\tau, \tag{10.8}$$

where

$$A_{o,j} = \int_0^1|D(sv_{n,j}(t+h)+(1-s)v_{n,j}(t))|^{p-2}ds.$$

If $\mathcal{K}$ is a compact subset of $B_j\times(s,T)$, we have

$$A_{o,j} \geq \gamma\|Dv_{n,j}\|^{p-2}_{\infty,\mathcal{K}}.$$

It follows from (10.8) that

$$\iint_{\mathcal{K}} |Dw|^2\, dxd\tau \le \frac{\gamma(N,p,T)\|Dv_{n,j}\|_{\infty,\mathcal{K}}^{2-p}}{s} \int_0^{T-h}\int_{B_j} \left|\frac{\partial}{\partial\tau} \fint_t^{t+h} v(x,\tau)d\tau\right|^2 dxd\tau.$$

The last integral is finite by virtue of (10.5) and the lemma follows.

11. More on the time–compactness

We record a simple consequence of Lemma 10.1. If $x\to\zeta(x)$ is the usual cutoff function in $K_{2\rho}$ that equals one on K_ρ, we find from the weak formulation $(8.1)_n$, $\forall\, 0<s<t\le T$,

$$\int_s^t\int_{K_{2\rho}} (u_t)^-\zeta\, dxd\tau - \int_s^t\int_{K_{2\rho}} (u_t)^+\zeta dxd\tau = -\int_s^t\int_{K_{2\rho}} u_t\zeta\, dxd\tau$$
$$= \int_s^t\int_{K_{2\rho}} |Du|^{p-2}DuD\zeta\, dxd\tau.$$

Therefore

$$\int_s^t\int_{K_{2\rho}} (u_t)^- dxd\tau \le \int_s^t\int_{K_{2\rho}} (u_t)^+ dxd\tau + \frac{1}{\rho}\int_s^t\int_{K_{2\rho}} |Du|^{p-1} dxd\tau.$$

The first integral on the right hand side is estimated by (10.2), i.e.,

$$\int_s^t\int_{K_{2\rho}} (u_t)^+ dxd\tau \le \frac{1}{s(2-p)}\int_s^t\int_{K_{2\rho}} u(x,\tau)dxd\tau$$
$$\le \gamma\frac{t-s}{s}\left\{\int_{K_{4\rho}} u_o\, dx + \left(\frac{t}{\rho^\lambda}\right)^{\frac{1}{2-p}}\right\}.$$

Estimating the second integral by Proposition 8.1 gives

LEMMA 11.1. *There exists a constant $\gamma=\gamma(N,p)$, such that*

(11.1) $\forall\, 0<s<t\le T,\quad \forall n=1,2,\dots,$

$$\int_s^t\int_{K_\rho} |u_{n,t}| dxd\tau \le \gamma\,\frac{t-s}{s}\left\{\int_{K_{4\rho}} u_o\, dx + \left(\frac{t}{\rho^\lambda}\right)^{\frac{1}{2-p}}\right\}.$$

12. The limiting process

By construction $u_n \nearrow u$ a.e. in Σ_T and by (8.3)

$$\sup_{0<\tau<t} \int_{K_\rho} u(x,\tau)\,dx \le \gamma \left\{ \int_{K_{2\rho}} u_o\,dx + \left(\frac{t}{\rho^\lambda}\right)^{\frac{1}{2-p}} \right\}, \tag{12.1}$$

for all $0<t\le T$. Moreover by (11.1)

$$u \in C(0,T; L^1_{loc}(\mathbf{R}^N)). \tag{12.2}$$

By Lemma 8.1 the sequence $\{u_n^{\frac{p-1-\alpha}{p}}\}$ is equibounded in

$$L^p\left(0,T; W^{1,p}(K_\rho)\right),\ \forall \rho>0, \quad \text{provided } \alpha \in (0,p-1).$$

Since the whole sequence $\{u_n\}_{n\in\mathbf{N}} \to u$ in $L^1_{loc}(\mathbf{R}^N)$,

$$u_n^{\frac{p-1-\alpha}{p}} \longrightarrow u^{\frac{p-1-\alpha}{p}} \quad \text{weakly in } L^p(0,T; W^{1,p}(K_\rho)),\ \forall \rho>0.$$

This implies that the sequences

$$u_{n,k} = u_n \wedge k = \min\{u_n, k\}$$

are equibounded in $L^p\left(0,T; W^{1,p}(K_\rho)\right)$, $\forall \rho>0$, and

$$u_{n,k} \longrightarrow u \wedge k \quad \text{weakly in } L^p\left(0,T; W^{1,p}(K_\rho)\right),\ \forall \rho>0,\ \forall k>0. \tag{12.3}$$

LEMMA 12.1. *$Du_{n,k} \to Du_k$ strongly in $L^p_{loc}(\Sigma_T)$.*

PROOF: In the weak formulation of $(8.1)_n$, take the testing function

$$(u_{n,k} - u_k)\,\varphi, \quad \varphi \in C_o^\infty(\Sigma_T), \qquad \varphi \ge 0$$

to obtain

$$\begin{aligned} \iint_{\Sigma_T} |Du_{n,k}|^p \varphi\,dxd\tau &= \iint_{\Sigma_T} |Du_n|^{p-2} Du_n \cdot Du_k \varphi\,dxd\tau \\ &+ \iint_{\Sigma_T} |Du_n|^{p-2} D(u+\nu)(u_k - u_{n,k}) D\varphi\,dxd\tau \\ &+ \iint_{\Sigma_T} u_{n,t}\,(u_k - u_{n,k})\varphi dxd\tau \equiv I_o + I_1 + I_2. \end{aligned} \tag{12.4}$$

We first estimate the integrals $I_i, i=1,2$. Let $\alpha \in (0,p-1)$ be so small that $(\alpha+1)(p-1) \le 1$. Then by Lemma 8.1

$$|I_1| \le \iint_{\Sigma_T} |Du_n|^{p-1} u_n^{-(\alpha+1)\frac{p-1}{p}} u_n^{(\alpha+1)\frac{p-1}{p}} (u_k - u_{n,k})\varphi \, dxd\tau$$

$$\gamma \le \left\| Du_n^{\frac{p-1-\alpha}{p}} \right\|_{p,\mathrm{supp}\{\varphi\}}^{p-1} \left(\iint_{\Sigma_T} u_n^{(\alpha+1)(p-1)} (u_k - u_{n,k})^p \varphi^p dxd\tau \right)^{\frac{1}{p}}$$

$$\le \gamma\,(\alpha, p, u_o, \varphi) \left(\iint_{\Sigma_T} (u+1)(u_k - u_{n,k})^p \varphi^p dxd\tau \right)^{\frac{1}{p}}$$

$$\longrightarrow 0 \quad \text{as } n \to \infty.$$

We estimate $|I_2|$ by making use of the regularising inequality (10.2).

$$|I_2| \le \frac{\gamma(\varphi)}{2-p} \iint_{\Sigma_T} u_n\,(u_{n,k} - u_k)\,\varphi dxd\tau$$

$$\longrightarrow 0 \quad \text{as } n \frac{2-p}{p} \infty.$$

We return to (12.4) and estimate

$$I_o = \iint_{\Sigma_T} |Du_{n,k}|^{p-1} |Du_k|\,\varphi\,dxd\tau$$

$$\le \frac{p-1}{p} \iint_{\Sigma_T} |Du_{n,k}|^p \varphi\,dxd\tau + \frac{1}{p} \iint_{\Sigma_T} |Du_k|^p \varphi\,dxd\tau.$$

Combining these calculations in (12.4) gives

$$\iint_{\Sigma_T} |Du_{n,k}|^p \varphi\,dxd\tau \le \iint_{\Sigma_T} |Du_k|^p \varphi\,dxd\tau + 0\left(\frac{1}{n}\right).$$

From this, by lower semicontinuity

$$\iint_{\Sigma_T} |Du_k|^p \varphi\,dxd\tau \le \liminf_{n\to\infty} \iint_{\Sigma_T} |Du_{n,k}|^p \varphi\,dxd\tau$$

$$\le \iint_{\Sigma_T} |Du_k|^p \varphi\,dxd\tau.$$

This proves the lemma.

Next, by Lemma 10.1 the sequence

$$\left\{ \frac{\partial}{\partial t} (u_n + 1)^{\frac{2-\theta}{2}} \right\}_{n\in\mathbf{N}}$$

is equibounded in $L^2_{loc}(\Sigma_T)$ for all $\theta \geq \alpha + 1$ and for all $\alpha \in (0, p-1)$. Therefore

$$\frac{\partial}{\partial t}(u_n+1)^{\frac{2-\theta}{2}} \longrightarrow \frac{\partial}{\partial t}(u+1)^{\frac{2-\theta}{2}} \text{ weakly in } L^2(s,t;L^2(K_\rho)),$$

for all $0<s<t\leq T$ and all $\rho>0$. This implies that

$$\left\{ \frac{\partial}{\partial t} u_{k,n} \right\} \in L^2_{loc}\Sigma_T$$

uniformly in n and

(12.5)
$$\frac{\partial}{\partial \tau} u_{k,n} \longrightarrow \frac{\partial}{\partial \tau}(u \wedge k) \quad \text{weakly in } L^2_{loc}(\Sigma_T),\ \forall k > 0.$$

Choose $\psi \in \overset{\circ}{X}_{loc}(\Sigma_T)$ and in $(8.1)_n$ consider the testing function $\varphi = (\psi - u)_+$. Fix $0<s<t\leq T$ and let

$$k = \|\psi\|_{\infty, \mathbf{R}^N \times (s,t)}.$$

Then

$$(\psi - u)_+ = (\psi - u \wedge k)_+ \in \overset{\circ}{X}_{loc}(\Sigma_T),$$

so that φ is an admissible testing function. It gives

(12.6)
$$\int_s^t\!\!\int_{\mathbf{R}^N} \left\{ \frac{\partial}{\partial \tau} u_n (\psi - u)_+ + |Du_n|^{p-2} Du_n \cdot D(\psi - u)_+ \right\} dx d\tau = 0.$$

Since $u_n \leq u$, $\forall n \in \mathbf{N}$, we have

$$\begin{aligned} \frac{\partial}{\partial t} u_n (\psi - u)_+ &= \frac{\partial}{\partial t} u_n (\psi - u \wedge k)_+ \\ &= \frac{\partial}{\partial t}(u_n \wedge k)(\psi - u)_+ \quad \text{a.e. } \Sigma_T. \end{aligned}$$

Therefore in view of (12.5)

$$\begin{aligned} \lim_{n\to\infty} \int_s^t\!\!\int_{\mathbf{R}^N} \frac{\partial}{\partial \tau} u_n (\psi - u)_+ dx d\tau &= \int_s^t\!\!\int_{\mathbf{R}^N} (u \wedge k)_t (\psi - u \wedge k)_+ dx d\tau \\ &\equiv \int_s^t\!\!\int_{\mathbf{R}^N} u_t (\psi - u)_+ dx d\tau. \end{aligned}$$

Analogously,

$$
\begin{aligned}
&|Du_n|^{p-2}Du_nD(\psi-u)_+\\
&=|D(u_n\wedge k)|^{p-2}D(u_n\wedge k)D(\psi-u\wedge k)_+\\
&=\left(|Du_{n,k}|^{p-2}Du_{n,k}-|Du_k|^{p-2}Du_k\right)\cdot D(\psi-u_k)\\
&+|Du|^{p-2}Du\cdot D(\psi-u)_+.
\end{aligned}
$$

By a calculation similar to that in Lemma 5.2 and leading to (5.6) we have

$$\left||Du_{n,k}|^{p-2}Du_{n,k}-|Du_k|^{p-2}Du_k\right|\le\left|D\left(u_{n,k}-u_k\right)\right|^{p-1}.$$

Therefore taking into account Lemma 12.1 and letting $n\to\infty$ in (12.6) gives

$$\int_s^t\int_{\mathbf{R}^N}\left\{u_t(\psi-u)_++|Du|^{p-2}Du\cdot D\left(\psi-u\right)_+\right\}dxd\tau=0,$$

for all $\psi\in\overset{\circ}{X}_{loc}\left(\Sigma_T\right)$. It remains to prove that u takes the initial datum u_o in the sense of $L^1_{loc}(\mathbf{R}^N)$ and that $u\in\mathcal{S}^*$.

12-(i). Continuity in $L^1_{loc}(\mathbf{R}^N)$ at $t=0$

Fix $\rho>0$ and let $u_{o,\varepsilon}$ be a net of functions satisfying

$$\begin{cases}u_{o,\varepsilon}\equiv0, & \text{for } |x|>4\rho\\ u_{o,\varepsilon}\longrightarrow u_o, & \text{in } L^1\left(K_{2\rho}\right).\end{cases}$$

Such a family can be constructed by first defining a function that coincides with u_o in $K_{3\rho}$ and zero otherwise and then by mollifying the function so obtained. Let also u_ε be the unique solution of (1.1) with initial datum $u_{o,\varepsilon}$. We take the difference of $(8.1)_n$ and the equation satisfied by u_ε. In the p.d.e. so obtained take the testing function

$$\varphi=[(u_n-u_\varepsilon)_++\delta]^\sigma\zeta$$

where $\sigma,\delta\in(0,1)$ and $x\to\zeta(x)$ is the usual cutoff function in $K_{2\rho}$ that equals one on K_ρ. We perform an integration by parts and let $\delta\to0$, $s\to0$, $\sigma\to0$, to obtain

$$
\begin{aligned}
\int_{K_\rho}(u_n(t)-u_\varepsilon(t))_+dx\le&\int_{K_{2\rho}}(u_{0,n}-u_{0,\varepsilon})_+dx\\
&+\frac{2\gamma}{\rho}\int_0^t\int_{K_{2\rho}}\left(|Du_n|^{p-1}+|Du_\varepsilon|^{p-1}\right)dxd\tau.
\end{aligned}
$$

We use (8.4), interchange the role of u_n and u_ε and, for $t>0$ fixed, let $n\to\infty$. This gives

$$\int_{K_\rho} |u(t) - u_\varepsilon(t)| dx \le \int_{K_{2\rho}} |u_o - u_{o,\varepsilon}|\, dx$$
$$+\gamma\left(\frac{t}{\rho^\lambda}\right)^{\frac{1}{p}}\left\{\int_{K_{4\rho}} u_o\, dx + \left(\frac{t}{\rho^\lambda}\right)^{\frac{1}{2-p}}\right\}^{\frac{2}{p}(p-1)}.$$

From this

$$\int_{K_\rho} |u(t) - u_o| dx \le 2\int_{K_{2\rho}} |u_o - u_{o,\varepsilon}| dx + \int_{K_\rho} |u_\varepsilon(t) - u_{o,\varepsilon}| dx + O\Big(t^{\frac{1}{p}}\Big).$$

Letting $t \searrow 0$

$$\lim - sup_{t\searrow 0} \int_{K_\rho} |u(t) - u_o)| dx \le 2\int_{K_{2\rho}} |u_o - u_{o,\varepsilon}| dx, \ \ \forall \varepsilon \in (0,1).$$

12-(ii). $u \in \mathcal{S}^*$

By (10.2), $\forall n \in \mathbf{N}$ and for all $k > 0$

$$\frac{\partial}{\partial t}(u_n \wedge k) \le \frac{1}{2-p}\frac{u_n}{t}.$$

As $n \to \infty$

$$(u \wedge k)_t \le \frac{1}{2-p}\frac{u}{t} \quad \text{a.e. in } \Sigma_T. \tag{12.7}$$

The limit is first taken in $\mathcal{D}'(0,T)$ and then (12.7) holds almost everywhere in Σ_T in view of (12.5). Next from Lemma 9.1 it follows that $\forall C > 1$

$$\int_s^t\int_{K_\rho} |Du_n|^p \frac{1}{u_n}\chi[k < u_n]\chi[u < Ck] dx d\tau = O\Big(\frac{1}{k}\Big).$$

Here we have used the fact that $u_n \nearrow u$ implies $[u_n < Ck] \supset [u < Ck]$. Letting $n \to \infty$ for $k > 0$ and $C > 1$ fixed yields by lower semicontinuity

$$\int_s^t\int_{K_\rho} |Du|^p \frac{1}{u}\chi[k < u < Ck] dx d\tau = O\left(\frac{1}{k}\right).$$

We conclude by remarking that the requirement $u \in \mathcal{S}^*$ is necessary and sufficient for uniqueness. Indeed, if solutions in $\mathcal{S}$ are unique, they can be constructed starting from their traces on $t = \tau \in (0,T)$ to yield $u \in \mathcal{S}^*$. Vice versa solutions in $\mathcal{S}^*$ are unique.

13. Bounded solutions. A counterexample

Let $r \geq 1$ satisfy $\lambda_r \equiv N(p-2)+rp>0$. If $u_o \in L^r_{loc}(\mathbf{R}^N)$, then by energy estimates, the sequence of approximating solutions of $(8.1)_n$ satisfies

$$\{u_n\} \in L^r_{loc}(\Sigma_T) \quad \text{uniformly in } n.$$

Therefore by Theorem 5.1 of Chap. V, $\{u_n\} \in L^\infty_{loc}(\Sigma_T)$ uniformly in n. It follows from the regularity results of Chaps. IV and IX that

$$\{u_n\},\ \{u_{n,x_j}\} \in C^\alpha_{loc}(\Sigma_T),\ j=1,2,\dots,N, \quad \text{uniformly in } n,$$

for some $\alpha \in (0,1)$ depending only upon N and p. This gives a regular solutions to the Cauchy problem (1.1)-(1.2). A similar analysis can be carried if the initial datum is a measure μ and $\lambda_1 > 0$, i.e.,

$$p > \frac{2N}{N+1}. \tag{13.1}$$

We show next that if (13.1) is violated, then initial data in $L^1_{loc}(\mathbf{R}^N)$ might produce unbounded solutions.

13-(i). A counterexample

Let $a \in (0,1)$ be a positive constant and let B_a denote the ball of radius a in $\mathbf{R}^N$ centered at the origin. Consider the functions

$$z = \frac{\left(a^2-|x|^2\right)_+^2}{|x|^N \left|\ln|x|^2\right|^\beta} \quad \text{and} \quad v = (1-ht)_+ z, \tag{13.2}$$

where $\beta, h > 1$ are to be chosen. One verifies that

$$z \in L^1(B_a), \quad \text{and} \quad z \notin L^{1+\varepsilon}(B_a),\ \forall \varepsilon > 0.$$

Consider also the Cauchy problem

$$\begin{cases} u_t - \operatorname{div}|Du|^{p-2}Du = 0, \text{ in } \Sigma_1 \equiv \mathbf{R}^N \times (0,1), \\ u(\cdot,0) = z. \end{cases} \tag{13.3}$$

The p.d.e. is meant in the sense of (2.1)–(2.2) and the initial datum is taken in the sense of $L^1_{loc}(\mathbf{R}^N)$.

LEMMA 13.1. *Assume that $N(p-2)+p=0$. The constants $a \in (0,1)$ and $\beta, h > 1$ can be determined a priori so that v is a non-negative, weak subsolution of (13.3) in Σ_1.*

PROOF: By calculation on the set $0<|x|<a$,

$$Dz = -\frac{z}{|x|^2} Fx,$$

where

$$F=\left\{N+\frac{2\beta}{\ln|x|^2}+\frac{4|x|^2}{a^2-|x|^2}\right\}.$$

We choose $a=e^{-k}$ and $k>1$ so large that $F>0$. Compute

$$|Dz|^{p-2}Dz=-\frac{z^{p-1}F^{p-1}}{|x|^p}x$$

$$\operatorname{div}(|Dz|^{p-2}Dz)=-(p-1)\frac{z^{p-2}F^{p-1}}{|x|^p}Dz\cdot x+p\frac{z^{p-1}F^{p-1}}{|x|^{p+1}}D|x|\cdot x$$
$$-N\frac{z^{p-1}F^{p-1}}{|x|^p}-(p-1)\frac{z^{p-1}F^{p-2}}{|x|^p}DF\cdot x.$$

Using the formulae

$$Dz\cdot x=-zF,\qquad D|x|\cdot x=|x|$$
$$DF\cdot x=\frac{-4\beta}{\ln^2|x|^2}+\frac{8|x|^2}{(a^2-|x|^2)}+\frac{8|x|^4}{(a^2-|x|^2)^2},$$

we obtain

$$\operatorname{div}|Dz|^{p-2}Dz=\frac{z^{p-1}F^{p-2}}{|x|^p}\{(p-1)F^2-(N-p)F-(p-1)DF\cdot x\}.$$

We calculate the expression in braces on the right-hand side using the definition of F and the fact that $N(p-2)+p=0$, to obtain

$$\operatorname{div}(|Dz|^{p-2}Dz)=2(p-1)\frac{z^{p-1}F^{p-2}}{|x|^p}\mathcal{H},$$

$$\mathcal{H}\equiv\frac{2\beta(\beta+1)}{\ln^2|x|^2}+\frac{4|x|^4}{(a^2-|x|^2)_+^2}+\frac{8|x|^2}{\ln|x|^2(a^2-|x|^2)_+}$$
$$+\frac{N\beta}{\ln|x|^2}+(N-2)\frac{2|x|^2}{(a^2-|x|^2)_+}$$
$$\geq\frac{4|x|^2}{(a^2-|x|^2)_+}\left[\frac{|x|^2}{(a^2-|x|^2)_+}+\frac{2\beta}{\ln|x|^2}\right]$$
$$+\frac{N\beta}{\ln|x|^2}.$$

Consider the sets

$$\mathcal{E}_k^{(1)}\equiv\left\{\frac{2}{3}e^{-2k}\leq|x|^2<e^{-2k}\right\},\quad \mathcal{E}_k^{(2)}\equiv\left\{|x|^2<\frac{2}{3}e^{-2k}\right\},\quad k>1.$$

One verifies that on $\mathcal{E}_k^{(1)}$ we have

$$\mathcal{H}\geq 8\left(2-\frac{\beta}{k}\right)-\frac{N\beta}{2k}.$$

Therefore $\mathcal{H}\geq 0$ on $\mathcal{E}_1$ if k is sufficiently large. On $\mathcal{E}_k^{(2)}$ we have $F\geq(N-\beta/k)>1$. Therefore

$$\operatorname{div}(|Dz|^{p-2}Dz)\geq\frac{z^{p-1}}{|x|^p}\frac{\gamma(N,p)}{\ln|x|^2}.$$

Finally we compute in $\{0<|x|<a\}$

$$\begin{aligned}\mathcal{L}(v)&\equiv v_t-\operatorname{div}(|Dv|^{p-2}Dv)\\&=-hz-(1-ht)_+^{p-1}\operatorname{div}(|Dz|^{p-2}Dz).\end{aligned}$$

On $\mathcal{E}_k^{(1)}$, $\mathcal{L}(v)\leq 0$ and on $\mathcal{E}_k^{(2)}$

$$\mathcal{L}(v)\leq z\left[-h-\frac{z^{p-2}}{|x|^p}\frac{\gamma(N,p)}{\ln|x|^2}\right].$$

By calculation on $\mathcal{E}_k^{(2)}$,

$$-\frac{z^{p-2}}{|x|^p}\frac{\gamma(N,p)}{\ln|x|^2}\leq\gamma(N,p)\frac{\left(a^2-|x|^2\right)^{2(p-2)}}{\left|\ln|x|^2\right|^{\beta(p-2)+1}},$$

where we have used the fact that $\lambda_1\equiv N(p-2)+p=0$. We select $\beta>1$ so that $\beta(p-2)+1>0$. This gives

$$-\frac{z^{p-2}}{|x|^p}\frac{\gamma(N,p)}{\ln|x|^2}\leq\gamma^*(N,p,k).$$

Therefore

$$\mathcal{L}(w)\leq z(-h+\gamma^*(k)).$$

Choosing $h=\gamma^*(k)$ proves that

$$\mathcal{L}(v)\leq 0\qquad\text{on }\{0<|x|<a\}\times(0,1).\tag{13.4}$$

To prove that indeed v is a weak subsolution in the whole Σ_1, multiply (13.4) by a non-negative function $x\to\varphi(x)\in C_o^\infty(\Sigma_1)$, and integrate over the cylindrical domain with annular cross section $Q_\varepsilon\equiv\{\varepsilon<|x|<a-\varepsilon\}\times(0,1)$. We obtain

$$\begin{aligned}&\iint_{\Sigma_1}\left\{v_t\varphi+|Dv|^{p-2}Dv\cdot D\varphi\right\}dxd\tau\\&=\lim_{\varepsilon\searrow 0}\iint_{Q_\varepsilon}\left\{v_t\varphi+|Dv|^{p-2}Dv\cdot D\varphi\right\}dxd\tau\\&\leq\lim_{\varepsilon\searrow 0}\int_0^1\int_{\{|x|=a-\varepsilon\}}|Dv|^{p-2}Dv\cdot\frac{x}{|x|}\varphi d\sigma d\tau\\&-\lim_{\varepsilon\to\searrow 0}\int_0^1\int_{\{|x|=\varepsilon\}}|Dv|^{p-2}Dv\cdot\frac{x}{|x|}\varphi d\sigma d\tau,\end{aligned}$$

where $d\sigma$ denotes the surface measure on $\{|x|=\varepsilon\}$ and on $\{|x|=\varepsilon\}$. The limits on the right hand side are zero. In particular we have

$$\forall\psi \in X_{loc}(\Sigma_1)\,,\quad \forall\zeta \in C_o^\infty(\mathbf{R}^N),\quad \zeta \geq 0,$$

$$\iint_{\Sigma_1} \{v_t(\psi - v)_+ + |Dv|^{p-2}Dv\cdot D(\psi - v)_+\}\, dxd\tau \leq 0. \tag{13.5}$$

One also verifies by direct calculation that v satisfies (5.1) and (5.2) and therefore is a subsolution of (13.3) in the class $\mathcal{S}^*$.

Next we return to (13.3). This problem has a unique solution $u \in \mathcal{S}^*$, by the construction of §§8–12 and the uniqueness theorem 7.1. By the comparison principle $u \geq v$ and therefore u is not bounded. The comparison principle here is applied as follows. By the definition of weak solution the truncated functions $u_k \equiv \min\{u; k\}$ are, for all $k > 0$ distributional subsolutions of (13.3). Setting

$$w \equiv v - u \quad \text{and} \quad w_{(k)} \equiv v - u_k$$

and using (13.5) we find

$$\forall 0 < s < t \leq T,\quad \forall\psi \in X_{loc}(\Sigma_1)\,,\quad \forall\zeta \in C_o^\infty(\mathbf{R}^N),\ \zeta \geq 0$$

$$\int_s^t\int_{\mathbf{R}^N}\Big\{\frac{\partial}{\partial t}w_{(k)}(\psi - v)_+$$

$$+ \left[|Dv|^{p-2}Dv - |Du_k|^{p-2}Du_k\right]\cdot D\left((\psi - v)_+\zeta\right)\Big\}dxd\tau \leq 0.$$

Observe that $w(t) \to 0$ as $t \searrow 0$ in $L^1_{loc}(\mathbf{R}^N)$. Therefore we may proceed as in the proof of the uniqueness theorem and establish an analog of Proposition 6.1, i.e.

$$\forall 0 < t \leq T,\quad \forall q > 1,\quad \forall\rho > 0,\quad \forall\sigma \in (0,1)$$

$$\int_{K_\rho} \left(w^+(t)\right)^q dx \leq \frac{\gamma}{(\sigma\rho)^p}\int_0^t\int_{K_{(1+\sigma)\rho}} \left(w^+\right)^{p-2+q} dxd\tau.$$

Proceeding as in the proof of Theorem 7.1 we find $w^+ = 0$.

Remark 13.1. If $N(p-2)+p > 0$ then v satisfies (13.4) but it is not a subsolution of (13.3) in the whole Σ_1. In particular it does not satisfy the requirement (5.2) of the class $\mathcal{S}^*$. If $N(p-2)+p < 0$ then v does not satisfy (13.4).

14. Bibliographical notes

Equations of the type of (1.1) arise in modelling of non-newtonian fluids (see Kalashnikov [57], Martinson–Paplov [74,75], Antonsev [5] and Joseph-Nield-Papanicolau [56]). Questions of solvability, even though in a different context, were first investigated by Brézis and Friedman [18]. The notion of weak solution introduced in §2 is taken from [42]. Bénilan has informed us of a more general notion of solution, introduced in [11], that would include solutions of variable sign. The remainder of the chapter is essentially taken from [42]. It would be of interest to investigate questions of existence/uniqueness for (1.1) in Σ_T when the initial datum is of variable sign or is a measure. Singular equations are little understood, mostly if p violates (13.1). Preliminary investigations seem to indicate questions of limiting Sobolev exponent (see [19]) and differential geometry.

Bibliography

[1] R. Adams, *Sobolev Spaces*, Academic Press, New York (1975).

[2] L. V. Alfors, *Lectures on quasiconformal mappings*, Wadsworth & Brooks/-Cole, Monterey CA (1987).

[3] D. Andreucci, A priori bounds for weak solutions of the filtration equation, SIAM J. Math. Anal. 22 # 1 (1991), pp. 138–145.

[4] D. Andreucci and E. DiBenedetto, A new approach to initial traces in non-linear filtration, Ann. Inst. H. Poincaré Analyse non Linéaire, 7 # 4 (1990), pp. 305–334.

[5] S.N. Antonsev, Axially symmetric problems of gas dynamics with free boundaries, Doklady Akad. Nauk SSSR 216 # 3 (1974), pp. 473–476.

[6] D.G. Aronson and L.A. Caffarelli, The initial trace of a solution of the porous medium equation, Trans. AMS 280 # 1 (1983), pp. 351–366.

[7] D.G. Aronson and J. Serrin, Local behaviour of solutions of quasilinear parabolic equations, Arch. Rational Mech. Anal. 25 (1967), pp. 81–123.

[8] G.I. Barenblatt, On some unsteady motions of a liquid or a gas in a porous medium, Prikl. Mat. Mech. 16 (1952), pp. 67–78.

[9] Ph. Bénilan and M.G. Crandall, The continuous dependence on ϕ of solutions of $u_t - \Delta\phi(u) = 0$, Indiana Univ. Math. J. 30 (1981), pp. 161–177.

[10] Ph. Bénilan, M.G. Crandall and M. Pierre, Solutions of the porous medium medium equation in R^N under optimal conditions on initial values, Indiana Univ. Math. J. 33 (1984), pp. 51–87.

[11] Ph. Bénilan and T. Gallouët (personal communication).

[12] S.N. Bernstein, *Collected works. III Differential equations, calculus of variations and geometry (1903–1947)*, Izdat. Akad. Nauk SSSR, Moscow (1960) (Russian).

[13] J.G.Berryman, Evolution of a stable profile for a class of nonlinear diffusion equations with fixed boundaries, J. Math. Phys. 18 # 11 (1977), pp. 2108–2115.

[14] J.G. Berryman and C.J. Holland, Stability of the separable solution for fast diffusion equation, Arch. Rational Mech. Anal. 74 (1980), pp. 379–388.

[15] L. Boccardo and T. Gallouët, Non linear elliptic and parabolic equations involving measure data, (to appear).

[16] L. Boccardo, F. Murat and J.P. Puel, L^∞ estimates for some non linear elliptic partial differential equations and applications to an existence result, SIAM J. Math. Anal.

[17] B. Bojarski and T. Iwaniec, p-harmonic equations and quasiregular Mappings, Inst. Angew. Universität Bonn (preprint 1983).

[18] H. Brézis and A. Friedman, Nonlinear parabolic equations involving measures as initial conditions, J. Math. Pures Appl. 62 (1983), pp. 73–97.

[19] H. Brézis, Elliptic equations with limiting Sobolev exponents–The impact of topology, Comm. Pure Appl. Math. XXXIX (1986), S17-S39.

[20] L.A. Caffarelli and L.C. Evans, Continuity of the temperature in the two phase Stefan problem, Arch. Rational Mech. Anal. 81 (1983), pp. 199–220.

[21] L.A. Caffarelli and A. Friedman, Regularity of the free-boundary of a gas in a n-dimensional porous medium, Indiana Univ. Math. J. 29, (1980), pp. 361–369 .

[22] S. Campanato, Equazioni ellittiche del II° ordine e spazi $\mathcal{L}^{2,\lambda}$, Ann. Math. Pura Appl. 69 (1965), pp. 321–381.

[23] S. Campanato, Equazioni paraboliche del secondo ordine e spazi $\mathcal{L}^{p,\theta}(\Omega, \delta)$, Ann. Math. Pura Appl. 73 (1966), pp. 55–102.

[24] Y.Z. Chen, Hölder estimates for solutions of uniformly degenerate quasilinear parabolic equations, Chin. Ann. Math. 5B (4) (1984), pp. 661–678.

[25] Y.Z. Chen, Hölder continuity of the gradients of solutions of non-linear degenerate parabolic systems, Acta Math. Sinica, New Series 2 # 4 (1986), pp. 309–331.

[26] Y.Z. Chen and E. DiBenedetto, On the local behaviour of solutions of singular parabolic equations, Arch. Rational Mech. Anal. 103 # 4 (1988), pp. 319–345.

[27] Y.Z. Chen and E. DiBenedetto, Boundary estimates for solutions of nonlinear degenerate parabolic systems, J. Reine Angew. Math. 395 (1989), pp. 102–131.

[28] Y.Z. Chen and E. DiBenedetto, Hölder estimates of solutions of singular parabolic equations with measurable coefficients, Arch. Rational Mech. Anal. 118 (1992), pp. 257–271.

[29] Y.Z. Chen and E. DiBenedetto, On the Harnack inequality for non-negative solutions of singular parabolic equations, Proc. of Conf. *Non-linear diffusion*, in Honour of J. Serrin, Minneapolis May 1990.

[30] H. Choe, Hölder regularity for the gradient of solutions of certain singular parabolic equations, Comm. Part. Diff. Equations 16 # 11 (1991), pp. 1709–1732.

[31] H. Choe, Hölder continuity of solutions of certain degenerate parabolic systems, Non-linear Anal. 8 # 3 (1992), pp. 235–243.

[32] G. DaPrato, Spazi $\mathcal{L}^{(p,\theta)}(\Omega,\delta)$ e loro proprietà, Ann. Math. Pura Appl. 69 (1965), pp. 383–392.

[33] E. DeGiorgi, Sulla differenziabilita' e l'analiticita' delle estremali degli integrali multipli regolari, Mem. Acc. Sci. Torino, Cl. Sc. Fis. Mat. Nat. (3) 3 (1957), pp. 25–43.

[34] E. DiBenedetto, Continuity of weak solutions to certain singular parabolic equations, Ann.Mat. Puta Appl. 4 # 130 (1982), pp.131–176.

[35] E. DiBenedetto, Continuity of weak solutions to a general porous medium equation, Indiana Univ. Math. J. 32 # 1 (1983), pp. 83–118.

[36] E. DiBenedetto and A. Friedman, Regularity of solutions of non-linear degenerate parabolic systems, J. Reine Angew. Math. 349 (1984), pp. 83–128.

[37] E. DiBenedetto and A. Friedman, Hölder estimates for non-linear degenerate parabolic systems, J. Reine Angew. Math. 357 (1985), pp. 1–22.

[38] E. DiBenedetto, A boundary modulus of continuity for a class of singular parabolic equations, J. Diff. Equations 6 # 3 (1986), pp. 418–447.

[39] E. DiBenedetto, On the local behaviour of solutions of degenerate parabolic equations with measurable coefficients, Ann. Sc. Norm. Sup. Pisa Cl. Sc. Serie IV, XIII 3 (1986), 487–535.

[40] E. DiBenedetto, Intrinsic Harnack type inequalities for solutions of certain degenerate parabolic equations, Arch. Rational Mech. Anal. 100 # 2 (1988), pp. 129–147.

[41] E. DiBenedetto and M.A. Herrero, On the Cauchy problem and initial traces for a degenerate parabolic equation Trans. AMS 314 (1989), pp. 187–224.

[42] E. DiBenedetto and M.A. Herrero, Non negative solutions of the evolution p-Laplacian equation. Initial traces and Cauchy problem when $1 < p < 2$, Arch. Rational Mech. Anal. 111 # 3 (1990), pp. 225–290.

[43] E. DiBenedetto, Y. C. Kwong and V. Vespri, Local space analiticity of solutions of certain singular parabolic equations, Indiana Univ. Math. J. 40 # 2 (1991), pp. 741–765.

[44] E. DiBenedetto and Y.C. Kwong, Intrinsic Harnack estimates and extinction profile for certain singular parabolic equations, Trans AMS 330 # 2 (1992), pp. 783–811.

[45] E. DiBenedetto, J. Manfredi and V. Vespri, Boundary gradient bounds for evolution $p-$laplacian equations. Proc. Int. Conf. of Evolution Equations and their Ground States, Gregynog, Wales, 1989

[46] E. DiBenedetto, J. Manfredi, On the local behaviour of solutions of degenerate elliptic systems, Amer. J. Math. (to appear)

[47] B. Fuglede, A criterion of non-vanishing differential of a smooth map, Bull. London Math. Soc. 14 (1982), pp. 98–102.

[48] M. Giaquinta, *Multiple integrals in the calculus of variations and nonlinear elliptic systems.* Ann. Math. Studies 105, Princeton Univ. Press., Princeton N.J. (1983).

[49] M. Giaquinta and E. Giusti, Global $C^{1,a}-$regularity for second order quasilinear elliptic equations in divergence form, J. Reine Angew. Math. # 351 (1984), pp. 55–65.

[50] J. Hadamard, Extension á l'équation de la chaleur d'un théoréme de A. Harnack, Rend. Circ. Mat. Palermo Ser. 23 (1954), pp. 337–346.

[51] A.M. Il'in, A.S. Kalashnikov and O.A. Oleinik, Linear equations of second order of parabolic type, Uspeki Matem. NAUK, 17 # 3 (1962), pp. 3–145.

[52] A.V. Ivanov, Uniform Hölder estimates for generalized solutions of quasilinear parabolic equations admitting a double degeneracy, Algebra Anal. 3 # 2 (1991), pp. 139–179.

[53] A.V.Ivanov, The classes $B_{m\ell}$ and Hölder estimates for quasilinear doubly degenerate parabolic equations. Zap. Nauchn. Sem. St. Petersburg Otdel. Math. Inst. Steklov (LOMI) 197 (1992), pp. 42–70 (Engl. transl: J. Soviet Math.).

[54] A.V. Ivanov and P.Z. Mkrtchen, On the regularity up to the boundary of weak solutions of the first initial–boundary value problem for quasilinear doubly degenerate parabolic equations. Zap. Nauchn. Sem. St. Petersburg, Otdel. Math. Inst. Steklov (LOMI) 196 (1991), pp. 3–28.

[55] T. Iwaniec, Projections onto gradient fields and L^p-estimates for degenerate elliptic equations. Studia Math. 75 (1983), pp. 293–312.

[56] D.D. Joseph, D.A. Nield and G. Papanicolau, Non linear equations governing flow in a saturated porous medium (preprint).

[57] A.S. Kalashnikov, On a heat conduction equation for a medium with non-uniformly distributed non-linear heat sources or absorbers, Bull. Univ. Moscow, Math. Mech. 3 (1983), pp. 20–24.

[58] A.S. Kalashnikov, Cauchy's problem in classes of increasing functions for certain quasi-linear degenerate parabolic equations of the second order, Diff. Uravneniya 9 # 4 (1973), 682–691.

[59] A.S. Kalashnikov, On uniqueness conditions for the generalized solutions of the Cauchy problem for a class of quasi-linear degenerate parabolic equations, Diff. Uravneniya 9 # 12 (1973), 2207–2212.

[60] S.N. Kruzkov, On the apriori estimation of solutions of linear parabolic equations and of solutions of boundary value problems for a certain class of quasi-linear parabolic equations, Dokl. Akad. NAUK SSSR # 138 (1961), pp. 1005–1008 (Engl. transl.: Soviet Math. Dokl. # 2 (1961), pp. 764–767).

[61] S.N. Kruzkov, A priori estimates and certain properties of the solutions of elliptic and parabolic equations of second order, Mat. Sbornik 65 # 107 (1964), pp. 522–570 (Engl. trans.: Amer. Math. Soc. Transl. 2 # 68 (1968), pp. 169–220).

[62] S.N. Kruzkov, Results concerning the nature of the continuity of solutions of parabolic equations and some of their applications, Math. Zametki 6 (1969), pp. 97–108 (Russian).

[63] N.V. Krylov, *Non-linear elliptic and parabolic equations of the second order*. D. Reidel, Dordrecht, Holland (1987).

[64] N.V. Krylov and M.V. Safonov, A certain property of solutions of parabolic equations with measurable coefficients, Math. USSR Izvestijia 16 # 1 (1981), pp. 151–164.

[65] O.A. Ladyzenskaja, New equations for the description of motion of viscous incompressible fluids and solvability in the large of boundary value problems for them, Proc. Steklov Inst. Math. # 102 (1967), pp. 95–118 (English transl.: Trudy Math. Inst. Steklov # 102 (1967), pp. 85–104).

[66] O.A. Ladyzenskaja, and N.N. Ural'tzeva, *Linear and quasilinear elliptic equations*, Academic Press, New York (1968).

[67] O.A. Ladyzenskaja, V.A. Solonnikov and N.N. Ural'tzeva, *Linear and quasilinear equations of parabolic type*. Transl. Math. Mono. Vol. 23 AMS, Providence, RI (1968).

[68] G.M. Lieberman, The first initial-boundary value problem for quasilinear second order parabolic equations, Ann. Scuola Norm. Sup. Pisa 13 (1986), pp. 347–387.

[69] G.M. Lieberman, Boundary regularity for solutions of degenerate elliptic equations, Non-linear Anal. 12 (1988), pp. 1202–1219.

[70] G.M. Lieberman, Boundary and initial regularity for solutions of degenerate parabolic equations, Nonlinear Anal. TMA 20 (1993), pp. 551–570. (to appear).

[71] G.M. Lieberman, Mean oscillation estimates and Hölder regularity for the gradients of solutions of degenerate parabolic systems (to appear).

[72] Lin Fan Hua, Boundary $C^{1,\beta}$–regularity of p–harmonic functions (preprint 1988).

[73] J.L. Lions, *Quelques methodes de resolution dés problemes aux limites nonlineaires*. Dunod, Paris (1969).

[74] L.K. Martinson and K.B. Paplov, The effect of magnetic plasticity in non-Newtonian fluids, Magnit. Gidrodinamika 3 (1969), pp. 69–75.

[75] L.K. Martinson and K.B. Paplov, Unsteady shear flows of a conducting fluid with a rheological power law, Magnit. Gidrodinamika # 2 (1970), pp. 50–58.

[76] V.G. Mazja, *Sobolev Spaces*. Springer-Verlag, New York, (1985).

[77] M. Meier, Boundedness and integrability properties of weak solutions of quasilinear elliptic systems. J. Reine Angew. Math. 333 (1982), pp. 191–220.

[78] G. Minty, Monotone (non-linear) operators in Hilbert spaces, Duke Math. J. 29 (1962), pp. 341–346.

[79] C.B. Morrey, *Multiple integrals in the calculus of variations*. Springer-Verlag, New York, (1966).

[80] C.B. Morrey, Partial regularity results for non-linear elliptic systems. J. Math. Mech. 17 (1968), pp. 649–670.

[81] J. Moser, A new proof of DeGiorgi's theorem concerning the regularity problem for elliptic differential equations, Comm. Pure Appl. Math. 13 (1960), pp. 457–468.

[82] J. Moser, On Harnack's theorem for elliptic differential equations, Comm. Pure Appl. Math. 14 (1961), pp. 577–591.

[83] J. Moser, A Harnack inequality for parabolic differential equations, Comm. Pure Appl. Math. 17 (1964), pp. 101–134.

[84] J. Nash, Continuity of solutions of parabolic and elliptic equations, Amer. J. Math. 80 (1958), pp. 931–954.

[85] M. Pierre, Uniqueness of the solutions of $u_t - \Delta\phi(u) = 0$ with initial datum a measure, Nonlin. Anal. 6, # 2 (1987), pp. 175–187.

[86] B. Pini, Sulla soluzione generalizzata di Wiener per il primo problema di valori al contorno nel caso parabolico, Rend. Sem. Math. Univ. Padova 23 (1954), pp. 422–434.

[87] M. Porzio, A priori bounds for weak solutions of certain degenerate parabolic equations, Nonlin. Anal. 20, # 11 (1991), pp. 1093, 1107.

[88] M. Porzio and V. Vespri, Hölder estimates for local solutions of some doubly non-linear degenerate parabolic equations, J. Diff. Equ. (to appear).

[89] Y.G. Resetniak, *Mappings with bounded distorsion in space*. 'Nauka' Novosibirski, Moscow (1982) (Russian).

[90] P.E. Sacks, Continuity of solutions of a singular parabolic equation, Nonlinear Anal. 7 (1983), pp. 387–409.

[91] M.V. Safonov, The Harnack inequality for elliptic equations and the Hölder continuity of their solutions, Boundary problems of mathematical physics and adjacent questions in the theory of functions, Zap. Nauchn. Sem. Leningrad Otdel. Math. Inst. Steklov (LOMI) 96 (1980), pp. 272–287 (Engl. transl.: J. Soviet Math. (LOMI) 20 (1983), pp. 851–863).

[92] J. Serrin, Local behaviour of solutions of quasilinear elliptic equations, Acta Math. 111 (1964), pp. 101–134.

[93] G. Stampacchia, *Equations elliptiques du second ordre a coefficients discontinues*. Sém. Math. Sup. 16, Les Presses de l'Université de Montréal, Montréal (1966).

[94] S. Tacklind, Sur les classes quasianalitiques des solutions des équations aux derivée partielles du type parabolique, Acta Reg. Soc. Sc. Uppsaliensis (Ser. 4) 10, # 3 (1936), pp. 3–55.

[95] P. Tolksdorff, Everywhere regularity for some quasi-linear systems with lack of ellipticity, Ann. Mat. Pura Appl. 4 # 134 (1983), pp. 241–266.

[96] N.S. Trudinger, On Harnack type inequalities and their application to quasilinear elliptic partial differential equations, Comm. Pure Appl. Math. 20 (1967), pp. 721–747.

[97] N.S. Trudinger, Pointwise estimates and quasilinear parabolic equations, Comm. Pure Appl. Math. 21 (1968), pp. 205–226.

[98] A.N. Tychonov, Thèoremes d'unicité pour l'equation de la chaleur Math. Sbornik 42 (1935), pp. 199–216.

[99] K. Uhlenbeck, Regularity for a class of non-linear elliptic systems, Acta Math. 138 (1977), pp. 219–240.

[100] N.N. Ural'tceva, Degenerate quasilinear elliptic systems, Zap. Nauk. Sem. Leningrad Otdel. Math. Inst. Steklov # 7 (1968), pp. 184–222 (Russian).

[101] V. Vespri, L^∞ estimates for non-linear parabolic equations with natural growth conditions, Rend. Sem. Mat. Univ. Padova (in press).

[102] V. Vespri, On the local behaviour of a certain class of doubly non-linear parabolic equations, Manuscripta Math. 75 (1992), pp. 65–80.

[103] V. Vespri, Harnack type inequalities for solutions of certain doubly nonlinear parabolic equations. J. Math. Anal. Appl. (in press).

[104] M. Wiegner, On C_α-regularity of the gradient of solutions of degenerate parabolic systems, Ann. Mat. Pura Appl. 4 # 145 (1986), pp. 385–405.

[105] D.V. Widder, Positive temperatures in an infinite rod, Trans. AMS, # 55 (1944), pp. 85–95.